Biotechnology: Beyond the Basics

Biotechnology: Beyond the Basics

Editor: Wendell Carter

Callisto Reference

www.callistoreference.com

Callisto Reference,
118-35 Queens Blvd., Suite 400,
Forest Hills, NY 11375, USA

Visit us on the World Wide Web at:
www.callistoreference.com

ISBN: 978-1-64116-074-2 (Hardback)

The publisher's policy is to use permanent paper from mills that operate a sustainable forestry policy. Furthermore, the publisher ensures that the text paper and cover boards used have met acceptable environmental accreditation standards.

Printed in China.

Cataloging-in-Publication Data

Biotechnology : beyond the basics / edited by Wendell Carter.
p. cm.
Includes bibliographical references and index.
ISBN 978-1-64116-074-2
1. Biotechnology. 2. Genetic engineering. I. Carter, Wendell.
TP248.2 .B56 2019
660.6--dc21

Table of Contents

Preface

Biotechnology involves the development of products by using living systems or organisms. It builds on the principles of biochemical engineering, molecular engineering, biomedical engineering and immunology. It diverges into diverse branches like bioinformatics for the analysis of biological data in complex systems; blue biotechnology for marine studies; green biotechnology for application in agriculture; red biotechnology for developing pharmaceutical products and white biotechnology for industrial applications. Modern developments in the field of biotechnology have applications in the pharmaceutical, agricultural and food industries. Most of the topics introduced in this book cover new techniques and their applications in a multidisciplinary manner. It highlights the researches that have transformed this discipline and aided its advancement. Comprehensive language and extensive use of examples make this book a complete reference text for all professionals and students involved in this area of study.

This book has been the outcome of endless efforts put in by authors and researchers on various issues and topics within the field. The book is a comprehensive collection of significant researches that are addressed in a variety of chapters. It will surely enhance the knowledge of the field among readers across the globe.

It gives us an immense pleasure to thank our researchers and authors for their efforts to submit their piece of writing before the deadlines. Finally in the end, I would like to thank my family and colleagues who have been a great source of inspiration and support.

Editor

1

Combinatorial metabolic engineering of industrial *Gluconobacter oxydans* DSM2343 for boosting 5-keto-D-gluconic acid accumulation

Jianfeng Yuan, Mianbin Wu, Jianping Lin* and Lirong Yang

Abstract

Background: L-(+)-tartaric acid (L-TA) is an important organic acid, which is produced from the cream of tartar or stereospecific hydrolysis of the *cis*-epoxysuccinate. The former method is limited by the availability of raw material and the latter is dependent on the petrochemical material. Thus, new processes for the economical preparation of L-TA from carbohydrate or renewable resource would be much more attractive. Production of 5-keto-D-gluconate (5-KGA) from glucose by *Gluconobacter oxydans* is the first step to produce L-TA. The aim of this work is to enhance 5-KGA accumulation using combinatorial metabolic engineering strategies in *G. oxydans*. The *sldAB* gene, encoding sorbitol dehydrogenase, was overexpressed in an industrial strain *G. oxydans* ZJU2 under a carefully selected promoter, P_{0169}. To enhance the efficiency of the oxidation by *sldAB*, the coenzyme pyrroloquinoline quinone (PQQ) and respiratory chain were engineered. Besides, the role in *sldAB* overexpression, coenzyme and respiratory chain engineering and their subsequent effects on 5-KGA production were investigated.

Results: An efficient, stable recombinant strain was constructed, whereas the 5-KGA production could be enhanced. By self-overexpressing the *sldAB* gene in *G. oxydans* ZJU2 under the constitutive promoter P_{0169}, the resulting strain, *G. oxydans* ZJU3, produced 122.48 ± 0.41 g/L of 5-KGA. Furthermore, through the coenzyme and respiratory chain engineering, the titer and productivity of 5-KGA reached 144.52 ± 2.94 g/L and 2.26 g/(L · h), respectively, in a 15 L fermenter. It could be further improved the 5-KGA titer by 12.10 % through the fed-batch fermentation under the pH shift and dissolved oxygen tension (DOT) control condition, obtained 162 ± 2.12 g/L with the productivity of 2.53 g/(L · h) within 64 h.

Conclusions: The 5-KGA production could be significantly enhanced with the combinatorial metabolic engineering strategy in *Gluconobacter* strain, including *sldAB* overexpression, coenzyme and respiratory chain engineering. Fed-batch fermentation could further enlarge the positive effect and increase the 5-KGA production. All of these demonstrated that the robust recombinant strain can efficiently produce 5-KGA in larger scale to fulfill the industrial production of L-TA from 5-KGA.

Keywords: 5-keto-D-gluconate, L-(+)-tartaric acid, Pyrroloquinoline quinone, Respiratory chain, Fed-batch fermentation

* Correspondence: linjp@zju.edu.cn
Key Laboratory of Biomass Chemical Engineering of the Ministry of Education,College of Chemical and Biological Engineering, Zhejiang University, Hangzhou 310027, China

Background

L-(+)-tartaric acid (L-TA), an important naturally existing hydroxyl carboxylic acid, is mainly used as antioxidant in food industry, as a chiral reagent in organic synthesis, as an acidic reducing agent in the textile industry and in galvanochemistry [1–3]. It is also an alternative to citric acid as an acidulant in food additives for its superior organoleptic properties [4]. Currently, the L-TA that is commercially available is produced exclusively through the stereospecific hydrolysis of *cis*-epoxysuccinate [5–7]. In this process, the reaction is catalyzed by the *cis*-epoxysuccinate hydrolase from *Rhodococcus rhodochrous*, *Nocardia tartaricans*, *Corynebacterium* sp. or *Pseudomonas* sp. [7]. However, *cis*-epoxysuccinate is derived from petrochemical-based precursor maleic anhydride, and this limits the production of L-TA. Therefore, development of the sustainable alternative solution for L-TA manufacture has recently attracted increasing attention.

A promising route has been employed for the production of L-TA by sequential whole-cell catalyzed oxidation and chemical catalysis, in which the glucose was first biologically converted to 5-keto-D-gluconate (5-KGA) by *Gluconobacter oxydans*, then to L-TA in the presence of ammonium vanadate as a trace element [4, 8]. Thus, the strategy towards an efficient synthetic route to L-TA was to optimize the enzymatic production of 5-KGA [4, 9].

Through genomic analyses, the genome sequence of *G. oxydans* 621H has been published, which leads to new insights into its metabolic pathway [10]. The essential genetic elements related to 5-KGA metabolism have been systematically identified (Fig. 1). The membrane-bound glucose dehydrogenase (mGDH, GOX0265), a quinoprotein containing pyrroloquinoline quinone (PQQ), oxidizes D-glucose to D-glucono-δ-lactone, which is subsequently converted to gluconic acid (GA) spontaneously or by gluconolactonase [11–13]. GA can be further oxidized to 5-KGA or 2-keto-D-gluconate (2-KGA) by PQQ-dependent sorbitol dehydrogenase (SLDH, GOX0854-0855) or FAD-dependent gluconate 2-dehydrogenase (GA2DH, GOX1230-1232), which transfer electrons from glucose to the respiratory chain ubiquinone and then to terminal ubiquinol oxidases to generate the proton motive force [10, 14]. PQQ, heme *c* or FAD serve as prosthetic groups [15, 16].

So far several attempts have been made to increase 5-KGA production, such as overexpression of gluconate: NADP 5-oxidoreductase [9], inactivation of GA2DH [3, 17] and overexpression of SLDH [18, 19] and optimization of the media conditions [20]. The highest yield achieved of 5-KGA was 240–295 mM (about 60 g/L) in a batch fermentation over periods as long as 72 h with a productivity of 0.83 g/(L · h) [18]. However, the methods used are

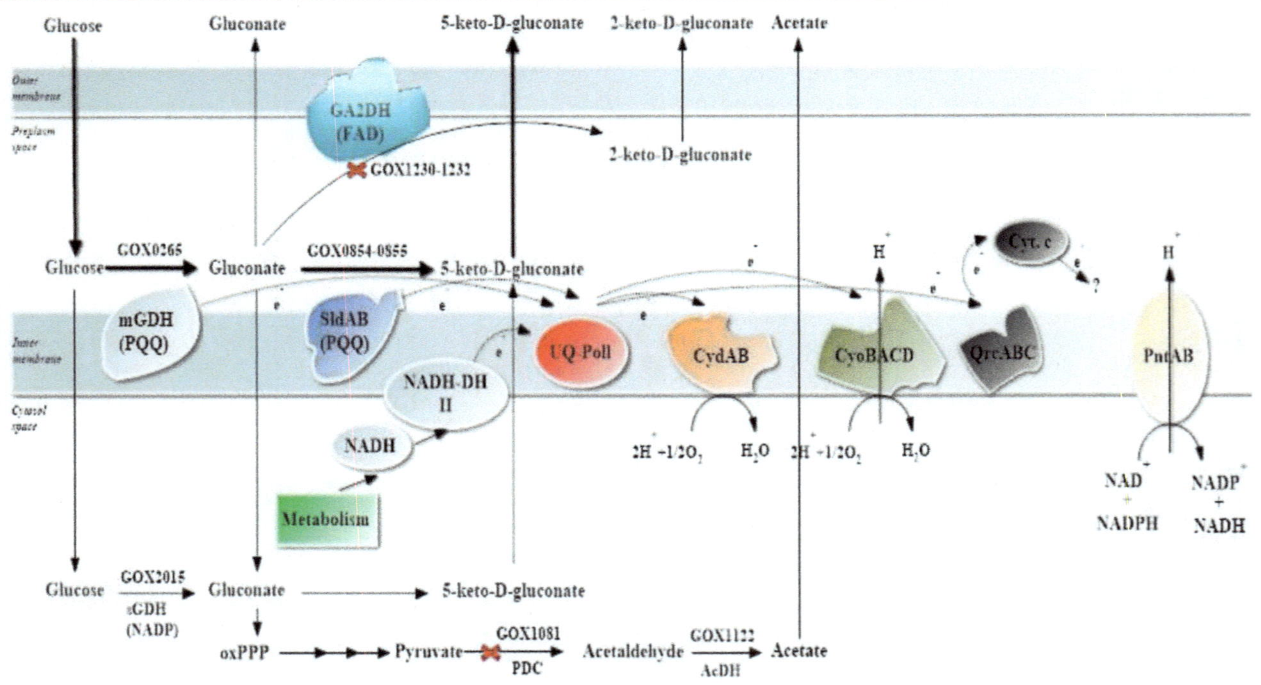

Fig. 1 Scheme of glucose metabolism and the respiratory chain in *Gluconobacter oxydans* DSM2343. Abbreviations: mGDH, membrane-bound PQQ-dependent glucose dehydrogenase; SldAB, membrane-bound PQQ-dependent sorbitol dehydrogenase; GA2DH, membrane-bound FAD-dependent gluconate 2-dehydrogenase; CydAB, cytochrome *bd* oxidase; CyoBACD, cytochrome bo_3 oxidase; QrcABC, cytochrome bc_1 complex; PntAB, transhydrogenase. the red-cross means the metabolic pathway cut off

individual engineering approaches and the efficiency of 5-KGA production was unsatisfactory, especially there was a large amount of residual GA left at the end of the biotransformation process.

In early work, we constructed the recombinant strain *G. oxydans* ZJU2, in which the GOX1231 and GOX1081 genes were markerless deleted [21]. The engineering of 2-KGA and acetic acid metabolism pathway could enhanced the 5-KGA production, but the titer and productivity of 5-KGA was unsatisfied. Hence, we set out to develop an efficient *Gluconobacter* cell factory to facilitate 5-KGA biosynthesis using combinatorial strategies. The *sldAB* gene, encoding the SLDH, was plasmid-based overexpressed with a strong promoter, P_{0169} [22]. Inspired by the mechanism from most of the *Gluconobacter*, it is proposed that the cofactor PQQ and respiratory chain engineering could be enhanced efficiency of the membrane-bound quinoaproteins. As a result, the genes involved with the PQQ cluster [23, 24] and terminal ubiquinol cytochrome bo_3 oxidase [25] were fused expression. The related specific enzyme activity, H^+/O ratio and 5-KGA titer were investigated. Base on the pH shift and dissolved oxygen tension (DOT) control, the robust *G. oxydans* cells were more facilitated 5-KGA accumulation by the fed-batch fermentation. This study represents combinatorial engineering approaches collectively increased the titer of 5-KGA in the *G. oxydans*, which can provide insights into devising engineering strategies to improve the object production.

Results and discussion

Promoter selection in *G. oxydans* strain

Increasing interest in *Gluconobacter* has led to investigations on strain improvement. The well characterized promoter is a prerequisite to the understanding of gene expression. So far, a few studies concerning the isolation and characterization of *G. oxydans* promoters, such as P_{0169} [22], P_{tufB} [26], P_{0264} [27], and P_{0452} [27], have been reported. However, which one is more suit for gene expression in *G. oxydans* is inconclusive. To select the appropriate promoter for gene expression in *G. oxydans*, we generated the different promoters in front of a *gfp* report gene in a pBBR1MCS5 vector. The promoter activity was indirectly determined by measuring the whole cell fluorescence intensity (RFU/OD_{600}). As shown in Fig. 2, the RFU/OD_{600} presented a linear increase with the cell growth until 24 h. Green fluorescent protein (GFP) contains a fluorescent cyclic tri-peptide, and oxygen is required for the final oxidation of the mature, cyclized fluorophore of GFP [22]. For this reason, the RFU was impacted by the level of the DO, which explained that the RFU/OD_{600} remained stable after 24 h. However, among this four different promoters, the cells expressing *gfp* under regulation of P_{0169} promoter exhibited outstanding effect in the *G. oxydans*. The observation elucidated that the P_{0169} promoter was evidently recognized by *G. oxydans* and could reliably drive heterologous gene expression in *G. oxydans*.

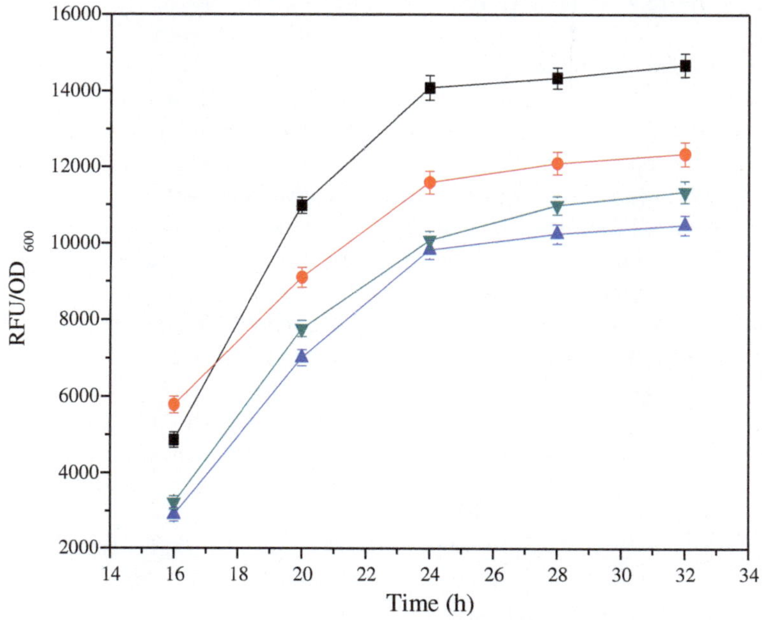

Fig. 2 Expression of green fluorescent protein in *G. oxydans* DSM2343 under the control of different promoter. The whole cell relative fluorescence unit (RFU) are the averages of three different experiments divided by the cell density at 600 nm. ■ P_{0169} promoter, ● P_{0264} promoter, ▲ P_{0452} promoter, ▼ P_{tufB} promoter

Enzyme activity and relative transcriptional level of *sldAB* overexpression strain

The glucose metabolism of the mutant *G. oxydans* ZJU2 strain showed the positive effect on the 5-KGA production [21], but it remained low productivity of 5-KGA. To boost the 5-KGA production, the *sldAB* gene was plasmid-based overexpressed in *G. oxydans* ZJU2 under the control of the selected P_{0169} promoter, generating *G. oxydans* ZJU3. The enzyme activity of SLDH of *G. oxydans* ZJU3 toward GA and the mRNA abundance of *sldAB* were investigated (Table 1). The SLDH activity was 2.55 ± 0.04 U/mg protein, which was more 3-fold higher than that of the reference strain. The *sldAB* expression data obtained were normalized to the transcriptional level of the 16S RNA. *G. oxydans* ZJU3 achieved the relative transcriptional level of 4.12 ± 0.04, which was 4-fold higher than those of the control. The results showed that the specific enzyme activities and transcription of the *sldAB* gene in *G. oxydans* could be markedly enhanced by *sldAB* overexpressed under the control of the selected P_{0169} promoter. In the literature [26], the stability and transcriptional level of the mRNA by adding poly (A/T) tails at the 3'-UTR of the *sldAB* were discussed. It was revealed that an artificial poly (A/T) tail was proposed to slow down the mRNA degradation process in bacteria and the high *sldAB* expression levels were achieved. This well demonstrated the importance of the mRNA stability on the gene expression, which should be considered in our late study.

Batch fermentation by *sldAB* overexpression strain

The batch fermentation by the recombinant *G. oxydans* ZJU3 strain was performed on a 15 L agitation tank under DOT rich condition. The results demonstrated that all tested strains had the similar trend of glucose consumption rate and cell growth (Fig. 3). The glucose was rapidly oxidized and exhausted at 30 h, while about 130 ± 1.76 g/L of GA was accumulated over the same time frame. The reference strain, *G. oxydans* ZJU2, accumulated 82.48 ± 1.10 g/L 5-KGA in the broth, but the residual GA was quite high at about 62.07 ± 1.04 g/L (Fig. 3a). In the fermentation process using recombinant *G. oxydans* ZJU3, 122.48 ± 0.41 g/L of 5-KGA with the productivity of 1.92 g/(L · h) were obtained at 64 h (Fig. 3b). Compared with the reference strain, the titer of 5-KGA was increased 48.99 % and the residual GA was 24.95 ± 0.76 g/L, decreased 59.80 %.

However, the DCW of the tested strains was about 3.58 ± 0.13 g/L, which was lower than other bacteria. In recent study, DNA microarray analysis and ^{13}C metabolic flux analysis (^{13}C -MFA) are used to characterize the two growth phases of *G. oxydans* in the presence of glucose [25]. In the first growth phase, 90 % of the glucose is oxidized by the mGDH to GA, accompanied by reasonable growth, high demand for oxygen and a low formation of CO_2. The onset of phase II results in reduced biomass and demand for oxygen while the GA was oxidized in the periplasm to 5-KGA (Fig. 3). Thus, only a small percentage of the carbon source is taken up by cell, resulting in modest cell growth and poor cell yield. *G. oxydans* IFO3293 and *G. oxydans* 621H cells cultivated on glucose medium provids a cell yield of 0.09 $g_{cdw}/g_{glucose}$ [28, 29]. In comparison to these values, *E. coli* reaches a value of 0.49 $g_{cdw}/g_{glucose}$ [30], and *Bacillus subtilis* reaches a yield of 0.32 $g_{cdw}/g_{glucose}$ [31]. In phase I and II, the cytoplasmic sugar catabolism proceeded predominantly via the PPP, particularly in phase II [28].

During the glucose metabolism by *G. oxydans*, the mGDH and SLDH serve as the main enzymes, and PQQ is the cofactor. The PQQ supplementation experiments showed that the control strain *G. oxydans* ZJU2 was not significantly influenced by the exogenous PQQ (Fig. 3c), but the *G. oxydans* ZJU3 strain showed the respond to PQQ supplementation in a similar manner, achieving the highest 5-KGA production when the added PQQ was up to 500 μg/L (Fig. 3d). A total of 131.92 ± 2.11 g/L of 5-KGA was produced by *G. oxydans* ZJU3. These results indicated that the addition of PQQ could enhance the production of 5-KGA in the SLDH overexpression strain, which was in agreement with a recent report that the overexpression of PQQ-dependent dehydrogenases could lead to imbalance between coenzyme PQQ level and the corresponding quinoproteins [24].

Enhanced PQQ biosynthesis to improve 5-KGA production

To further augment the 5-KGA production, the PQQ biosynthesis gene, *pqqABCDE,* was engineered and the biomass growth, PQQ concentration and 5-KGA production were shown in Table 2. In *G. oxydans* ZJU4, overexpression of *pqqABCDE* gene cluster under the control of P_{0169} promoter led to the excretion of 674.82 ± 4.12 μg/L of PQQ into the supernatant, which was enhanced by 383.53 % compared with that by the parent strain while the 5-KGA concentration was 131.76 ± 1.89 g/L, an increase of 7.58 %. This also confirmed the results from the PQQ supplement experiment (Fig. 3) that the coenzyme PQQ was the key factor driving the

Table 1 Enzyme activities and relative transcriptional levels of the membrane-bound SLDH in *G. oxydans* strains

Strains	Specific SLDH activity (U/mg protein)	Relative transcriptional levels of *sldAB* gene
G. oxydans ZJU2	0.75 ± 0.02	1.03 ± 0.02
G. oxydans ZJU2/ pBB5-P_{0169}	0.74 ± 0.01	1.02 ± 0.03 4.12 ± 0.04
G. oxydans ZJU2/ pBB5-P_{0169}-*sldAB*	2.55 ± 0.04	

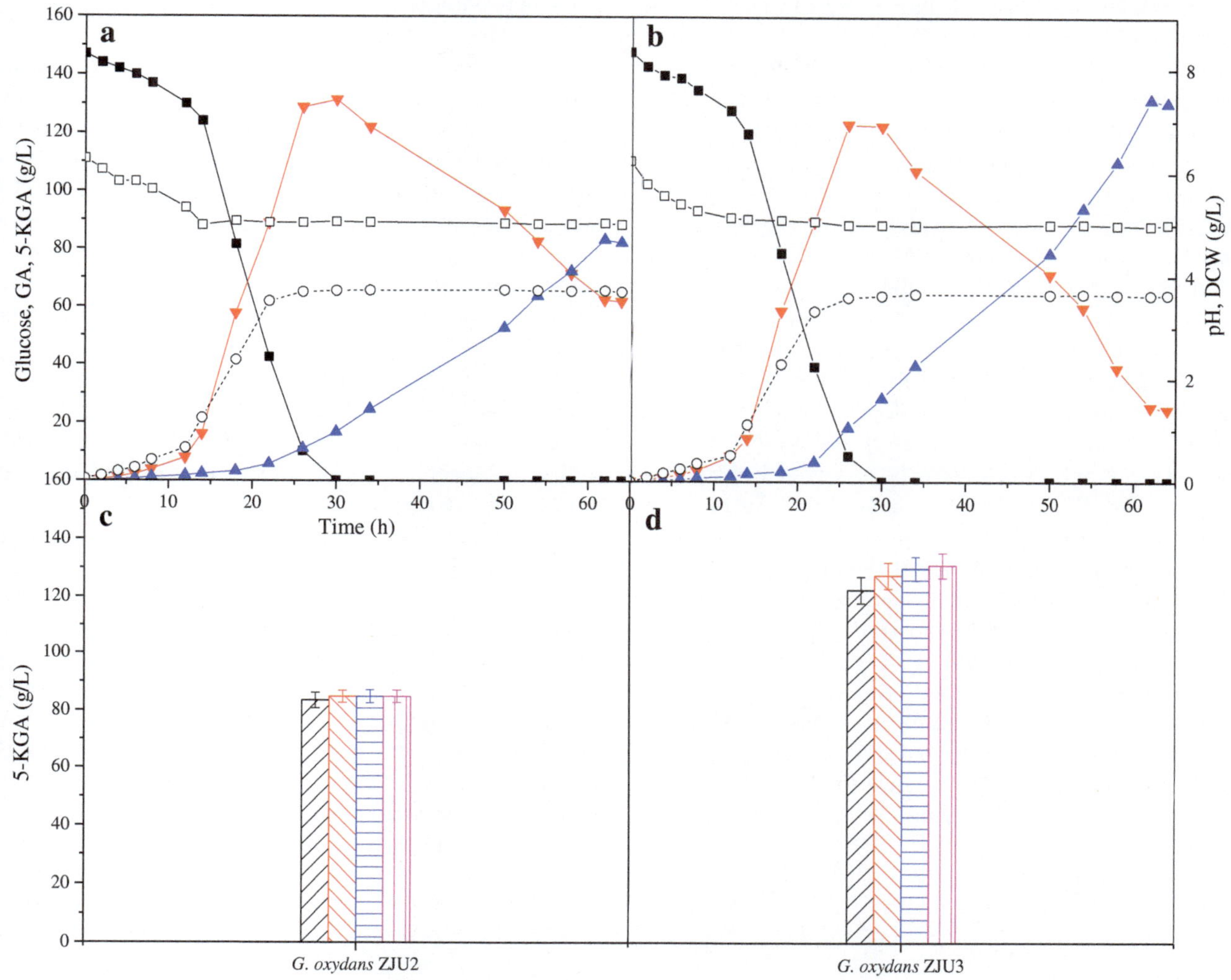

Fig. 3 Time-course of the oxidative fermentation in a 15-L fermentation tank and coenzyme PQQ complement study. **a**, **c** *G. oxydans* ZJU2, **b**, **d** *G. oxydans* ZJU3 strains. ■ Glucose, ▼ GA, ▲ 5-KGA, ○ DCW, □ pH value, control, 100 μg/L, 200 μg/L, 500 μg/L

5-KGA production. In *G. oxydans* ZJU5, simultaneous expression of the PQQ gene cluster and an associated *tldD* gene produced 757.83 ± 2.43 μg/L of PQQ, which was increased by 12.30 % compared with that from *G. oxydans* ZJU4. These results were consistent with the literature finding that the *tldD* gene was related to the PQQ biosynthesis [32]. The 5-KGA production by *G. oxydans* ZJU5 was 134.88 ± 2.16 g/L. In addition, the cell growth of *G. oxydans* ZJU5 was comparable to that of *G. oxydans* ZJU4. This finding implied that *tldD* gene expression did not affect the cell growth.

Overexpression of metabolic pathways involving redox reactions may lead to cofactor imbalances, thus impairing the yield of products. Cofactor engineering approaches are often adopted to compensate for imbalance of cofactors to improve product biosynthesis [33], for example, manipulating the availability of intracellular NADH [34] and NADPH [35]. In *G. oxydans*,

Table 2 Effects of overexpression *pqqABCDE* cluster and *tldD* genes in *G. oxydans* strains

Strains	Max. DCW (g/L)	PQQ concentration (μg/L)	5-KGA concentration (g/L)
G. oxydans ZJU3	3.58 ± 0.13	139.56 ± 1.87	122.48 ± 0.41
G. oxydans ZJU3/pUCpr	3.39 ± 0.09	137.74 ± 2.24	118.89 ± 1.28
G. oxydans ZJU3/pUCpr-T1	3.41 ± 0.11	674.82 ± 4.12	131.76 ± 1.89
G. oxydans ZJU3/pUCpr-T2	3.40 ± 0.08	757.83 ± 2.43	134.88 ± 2.16

quinoproteins and their cofactor PQQ can catalyze the oxidation of substrates by PQQ regulated electron transfer in the respiratory chain. However, the biosynthesis of quinoproteins and their cofactor PQQ are usually independent [24]. Therefore, the PQQ gene cluster was overexpressed in *Gluconobacter oxydans* WSH-003 [23] and *Ketogulonigenium vulgare* [24], which demonstrated to improve 2-keto-L-gulonic acid (2-KLG) production by cofactor engineering. These advance significantly facilitated the development of the efficient strains to produce 5-KGA.

Notably, it was reported that disruption of *tldD* gene in *G. oxydans* led to a drop of PQQ excretion below the detection limit and a decrease in cell growth, indicating that the related *tldD* gene was essential for PQQ biosynthesis [32]. The TldD protein of *G. oxydans* 621H, related to the *E. coli* TldD, is a peptidase involved in processing of small peptides. In other PQQ-producing bacteria, the peptidase-like protein PqqF is required for PQQ synthesis and has a similar function as the TldD in *G. oxydans* [32]. Therefore, our study demonstrated that overexpression of the *tldD* gene could increase PQQ level by 12.30 % in *G. oxydans* ZJU5 compared with that of *G. oxydans* ZJU4 (Table 2). This cofactor engineering was adopted to compensate the shortage of cofactors in the PQQ-dependent SLDH overexpression strain and to improve the 5-KGA production.

The engineering of the respiratory chain

In an earlier study, the cytochrome bo_3 oxidase (*cyoBACD*, GOX1911-1914) was found to be the main component for proton extrusion via the respiratory chain in *G. oxydans* 621H [25]. To enhance the respiratory chain activity and the 5-KGA production, the cytochrome bo_3 oxidase was engineered. The *cyoBACD* genes were fused into the PQQ overexpression plasmid (pUCpr-T1 and pUCpr-T2) under the P_{0169} promoter control, generating pUCpr-T3 and pUCpr-T4. The corresponding plasmids were electrotransferred into *G. oxydans* ZJU3, resulting in *G. oxydans* ZJU6 and *G. oxydans* ZJU7, respectively.

The batch fermentations of *G. oxydans* ZJU4, *G. oxydans* ZJU5, *G. oxydans* ZJU6, and *G. oxydans* ZJU7, under an excess of oxygen, showed that all recombinant *G. oxydans* strains reached their maximal OTR. The time points of maximal OTR correlated with the time points at which about 60 % glucose had been predominantly oxidized to GA in the periplasm (Fig. 4). *G. oxydans* ZJU4 and *G. oxydans* ZJU5 reached their maximal OTR of 26.52 mmol/L · h and 29.85 mmol/L · h, respectively at 19 h. After complete consumption of glucose, the OTR decreased to 1.5 to 1.8 mmol/L · h. The specific maximal CO_2 production rate (CTR) of both strains was 7.5 mmol/L · h at 32 h, then it began to decrease (Fig. 4a, b). The CTR decreased below 5.0 mmol/L · h at 57 h for

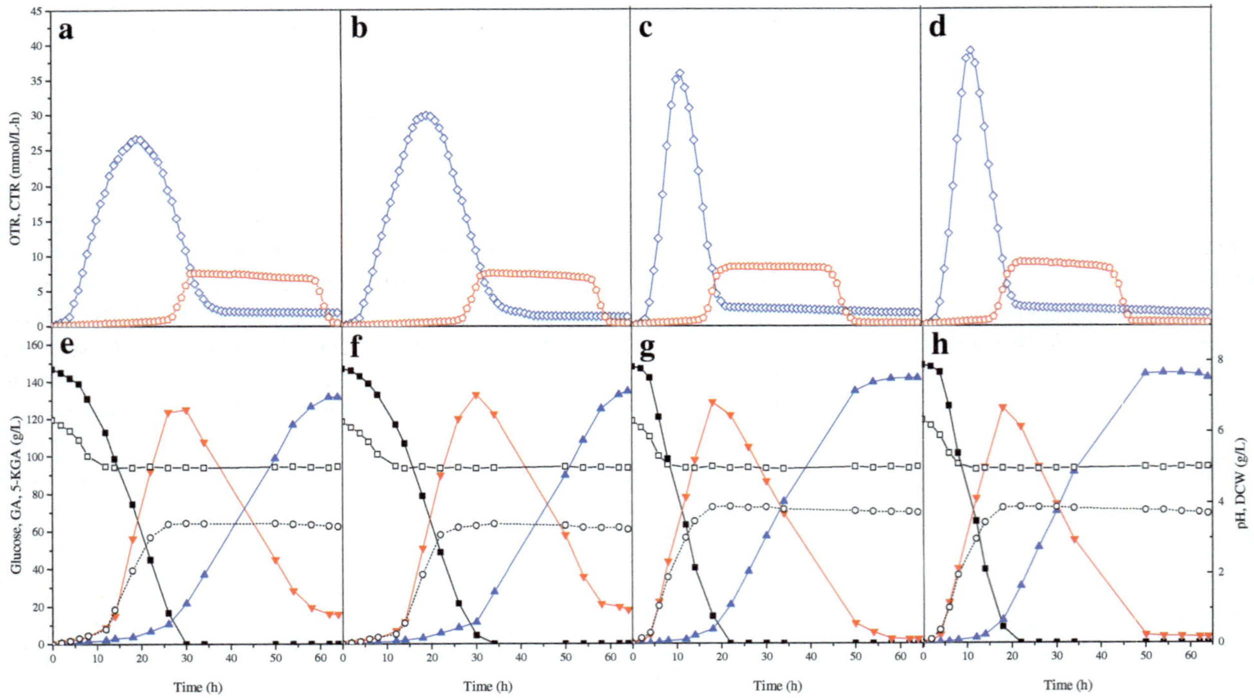

Fig. 4 The OTR and CTR value and the glucose metabolism. **a**, **e** *G. oxydans* ZJU4, **b**, **f** *G. oxydans* ZJU5, **c**, (g) *G. oxydans* ZJU6, **d**, **h** *G. oxydans* ZJU7. ■ Glucose, ▼ GA, ▲ 5-KGA, ○ DCW, □ pH value, ◇ OTR, and ⬠ CTR

the *G. oxydans* ZJU5 strain, but it decreased below this level at 60 h for *G. oxydans* ZJU4. This meant that the *tldD* gene overexpression could increase the PQQ biosynthesis and the oxidation rate of SLDH. The situations of *G. oxydans* ZJU6 and *G. oxydans* ZJU7 were remarkable different with those of *G. oxydans* ZJU4 and *G. oxydans* ZJU5, where 141.86 ± 2.89 g/L and 144.52 ± 2.94 g/L 5-KGA was accumulated, respectively. The glucose oxidation rate was accelerated, and glucose was exhausted at 20 h. The maximal OTR of 35.83 mmol/L · h and 38.97 mmol/L · h were achieved at 11 h, respectively. After 22 h, the CTRs of ZJU6 and ZJU7 reached the maximal values of 8.38 mmol/L · h and 8.98 mmol/L · h, respectively. The CTR of *G. oxydans* ZJU7 decreased to 0.5 mmol/L · h at 47 h, which was 13 h and 3 h short than those of *G. oxydans* ZJU5 and *G. oxydans* ZJU6, respectively (Fig. 4c, d). This indicated that the oxidation of GA to 5-KGA by *G. oxydans* ZJU7 had completely finished at 47 h.

Using the constructed *G. oxydans* strains, the H^+/O ratio and terminal ubiquinol bo_3 oxidase activity were also measured when the cells were in the logarithmic growth phase (DCW of 1.5). The results were shown in Table 3. The average H^+/O ratio and terminal ubiquinol oxidase activity of the control strain *G. oxydans* ZJU3 were 1.24 ± 0.11 and 0.31 ± 0.04 μmol/min · mg, respectively. The *G. oxydans* ZJU4 and *G. oxydans* ZJU5 had an H^+/O ratio and terminal ubiquinol oxidase activity comparable to those of the control strain, implying the mutant growth was not impaired under conditions of PQQ cluster overexpression. However, the recombinant *G. oxydans* ZJU6 and *G. oxydans* ZJU7 showed a 64 % increase of the H^+/O ratio (1.95 ± 0.23 and 2.01 ± 0.16) and 1.5-fold of the ubiquinol oxidase activity (0.78 ± 0.05 μmol/min · mg and 0.80 ± 0.06 μmol/min · mg) compared with the control strain, respectively. A total of 141.86 ± 2.89 g/L and 144.52 ± 2.94 g/L 5-KGA was accumulated by *G. oxydans* ZJU6 and *G. oxydans* ZJU7, respectively. This indicated the expression of ubiquinol bo_3 oxidase could enhance respiratory proton translocation and increase the 5-KGA production in *G. oxydans.*

G. oxydans possesses a branched respiratory chain consisting of two terminal ubiquinol oxidases, cytochrome bo_3 oxidase and cytochrome *bd* oxidase (*cydAB*, GOX0278-0279) (Fig. 1). The genome sequence also revealed genes for a cytochrome bc_1 complex (*qrcABC*, GOX0565-0567) and a soluble cytochrome *c* (GOX0258) [10]. The absence of cytochrome *bd* oxidase did not affect the cell growth or proton extrusion via the respiratory chain, whereas absence of the genes encoding cytochrome bo_3 oxidase caused a severe growth defect [25]. Plasmid-based overproduction of cytochrome bo_3 oxidase under the P_{0169} promoter control increased the respiration-driven proton extrusion by 66.7 % in *G. oxydans* ZJU6 and *G. oxydans* ZJU7, compared with the control strains (Table 3). The increase in the H^+/O ratio might be owing to the presence of more bo_3 quinol oxidase and its proposed high oxygen affinity, which should favor oxygen reduction under the experimental conditions of an oxygen pulse [25]. The H^+/O ratio for the reference strain, *G. oxydans* ZJU3, was measured as 1.21 ± 0.11. In the literature, H^+/O ratios reported for *E. coli* vary between 3.4 and 4.5 [36, 37], which was more than 2-fold higher than that of the *G. oxydans* ZJU3 strain. A major difference of the respiratory chain between the two species is the lack of the multisubunit proton-pumping NADH dehydrogenase I (NDH-I) in the *G. oxydans* [10, 25]. However, NDH-I is preferentially synthesized during anaerobic growth in the presence of alternate electron acceptors [38]. Therefore, what extent NDH-I contributes to the H^+/O ratio is unclear. As the *bd* oxidase was not relevant for proton translocation, an H^+/O ration of 4 might be assumed for *G. oxydans* [39]. However, our results were much lower than a value of 4. Richhardt explained that cytochrome bo_3 oxidase might not function as a primary proton pump but as a Na^+ pump and the respiratory chain could involve a reverse electron transfer coupled to an influx of protons [25].

Interestingly, an increased 15.48 % cell yield was observed when the bo_3 oxidase gene was overexpressed (Fig. 4 g, h). The possibility of an increased bo_3 oxidase level in *G. oxydans* could cause a shift of the electron flux from the non-proton pumping *bd* oxidase to the proton pumping bo_3 oxidase [25]. Another explanation for the increased cell growth is the assumption that the

Table 3 H^+/O ratio and ubiquinol oxidase activity of recombinant *G. oxydans* strains

Strains	H^+/O ratio[a] (No. of experiments)	Ubiquinol oxidase activity (μmol/min · mg)	5-KGA production (g/L)
G. oxydans ZJU3	1.21 ± 0.11 (8)	0.31 ± 0.04	122.48 ± 0.41
G. oxydans ZJU4	1.19 ± 0.10 (8)	0.32 ± 0.05	131.76 ± 1.89
G. oxydans ZJU5	1.20 ± 0.11 (8)	0.30 ± 0.11	134.88 ± 2.16
G. oxydans ZJU6	1.95 ± 0.23 (8)	0.78 ± 0.05	141.86 ± 2.89
G. oxydans ZJU7	2.01 ± 0.16 (8)	0.80 ± 0.06	144.52 ± 2.94

[a]the H^+/O measured by the oxygen pulse method

activity of the bo_3 oxidase is limited and can be increased by its overproduction. This was confirmed in the experiments that the ubiquinol oxidase activity increased 2.5-fold (Table 3) and the OTR enhanced 9.12 mmol/L · h (Fig. 4) by overexpressed the *cyoBACD* genes, indicating that cytochrome bo_3 quinol oxidase played the positive role in the cell growth and substrate oxidation. Recently, a limitation of oxygen consumption by the activity of the terminal oxidase was reported after overexpressing the membrane-bound glucose dehydrogenase (mGDH, GOX0265) [40]. This implied that the activity of bo_3 quinol oxidase might be high enough to satisfy the increased capacity for oxygen reduction.

Process optimization and fed-batch fermentation

The constructed *G. oxydans* ZJU7 showed good potential for 5-KGA production. To fulfill the requirement of industrial production of 5-KGA, the fermentation process was optimized, including the two-stage pH control, DO control and glucose feed. As shown in Fig. 5a, the glucose was oxidized faster and exhausted at 14 h when the pH was controlled at 5.5. This result was consistent with the previous results where a pH of 5.5 was suitable for the activity of membrane-bound glucose dehydrogenase [41]. However, a pH of 4.5 was more suitable for synthesis of 5-KGA by SLDH. In the early study, the pH of the culture medium was an important factor

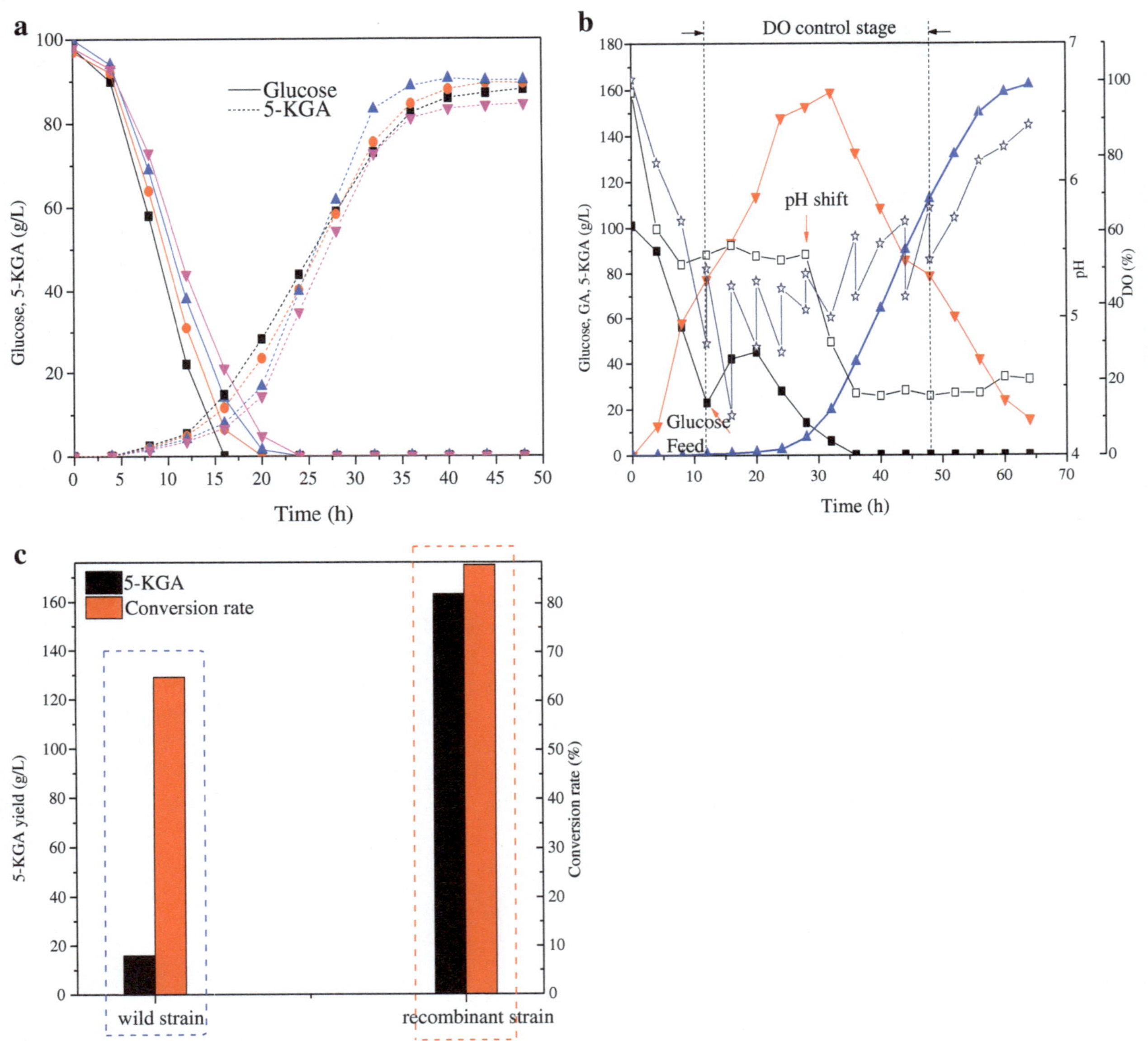

Fig. 5 a Effect of pH on the glucose consumption and 5-KGA production by *G. oxydans* ZJU7. ■ pH 5.5, ● pH 5.0, ▲ pH 4.5, ▼ pH 4.0. **b** The glucose fed-batch fermentation by *G. oxydans* ZJU7 under DO control and pH shift condition. ■ Glucose, ▼ GA, ▲ 5-KGA, □ pH value and ☆ DO. **c** The 5-KGA production and conversion rate between *G. oxydans* ZJU7 and wild-type strains. black-bar, 5-KGA, red-bar, conversion rate

for selective production of 5-KGA by *G. suboxydans* IFO 12528, in which an 87 % glucose conversion rate could be achieved by controlling the medium pH in a range of 3.5-4.0 [20]. According to the experimental results, we have established a two-stage pH control strategy, whereas the pH is controlled as 5.5 in the first stage to let the glucose oxidation and cell growth, then it is shifted to 4.5, facilitating 5-KGA formation.

In its natural habitat, *G. oxydans* is likely subjected to low-oxygen stress conditions because of the rapid oxygen consumption as a result of its own metabolism. The OTR and CTR were previously investigated during the 5-KGA production, in which the high OTR was observed when the glucose was oxidized to GA in the periplasm (Fig. 4a, b, c, d). It was reported that oxygen limitation could cause expression changes of 486 genes, representing 20 % of the chromosomal genes [42]. Therefore, during this stage, it must ensure the adequate DOT above 20 % by increasing the agitation speed. The onset of 5-KGA formation decreased the demand for DOT but increased the CTR (Fig. 4a, b, c, d). Hence, to improve the cost performance of the industrial production of 5-KGA, DOT was continuously controlled by adjusting the agitation speed according to OTR and CTR.

Additionally, to achieve the hyper-production of 5-KGA by the recombinant *G. oxydans* ZJU7 strain, the glucose fed-batch fermentation was carried out under the DOT and pH control. A glucose feed of about 1400 g was started at 12 h when the glucose concentration was below 40 g/L and was used to maintain the glucose at 40 ~ 50 g/L (Fig. 5b). The GA quickly accumulated during this stage. When the glucose concentration dropped to 14 g/L at 28 h, the 5-KGA began to quickly be formed and the pH shifted from 5.5 to 4.5. Concurrently, the DOT was controlled above 20 %. Using glucose fed-batch fermentation, DOT control, and a pH shift strategy, the 5-KGA titer was increased to 162 ± 2.12 g/L with the 2.53 g/(L · h) productivity by the recombinant *G. oxydans* ZJU7 strain, which was increased by 10-fold compared to the wild-type strain (Fig. 5c). This indicated that the PQQ overexpression and respiratory chain modification could efficiently enhance 5-KGA accumulation. Furthermore, the problem of residual GA was well solved, and the final concentration of GA was reduced to 15.38 g/L, which should decrease the separation difficulty. These results illustrated that the supplement of co-enzyme to the membrane-bound dehydrogenases was increased by overexpression of the PQQ cluster genes. The efficiency of electron transfer to O_2 was enhanced by overexpression of the genes for the bo_3 oxidase, which increased the activity of respiratory chain.

Conclusion

Bacteria of the genus *Gluconobacter* exhibit so-called oxidative fermentation, or incomplete oxidation, a highly unusual metabolic feature that has been exploited industrially for decades in the production of vitamin C, GA and dihydroxyacetone [43]. In this study, the role of the combinatorial metabolic engineering of the industrial *G. oxydans* for the boosting of 5-KGA accumulation was investigated. In summary, the *sldAB* was overexpressed under the selected P_{0169} promoter. In addition, the PQQ gene cluster and terminal ubiquinol bo_3 oxidase were fused overexpression to strengthen the glucose oxidation. Under the optimized culture conditions of the fed-batch process, the combinatorial approaches collectively increased the 5-KGA titer 10-fold, reached 162 ± 2.12 g/L. The results showed great potential for optimizing the current producer strain of *G. oxydans* used in industrial biotechnology, which presents the first report of 5-KGA production by combinatorial metabolic engineering approaches in *G. oxydans*. We envision that these approaches could provide framework for devising engineering strategies to improve the production of biochemicals in *G. oxydans*. Nevertheless, some issues are still worth further study, for example, how many copies of *pqq* cluster and bo_3 oxidase genes overexpression can be matched the quinoproteins requirements, especially in the quinoproteins overexpression strains. The genome sequence has revealed that the TCA cycle is incomplete as genes for succinate dehydrogenase and succinyl CoA synthetase were absent [10]. To develop more robust strains, the strategy involving direct repair of this defective metabolic pathway by genomic integration of heterologous genes should be investigated, while considering the *G. oxydans* as a broadly applicable host for oxidative industrial bioconversions.

Methods

Bacterial strains, plasmids, and media

The bacterial strains, and plasmids used in this study are listed in Table 4. *Escherichia coli* strains were cultivated in Luria–Bertani (LB) medium at 37 °C. The 50 μg/mL gentamicin or 100 μg/mL ampicillin were used whenever required. Agar (1.5 %) was added to obtain solid media. *Gluconobacter oxydans* DSM2343 strains were cultivated on mannitol medium (MP) containing 5 g/L yeast extract, 3 g/L peptone, and 25 g/L mannitol at 30 °C. For *G. oxydans* possesses a natural resistance against cefoxitin, thus, as a precaution to prevent bacterial contamination, 50 μg/mL cefoxitin was added.

Construction of shuttle vector

A shuttle vector pUCpr, compatible to the broad-host-plasmid pBBR1MCS5, was constructed. The 2446-bp *par-rep* gene fragment of cryptic plasmid pGOX3 from *G. oxydans* DSM2343 was amplified with primers pr_PstI_F / pr_SalI_R (Table 5). The sequenced PCR

Table 4 Bacterial strains and plasmids used in this work

	Properties	Source
Strains		
E. coli DH5α	F^-, *endA1*, *hsdR17* (rk-mk-),*supE44*, *thi1*, *recA1*, *gyrA*, (Nalr), *relA1*, D(*lacZYAargF*), U169, and F80*lacZ*DM15	Invitrogen
G. oxydans DSM2343	Wild-type, Cef^R	DSMZ[a]
G. oxydans ZJU2	Gluconate 2-dehydrogenase and pyruvate decarboxylase deletion strain derived from *G. oxydans* DSM2343, Cef^R	[21]
G. oxydans ZJU3	*G. oxydans* ZJU2 harboring pBB5-P_{0169}-*sldAB*, Cef^R, Gm^R	This work
G. oxydans ZJU4	*G. oxydans* ZJU3 harboring pUCpr-T1, Cef^R, Gm^R, Amp^R	This work
G. oxydans ZJU5	*G. oxydans* ZJU3 harboring pUCpr-T2, Cef^R, Gm^R, Amp^R	This work
G. oxydans ZJU6	*G. oxydans* ZJU3 harboring pUCpr-T3, Cef^R, Gm^R, Amp^R	This work
G. oxydans ZJU7	*G. oxydans* ZJU3 harboring pUCpr-T4, Cef^R, Gm^R, Amp^R	This work
Plasmids		
pUC19	Cloning vector, ColE1 *ori*, Amp^R	Invitrogen
pET28 (a)-GFP	*gpf* gene expressed vector, Km^R	Laboratory preservation
pBBR1MCS-5	Broad-host-range (bhr) expression vector, Gm^R	[45]
pBB5-P_{tufB}	Insert P_{tufB} promoter vector derived from pBBR1MCS-5, Gm^R	This work
pBB5-P_{0264}	Insert P_{0264} promoter vector derived from pBBR1MCS-5, Gm^R	This work
pBB5-P_{0452}	Insert P_{0452} promoter vector derived from pBBR1MCS-5, Gm^R	This work
pBB5-P_{0169}	Insert P_{0169} promoter vector derived from pBBR1MCS-5, Gm^R	This work
pBB5-P_{0169}-*sldAB*	*sldAB* gene overexpression vector derived from pBBR1MCS-5, inserted P_{0169} promoter, Gm^R	This work
pUCpr	Constructed expression vector derived from pUC19, *par-rep*, Amp^R	This work
pUCpr-T1	pUCpr-P_{0169}-*pqqABCDE* vector derived from pUCpr, Amp^R	This work
pUCpr-T2	pUCpr-P_{0169}-*pqqABCDE-tldD* vector derived from pUCpr, Amp^R	This work
pUCpr-T3	pUCpr-P_{0169}-*pqqABCDE*-P_{0169}-*cyoBACD* vector derived from pUCpr, Amp^R	This work
pUCpr-T4	pUCpr-P_{0169}-*pqqABCDE-tldD*-P_{0169}-*cyoBACD* vector derived from pUCpr, Amp^R	This work

[a]DSMZ, Deutsche Sammlung von Mikroorganismen und Zellkulturen, Braunschweig, Germany

product was digested and inserted into the *Pst*I/*Sal*I site of the pUC19, resulting in pUCpr [44].

Promoter selection

Four different promoters from *G. oxydans* DSM2343 were carefully selected. A putative promoter sequence, P_{0169} [22], the promoter of elongation factor *Tu*, P_{tufB} [26], and ribosomal proteins L35 and L13, P_{0264} and P_{0452} [27], were amplified by PCR with primers as listed in Table 5. The resulting DNA fragments P_{tufB}, P_{0264}, P_{0452} and P_{0169} were digested with restriction enzymes *Sac*I and *Xba*I, then were ligated into pBBR1MCS-5 [45], generated the vector pBB5-P_{tufB}, pBB5-P_{0264}, pBB5-P_{0452} and pBB5-P_{0169}, respectively. The report gene *gfp* was amplified from the cloning vector pET28 (a)-GFP. The resulting product was digested with *Xba*I and *Hind*III, and then cloned into the *Xba*I and *Hind*III site of pBB5-P_{tufB}, pBB5-P_{0264}, pBB5-P_{0452} and pBB5-P_{0169} to generate the corresponding promoter strength reporter plasmids.

The reporter plasmids were first transformed into *E. coli* DH5α, analyzed for the correct insert by DNA sequencing and then transformed into *G. oxydans* DSM2343 by electroporation (2000 V, 200 Ω and 25 μF) in a 2-mm cuvette using a Gene Pulser II (Bio-Rad, München, Germany) as described previously [46]. The transformants were selected by cefoxitin and gentamicin. The whole cell fluorescence intensity (RFU/OD, the relative fluorescence unit divided by the corresponding cell density) was measured [22]. Cells were harvested and washed twice with KPB buffer (pH 7.0), and then photographed by using a confocal laser scanning microscope. The cell density was measured by the absorbance at 600 nm using the spectrophotometer (UVmini-1240, SHIMADZU®), which determined the strengths of the different promoters.

Table 5 The oligonucleotides primers used in this work

Primer	Sequence (5′ → 3′)	Usage
pr_PstI_F	AACTGCAGgtttatcggccgttgaatat	Amplify the *par-rep* gene
pr_SalI_R	ACGCGTCGACggtgtttaaacagtgttacggt	
0169_SacI_F	ATAGAGCTCtgaaagcggctggcgcgt	Amplify the 5′-UTR of GOX0169 promoter
0169_XbaI_R	GCTCTAGAgcggaaggcgttataccctga	
0264_SacI_F	ATAGAGCTCgttgcgcctgaatgagagg	Amplify the 5′-UTR of GOX0264 promoter
0264_XbaI_R	GCTCTAGAttcggtctccctcgccgtaa	
0452_SacI_F	ATAGAGCTCggcttcgtggtgaacgcc	Amplify the 5′-UTR of GOX0452 promoter
0452_XbaI_R	GCTCTAGAtagtgacattccagcttggg	
tufB_ SacI_F	ATAGAGCTCcgatggtaagaaatccactgc	Amplify the *tufB* promoter
tufB_ XbaI_R	ATATCTAGAccaaaaccccgctccacc	
GFP_XbaI_F	ATATCTAGAatggtgagcaagggc	Amplify the *gfp* reported gene
GFP_HindIII_R	CCCAAGCTTctacttgtacagctc	
SLDH_XbaI_F	GCTCTAGAggactttcagttctggaggctttcacca	Amplify the *sldAB* gene
SLDH_EcoRI_R	CGGAATTCtcccacccgaaaaatggaaaaaacg	
ADD_0169_F	acactgtttaaacaccgtgaaagcggctggcgc	Amplify the fuse fragments
pQQ_Fuse0169_R	acatccgcgcggaaggcgttatac	
pQQ_Fuse0169_F	ccttccgcgcggatgttcagg	
tldD_FusepQQ_R	ccggctagaagatggcctctc	
tldD_FusepQQ_F	gccatcttctagccggtctgttc	
0169_FusetldD_R	ctttcaggatcttcttcatg	
0169_FusetldD_F	tcgcgactgaaagcggctggc	
ADD_0169_R	cggtacccggggatcctgcggaaggcgttatac	
cyoBACD_XbaI_F	CGATTCTAGAactactgcaagccggaacgg	Amplify the terminal cytochrome bo_3 oxidase
cyoBACD_SacI_R	ACTGGAGCTCaagggctggcaggatttctc	
RT16S_F	gcggttgttacagtcagatg	-
RT16S_R	gcctcagcgtcagtatcg	-
RTsldh_F	atcatgccgaccaagcgtggc	-
RTsldh_R	cgtcggcgaacgcggatcg	-

[a]The capital and underlined sequences indicate the restriction enzyme sites

Overexpression of *sldAB* gene in *G. oxydans* ZJU2

Based on the bioinformatics analysis of *G. oxydans* 621H genome sequence [10], SLDH_XbaI_F / SLDH_EcoRI_R primers were designed and the open reading frames (ORFs) of SLDH (*sldAB*, GOX0854-0855) was PCR-amplified with the primers. The genomic DNA of *G. oxydans* DSM 2343 was used as a template. The sequenced amplicon was digested with the restriction endonucleases *Xba*I and *EcoR*I and cloned into pBB5-P_{0169} vector restricted with the same enzymes, resulting in plasmids pBB5-P_{0169}-*sldAB*. This plasmid and as a control pBB5-P_{0169} vector were transferred into the desired *G. oxydans* ZJU2 strain by electroporation and were selected for a gentamycin-resistant phenotype. The correct strain *G. oxydans* ZJU2/pBB5-P_{0169}-*sldAB* was named *G. oxydans* ZJU3.

Co-expression of the PQQ biosynthesis genes and the terminal ubiquinol cytochrome bo_3 oxidase genes in *G. oxydans* ZJU3

In the oxidation process by the membrane-bound SLDH, the PQQ serves as prosthetic groups [10, 15, 16], which transfer electrons to the respiratory chain [10]. To enhance the 5-KGA production, the genes encoding the cofactor PQQ and the terminal ubiquinol cytochrome bo_3 oxidase of the respiratory chain were reinforced. To fulfill the experiments, the PQQ gene cluster (GOX0983-0987) and the related *tldD* gene (GOX1104) were cloned and the generated plasmids pUCpr-T1 and pUCpr-T2. The constructed plasmids were transferred into *G. oxydans* ZJU3, resulting the recombinant strains *G. oxydans* ZJU4 and *G. oxydans* ZJU5, respectively. The PQQ and terminal ubiquinol cytochrome bo_3 oxidase fused expression plasmid was constructed by the *pEASY*-Uni seamless cloning and assembly kit CU101 (TransGen, China). The PCR-amplified promoter P_{0169}, *pqqABCDE*, and *tldD* were cloned into the *Xba*I site of the shuttle vector pUCpr, which resulted in the plasmid pUCpr-T1-P_{0169} and pUCpr-T2-P_{0169}. Then the *cyoBACD* (GOX1911-1914) was inserted the *Xba*I and *Kpn*I site of pUCpr-T1-P_{0169} and pUCpr-T2-P_{0169} to generate the fusion expressed plasmid pUCpr-T3 and pUCpr-T4, respectively. After verification of the accuracy of the plasmid by the sequencing, the fusion plasmid was transferred in the recombinant strain *G. oxydans* ZJU3 by electroporation [46] and selected for a gentamycin-kanamycin resistant phenotype, generated *G. oxydans* ZJU6 and *G. oxydans* ZJU7.

Measurements of enzyme activity and protein concentration

For the purpose of preparation of the cell crude extract and membrane fraction, a single colony of the *G. oxydans* strain was pre-incubated in MP medium. The *G. oxydans* cells were harvested by centrifugation (10,000 × g, 5 min, 4 °C) and resuspended in 20 mL 50 mM sodium phosphate buffer (pH 6.0). The cells were disrupted with an ultrasonifier (JY92–2D, Xinzhi, NingBo) for 50 cycles (2000 w, 3 s sonication, 5 s pause) on ice. The cell debris were removed by centrifugation at 5500 × g for 20 min at 4 °C and the supernatant was used as the crude extract. For the preparation of membranes, the supernatant was centrifuged for 60 min at 180,000 × g at 4 °C. The resulting sediments were collected and resuspended into 50 mM sodium phosphate buffer (pH 6.0), and used as the membrane fraction.

Enzyme activities were determined using a spectrophotometer (Uvmini-1240 SHIMADZU®). Substrate-dependent changes of redox states of artificial electron acceptors (2, 6-dichlorophenolindophenol, DCPIP, Sigma) were determined at 600 nm and 30 °C. The basal reaction mixture contained 50 mM PBS pH 6.0, 0.25 mM DCPIP, and 0.325 mM phenazine methosulphate (PMS, Sigma), which was prepared and pre-warmed to 30 °C before the assay. Measurements were performed in a cuvette with a 1-cm light path containing a 0.8-mL basal reaction mixture and 10 μL enzyme pre-incubated at 30 °C for 5 min. The reaction was started by adding 20 μL of a 2.0 M gluconate solution. One unit of enzyme activity (U) was defined as the amount of enzyme that can catalyze the conversion of 1 μM DCPIP per min at 30 °C. The concentration of proteins was determined with the Pierce™ BCA assay kit.

Quantitative real-time PCR (RT-PCR)

Cells were harvested at an OD_{600} of 2.5 at room temperature and immediately frozen in liquid nitrogen. Cells were then stored at −80 °C until RNA extraction. Total RNA was extracted with RNAiso™ Plus from Takara (Dalian, China) and treated with RNase-free DNase. Following chloroform extraction step, RNA was precipitated with isopropanol and the pellet washed twice in 75 % ethanol. After air-drying, RNA was resuspended in RNase-free water. The quantity of total RNA was verified using an Eppendorf Biophotometer (Eppendorf, Hamburg, Germany). The cDNA was synthesized from the total RNA using a PrimeScript RT Reagent Kit (Perfect Real Time) (Takara) according to the manufacturer's protocol. The products were quantified via real-time PCR with StepOnePlus™ Real-Time PCR System (Applied Biosystems, USA) using primer RTsldh_F/RTsldh_R. The 16S rRNA gene was used as internal standard based on the primer RT16S_F/RT16S_R.

Characterization of respiration activity and H^+/O measurements

Cultivation of *G. oxydans* strains was performed and the respiration activity was measured by the exhaust gas analysis system, such as the oxygen transfer rate (OTR), carbon dioxide transfer rate (CTR). The number of H^+ moved upon respiration (H^+/O ratio) is principally important for the efficiency of cellular ATP production. Hence the H^+/O ratio of bacterial cells have been measured to evaluate the efficiency of the respiratory chain. The H^+/O and terminal ubiquinol oxidase activity were measured using a method previously reported in the literature [36].

Batch fermentations

Batch and fed-batch fermentations were conducted in a 15 L stirred tank bioreactor (Fus-D; Guoqiang Bioengineering Equipment Co., Ltd, Shanghai, China) with 9 L of the initial medium, which was composed of 0.41 g/L $(NH_4)_2SO_4$, 0.1 g/L $(NH_4)_2HPO_4$, 0.01 g/L $MgSO_4 \cdot 7H_2O$, 3.0 g/L corn steep liquor paste and 100–150 g/L glucose (depending on the experiment) [19]. The $CaCO_3$ (20 g/L) was sterilized separately and then added to the medium. The seed culture was prepared by inoculation of a single colony into a 5-mL MP medium tube, then into 200 mL of fresh seed medium in 500-mL flasks and cultivated on a rotary shaker at 220 rpm for 16 h. The seed culture (5 %, v/v) was then inoculated into the fermentation medium and the fermentation was carried out at 30 °C and pH was controlled at 5.0 by automatic addition of 5 M NaOH. In order to examine whether PQQ could enhance the 5-KGA production in engineered strains, different amount of PQQ (0 μg/L, 100 μg/L, 200 μg/L, 500 μg/L) was exogenously added to the mixed cultures at the beginning of the fermentation process, and the 5-KGA concentrations were measured after 64 h.

Fed-batch fermentation

In this study, optimization of the pH condition was examined in a medium containing 100 g/L glucose as the initial substrate. The pH was controlled as 5.5, 5.0, 4.5, and 4.0 and the glucose and 5-KGA were detected. Hence, a two-stage pH control strategy was employed. In the first stage of glucose oxidized to GA, then shifted to 4.5 in the process of 5-KGA production. In further experiment, fed-batch culture was performed. When the fermentation was began, the initial volume was 6 L and the glucose concentration was 100 g/L, then 2 L of the feed medium containing 1400 g glucose was added when the concentration of glucose turned to 30 g/L and maintained between 30 and 40 g/L.

Analysis

Samples were centrifuged at 12,000 g for 2 min, and the supernatant was passed through a 0.22 μm filter. The residual glucose concentration was determined by a bioanalyzer (SBA-40D, Shandong Academy of Sciences, China) after dilution to an appropriate concentration. The GA and 5-KGA in the fermentation broth were analyzed by high-performance liquid chromatography (HP1100, Agilent 1100 series) using a RSpak DE-613 column (Shodex, Japan), with 2 mM $HClO_4$ as the mobile phase at a flow rate of 0.5 mL/min and a UV absorption of 210 nm. Acetic acid was detected by GC (Agilent 6820 series). The biomass dry cell weight was determined by applying membrane filtration. Before filtration, the $CaCO_3$ was removed by reaction with HCl. The concentration of PQQ in the culture supernatants was measured according to the literature [24]. All experiments were performed in triplicate and the Origin 8.0 software package was used for statistical analysis. Analysis of variance was performed. Each data point represents the mean ± SD from triplicate experiments.

Abbreviations used

2-KGA: 2-keto-D-gluconate; 5-KGA: 5-keto-D-gluconate; DCPIP: 2, 6-dichlorophenolindophenol; DCW: dry cell weight; DOT: dissolved oxygen tension; GA: gluconic acid; PQQ: pyrroloquinoline quinone.

Competing interests

All authors declare that they have no competing interests.

Authors' contributions

JY, MW and JL conceived and designed the study. JY and JL drafted the manuscript. LY participated in the project design and manuscript preparation. All authors read and approved the final manuscript.

Authors' information

JY, Ph.D. student of Zhejiang University. MW, JL, and LY are staffs of Zhejiang University. The address is as follow: Key Laboratory of Biomass Chemical Engineering of Ministry of Education, Department of Chemical and Biological Engineering, Zhejiang University, Hangzhou 310027, China

Acknowledgments

This study was financially supported by the National High Technology Research and Development Program of China (863 Program, No. 2014AA022105) and the Zhejiang Leading Team of S & T Innovation Program (2011R50002).

References

1. Chandrashekar K, Felse PA, Panda T. Optimization of temperature and initial pH and kinetic analysis of tartaric acid production by *Gluconobacter suboxydans*. Bioprocess Eng. 1999;20:203–7.
2. DeBolt S, Cook DR, Ford CM. L-tartaric acid synthesis from vitamin C in higher plants. Proc Natl Acad Sci U S A. 2006;103:5608–13.
3. Elfari M, Ha SW, Bremus C, Merfort M, Khodaverdi V, Herrmann U. A *Gluconobacter oxydans* mutant converting glucose almost quantitatively to 5-keto-D-gluconic acid. Appl Microbiol Biotechnol. 2005;66:668–74.
4. Matzerath I, Kläui W, Klasen R, Sahm H. Vanadate catalyzed oxidation of 5-keto-D-gluconic acid to tartaric acid: the unexpected effect of phosphate and carbonate on rate and selectivity. Inorg Chim Acta. 1995;237:203–5.
5. Li X, Xu TC, Ma XH, Guo KP, Kai L, Zhao YH. Optimization of culture conditions for production of *cis*-epoxysuccinic acid hydrolase using response surface methodology. Bioresource Technol. 2008;99:5391–6.
6. Orru RVA, Archelas A, Furstoss R, Faber K. Epoxide hydrolases and their synthetic applications. Adv Biochem Engin/Biotechnol. 1999;63:145–67.
7. Willaert R, Vuyst LD. Continuous production of L-(+)-tartaric acid from *cis*-epoxysuccinate using a membrane recycle reactor. Appl Microbiol Biotechnol. 2006;71:155–63.
8. Klasen R, Bringer-Meyer S, Sahm H. Incapability of *Gluconobacter oxydans* to produce tartaric acid. Biotechnol Bioeng. 1992;40:183–6.
9. Klasen R, Bringer-Meyer S, Sahm H. Biochemical characterization and sequence analysis of the gluconate: NADP 5-oxidoreductase gene. J Bacteriol. 1995;177:2637–43.
10. Prust C, Hoffmeister M, Liesegang H, Wiezer A, Fricke WF, Ehrenreich A. Complete genome sequence of the acetic acid bacterium *Gluconobacter oxydans*. Nat Biotechnol. 2005;23:195–200.
11. Kataoka N, Matsutani M, Yakushi T, Matsushita K. Efficient production of 2, 5-diketo-D-gluconate via heterologous expression of 2-ketogluconate dehydrogenase in *Gluconobacter japonicas*. Appl Environ Microbiol. 2015;81:3552–60.
12. Matsushita K, Shinagawa E, Adachi O, Ameyama M. Reactivity with ubiquinone of quinoprotein $_D$-glucose dehydrogenase from *Gluconobacter suboxydans*. J Biochem. 1989;105:633–7.
13. Shinagawa E, Ano Y, Yakushi T, Adachi O, Matsushita K. Solubilization, purification, and properties of membrane-bound D-glucono-δ-lactone hydrolase from *Gluconobacter oxydans*. Biosci Biotechnol Biochem. 2009;73:241–4.
14. Toyama H, Furuya N, Saichana I, Ano Y, Adachi O, Matsushita K. Membrane bound, 2-keto-D-gluconate yielding D-gluconate dehydrogenase from *Gluconobacter dioxyacetonicus* IFO 3271: molecular properties and gene disruption. Appl Environ Microbiol. 2007;73:6551–6.
15. Shinagawa E, Matsushita K, Adachi O, Ameyama M. Evidence for electron transfer via ubiquinone between quinoproteins D-glucose dehydrogenase and alcohol dehydrogenase of *Gluconobacter suboxydans*. J Biochem. 1990;107:863–7.
16. Toyama H, Mathews FS, Adachi O, Matsushita K. Quinohemoprotein alcohol dehydrogenase: structure, function, and physiology. Arch Biochem Biophys. 2004;428:10–21.
17. Saichana I, Moonmangmee D, Adachi O, Matsushita K, Toyama H. Screening of thermotolerant *Gluconobacter* strains for production of 5-keto-D-gluconic acid and disruption of flavin adenine dinucleotide-containing D-gluconate dehydrogenase. Appl Environ Microbiol. 2009;75:4240–7.
18. Merfort M, Herrmann U, Ha SW, Elfari M, Bringer-Meyer S, Görisch H. Modification of the membrane-bound glucose oxidation system in *Gluconobacter oxydans* significantly increases gluconate and 5-keto-D-gluconic acid accumulation. Biotechnol J. 2006;5:556–63.
19. Merfort M, Hermann U, Bringer-Meyer S, Sahm H. High-yield 5-keto-D-gluconic acid formation is mediated by soluble and membrane-bound gluconate-5-dehydrogenase of *Gluconobacter oxydans*. Appl Microbiol Biotechnol. 2006;73:443–51.

20. Ano Y, Shinagawa E, Adachi O, Toyama H, Yakushi T, Matsushita K. Selective, high conversion of D-glucose to 5-keto-D-gluconate by *Gluconobacter suboxydans*. Biosci Biotechnol Biochem. 2011;75:586–9.
21. Yuan JF, Wu MB, Lin JP, Yang LR. Enhancement of 5-keto-D-gluconate production by a recombinant *Gluconobacter oxydans* using a dissolved oxygen control strategy. *J Biosci Bioeng* 2015, http://dx.doi.org/10.1016/j.jbiosc.2015.12.006.
22. Shi LL, Li KF, Zhang H, Liu X, Lin JP, Wei DZ. Identification of a novel promoter *gHp0169* for gene expression in *Gluconobacter oxydans*. J Biotechnol. 2014;175:69–74.
23. Gao LL, Hu YD, Liu J, Du GC, Zhou JW, Chen J. Stepwise metabolic engineering of *Gluconobacter oxydans* WSH-003 for the direct production of 2-keto-L-gulonic acid from D-sorbitol. Metab Eng. 2014;24:30–7.
24. Du J, Bai W, Song H, Yuan YJ. Combinational expression of sorbose/ sorbosone dehydrogenases and cofactor pyrroloquinoline quinone increases 2-keto-L-gulonic acid production in *Ketogulonigenium vulgare–Bacillus cereus* consortium. Metab Eng. 2013;19:50–6.
25. Richhardt J, Luchterhand B, Bringer S, Büchs J, Bott M. Evidence for a key role of cytochrome bo_3 oxidase in respiratory energy metabolism of *Gluconobacter oxydans*. J Bacteriol. 2013;195:4210–20.
26. Xu S, Wang XB, Du GC, Zhou JW, Chen J. Enhanced production of L-sorbose from L-sorbitol by improving the mRNA abundance of sorbitol dehydrogenase in *Gluconobacter oxydans* WSH-003. Microb Cell Fact. 2014;13:146–52.
27. Kallnik V, Meyer M, Deppenmeier U, Schweiger P. Construction of expression vectors for protein production in *Gluconobacter oxydans*. J Biotechnol. 2010;150:460–5.
28. Richhardt J, Bringer S, Bott M. Role of the pentose phosphate pathway and the entner-doudoroff pathway in glucose metabolism of *Gluconobacter oxydans* 621H. Appl Microbiol Biotechnol. 2013;97:4315–23.
29. Krajewski V, Simić P, Mouncey NJ, Bringer S, Sahm H, Bott M. Metabolic engineering of *Gluconobacter oxydans* for improved growth rate and growth yield on glucose by elimination of gluconate formation. Appl Environ Microbiol. 2010;76:4369–76.
30. Soini J, Ukkonen K, Neubauer P. High cell density media for *Escherichia coli* are generally designed for aerobic cultivations consequences for large-scale bioprocesses and shake flask cultures. Microb Cell Fact. 2008;7:26–36.
31. Dauner M, Sonderegger M, Hochuli M, Szyperski T, Wűthrich K, Hohmann HP. Intracellular carbon fluxes in riboflavin producing *Bacillus subtilis* during growth on two-carbon substrate mixtures. Appl Environ Microbiol. 2002;68: 1760–71.
32. Hölscher T, Görisch H. Knockout and overexpression of pyrroloquinoline quinone biosynthetic genes in *gluconobacter oxydans* 621H. J Bacteriol. 2006;188:7668–76.
33. San KY, Bennett GN, Berrios-Rivera SJ, Vadali RV, Yang YT, Horton E. Metabolic engineering through cofactor manipulation and its effects on metabolic flux redistribution in *Escherichia coli*. Metab Eng. 2002;4:182–92.
34. Sánchez AM, Bennett GN, San KY. Effect of different levels of NADH availability on metabolic fluxes of *Escherichia coli* chemostat cultures in defined medium. J Biotechnol. 2005;117:395–405.
35. Chemler JA, Fowler ZL, McHugh KP, Koffas MAG. Improving NADPH availability for natural product biosynthesis in *Escherichia coli* by metabolic engineering. Metab Eng. 2010;12:96–104.
36. Minohara S, Sakamoto J, Sone N. Improved H^+/O ratio and cell yield of *Escherichia coli* with genetically altered terminal quinol oxidases. J Biosci Bioeng. 2002;93:464–9.
37. Lawford HG, Haddock BA. Respiration-driven proton translocation in *Escherichia coli*. Biochem J. 1973;136:217–20.
38. Tran QH, Bongaerts J, Vlad D, Unden G. Requirement for the proton-pumping NADH dehydrogenase I of *Escherichia coli* in respiration of NADH to fumarate and its bioenergetic implications. Eur J Biochem. 1997;244:155–60.
39. Puustinen A, Finel M, Virkki M, Wikström M. Cytochrome *o* (*bo*) is a proton pump in *Paracoccus denitrificans* and *Escherichia coli*. FEBS Lett. 1989;249:163–7.
40. Meyer M, Schweiger P, Deppenmeier U. Effects of membrane-bound glucose dehydrogenase overproduction on the respiratory chain of *Gluconobacter oxydans*. Appl Microbiol Biotechnol. 2013;97:3457–66.
41. Qizi GN, Parshad R, Verma V, Chopra CL, Buse R, Träger M. Diketo-gluconate fermentation by *Gluconobacter oxydans*. Enzyme Microb Technol. 1991;13: 504–7.
42. Hanke T, Richhardt J, Polen T, Sahm H, Bringer S, Bott M. Influence of oxygen limitation, absence of the cytochrome *bc*(1) complex and low pH on global gene expression in *Gluconobacter oxydans* 621H using DNA microarray technology. J Biotechnol. 2012;157:359–72.
43. McNeil B, Harvey L. Energy well spent on a prokaryotic genome. Nat Biotechnol. 2005;23:186–7.
44. Zhang L, Lin JP, Ma YS, Wei DZ, Sun M. Construction of a novel shuttle vector for use in *Gluconobacter oxydans*. Mol Biotechnol. 2010;46:227–33.
45. Kovach ME, Elzer PH, Hill DS, Robertson GT, Farris MA, Roop RM. Four new derivatives of the broad-host-range cloning vector pBBR1MCS carrying different antibiotic resistance cassettes. Gene. 1995;166:175–6.
46. Mostafa HE, Heller KJ, Geis A. Cloning of *Escherichia coli lacZ* and *lacY* genes and their expression in *Gluconobacter oxydans* and *acetobacter liquefaciens*. Appl Environ Microbiol. 2002;68:2619–23.

2

The effects of TORC signal interference on lipogenesis in the oleaginous yeast *Trichosporon oleaginosus*

Felix Bracharz, Veronika Redai, Kathrin Bach, Farah Qoura and Thomas Brück*

Abstract

Background: Oleaginous organisms are a promising, renewable source of single cell oil. Lipid accumulation is mainly induced by limitation of nutrients such as nitrogen, phosphorus or sulfur. The oleaginous yeast *Trichosporon oleaginosus* accumulates up to 70% w/w lipid under nitrogen stress, while cultivation in non-limiting media only yields 9% w/w lipid. Uncoupling growth from lipid accumulation is key for the industrial process applicability of oleaginous yeasts. This study evaluates the effects of rapamycin on TOR specific signaling pathways associated with lipogenesis in *Trichosporon oleaginosus* for the first time.

Results: Supplementation of rapamycin to nutrient rich cultivation medium led to an increase in lipid yield of up to 38% g/L. This effect plateaued at 40 μM rapamycin. Interestingly, the fatty acid spectrum resembled that observed with cultivation under nitrogen limitation. Significant changes in growth characteristics included a 19% increase in maximum cell density and a 12% higher maximum growth rate. *T. oleaginosus* only has one Tor gene much like the oleaginous yeast *Rhodosporidium toruloides*. Consequently, we analyzed the effect of rapamycin on *T. oleaginosus* specific TORC signaling using bioinformatic methodologies.

Conclusions: We confirm, that target of rapamycin complex 1 (TORC1) is involved in control of lipid production and cell proliferation in *T. oleaginosus* and present a homology based signaling network. Signaling of lipid induction by TORC1 and response to carbon depletion to this complex appear to be conserved, whereas response to nitrogen limitation and autophagy are not. This work serves as a basis for further investigation regarding the control and induction of lipid accumulation in oil yeasts.

Keywords: *Trichosporon oleaginosus*, Rapamycin, TORC, Single cell oil, Oil yeast

Background

Production volumes of bio-based lipids have experienced a 65% increase over the last decade due to increasing demands by food, chemical- and pharmaceutical industries [1, 2]. Especially the application of natural oils for biofuels, oleochemicals and bioactive substances is expanding rapidly [3]. However, the application of plant oils for non-food use accelerates land use and climate change, which in turn negatively impacts on biodiversity. Further, the majority of pharmaceutically active high value lipids such as long chain polyunsaturated fatty acids are still sourced from fish and crustaceans, which negatively affects sensitive marine ecosystems.

Microbial oils have been designated as a sustainable alternative to plant and animal based lipids. In particular, oleaginous yeast, which can accumulate between 20 and 70% w/w lipids [4, 5] have gained increasing interest as providers for sustainable oleochemical building blocks in the biofuel, lubricant, food and cosmetics industry. However, induction of lipogenesis in microorganisms is linked to nutrient restriction (N or P), which results in significant reductions in cell growth. In order to design an economically relevant continuous production process that provides for high lipid and biomass formation the metabolic uncoupling of cell growth from intracellular lipid accumulation is a prerequisite. This aim however

* Correspondence: brueck@tum.de
Industrial Biocatalysis Group, Technische Universität München, Lichtenbergstraße 4, 85748 Garching, Germany

requires in depth knowledge of the cells regulatory mechanisms involved in lipogenesis.

The target of rapamycin complexes (TORCs) are central, highly conserved regulators in the control of cell proliferation, sexual development, cell skeleton organization, lipogenesis and other essential functions. TORCs draw their name from the TORC1 inhibiting heterocyclic makrolid rapamycin, derived from *Streptomyces hygroscopicus*. The complexes consist of different components depending on the organism. The key component is Tor1, a phosphatidylinositol 3 kinase-like serine/threonine kinase (mTOR in mammals).

Whereas some involvement of TORCs, such as TORC1 activation of translation and transcription are conserved from yeast over plants to mammals [6], other downstream processes are vastly different depending on the organism. It is reported that the signaling network plays a crucial role in the control of lipid homeostasis [7, 8]. Understanding the regulatory system of lipid accumulation would therefore contribute to the ability for targeted, genetic modification of single cell oil production strains.

Recently, the effects of rapamycin on the lipid accumulating microalgae *Euglena gracilis* in comparison to model algae *Chlamydomonas reinhardtii* and *Cyanidioschyzon merolae* were characterized [9]. Lipid content was significantly increased when the algae were exposed to minor concentrations of rapamycin. By contrast, higher rapamycin concentrations resulted in growth inhibition. While the effect of rapamycin has been well described in model yeasts, it has as of now not been evaluated in non-conventional oil forming yeasts strains. Therefore, this study evaluates the significance of TORC signaling pathways on lipogenesis in oleaginous yeast for the first time.

Methods

Strains and media

Wild type *Trichosporon oleaginosus* ATCC 20509 (DSM-11815), obtained from the "Deutsche Sammlung von Mikroorganismen und Zellkulturen" (DMSZ) (Braunschweig, Germany) was used for all experiments. Cultivation was done in YPD medium (glucose, 20 g/L; tryptone, 20 g/L; yeast extract, 10 g/L), nitrogen limitation medium [10] (glucose, 30 g/L; yeast extract, 0.5 g/L; (NH4)2SO4, 0.3 g/L; $MgSO_4{\cdot}7H_2O$, 1.5 g/L; KH_2PO_4, 2.4 g/L; Na_2HPO_4 0.91 g/L; $CaCl_2{\cdot}H_2O$, 0.22 g/L; $ZnSO_4{\cdot}7H_2O$, 0.55 μg/L; $MnCl_2{\cdot}4H_2O$, 22.4 μg/L; $CuSO_4{\cdot}5H_2O$, 25 μg/L; $FeSO_4{\cdot}7H_2O$, 25 μg/L, pH 6.1). Rapamycin (Tecoland, CA, USA) was solved in DMSO (10 mM stock) and added directly to the media in different concentrations after the inoculation.

Cultivation conditions

T. oleaginosus was cultivated as triplicate in yeast extract peptone dextrose medium (YPD) with different concentrations of rapamycin solution for 7 days at 28 °C in 500 mL baffled shake flasks. Cultivation was carried out in 100 mL YPD and nitrogen limitation medium (MNM) with glucose. Cells from an overnight culture grown in YPD medium under the same cultivation conditions were washed in ddH_2O and used to inoculate all cultivations at $OD_{600} = 0.5$. Where applicable, rapamycin was added 8 h after inoculation and adjusted to varying concentrations. 6 mL samples were taken daily for analysis of cell-dry weight, lipid content and fatty acid distribution.

Biomass and lipid determination

Determination of cellular dry weight occurred by pelleting 2 mL samples (12,000 g for 10 min), washing cells with ddH_2O and freeze drying at −80 °C for 24 h in pre-weighed microtubes. Cellular total lipid was obtained by extraction with chloroform and methanol by Folch et al. [11] (adapted). Cell pellets from 12 mL culture were washed with ddH_2O twice and disrupted four times by french press (EmulsiFlex®-B15, Avestin) at 2400 bar. A triplicate of 4 mL cell lysate was transferred to glass vials with screw caps and mixed with 6 mL of Folch reagent (2:1 chloroform/methanol) each. After extraction by shaking at 900 rpm and room temperature for 1 h, 1 mL 0.9% NaCl was added to aid phase separation. Samples were vortexed, centrifuged at 1000 g and the chloroform phase was transferred to pre-weighed glass vials. After evaporation of the solvent und a constant stream of dried nitrogen, vials were weighed and lipid content was calculated per dry weight in % g/g.

Analysis of fatty acid composition

Triplicates of 2 mL culture were pelleted by centrifugation, washed with ddH_2O and freeze dried at −80 °C. Between 10 and 20 mg were used for the fatty acid analysis. Fatty acid methyl esters (FAMES) were obtained by direct conversion of cell biomass by methanol transesterification [12].

FAMEs were analyzed on a GC-2025 gas chromatograph from Shimadzu (Nakagyo-ku, Kyōto, Japan) with flame ionisation detector and an AOC-20i auto injector (Shimadzu). 1 μl sample was applied onto a ZB-WAX column (30 m, 0.32 mm ID; 0.25 μm df; Phenomenex (Torrance, CA, USA)) with an initial column temperature of 150 °C (maintained for 1 min). A temperature gradient was applied from 150–240 °C (5 °C/min), followed by 6 min maintenance at 240 °C. Fatty acids were identified according to retention times of authentic standards.

Nile red assay

Semi-quantitative estimation of intracellular lipids was done by Nile Red staining according to Sitepu et al. [13]. 200 μL of culture sample was adjusted to $OD_{600} = 1$ and transferred to MaxiSorp F96 plates (Thermo Scientific Waltham, MA, USA) as 5 replicates. After addition of 25 μL DMSO, the blank measurement was taken. Subsequently, 25 μL of Nile Red staining solution (0.1 mg/mL in DMSO) were added and Nile Red fluorescence intensity was measured at 590 nm (excitation 530 nm) on an EnSpire 2 plate reader from Perkin Elmer (Waltham, MA, USA). Lipids were estimated by correcting FI (fluorescence intensity) by OD_{600}.

Online OD measurements

Online OD measurements were conducted by measuring real-time backscatter at 525 nm with a Cell Growth Quantifier (Aquila biolabs - Baesweiler) using 100 mL YPD in 250 mL shake flask without baffles for 72 h. Rapamycin concentration was adjusted to 5 μM 8 h after inoculation. The backscatter signal was calibrated with a manual 2 point OD_{600} measurement by an HP 8453 photometer.

Statistical analysis

Statistical analysis and data visualization was done in R [14]. Data were fitted to a Richards' growth curve using the grofit package [15]. All error bars show standard deviations. Stars show significance with $p = 0.05$ in comparison to untreated cultures.

Assembly of TORC-network

Previously published genomic and transcriptomic data of *T. oleaginosus* [16] were searched for TORC homologues using sequences from *Schizosaccharomyces pombe* and *Saccharomyces cerevisiae, Candida curvata* and *Cryptococcus neoformans*.

Results and discussion

Effect of rapamycin on nile red fluorescence

Cultivation of *T. oleaginosus* in full YPD medium does not lead to nutrient limitation and is associated with a low accumulation of intracellular lipids. Supplementation of this medium with rapamycin (20 μM, [9]) after 8 h cultivation time resulted in significantly higher nile red fluorescence (corrected for OD) signal indicating an increased lipid production (Fig. 1). The increase of 66% FI/OD at 72 h subsequently decreased to 44% (92 h) and 40% after 116 h respectively. During cultivation, no decrease in OD was observed, indicating that the cell growth was not affected by the applied rapamycin concentrations.

It is reported that Nile Red is a semiquantitative lipid stain [17], as its specific fluorescent signal is dependent on the fatty acid profile, the type of lipid (phospho-, triacylglyceride or steran) and the protein content within intracellular lipid bodies.

Effect of rapamycin on lipid accumulation

To confirm an increase in lipid yield, a gravimetric biomass and lipid determination was conducted using cells grown in cultures containing between 0 and 40 μM rapamycin. The addition of 40 μM rapamycin resulted in a maximal lipid increase of 38% compared to controls. Moreover, at 5 μM rapamycin the total biomass production is significantly increased. This data contrasts reports on the effects of rapamycin on algae and bacterial growth, where supplementation of the compound resulted in decreased biomass formation in line with its established cell cycle inhibition effects [8, 18].

Due to the expected logarithmic dependency (Additional file 1), effects of rapamycin on lipid content (Fig. 2) saturate at low concentrations. To confirm the effect of rapamycin on intracellular lipogenesis, we applied a one tailed Welch's t-test between samples in the absence and presence of rapamycin. The null hypothesis of both sample sets being of the same distribution is rejected with $p = 0.003$ and a confidence interval of 1.4–5% g/g increase in the absence and presence of rapamycin.

Impact of rapamycin on *T. oleaginosus* growth kinetics

As our initial data suggested that at 5 μM rapamycin (at t: 8 h cultivation time) cell growth was enhanced, we applied a real-time backscatter measurement in order to compare cellular growth in the absence and presence of rapamycin (Fig. 3). Cell growth could be sufficiently described by a fit to a Richards' curve [19], which allowed extraction of μ_{max} (maximum growth rate), lambda (lag phase duration) and A (maximum OD_{600}) values (Table 1).

Interestingly, we observed significant differences in the growth parameters of each culture. For the rapamycin treated culture, the maximum growth rate μ_{max} (12% increase) and the maximum optical density (19% increase) were elevated compared to controls.

This translates to a change in the maximum growth rate from a nominal 1.2 to 1.35 OD_{600}/h and an increase in the maximum cell density from OD 30.8 (61 g/L) to OD 36.6 (71 g/L) respectively. Hence, in the presence of 5 μM rapamycin the cell density increased by 19% concomitantly with a 25% lipid increase. These cumulative values translate to a 49% improved space time yield compared to controls in the absence of rapamycin.

The improved growth rate may be attributed to the upregulation of pathways relating to alternative nitrogen sources. A simultaneous assimilation of many different nitrogen sources could be advantageous in a high nutrient environment such as YPD. Furthermore, TORCs are known to affect cell cycle progression. Shortening of the G2 phase could lead to an increased growth rate, while sacrificing replication fidelity and long term offspring survival.

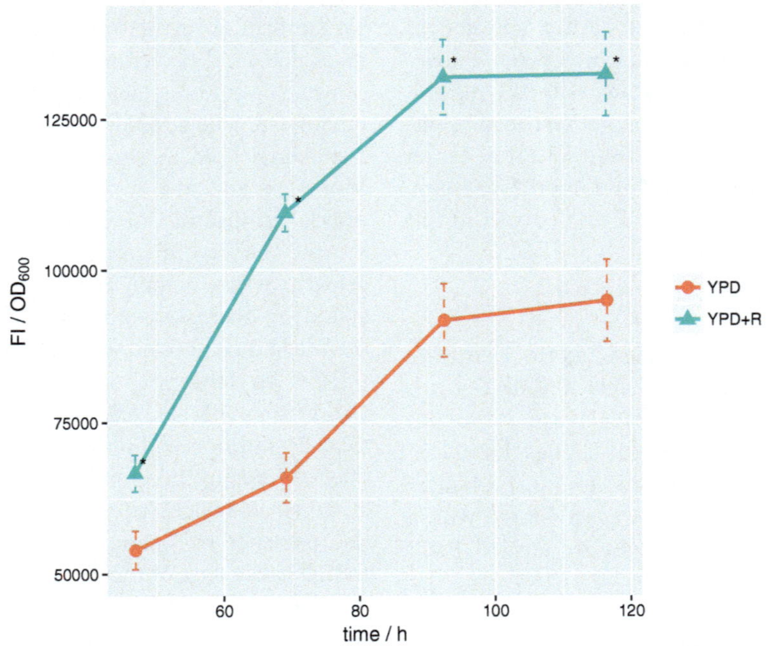

Fig. 1 Effect of rapamycin on nile red fluorescence: *T. oleaginosus* grown in YPD without (YPD) and with 20 μM rapamycin (YPD + R) supplementation. At different time points, nile red fluorescence (FI) and OD_{600} were measured. The ratio between the latter is plotted on the y axis and is a semiquantitative indicator of lipid content. Stars show significance at $p = 0.05$ and error bars show standard deviation

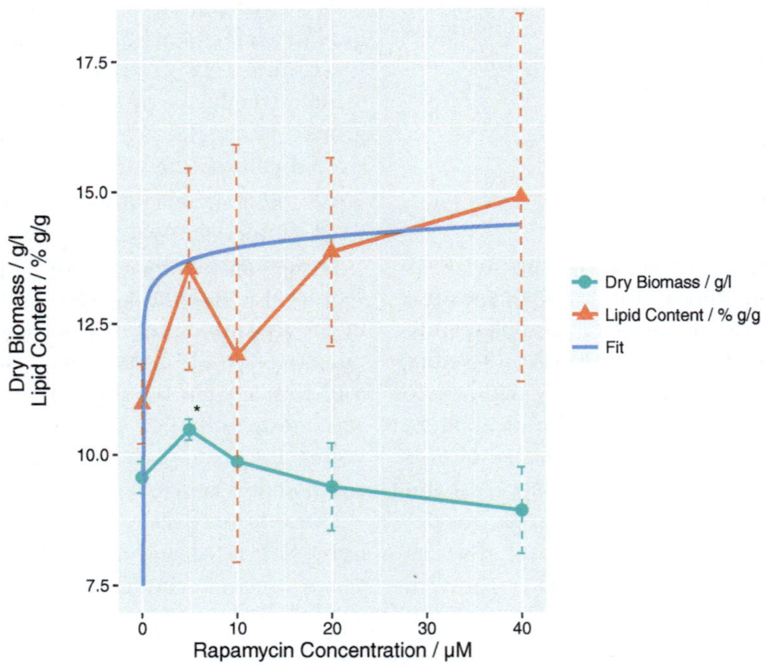

Fig. 2 Effect of rapamycin on lipid content and biomass: Total lipid content (*triangle*) and dry biomass (*circle*) of *T. oleaginosus* after 72 h of cultivation in YPD are measured with different concentrations of rapamycin supplementation between 0 and 40 μM. Error bars show standard deviation and the star shows significance at $p = 0.05$ in comparison to culture without rapamycin supplementation. The *blue line* shows a robust logarithmic fit of the lipid content in dependence of rapamycin concentration as described in Additional file 1

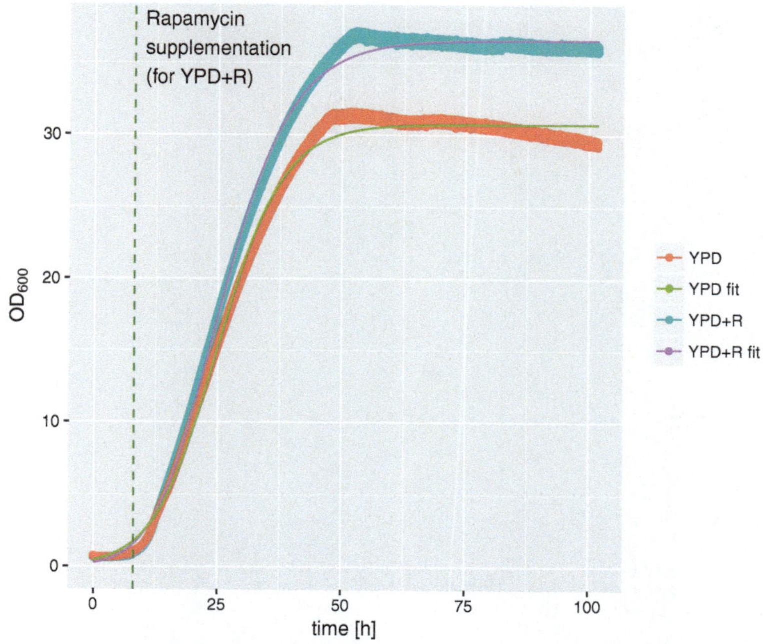

Fig. 3 Online-OD measurement: Growth curves of *T. oleaginosus* without (YPD) and with 5 μM rapamycin (YPD + R) obtained by measuring backscatter with an online OD system. *Thin lines* (YPD fit, YPD + R fit) show fit of Richards' growth curves

Fatty acid profile

Supplementation of rapamycin caused a non-concentration dependent shift in fatty acid spectrum (Fig. 4). Under these conditions, a major decrease of C18:0 in favor of C18:1 fatty acids was observed. Additionally, a minor decrease of C16:0 and a minor increase in C18:3 could be detected. These changes resemble the fatty acid profile obtained by cultivation in nitrogen limiting medium, thereby supporting the notion that rapamycin is at least partially simulating a low nutrient environment to the cells regulatory system. Fatty acid spectra show, that the effect of rapamycin saturates at comparatively low concentrations, confirming the fit in Fig. 2.

In *Euglena gracilis* addition of rapamycin led to an increase in lipid amount but almost no change in fatty acid profile was reported [9]. The reason for this appears to be the different mechanism of rapamycin response between this algae and previously identified yeasts or animal cells [9]. This observation in this study motivated us to use a bioinformatics approach to investigate the effects of rapamycin on the Trichosporon cell signaling network and lipogenesis.

TORC-network

A homology-based TORC signaling network, including upstream and downstream elements (Fig. 5), was assembled. A table of all proposed pathway components can be found in Additional file 2.

The SNF1/AMPK pathway is highly conserved from yeast to mammals, and homologues of its main components could also be detected in *T. oleaginosus*. These homologues termed Elm1, Sak1 and Tos3 are kinases which phosphorylate and activate Snf1 (orthologue to the mammalian AMPK), which in turn is a central regulator required for energy homeostasis. In *S. cerevisiae* Snf1 is mainly responsible for adaption to a glucose limiting environment [20]. Inactivation of Snf1 is caused by dephosphorylation initiated by the Reg1/Glc7 complex [20]. In a low carbon environment, phosphorylated Snf is translocated to the nucleus, where it phosphorylates the transcriptional activator Sip4. Subsequently, Sip4 activates the transcription of glucose-repressed genes [20]. This process is conserved in ascomycetes *S. cerevisiae* and *S. pombe*

Table 1 Comparison of growth curve parameters extracted from Richards' fit: Maximum growth rate (μ_{max}), lag phase (λ) and maximum cell density (A) of *T. oleaginosus* cultivated in YPD with and without rapamycin

	YPD	YPD + 5 μM rapamycin	Change / %
μ_{max} / OD_{600}/h	$1.2068 \pm 1.779*10^{-3}$	$1.349 \pm 1.362*10^{-3}$	$11.79 \pm 2.6*10^{-3}$
λ / min	$12.433 \pm 23.104*10^{-3}$	$12.151 \pm 5.776*10^{-3}$	$-2.27 \pm 3.13*10^{-3}$
A /OD_{600}	$30.754 \pm 7.421*10^{-3}$	$36.578 \pm 6.73*10^{-3}$	$18.93 \pm 0.46*10^{-3}$

All values are given with standard deviations

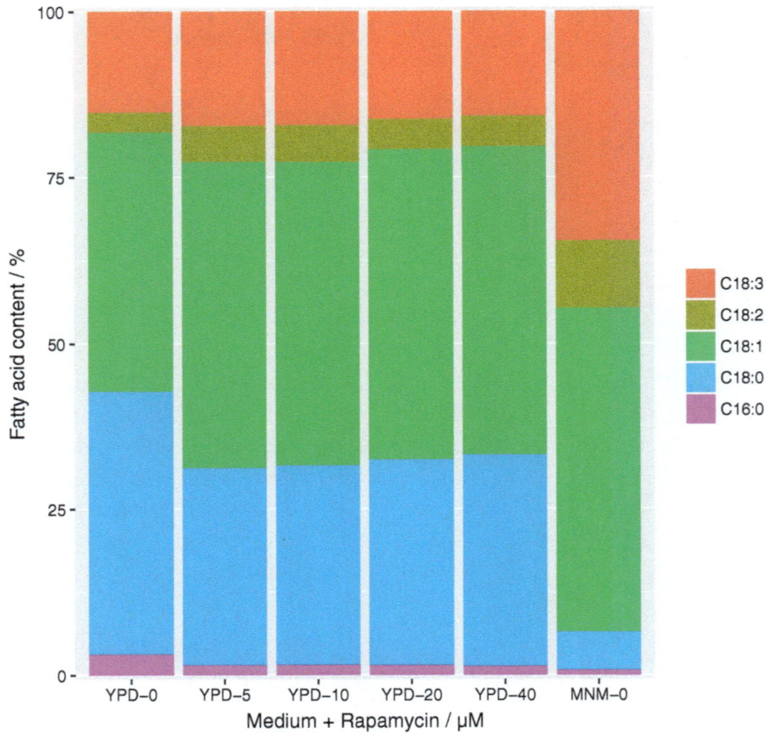

Fig. 4 Effect of rapamycin on fatty acid content: Profile of the main fatty acids of *T. oleaginosus* after 72 h of cultivation in Minimal N Medium (MNM-0) or YPD with different rapamycin concentrations (YPD-0 – YPD-40)

alike. Interestingly, activated Snf1 also inhibits Acc1p, which is responsible for catalysis of acetyl-CoA to malonyl-CoA, a main precursor for the production of fatty acids [21]. This indicates a direct link between metabolic signaling under nutrient stress conditions and the regulation of cellular lipid biosynthesis.

Furthermore, Snf1 is reported to inhibit the Tsc1/Tsc2 complex in model yeasts and in mammals. Moreover, the Tsc1/Tsc2 complex further integrates signals from other pathways, including the MAPK/ERK pathway [22], cytokines, hypoxia signals and Wnt signaling respectively. More recently, it has been suggested that in yeast,

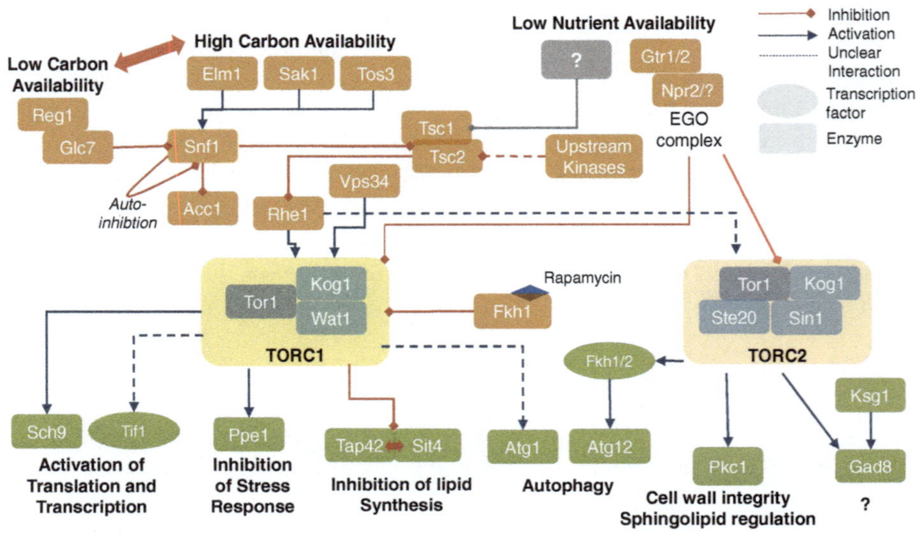

Fig. 5 TORC signaling network in *T. oleaginosus:* Proposed signaling network surrounding TORC in *T. oleaginosus* inferred by homology and assembled using data obtained from *S. pombe* and *S. cerevisiae*

involvement of other factors, especially in reaction to low nutrient content is feasible [23]. However, a direct inhibition of TORC1 by Vps34, an essential gene which channels amino acid availability to the Tsc1/Tsc2 complex, has been reported in yeasts and mammals [22, 24].

More specifically, TORC1 itself receives inputs from RheB over the Tsc1/2 axis and directly from the EGO complex. In model yeasts this EGO is composed of the kinases Gtr1/2 and Npr2/3 respectively. However in *T. oleaginosus*, we could not detect homologous of the Npr3 or Ego1-Ego3 complexes, which indicates that the absence of amino acids in the medium are not sensed via the EGO involved signaling. Inhibition of the TORC1 complex by rapamycin occurs via initial formation of an Fkh1 protein-rapamycin complex (mammalian homologue Fkbp12), which then binds to TORC1. This mechanism is highly conserved throughout the microbial and animal kingdom. Expectantly, *T. oleaginosus* Tor1 contained the characteristic rapamycin binding motif [25]. In nitrogen limitied media, the addition of rapamycin has no effect on *T. oleaginosus* biomass formation, growth kinetics and intracellular lipid content. This indicates that TORC1 may already be blocked under nutrient limiting cultivation conditions. Consequently, rapamycin addition would have no effects on biomass or lipid formation in the non-conventional yeast *T. oleaginosus*.

In consensus with data from model yeasts, we could identify homologues components of the TORC1 complex, namely Tor1 (with strong similarity to Tor2 of *S. pombe*), Kog1 and Wat1 respectively. Wat1 is a scaffold protein facilitating the connection between Tor1 and downstream substrates, like Ppe1 and Sch9 [26]. Analogous to model yeasts and mammals the Wat1 protein in *T. oleaginosus* mainly contains 7 WD40 repeats. Specifically in mice, Wat1 was required for TORC2 but not TORC1 activity. Most interestingly, we could not detect any Tco89 homologue in *T. oleaginosus*, which indicates that a component of the TORC complexes found in model yeast *S. cerevisiae* is absent. Therefore, the TORC1 complex of *T. oleaginosus* more closely resembles the situation reported for *S. pombe* or the oleaginous yeast *R. toruloides*. Indeed, this resemblances is confirmed by the absence of a second Tor gene for TORC2. Furthermore, we could not identify any Avo2 or Bit61 homologues in *T. oleaginosus*, both of which are non-functional TORC2-binding structures [27]. Transcripts for the TORC2 component Sin4 could be identified in different splicing isoform, which is consistent with previous findings in *S.cerevisiae* [24]. Notably, in *S. cerevisiae*, mutation of Sin4 leads to rapamycin resistance [28]. In analogy, the presence of Sin4 isoforms detected in *T. oleaginosus* could render the TORC2 complex resistant to rapamycin.

Downstream of primary TORC effects, we could identify significant differences between signals in model yeasts and *T. oleaginosus* affecting autophagy. Particularly, genes essential to the autophagy signaling pathway, namely Atg13, Atg17, Atg31, Atg29 could not found in *T. oleaginosus*. In this respect, the *T. oleaginosus* system may resemble the regulatory system of *Drosophila melanogaster*, in which hyperphosphorylated Atg1 in conjunction with Atg13 are sufficient to inhibit autophagy. In the model yeast S. *cerevisiae* the autophagy signaling is by far more complex and therefore may not apply to *T. oleaginosus* [29]. Further, individual autophagy related homologues (Atg5, Atg6, Atg16) were found, but the actual signaling pathway appears to differ significantly from other yeast systems.

The other main signaling pathways for regulation of lipid biosynthesis, transcriptional and translational initiation appear to be conserved with high similarity.

In *S. cerevisiae*, TORC1 inhibition liberates Tap42 and Sit4 from being bound to each other. This in turn activates the downstream transcription factors, Gat1 and Gln3. After Gat1 and Gln3 transport to the nucleus, these transcription factors induce the accumulation of lipids. Our bioinformatics analysis indicates that a similar mechanism is likely for *T. oleaginosus*. The highly conserved Sch9 is homologous to the mammalian S6K, which is responsible for activation of ribosomal protein S6 and therefore directly controls translation. By contrast, no homologue to Gaf1, which in *S. pombe* is central for the response to nitrogen stress [30], could be identified. The absence of Gaf1 therefore may modulate the cell cycle in *T. oleaginosus*. Nonetheless, Ppe1 homologue a kinase acting within the *S. pombe* stress response which also affects the cell cycle [31] could be identified.

TORC2 activation, especially in yeast, remains elusive. In *S. pombe*, interaction with Rhe1 is confirmed [32]. Furthermore, it was reported in mammals, that activation can be achieved by growth factors (PI3K axis). Furthermore, ribosomal association of the complex suggests its activation by nutrients. Especially the latter is also likely for yeast [33], considering the effect on growth in *S. pombe* [34]. In *T. oleaginosus* however, little can be reported about the effects of downstream TORC2 elements due to the absence of detailed cell biology studies.

Active TORC2 activates Fkh1/2 in *S. cerevisiae* (FOXO genes in mammals), which affects autophagy related genes, life span and stress response [27, 35, 36]. TORC2 as well as the highly conserved kinase Ksg1 activate Gad8 by phosphorylation [37]. The subsequent cellular effects strongly depend on the organism. For *S. cerevisiae*, this impacts on actin organization and cell wall synthesis, whereas for *C. elegans* mainly lipid metabolism and growth are affected. For the related and well described yeast, *S. pombe*, changes in amino acid uptake and general changes in stress response are described [27]. Lack of Rho/Rac homologues indicates a strong difference in the regulation of actin organization from *S. pombe.*

A confirmed element of TORC2 is Pkc activation, as was described for the closely related and pathogenic *Cryptococcus curvatus*. Pkc itself is involved in regulation of spingholipid biosynthesis, which impacts the structural integrity of the cell wall [38]. Most recently, the GATA transcription factor Gafl was reported to be responsible for sexual development in yeast and upregulation of amino acid transporters [30]. It is activated as response to nitrogen stress about 10 to 120 min after the onset of nitrogen stress. Therefore it can be hypothesized that it is part of a first, reversible response to nutrient stress. Persisting lack of nitrogen would then trigger the second, delayed phase which includes elevated mating in *S.pombe*.

Two factors indicate, that the observed effects are not due to rapamycin involvement with TORC2: Rapamycin resistance of TORC2 could be structurally substantiated by Avo3 (Ste20), which wraps around the Fkpb-binding domain of Tor1/2 [39]. Prolonged exposure of certain mammalian cell types to rapamycin showed inhibited assembly of TORC2 [40], however this was not observed in unicellular organisms [27] and therefore appears unlikely for *T. oleaginosus*. Secondly, no obvious differences in cell morphology were observed using microscopy (Additional file 3) and FACS (data not shown), indicating that for *T. oleaginosus*, rapamycin does not impact on cytoskeleton and actin organization, which are commonly affected by TORC2.

Conclusions

For the first time, lipogenesis could be induced in an oleaginous yeast without compromising on growth, resulting in a 1.4 fold increase in total lipid yield. We observed an increase in growth as well as lipid content in the absence of nutrient limitation, using YPD as model substrate with high nutrient content. TORC1 in *Trichosporon oleaginosus* can be inhibited by rapamycin, impacting on growth characteristics and lipid accumulation. However, considering a lack of reduction in growth and comparatively minor increases in lipid accumulation, inactivation of TORC1 is not sufficient to induce a cell state resembling nitrogen starvation.

It is possible, that *T. oleaginosus* either relies on TORC2 inhibition, requires additive signals of both complexes or employs another unknown pathway for full activation of nitrogen stress response and associated lipid accumulation. However, Torc2 regulation of lipid synthesis and its strong involvement in the upregulation of aminoacid transporters [27], one of the defining features of lipid accumulation in *T. oleaginosus*, make TORC2 involvement in nutrient limitation response in this yeast likely.

Proteomic and transcriptomic approaches are excellent tools for elucidating how rapamycin impacts on *T. oleaginosus* physiology. Comparing these data with previously obtained information about transcriptomic changes [16] in the presence of nitrogen stress will allow for pinpointing more clearly the relevance of TORC1 for lipid accumulation. Using a metabolomic strategy is a promising approach for a more in-depth study of key intermediates, such as glutamine and glutamate as components of central nitrogen metabolism.

Additional files

Additional file 1: Trend analysis: Analytical plots of robust regression using M-estimation of yeast lipid content in dependence of rapamycin concentration in cultivation medium (Fig. 2). Data points 9 and 7, both of which outliers causing deviation at [rapamycin] = 10 were excluded due to high Cook's distance. The resulting fit was plotted in Fig. 2 and was based on the following formula: Lipid content = 13.1996 + 0.3197*log ([Rapamycin]).

Additional file 2: Table of *T. oleaginosus* homologues in TORC signaling network: Abbreviations: *Trichosporon oleaginosus*, TO; *Schizosaccharomyces pombe*, SP; Nitrogen Catabolite Repression, NCR; *Saccharomyces cerevisiae*, SC; *Dictyostelium discodeum*, DD.

Additional file 3: Fluorescence microscopy: 200 μL *T. oleaginosus* cells grown for 72 h in YPD with and without 5 μM rapamycin supplementation were pelleted, washed with ddH_2O and resuspended in the same amount of water. 25 μL DMSO and 25 μL nile red (50 mg/ml) in DMSO were added and incubated in darkness for 10 min. Images were taken on a Zeiss Axio Lab A1 with an Axio Cam ICm1 (Oberkochen, Germany). Fluorescence was measured with a 525/25 filter with an exposure time of 500 ms.

Abbreviations

FACS: Fluorescence activated cell sorting; FI: Fluorescence intensity; MNM: Minimal nitrogen medium; N: Nitrogen; OD: Optical density; P: Phosphorus; S: Sulfur; TORC: Target of rapamycin complex; YPD: Yeast extract pepton dextrose medium

Acknowledgements

FB, FQ and TB would like to thank the German Federal Ministry of Education and Research for supporting the "Advanced Biomass –Value" research project (grant number: 03SF0446A). Information concerning the project is stated under: http://www.ibc.ch.tum.de/index.php?id=2042&L=1. Finally, the authors would like to express their gratitude to Dipl. Ing. (FH) Martina Haack for her kind help addressing analytical questions.

Funding

This work was supported by the BMBF (German Federal Ministry of Education and Research, 03SF0446A).

Authors' contributions

FB, VR and TB conceived the study and participated in its design and coordination. FB, VR, KB carried out the experimental work relating to the effects of rapamycin on *Trichsporon oleaginousus* growth. FQ supervised the experimental work and prepared scientific reports. FB carried out detailed bioinformatic studies to reconstruct the cell signaling pathways involved in rapamycin addition to the cultivation medium. The manuscript was drafted by FB and finalized by TB. The final manuscript version was approved by all authors.

Competing interests
The authors declare that they have no competing interests.

References

1. Lu C, Napier JA, Clemente TE, Cahoon EB. New frontiers in oilseed biotechnology: meeting the global demand for vegetable oils for food, feed, biofuel, and industrial applications. Curr Opin Biotechnol. 2011;22:252–9.
2. Pfleger BF, Gossing M, Nielsen J. Metabolic engineering strategies for microbial synthesis of oleochemicals. Metab Eng. 2015;29:1–11.
3. Fritsche UR, Sims REH, Monti A. Direct and indirect land-use competition issues for energy crops and their sustainable production - an overview. Biofuels Bioprod Biorefin. 2010;4:692–704. John Wiley & Sons, Ltd.
4. Papanikolaou S, Aggelis G. Lipids of oleaginous yeasts. Part I: Biochemistry of single cell oil production. Eur J Lipid Sci Technol. 2011;113:1031–51.
5. Ageitos JM, Vallejo JA, Veiga-Crespo P, Villa TG. Oily yeasts as oleaginous cell factories. Appl Microbiol Biotechnol. 2011;90:1219–27.
6. de Virgilio C, Loewith R. The TOR signalling network from yeast to man. Int J Biochem Cell Biol. 2006;38:1476–81.
7. Aggarwal D, Fernandez ML, Soliman GA. Rapamycin, an mTOR inhibitor, disrupts triglyceride metabolism in guinea pigs. Metabolism. 2006;55:794–802.
8. Madeira JB, Masuda CA, Maya-Monteiro CM, Matos GS, Montero-Lomelí M, Bozaquel-Morais BL. TORC1 inhibition induces lipid droplet replenishment in yeast. Mol Cell Biol. 2015;35:737–46.
9. Mukaida S, Ogawa T, Ohishi K, Tanizawa Y, Ohta D, Arita M. The effect of rapamycin on biodiesel-producing protist Euglena gracilis. Biosci Biotechnol Biochem. 2016;80:1223–9.
10. Görner C, Redai V, Bracharz F, Schrepfer P, Garbe D, Brück T. Genetic engineering and production of modified fatty acids by the non-conventional oleaginous yeast Trichosporon oleaginosus ATCC 20509. Green Chem. 2016;18:2037–46. The Royal Society of Chemistry.
11. Folch J, Lees M, Sloane SGH. A simple method for the isolation and purification of total lipides from animal tissues. J Biol Chem. 1957;226:497–509.
12. Griffiths MJ, van Hille RP, Harrison STL. Selection of direct transesterification as the preferred method for assay of fatty acid content of microalgae. Lipids. 2010;45:1053–60.
13. Sitepu IR, Ignatia L, Franz AK, Wong DM, Faulina SA, Tsui M, et al. An improved high-throughput Nile red fluorescence assay for estimating intracellular lipids in a variety of yeast species. J Microbiol Methods. 2012;91:321–8.
14. R Core Team. R: A language and environment for statistical computing. Vienna: R Foundation for Statistical Computing. 2013. http://www.R-project.org/. ISBN 3-900051-07-0.
15. Kahm M, Hasenbrink G, Lichtenberg-frate H, Ludwig J, Kschischo M. Grofit: fitting biological growth curves. J Stat Softw. 2010;33:1–21.
16. Kourist R, Bracharz F, Lorenzen J, Kracht ON, Chovatia M, Daum C, et al. Genomics and transcriptomics analyses of the oil-accumulating basidiomycete yeast Trichosporon oleaginosus: Insights into substrate utilization and alternative evolutionary trajectories of fungal mating systems. MBio. 2015;6:1–13.
17. Rumin J, Bonnefond H, Saint-Jean B, Rouxel C, Sciandra A, Bernard O, et al. The use of fluorescent Nile red and BODIPY for lipid measurement in microalgae. Biotechnol Biofuels. 2015;8:42.
18. Brown NF, Stefanovic-Racic M, Sipula IJ, Perdomo G. The mammalian target of rapamycin regulates lipid metabolism in primary cultures of rat hepatocytes. Metabolism. 2007;56:1500–7.
19. Richards FJ. A flexible growth function for empirical use. J Exp Bot. 1959;10: 290–301.
20. Hedbacker K, Carlson M. SNF1/AMPK pathways in yeast. Front Biosci. 2008;13:2408–20.
21. Zhang M, Galdieri L, Vancura A. The yeast AMPK homolog SNF1 regulates acetyl coenzyme A homeostasis and histone acetylation. Mol Cell Biol. 2013;33:4701–17.
22. Wang R, Levine B. Autophagy in cellular growth control. FEBS Lett. 2010;584:1417–26.
23. Huang J, Brendan DM. The TSC1–TSC2 complex: a molecular switchboard controlling cell growth. Biochem J. 2009;290:1717–21.
24. Jacinto E. What controls TOR? IUBMB Life. 2008;60:483–96.
25. Choi J, Chen J, Schreiber SL, Clardy J. Structure of the FKBP12-rapamycin complex interacting with the binding domain of human FRAP. Science. 1996;273:239–42.
26. Chantaravisoot N, Tamanoi F. mTOR signaling and human cancer. Struct. Funct. Regul. TOR complexes from Yeasts to Mamm. Part B. Cambridge: Academic Press; 2010. p. 301–16.
27. Cybulski N, Hall MN. TOR complex 2: a signaling pathway of its own. Trends Biochem Sci. 2009;34:620–7. Elsevier Ltd.
28. Zhang L, Liu N, Ma X, Jiang L. The transcriptional control machinery as well as the cell wall integrity and its regulation are involved in the detoxification of the organic solvent dimethyl sulfoxide in *Saccharomyces cerevisiae*. FEMS Yeast Res. 2013;13:200–18.
29. Chen Y, Klionsky DJ. The regulation of autophagy - unanswered questions. J Cell Sci. 2011;124:161–70.
30. Laor D, Cohen A, Kupiec M, Weisman R. TORC1 regulates developmental responses to nitrogen stress via regulation of the GATA transcription factor Gaf1. MBio Am Soc Microbiol. 2015;6:e00959–15.
31. Yanagida M, Ikai N, Shimanuki M, Sajiki K. Nutrient limitations alter cell division control and chromosome segregation through growth-related kinases and phosphatases. Philos Trans R Soc Lond B Biol Sci. 2011;366: 3508–20. The Royal Society.
32. Matsuo T, Otsubo Y, Urano J, Tamanoi F, Yamamoto M. Loss of the TOR kinase Tor2 mimics nitrogen starvation and activates the sexual development pathway in fission yeast. Mol Cell Biol. 2007;27:3154–64.
33. Weisman R, Cohen A, Gasser SM. TORC2–a new player in genome stability. EMBO Mol Med. 2014;6:995–1002. EMBO Press.
34. Ikai N, Nakazawa N, Hayashi T, Yanagida M. The reverse, but coordinated, roles of Tor2 (TORC1) and Tor1 (TORC2) kinases for growth, cell cycle and separase-mediated mitosis in Schizosaccharomyces pombe. Open Biol. 2011;1:110007.
35. Pic A, Lim FL, Ross SJ, Veal EA, Johnson AL, Sultan MR, et al. The forkhead protein Fkh2 is a component of the yeast cell cycle transcription factor SFF. EMBO J. 2000;19:3750–61. EMBO Press.
36. Hollenhorst PC, Bose ME, Mielke MR, Müller U, Fox CA. Forkhead genes in transcriptional silencing, cell morphology and the cell cycle: overlapping and distinct functions for FKH1 and FKH2 in Saccharomyces cerevisiae. Genetics. 2000;154:1533–48.
37. Du W, Forte GM, Smith D, Petersen J. Phosphorylation of the amino-terminus of the AGC kinase Gad8 prevents its interaction with TORC2. Open Biol. 2016;6:150189.
38. Lee H, Khanal Lamichhane A, Garraffo HM, Kwon-Chung KJ, Chang YC. Involvement of PDK1, PKC and TOR signalling pathways in basal fluconazole tolerance in Cryptococcus neoformans. Mol Microbiol. 2012;84:130–46. NIH Public Access.
39. Gaubitz C, Oliveira TM, Prouteau M, Leitner A, Karuppasamy M, Konstantinidou G, et al. Molecular basis of the rapamycin insensitivity of target of rapamycin complex 2. Mol Cell. 2015;58:977–88.
40. Sarbassov DD, Ali SM, Sengupta S, Sheen J-H, Hsu PP, Bagley AF, et al. Prolonged rapamycin treatment inhibits mTORC2 assembly and Akt/PKB. Mol Cell. 2006;22:159–68. Elsevier.

3

Knockdown of five trehalase genes using RNA interference regulates the gene expression of the chitin biosynthesis pathway in *Tribolium castaneum*

Bin Tang, Ping Wei, Lina Zhao, Zuokun Shi, Qida Shen, Mengmeng Yang, Guoqiang Xie and Shigui Wang*

Abstract

Background: RNA interference is a very effective approach for studies on gene function and may be an efficient method for controlling pests. Trehalase is a key gene in the chitin biosynthesis pathway in insects. Five trehalase genes have been cloned in *Tribolium castaneum*, though it is not known whether the detailed functions of these trehalases can be targeted for pest control.

Results: The functions of all five trehalase genes were studied using RNAi, and the most important results showed that the expression of all 12 genes decreased significantly from 12 to 72 h compared with the control groups, except *GP1* at 72 h, when the expression of the *TcTre2* gene was suppressed. The results also revealed different abnormal phenotypes, and the observed mortality rates ranged from 17 to 42 %. The qRT-PCR results showed that the expression of *TPS*, *GS*, two *GP*, *CHS1a* and *CHS1b* genes decreased significantly, while that of the *CHS2* gene decreased or increased after RNAi after the five trehalases were silenced at 48 h. In addition, *TPS* gene expression decreased from 12 to 72 h after dsTcTre injection.

Conclusions: These results demonstrate that silencing of any individual trehalase gene, especially *Tre1-4* and *Tre2* gene can lead to moulting deformities and a high mortality rate through the regulation of gene expression in the chitin biosynthesis pathway and may be a potential approach for pest control in the future.

Keywords: *Tribolium castaneum*, Trehalase, RNA interference, Chitin biosynthesis pathway, Gene function

Abbreviations: cDNA, Complementary DNA; CHS, Chitin synthase; CK, Control group; Cu, Cuticle; DMRT, Duncan's new multiple range test; dsGFP, Double-stranded GFP RNA; dsRNA, Double-stranded RNA; dsTcTre, Double-stranded trehalase RNA of *Tribolium castaneum*; Fb, Fat body; GFAT, Glutamine, fructose-6-phosphate transaminase; GFP, Green fluorescent protein; GP, Glycogen phosphorylase; GPI, Glucose-6-phosphate isomerise; GS, Glycogen synthase; h, Hour; HK, Hexokinase; Mg, Midgut; Mt, Malpighian tubules; PCR, Polymerase chain reaction; PFK, Phosphofructokinase; PM, Peritrophic membrane; qRT-PCR, Quantitative real-time polymerase chain reaction; RNAi, RNA interference; RT-PCR, Reverse transcription polymerase chain reaction; SEs, Standard errors; Tc, *Tribolium castaneum*; TPS, Trehalose-6-phosphate synthase; Tre, Trehalase; µg, Microgramme; µl, Microliter

* Correspondence: sgwang@mail.hz.zj.cn
Hangzhou Key Laboratory of Animal Adaptation and Evolution, College of Life and Environmental Sciences, Hangzhou Normal University, Hangzhou, Zhejiang, China

Background

RNA interference (RNAi), which is a robust and powerful experimental tool, has been widely used to study gene function through the suppression of gene expression [1]. Many reports on the use of double-stranded RNA (dsRNA) techniques to silence target genes that are resistant to insecticides have indicated that RNAi may be an efficient method for controlling insect pests, and novel pest management strategies involving the modification of plants are being developed [2–4]. In addition, RNAi technology has been widely used to investigate the functions of essential genes in insects [5]. RNAi has also been employed for the silencing of vital genes in *Tribolium castaneum*, *Nilaparvata lugens* and other insects [4, 6]. Furthermore, RNAi could be used in novel methods of pest management and may be more effective than other pest control measures once the problems of the efficiency and specificity of gene silencing are solved [7].

The chitin biosynthesis pathway is crucially important for insect life; the pathway regulates the main component of the cuticle of most insects and the peritrophic membranes of arthropods [8–13]. The first enzyme involved in the chitin biosynthesis pathway is trehalase (Tre, TRE or Treh), which hydrolyses trehalose into glucose as a source of the substrate for hexokinase (HK), while the last enzyme in the pathway is chitin synthase (CHS), which consists of two classes, CHS1 and CHS2, in the majority of insects [14, 15]. Glycogen phosphorylase (GP) in the decomposition pathway together with glycogen synthase (GS) in the synthetic pathway regulates glycogen levels. Glucose-6-phosphate generated through the hydrolysis of trehalose and degradation of glycogen is transformed into fructose-6-phosphate by glucose-6-phosphate isomerase (GPI), which plays different roles in the chitin biosynthesis pathway after the enzymatic action of glutamine: fructose-6-phosphate transaminase (GFAT) and in the glycolysis pathway after the enzymolysis of phosphofructokinase (PFK) (Additional file 1: Figure S1) [16–20].

In *T. castaneum*, there is one trehalose 6-phosphate synthase (TPS) that catalyses the synthesis of a trehalose from two glucose molecules, in addition to HKs (HK1 and HK2), one glycogen synthase (GS), two glycogen phosphorylases (GP1 and GP2), one glucose-6-phosphate isomerase (GPI), one fructose-6-phosphate transaminase (GFAT) and one phosphofructokinase (PFK). The two classes of chitin synthase (CHS1 and CHS2) are found in *T. castaneum*, and the CHS1 yields two alternatively spliced transcripts and the corresponding isoforms of the enzyme proteins, namely CHS1a and CHS1b [9]. CHS1 is responsible for moulting during the entire lifecycle of *T. castaneum*, and CHS2 is required for the formation of chitin in the peritrophic membrane (PM) and RNAi of this gene leads to reduced food uptake, growth arrest and failure to molt to the next instar insect. In addition, CHS1a is responsible for both the larval-pupal and pupal-adult moults, while CHS1b is only responsible for the latter [9].

Trehalase (a-glucoside-1-glucohydrolase, EC 3.2.1.28) is an anomer-inverting α-trehalose-1-D-glucosidase that hydrolyses a trehalose molecule into two glucose molecules. Two forms of trehalase, soluble trehalase (Tre1) and membrane-bound trehalase (Tre2), have been identified to date and cloned in several insect species, including *Tenebrio molitor* (*T. molitor* Coleoptera), *Pimpla hypochondriaca* (Hymenoptera), *Bombyx mori* (*B. mori* Lepidoptera), *Apis mellifera* (Hymenoptera), and *Spodoptera exigua* (*S. exigua* Lepidoptera) [21–27]. The first soluble and membrane-bound trehalases were found in *T. molitor* and *B. mori*, respectively [21, 23]. Previous research has shown that two types of trehalases regulate different CHS enzymes [28]. It has been reported that trehalase regulates the expression of chitin synthase in the insect cuticle and midgut, and inhibiting chitin biosynthesis kills pests when trehalase gene expression is suppressed or knocked down [28, 29]. However, the functions and relationships of many TcTres and whether and how they regulate different CHS enzymes are unknown, as indicated by Chen and Arakane [9, 28]. Trehalase has become an important target in pest management and control, and new and effective trehalase inhibitor products with practical significance as pesticides are being developed [30].

In the present study, five *T. castaneum* trehalase (*TcTre*) genes were identified and cloned, including four soluble trehalase genes (*TcTre1-1*, *TcTre1-2*, *TcTre1-3* and *TcTre1-4*) and one membrane-bound trehalase gene (*TcTre2*); the results indicated that these trehalases may have different functions. The flour beetle, *T. castaneum*, is a member of the coleopteran order of insects. *T. castaneum* is a holometabolous insect in the larva, pupa and adult stages and represents a serious pest that damages grain in storage. It is thought to be an effective and reliable insect for conducting RNAi research [7, 31, 32]. It has also been reported that RNAi is highly effective in *Tribolium*, even in the postembryonic stages of *Tribolium* [9, 33]. Therefore, it will be useful to study the characteristics and functions of the five *TcTre* genes using RNAi. The transcription levels of the five *TcTre* genes were found to be substantially down-regulated after specific dsRNA injection targeting every *TcTre*. The present study revealed the specific functions of the five *TcTres* genes in chitin biosynthesis in insects as well as some functions that are common to all of them.

Methods

Insect cultures

The *T. castaneum* insects were kindly provided by Professor Chao-Ming Wei of Shanxi Normal University.

The experimental beetle populations were raised in whole wheat flour with 5 % yeast powder at 30 ± 1 °C [7, 34]. The developmental stages were synchronised by providing fresh coarse wheat bran to replace the old coarse wheat bran every day, and the old coarse wheat bran mixture containing *T. castaneum* eggs was cultured for future experiments.

RNA extraction, cDNA synthesis and rapid amplification of full-length cDNA (PCR)

Total RNA was extracted from the whole body of *T. castaneum* larvae or adults using the TRIzol (Invitrogen, Carlsbad, California, USA) method. The RNA concentration was then determined by measuring the absorbance at 260 nm with a spectrophotometer [35]. The purified RNA was stored at -80 °C for future experiments. First-strand cDNA synthesis was performed using the PrimeScript® RT reagent Kit with gDNA Eraser (Takara). The obtained first-strand cDNA (1 μl) was used as the template for PCR.

Cloning of the fragments of five trehalase genes with the protein coding regions

Five pairs of specific primers based on the whole gene sequences published in GenBank were designed for *TcTre1-1* cDNA (XM968798), *TcTre1-2* cDNA (XM968883), *TcTre1-3* cDNA (XM968859), *TcTre1-4* cDNA (XM968826) and *TcTre2* cDNA (EFA11183) (Table 1). The components of the PCR mixture included PCR buffer containing 0.1 mM dNTPs, 0.2 μM each primer, and 0.5 U of HiFi-Taq DNA polymerase (Transgene, China) in a total volume of 25 μl. The amplification

Table 1 Primers used in present study

Gene	Application type	Primer set	Forwad (5′-3′)	Reverse (5′-3′)	Length of target fragment (bp)
TcTre1-1	cDNA cloning	TcTre1-1	GAACCACTTGTCCCTTGTC	TGCTAATCAAGTGGGCTT	1626
	qRT-PCR	QTcTre1-1	AACGACTCGCAATGGCTGG	CGGAGGCGTAGTGGAATAGAG	127
	dsRNA synthesis	dsTcTre1-1	CGACCTGAAATTAGCCCAGAAG	GGCTCCCCCACTCTTTC	357
TcTre1-2	cDNA cloning	TcTre1-2	CGTTTCATGTCGTTGTG	GAACACCACAGAGACGTAA	1692
	qRT-PCR	QTcTre1-2	GTGCCCAATGGGTTTATCG	CAACCACAACACTTCCTTCG	261
	dsRNA synthesis	dsTcTre1-2	GGACAATGAGTTGGGGCAG	AACCGCGTAAAATCGCTCG	493
TcTre1-3	cDNA cloning	TcTre1-3	ATGCCTTTCCTCCTACTTCTCACCC	GGTACTACGGCCGGGAAAAATAA	1647
	qRT-PCR	QTcTre1-3	CCTCTCATTCGTCACAAGCG	AAGCGTTTGATTTCTTTGCG	205
	dsRNA synthesis	dsTcTre1-3	GGGACCGTGGGTTAAACCAAC	CTGTGGCCGCTGAACTGAA	376
TcTre1-4	cDNA cloning	TcTre1-4	ATGAAGTCTCTCCTCGTTTCG	GATTTTACGCGGTTTCGG	1662
	qRT-PCR	QTcTre1-4	ACGGTGCCCGCATCTACTA	GTGTAGGTGGTCCCGTTCTTG	187
	dsRNA synthesis	dsTcTre1-4	ACGATATCGAGCTCAGAGTCC	TACGTCTCTGTGGTGTTCGC	462
TcTre2	cDNA cloning	TcTre2	TTAGTGTGTTGTGCGTTTGT	TGCGAAGAGCACGAGATAA	1929
	qRT-PCR	QTcTre2	CTCAGCCTGGCCCTTAGTTG	GGAGTCCTCGTAGATGCGTT	120
	dsRNA synthesis	dsTcTre2	GGTGCGCTCCAACTACAAAG	GCTCCGCAGTTCCGTGTAG	399
TcTPS	qRT-PCR	QTcTPS	CCTTGTTCCACTCAATGCCCGAC	GTCATGTATCCAGATCAAGGGAAC	105
TcHk1		QTcHk1	CGCACCGAATGCCAGAATC	GACCCACCCGACATCGATT	141
TcHk2		QTcHk2	CGAATCGGCCTAATAGTTGGC	GACGGAGCCCTCGATTTCAT	155
TcCHS1a		QTcCHS1a	CGTATAGCCGCCGACTTGAA	CCTATGACGAGAGCACCCAAGA	173
TcCHS1b		QTcCHS1b	CCAGGATTGCAAAGGAGTTG	TTGAGGAAACGTCCGAGGTC	179
TcCHS2		QTcCHS2	GAGTTGTGGCAATGTTCTCGC	GTGGTGTGGCCCTTGGTT	116
TcGS		QTcGS	GGAGTTTCATGCACGGACTG	GTTGGCTTGTCGATGGGAA	105
TcGP1		QTcGP1	GCGAGCGACTATGAACTGATG	TTGCGTGTCGGAAACTACAT	104
TcGP2		QTcGP2	GGACGAATATTACACAAGTTACGAC	CAGACCCACGATTACGCTT	119
TcGPI		QTcGPI	GTGATGCCGGAGGTGAAT	CACGTCGGTGATGGGCTT	112
TcGFAT		QTcGFAT	GGAACTGGACATGGACCGTA	GAACGGTGAGGATGCGAGTT	126
TcPFK		QTcPFK	GAGCAAGGACATGGAAGGGAA	CCAACCCAACCAGCCACTT	158
Tcβ-actin		QTcactin	AGGGCGTCATGGTCGGTAT	TTCTACAACGAGCTCCGCG	165
GFP	dsRNA synthesis	dsGFP	AAGGGCGAGGAGCTGTTCACCG	CAGCAGGACCATGTGATCGCGC	206

reactions were performed with the specific primers according to the following conditions: 10 min at 94 °C, followed by 30 cycles of 30 s at 94 °C, 30 s at 55 °C and 120 s at 72 °C, and then 10 min at 72 °C [25].

The products were subjected to agarose gel electrophoresis. The true electrophoretic DNA bands corresponding to the expected size of approximately 1600–2000 bp were excised from the agarose gel and purified using a DNA gel extraction kit (OMEGA, USA). The purified DNA was ligated into the pMD18-T vector (TaKaRa, Japan) and sequenced using the dideoxynucleotide method.

Analysis of the *TcTres* tissue distribution using semi-quantitative PCR

cDNA synthesis and RT-PCR were performed to analyse the tissue distribution of *TcTres* using gene-specific primers [19]. The midgut, fat body, cuticle and Malpighian tubules from mixed larvae including the penultimate larval instar and last-instar larvae were dissected in a saline solution (0.75 % NaCl) and stored at -80 °C for future analysis. Total RNA was isolated from these tissues and reverse transcribed into cDNA. Five pairs of specific primers based on the previously cloned *TcTre* genomic DNAs and one pair of specific of *Tcβ-actin* primers were designed (Table 1), which were the same as the primers used in qRT-PCR. First, the content of cDNA in each tissue was adjusted to ensure that the brightness of every electrophoresis band was the same, based on the cloning of *β-actin*, and the corresponding amount of cDNA was then added to clone every *TcTre* under the same PCR conditions indicated above to produce specific expression in the four tissues.

RNAi targeting the five trehalase genes

The cDNA regions of the five *TcTre* genes showing the greatest sequence divergence were targeted for dsRNA production. One region was the target for the *TcTre1-1*, *TcTre1-2*, *TcTre1-3* and *TcTre1-4* cDNA regions, as shown in Additional file 1: Figure S2. The lengths of the homologous dsRNAs for *TcTre1-1*, *TcTre1-2*, *TcTre1-3* and *TcTre1-4* used in this study were 357, 493, 376 and 462 bp, respectively. The other region was the target for TcTre2, for which the dsRNA contained a unique sequence near the 3'end, and its length was 399 bp (Additional file 1: Figure S2). The highest and lowest nucleotide sequence identities between the sequences were 53.2 and 39.2 %, respectively. Five pairs of primers with the T7 RNA promoter sequence flanking the 5'-ends were designed according to these regions (Table 1), which were synthesized using the Promega T7 Expression kit. Templates for in vitro transcription were prepared via PCR using gene-specific primers with the T7 polymerase promoter sequence at both ends [7].

The NT-88-V3 series (Nikon) micromanipulator system was used. The dorsal side of the first abdominal segment of *T. castaneum* was selected as the injection point, and 0.5–0.7 μg of the dsRNAs (3 μg/μl) targeting the five *TcTres* sequences and *GFP* was injected into each insect [32].

Sample collection and phenotype observations

After the dsRNAs were injected into last-instar insects, we collected 12 larvae (showing malformation at best) and divided them into three tubes every 12 h for the first 48 h, which were then stored at -80 °C. Most of the larvae developed into the prepupa stage by 48 h after RNAi treatment. The abnormal pupae were collected according to the above method at 72 h after RNAi treatment, when most of the insects had formed pupae. We observed the phenotypes of the insects every 12 h. When the dsRNAs were injected into penultimate-instar insects, samples were collected at 48 and 72 h, and the penultimate-instar insects started to moult to become last-instar insects at 72 h.

qRT-PCR analyses of *TcTres* and the expression levels of related genes

Total RNA was isolated from *T. castaneum* on each day of the insect's life cycle, including the larval, pupal and adult stages, and 1 μg of total RNA was extracted and used as the template to demonstrate the stability of *Tcβ-actin*. PCR was performed with the *QTcactin*-F and *QTcactin*-R primers under the following conditions: 95 °C for 5 min, followed by 28 cycles of 95 °C for 30 s, 55 °C for 30 s and 72 °C for 30 s, with a final extension at 72 °C for 10 min. The expression of *TcTres* was estimated through qRT-PCR using the Bio-Rad CFX96^{TM} system (Bio-Rad, America) and SsoFast^{TM} EvaGreen Supermix (Bio-Rad, America). All primers were designed to determine the expression of the corresponding homologous gene sequences, which included *TcTre1-1, TcTre1-2, TcTre1-3, TcTre1-4, TcTre2, CHS1a, CHS1b, CHS2, GS, GP1, GP2, HK1, HK2, GPI, GFAT* and *PFK* (Table 1); these primers were designed based on the unique regions of the sequences, which were identified in the alignment using Vector NTI Suite 7. Each reaction was performed in a final volume of 20 μl containing 1 μl of the cDNA sample (or standard), 1 μl (10 μmol/μl) of each primer, 7 μl of RNAase-free and DNAase-free water and 10 μl of SsoFast^{TM} EvaGreen Supermix. After 3 min of initial denaturation at 95 °C, the cycling protocol consisted of 40 cycles of denaturation at 95 °C for 5 s, annealing at 55 °C -62.5 °C for 20 s and, finally, the melting curve was performed at 65–95 °C (according to the instructions for SsoFast^{TM} EvaGreen Supermix). Standard curves were obtained using ten-fold serial dilutions of pooled total RNA.

Statistical analyses

The mRNA expression levels in the non-injected group and the dsGFP-injected group were designated as the controls. All of the data obtained in this study are presented as the means ± standard errors (SEs) of 3–6 replicates and were analysed through one-way ANOVA. A *P*-value of less than 0.05 or 0.01 was considered significant or extremelysignificant in Duncan's new multiple range test (DMRT) [36]. An asterisk indicates a significant difference in mRNA levels between the dsGFP group and each of the dsTcTres-injected groups measured at the same time (p, 0.05, *T* test), and a double asterisk indicates a highly significant difference (p, 0.01, *T* test).

Results

Analysis of the tissue distribution of the five *TcTres*

The tissue-specific expression of *TcTres* was determined through semi-quantitative reverse transcription PCR (RT-PCR). All five *TcTres* transcripts were detected in the midgut, fat body, cuticle and Malpighian tubules (Fig. 1). *TcTre1-1* expression was higher in the cuticle, fat body and Malpighian tubule and lower in the midgut. The highest and lowest *TcTre1-2* expression levels were observed in the fat body and Malpighian tubules, respectively. *TcTre1-3* expression was higher in the midgut, fat body and Malpighian tubules than in the cuticle. The expression of *TcTre1-4* was high in the midgut, fat body, cuticle and Malpighian tubules, whereas that of *TcTre2* was low; the negative control exhibited no expression.

Efficiency of *TcTre* RNAi in penultimate-instar larvae

RNAi targeting each *TcTre* was successful in each group, and the expression of the five trehalase genes had decreased significantly at 48 and 72 h after the corresponding dsRNA injections (Fig. 2). In addition, the transcript levels for any of these Tre genes are suppressed not only by the dsRNA for that gene, but also by dsRNA's for other Tre genes as well (Fig. 2a, b, c and d). All of the soluble *TcTre* transcripts decreased significantly after dsTcTre2 injection, except for *TcTre1-4* at 48 h. *TcTre2* transcript levels increased after RNAi targeting all of the soluble dsTcTres (Fig. 2e). Following RNAi treatment, most of the penultimate-instar larvae exhibited sloughing and became last-instar larvae at 72 h.

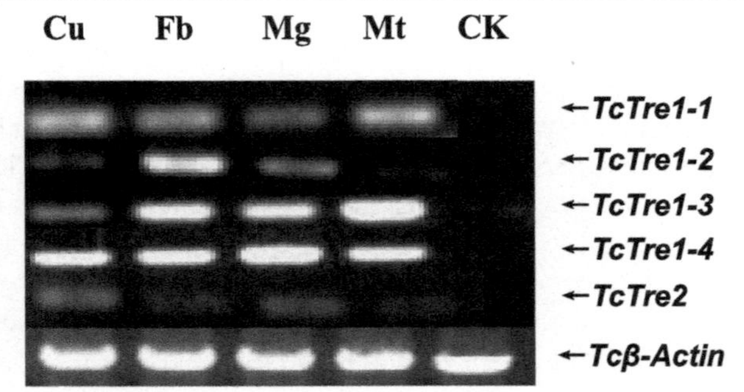

Fig. 1 Analysis of the five *TcTre* transcripts in different tissues of last-instar larvae using semi-quantitative PCR (RT-PCR). Total RNA was extracted from various tissues: fat body (Fb), midgut (Mg), cuticle (Cu) and Malpighian tubules (Mt). The expression of *Tcβ-actin* in the four tissues was used to standardize the relative expression levels of the five *TcTre* genes. CK represents PCR products without a template used as the controls

Efficiency and specificity of RNAi targeting *TcTres* in last-instar larvae

To verify the specificity of RNAi targeting the *TcTre* genes, the alignment of the five dsTcTres was assessed, and no consecutive identical 19-bp sequences between pairs of fragments were found (Additional file 1: Figure S2). All of the *TcTres* decreased significantly from 12 to 72 h after dsRNA injection compared with the control group, and *TcTre2* decreased most significantly after dsTcTre2 RNAi (Fig. 3). The corresponding RNAi efficiencies for *TcTre1-1*, *TcTre1-2* and *TcTre1-3* were 68, 56 and 93 % after 72 h, whereas those of *TcTre1-4* and *TcTre2* were 89 and 96 %, respectively, after just 48 h. The efficiencies obtained for *TcTre1-3*, *TcTre1-4* and *TcTre2* were higher than for the other *TcTres*.

The levels of the different *TcTres* showed both increases and decreases in every *TcTre* RNAi group (Fig. 3), whereas all of the *TcTres* showed decreases in the *TcTre2* RNAi group. Also the transcript level of TcTre1-2 is suppressed not only by the conrresponding dsRNA, but also by other Tre dsRNA's genes as well (Fig. 3b), the transcript level of *TcTre2* increased after RNAi targeting the other soluble *TcTres*, except after 36 and 72 h in the *TcTre1-1* RNAi group and 48 h in all of the soluble *TcTre* RNAi groups (Fig. 3e). The expression of *TcTre1-1* increased and decreased after RNAi targeting the other soluble *TcTres*; however, it increased rapidly in the *TcTre1-3* RNAi group at 24 h and in the *TcTre1-4* RNAi group at 72 h (Fig. 3a). *TcTre1-2* only showed an obvious increased at 48 and 72 h after *TcTre1-4* RNAi (Fig. 3b). *TcTre1-3* increased slightly at 24 h after *TcTre1-1* RNAi and at 48 h after *TcTre1-2* RNAi, although a significant reduction was detected in the other groups (Fig. 3c). There was variation in the expression of *TcTre1-4* in the other soluble RNAi groups, and it peaked at 36 h in the *TcTre1-3* RNAi group (Fig. 3d).

Analyses of phenotypes, aberrations and survival rates after RNAi

With the successful silencing of every *TcTre*, the insects subjected to RNAi exhibited a large number of deaths and presented various abnormal phenotypes (Fig. 4). The mortality rates recorded after RNAi targeting the five *TcTres* were as follows: *TcTre1-1* (35 %), *TcTre1-2* (17 %), *TcTre1-3* (23 %), *TcTre1-4* (42 %) and *TcTre2* (40 %) (Table 2). These values corresponded to 35, 12,

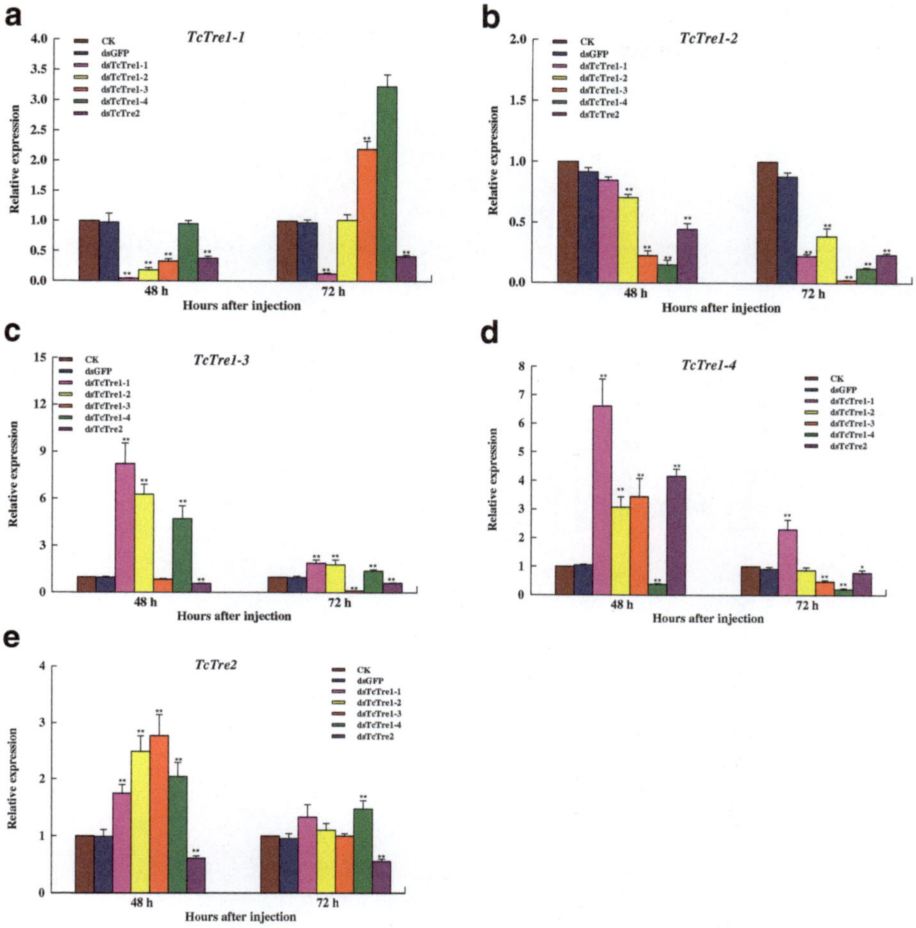

Fig. 2 Changes in the mRNA levels of the five *TcTres* after specific RNAi in penultimate-instar larvae. The penultimate-instar larvae were selected as the targets for dsRNA injection. **a**, **b**, **c**, **d** and **e** represent the mRNA levels of *TcTre1-1*, *TcTre1-2*, *TcTre1-3*, *TcTre1-4* and *TcTre2* relative to *Tcβ-actin* mRNA level after RNAi targeting all five of the *TcTres*, analysed via qRT-PCR

18, 35 and 28 insects and were higher than the death rates of the naive insects (0 %) and the insects injected with dsGFP (7 %). The death rates differed in different insect stages. The following percentages were obtained for the number of deaths observed among every type of abnormal insect divided by the total number of abnormal insects in the same stage (larva, pupa or adult stages). From larva to pupa, the percentages were *TcTre1-1* (20 %), *TcTre1-2* (14 %), *TcTre1-3* (13 %), *TcTre1-4* (24 %) and *TcTre2* (8 %), which corresponded to 20, 11, 10, 24 and 6 larvae that died before the pupal stage. During the pupa-adult stage, the percentages were *TcTre1-1* (8 %), *TcTre1-2* (0), *TcTre1-3* (4 %), *TcTre1-4* (13 %) and *TcTre2* (24 %), which corresponded to 6, 0, 3, 7 and 16 pupae that died before becoming adults, and most of these numbers were higher than in the two control groups. There were some deaths recorded during the adult stage, when the following percentages were determined: *TcTre1-1* (12 %), *TcTre1-2* (2 %), *TcTre1-3* (8 %), *TcTre1-4* (6 %) and *TcTre2* (12 %), which corresponded to just 9, 2, 5, 3 and 6 more adults than in the two control groups. Additional details are presented in Table 2, including the aberration rates, survival rates, pupation rates and eclosion rates (Table 2).

Insects in the larval, pupal and adult stages were present after RNAi targeting dsGFP as well in every *TcTre* RNAi group (Fig. 5). Three abnormal phenotypes were observed in the larval stage, and the insects died with the abnormal phenotype: the body size of the larvae decreased (i), or the size of the larvae abdomen decreased (ii) (both of which were associated with a reduced food intake that mainly occurred from 24 to 36 h after *TcTre* RNAi), and some larvae could not exuviate the larval cuticle to become pupae (iii). During the pupal stage, five malformations were present, which included shrunken pupae that were much smaller than normal pupae (iv); pupae with one set or both sets of wings together with membranous bulging wings (v); pupae with one set or both sets of wings bulging, leaving normal membranous wings (vi); colossal pupae that were larger

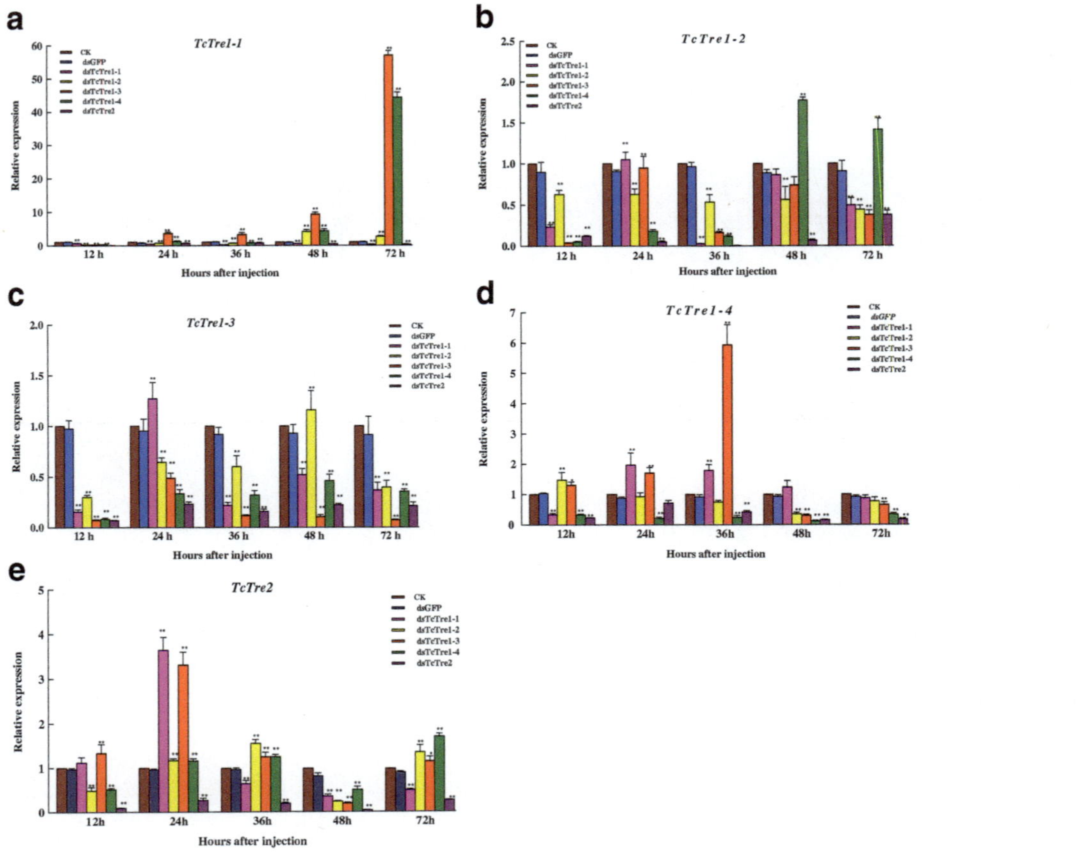

Fig. 3 Changes in the mRNA transcripts of the five *TcTre* genes after specific RNAi in last-instar larvae. **a**, **b**, **c**, **d** and **e** represent the mRNA levels of *TcTre1-1*, *TcTre1-2*, *TcTre1-3*, *TcTre1-4* and *TcTre2* relative to the *Tcβ-actin* mRNA level after RNAi targeting all five of the *TcTres*, analysed via qRT-PCR. Last-instar larvae were chosen as the targets for dsRNA injection

than normal pupae (vii); and disabled pupae that lacked some body parts (viii) (Fig. 4). All vii and viii pupae died with the abnormal phenotype. Three types of abnormal adults were observed: adults that survived with both sets of elytra slightly opened (ix); adults that partially shed their pupal cuticle with both sets of elytra open and membranous wings that were dislocated, leading to death in all of these individuals (x); and adults that showed combined characteristics of both pupae and adults, which were unable to extricate their appendages and died without completing adult eclosion (xi). All of the ix adults developed from "tiny abnormal" larvae that appeared abnormal in the larval stage and x adults that developed from most of the vi and v larvae; a small number of the iv pupae, vi pupae and v pupae became xi adults, and most of the iv larvae died with a burnt coloration on their body.

Effects of *TcTre* RNAi on the expression of the TPS, GS and GP genes

TPS, *GS* and *GP* are the key genes involved in energy metabolism during insect growth and development. qRT-PCR was used to detect mRNA levels after RNAi targeting the *TPS*, *GS* and two *GP* genes. The expression of *TPS* decreased from 12 to 72 h after *TcTre* RNAi, and its expression was decreased significantly at 12, 36 and 48 h (Fig. 5). In addition, *TPS* mRNA expression decreased significantly during the 24 h after *TcTre1-2*, *TcTre1-4* and *TcTre2* were knocked down, and its expression decreased significantly during the 72 h after *TcTre1-3* and *TcTre2* were silenced.

The results showed that the expression of the *TPS*, *GS* and the two *GP* genes was decreased significantly at 48 h after dsTcTre injection (Figs. 5 and 6a). The transcript levels of *TPS*, *GS*, *GP1* and *GP2* showed consecutive variation after RNAi targeting the five dsTcTres at 72 h. In addition, *GP1* and *GP2* exhibited high transcript levels at 72 h after dsTcTre1-2, dsTcTre1-3 and dsTcTre1-4 RNAi. The highest transcript levels of *GP1* and *GP2* were 79.6 and 6.6 times those in the respective control group injected with dsGFP (Fig. 6a). Moreover, the ascensional range of *GP1* gene expression was much higher prior to *GP2* gene expression.

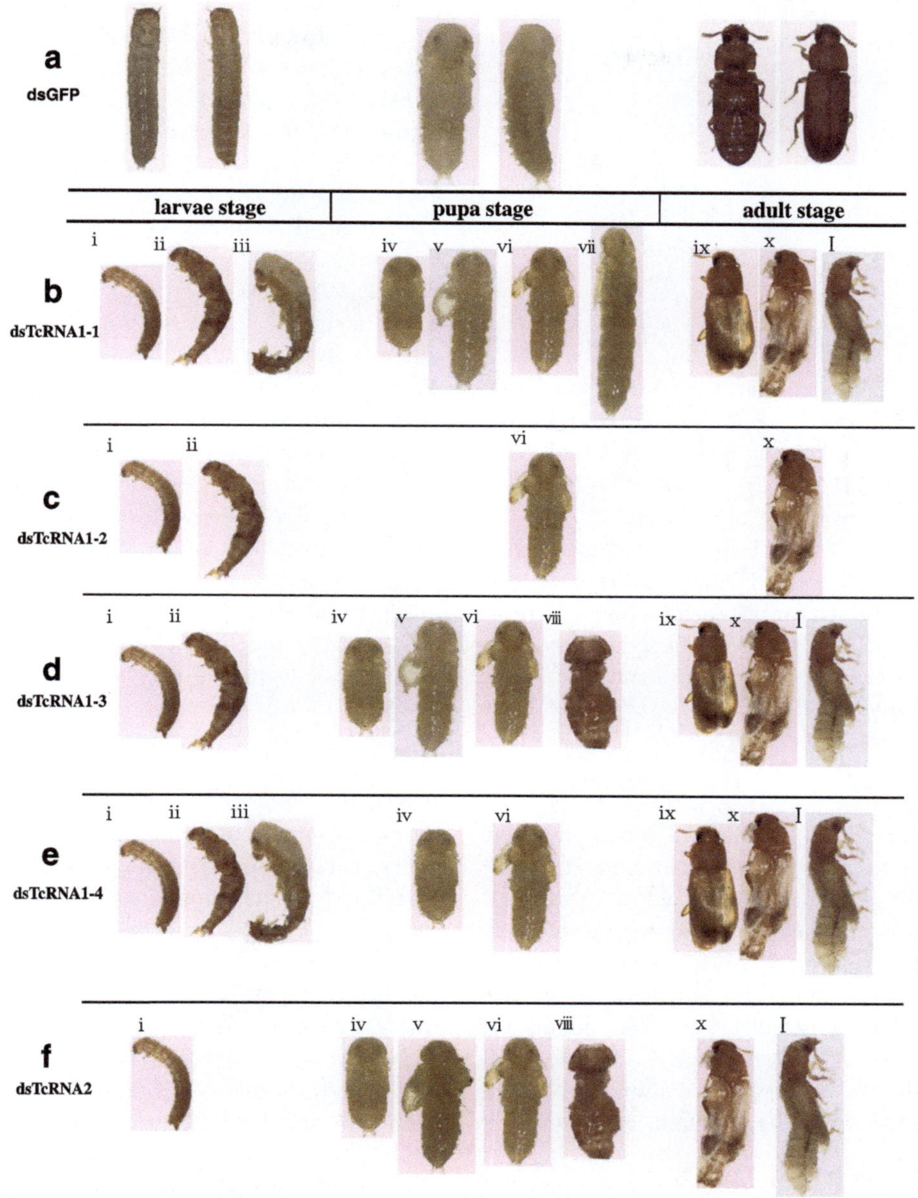

Fig. 4 Abnormal phenotypes were caused by RNAi targeting the five dsTcTres. All abnormal insects in the three stages (larva-pupa stage, pupal stage, and adult stage) are displayed and are divided into eleven types, from i to xi, after RNAi. **a** dsGFP-injected group. **b** dsTcTre1-1-injected group. **c** dsTcTre1-2-injected group. **d** dsTcTre1-3-injected group. **e** dsTcTre1-4-injected group. **f** dsTcTre2-injected group. The percentage below each type of abnormal insect is the corresponding proportion among all of the abnormal insects in that stage, such as the larva-pupa stage

Knockdown of *TcTres* and the expression of genes in the chitin biosynthesis pathway

qRT-PCR was used to detect mRNA levels after RNAi targeting all of the *TcTres*, including eight genes in the chitin biosynthesis pathway (*HK1, HK2, GPI, PFK, GFAT, CHS1a, CHS1b, CHS2*). The expression of *HK1* and *HK2* decreased significantly in the 48 h after the injection of the *TcTre1-1*, *TcTre1-3* and *TcTre2* dsRNAs, whereas their expression increased after dsTcTre1-2 injection (Fig. 6b). The expression of *HK1, HK2, CHS1a, CHS1b* and *CHS2* was decreased substantially after dsTcTre2 RNAi at both 48 and 72 h (Fig. 6b, c). The expression of *CHS1a* and *CHS1b* decreased slightly after RNAi targeting the four soluble dsTcTres at 48 h (Fig. 6c), and the expression of *CHS2* either increased or decreased in the various soluble *TcTre* RNAi groups. The transcript levels of *HK1, HK2, CHS1a, CHS1b* and *CHS2* exhibited consecutive variations after RNAi targeting the five dsTcTres at 72 h. The expression of *HK2, CHS1a* and *CHS1b* increased significantly at 72 h after dsTcTre1-2, dsTcTre1-3 and dsTcTre1-4 RNAi (Fig. 6b, c).

Table 2 *Tribolium castaneum* phenotype rate after RNAi

dsRNA	aberration rate (%)	survival rate (%)	pupation rate (%)	eclosion rate (%)
dsTcTre1-1	38	65	92	93
dsTcTre1-2	17	83	100	100
dsTcTre1-3	24	77	100	96
dsTcTre1-4	45	58	95	88
dsTcTre2	40	61	100	76
dsGFP	0	94	92	90
no-injection	0	100	100	100

The results also showed that *GPI, GFAT* and *PFK* gene expression decreased significantly in the 24 and 36 h after TcTre2 gene expression was inhibited through RNAi (Fig. 7). The expression of *GPI, GFAT* and *PFK* was increased at 24 h after *TcTre1-1, TcTre1-2* and *TcTre1-3* RNAi but was decreased at 24 h after *TcTre1-4, TcTre2* RNAi, with the exception of a decrease in *PFK* transcript levels after *TcTre1-3* RNAi observed at 24 h (Fig. 7). In addition, three gene transcripts were decreased significantly at 36 h after RNAi targeting the five *TcTres*, except that there was an increase after *TcTre1-2* RNAi.

Discussion

RNAi has been extensively studied in insects in recent years and applied in pest control management [7, 37, 38]. In most cases, RNAi results in a low rate of deformed phenotypes in insects and shows varying efficiency among insect tissues [5, 39, 40]. A previous study revealed that dsRNA is more efficient in *T. castaneum*, maintaining the silencing effect for a longer period and inducing a higher rate of deformations [7]. In this study, the dsRNAs for the five *TcTres* were demonstrated to be highly selective and clearly discriminant, and they were verified to be accurate after cloning; the levels of the targeted mRNAs were greatly reduced, along with compensatory expression of the other *TcTres* transcripts after RNAi targeting any of the *TcTres* (Figs. 2 and 3). For example, *TcTre1-1* transcripts are reduced not only by dsRNA for this gene but also by dsRNAs for *Tre1-2, Tre1-3* and by *Tre-2* genes at 48 h and by dsRNA for *Tre-2* gene at 72 h (Fig. 2a). It is clear that there is compensating over-expression of *Tre1-3* and *Tre1-4* genes at 72 h (Fig. 2a). Similarly the *Tre1-2* transcript suppression is not restricted to dsRNA for this gene but also by all other dsRNAs including those of the *Tre-2* gene (Fig. 2b). The results indicate that one of *Tre1-1, Tre1-2* and *Tre1-3* was inhibited, and the other genes were also lower expression or compensating over-expression, while other Tre transcripts compensating over-expression after *Tre1-4* or *Tre2* knockdown and the aberration and mortality rates was more higher than other three *Tre* genes (Table 2). These results are consistent with other observations showing that RNAi is effective in *Tribolium* larvae, pupae and adults, and even in embryos [41, 42].

There are many reports describing the characteristics of the two forms of trehalases. The tissue distribution of the two forms of trehalases differs. Tre1 is mainly expressed in the midgut and Malpighian tubules in *B. mori* and is not expressed in the fat body, silk gland, ovary, trachea, or brain [23, 43–45]. Tre2 is expressed in ovary cells, the spermatophore and the midgut in locusts and in the visceral muscle surrounding the midgut in *B. mori*; Tre2 exhibits both inactive and latent forms that can be activated by being released after destruction of membrane integrity [23, 46, 47]. Zhang also reported that Tre1 is expressed in the midgut, Malpighian tubules and ovary cells, whereas Tre2 is expressed in the fat body, midgut and Malpighian tubules [48]. However, both Tre1 and Tre2 are expressed in the cuticle, midgut,

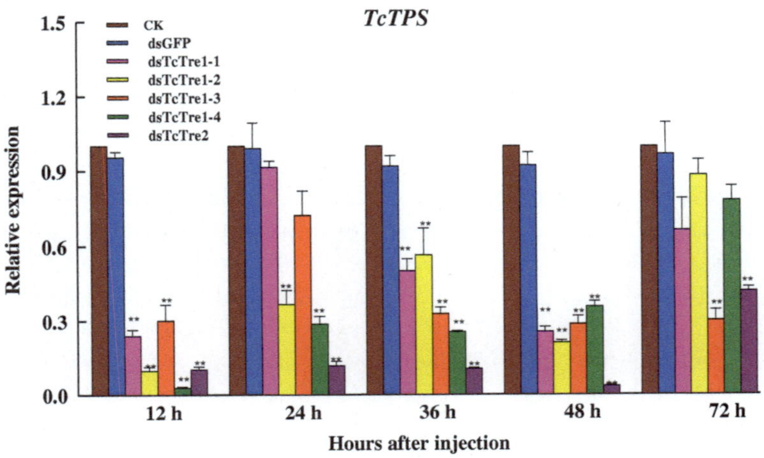

Fig. 5 Effects of RNAi targeting the five *TcTres* on trehalose-6-phosphate synthase (TPS) in last-instar larvae. The levels of *TPS* mRNA relative to the *Tcβ-actin* mRNA level were measured via qRT-PCR

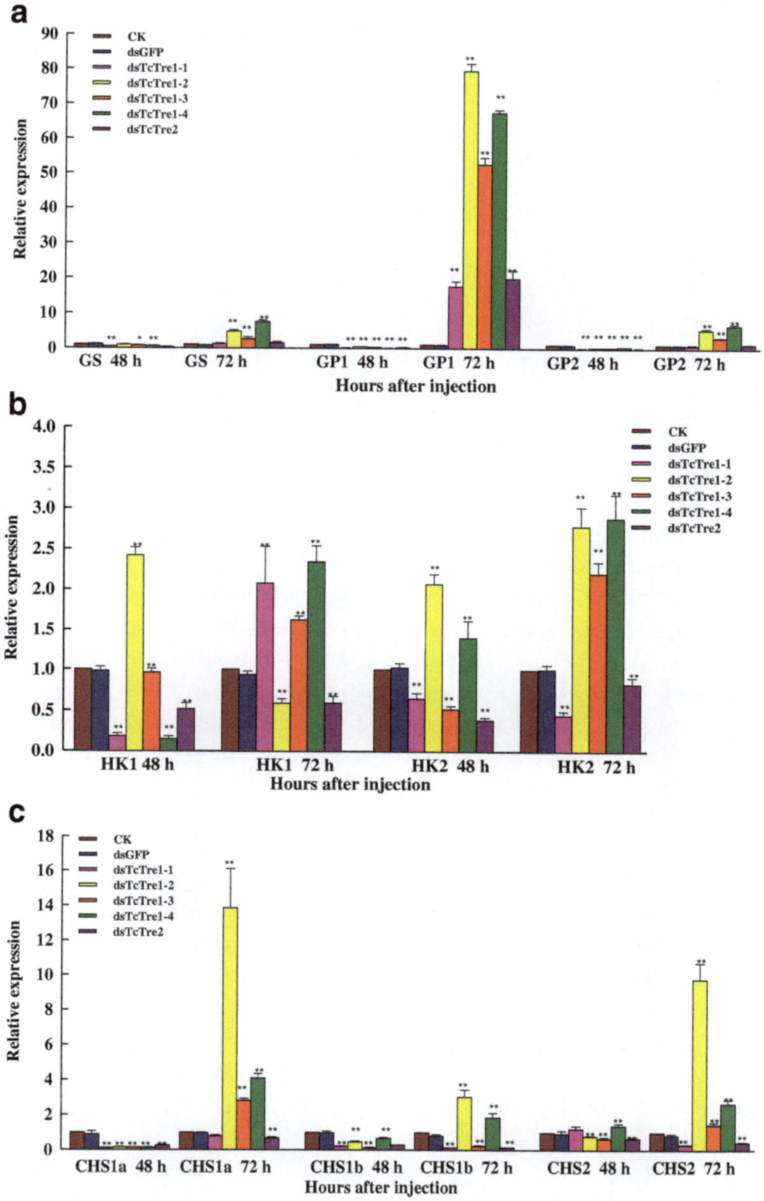

Fig. 6 Effects on key genes in the chitin biosynthesis pathway after specific RNAi in last-instar larvae at 48 and 72 h. The levels of these genes relative to the *Tcβ-actin* mRNA level were measured via qRT-PCR. **a** mRNA levels of glycogen synthase (*GS*) and glycogen phosphorylase (*GP*). **b** mRNA levels of two hexokinases (*HK1* and *HK2*). **c** mRNA levels of three chitin synthases (*CHS1a*, *CHS1b* and *CHS2*)

Malpighian tubules, tracheae and fat body of *S. exigua* [28], which is very similar to *T. castaneum*, and all five Tres genes are expressed in the cuticle, fat body, midgut and Malpighian tubules (Fig. 1).

Trehalase participates in both homeostasis and development and is involved in blood sugar metabolism in insects as well as flight metabolism, chitin synthesis during moulting and cold tolerance [26, 27, 49–53]. In addition, Tre2 is expressed as a transmembrane enzyme with an active site on the outside of the cell membrane and mainly hydrolyses extracellular trehalose, whereas Tre1 is located within the cell and hydrolyses intracellular trehalose [44, 54]. Some studies based on injection of the hormone 20-hydroxyecdysone (20E) into *B. mori* have observed an increase in *Tre1* transcript levels and no effect on the expression of *Tre2*, which indicates that insect moults are more closely associated with *Tre1* [44]. Many studies have also shown that soluble *Tre* levels increase before insect moults and after injecting 20-hydroxyecdysone (20E) [19, 28, 44]. In the present study, *GP1* and *GP2* increased rapidly at 72 h (Fig. 5b), as did *HK2* (Fig. 6a), most likely to compensate for the

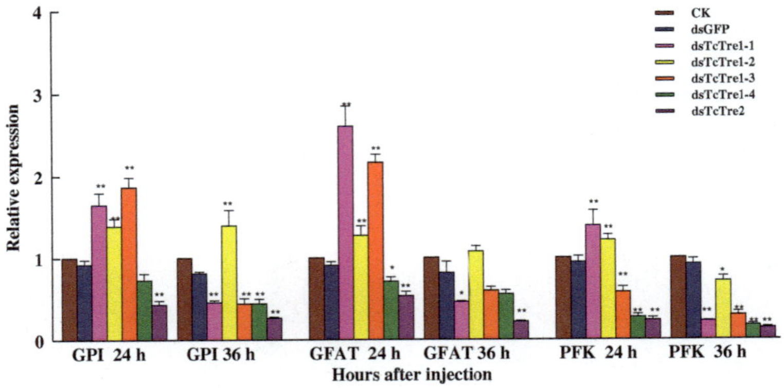

Fig. 7 Effects on other genes in the chitin biosynthesis pathway after specific RNAi in last-instar larvae at 24 and 36 h. The mRNA levels of glucose phosphate isomerase (*GPI*), fructose phosphate transaminase (*GFAT*) and phosphofructokinase (*PFK*) relative to the *Tcβ-actin* mRNA level were measured via qRT-PCR

reduced level of the *TcTre1-2* transcript, ensuring that the downstream *CHS* gene transcripts increased significantly (Fig. 6c). This result is consistent with the idea that conversion of glycogen into trehalose occurs in last-instar larvae [55]. At the same time, the *CHS1b* transcript decreased to a lower level, although its upstream gene transcripts *GP1*, *GP2*, *HK1* and *HK2* were increased substantially at 72 h after dsTcTre1-3 RNAi in the last-instar larvae (Figs. 5b and 6a, c). These results implied that the *TcTre1-3* gene plays a role in energy metabolism by hydrolysing trehalose into glucose and regulating the expression of the *CHS1b* gene, which was demonstrated indirectly based on the 33 % yield of xi adults; the xi adults exhibited a phenotype similar to that resulting from loss of the *CHS1b* gene in *Tribolium* after *CHS1b* RNAi [22]. We found that after dsTcTre1-4 injection, the *GFAT* transcript level was decreased to 0.71-fold of that in the control group injected with dsGFP at 24 h and to 0.55-fold at 36 h, whereas the *PFK* transcript level was decreased more strongly, to 0.27-fold at 24 h and 0.17-fold at 36 h (Fig. 6b). However, 90 % (nine insects) of the iv pupae (a greater percentage than for the other *TcTre* RNAi groups) showed a new phenotype that did not appear in Arakane's study due to knockdown of the *CHS* genes, which implied that this abnormal phenotype may be due to a lack of energy [9].

In addition, the 40 % abnormal insects and 39 % mortality recorded appeared to demonstrate that *TcTre2* is important, and the uniform distribution of many abnormal phenotypes indicated that *TcTre2* can adjust the expression of other *TcTres* (Fig. 4 and Table 2). All of the *TcTres* in the last-instar larvae were decreased sharply from 12 to 72 h after dsTcTre2 injection compared with the two control groups (Fig. 3); in contrast, *TcTre2* expression increased after RNAi targeting the four soluble dsTcTres, except for *TcTre1-2* and *TcTre1-4* at 12 h, *TcTre1-1* at 36 and 72 h and all of the soluble *TcTres* at 48 h (Fig. 3f). The same phenomenon was observed in the penultimate-instar larvae from 48 to 72 h after dsTcTre2 injection (Fig. 2), and a possible explanation for these findings is that the shortage of the *TcTre2* transcript resulted in a substantial loss of intracellular glucose following the decline of intracellular trehalose. This explanation is consistent with the finding that the expression of *TPS* decreased from 12 to 72 h after dsTcTre2 RNAi (Fig. 5a), and all of the *TcTres* therefore decreased as a corresponding response. These results led to the hypothesis that TcTre2 is a trehalase that mainly hydrolyses extracellular trehalose into glucose flowing into the cell, which is not inconsistent with the idea that Tre2 is a transmembrane enzyme with an active site on the outside of the cell membrane in *B. mori* [44]. Tre2 protein expression was shown to increase substantially, and there was no Tre1 protein present due to the hydrolysis of trehalose into glucose for glycogen accumulation in *B. mori* injected with diapause hormone, which revealed that *Tre2* plays a role in energy metabolism, similar to the finding that *Tre2* targets the hydrolysis of trehalose into glucose that will flow into the oocytes of *B. mori* and *Rhodnius prolixus* [56–58]. In addition, Becker found that the relationship between *GP* and *Tres* in non-feeding insects shows a positive correlation, involving increased expression and activity of *GP* along with a decrease in *Tres*. Our qRT-PCR results revealed that some gene transcripts in both the chitin synthesis pathway and the energy synthesis pathway decreased, including *HK*, *CHS*, *GPI*, *GFAT*, and *PFK*, although *GP1* increased substantially at 72 h and did not sufficiently compensate for the decline in glucose resulting from *TcTre2* RNAi (Figs. 5b and 6). This finding is consistent with Becker's observation that the insects lacked energy after dsTcTre2 injection [55, 59, 60].

Chitin biosynthesis is well established as a key target for pest control, and trehalase inhibitors are being developed for future pest management. It was reported that inhibition of the expression of the *Laodelphax stritellus* soluble and membrane-bound trehalase genes via the dsRNA feeding method resulted in mortality rates of 38.89 and 27.72 %, respectively [29]. In the pupal stage, the appearance of 60 % v and vi pupae among all of the abnormal pupae may also have resulted from the severe decline in the *CHS1a* transcript and *CHS1b* transcript, which caused a higher death rate (12 %) in the adult stage compared with the other *TcTre* RNAi groups (Fig. 4b). The abnormal pupal and adult phenotypes were consistent with Arakane's findings that abnormal iii larvae and adults emerged after RNAi targeting *CHS1a*, and vi adults emerged after RNAi targeting *CHS1b* or other genes [9, 61, 62].

As it known that the same family genes which have relative high protein sequence identity, so the compensatory function or cross silencing have been found in gene function study by RNAi. In the study of *Nilaparvata lugens TPS* genes function, *NLTPS2* gene's expression followed decreased when *NLTPS1* gene's expression inhibited by RNAi [63]. In our study, the expression of *TcTre1-1* also decreased when dsTcTre1-1, dsTcTre1-2, dsTcTre1-4 and dsTcTre2 were injected from 12 to 36 h, but it increased when dsTcTre1-3 injected from 24 to 72 h (Fig. 3a). As well as *TcTre1-2* (Fig. 3b), *TcTre1-4* (Fig. 3d) and *TcTre2* (Fig. 3e) have compensatory function at some extent when other *TcTre* was knockdown. The seeming cross silences similar with *TPS* genes function study also been found [63], while it is also reported that *N. lugens* soluble trehalases (*NlTre1-1* and *NlTre1-2*) and membrane-bound trehalase (*NlTre2*) also have the compensatory function [64]. The good inhibited effects and stronger phenotype were found when two *Tre1* and one *Tre2* gene in *N. lugens* was knockdown simultaneously [64]. However, double knockdown *vti1a* and *vti1b* did not produce a strong phenotype and it suggesting a compensatory mechanism exists within the secretory system though *vti1a* and *vti1b* by RNAi resulted in a significant decrease in the study of Vti family of SNARE proteins function [65]. Therefore, the crossing silencing and compensatory mechanism of the same gene family need more studied in the future.

Conclusions

These results demonstrate that the five trehalase genes exhibit different functions in chitin biosynthesis and energy metabolism by the way of RNAi. In particular, a mortality rate close to 40 % resulted from the silencing of TcTre1-4 or TcTre2 gene expression through RNAi, and it could be effective for pest control by a potential trehalase inhibitor [66]. Moreover, the development of anti-trehalase agents that target the chitin biosynthesis pathway or energy metabolism is an alternative method for exploring potential chemical insecticides.

Additional files

Additional file 1: Figure S1. Synthesis and degradation of trehalose and glycogen in insects. Adapted from Rockstein [16], Elbein et al. [17], Montooth et al. [18], Kunieda et al. [19], and Tang et al. [20]. The red arrow and dotted arrow indicate that the enzyme in this pathway was analysed and that this pathway is unknown, respectively. **Figure S2:** Alignment of the nucleotide sequences of five TcTre genes in the region of the dsRNAs.

Acknowledgements
We thank Dr. Chao-Ming Wei (College of Life Sciences, Shaanxi Normal University, Xi'an, China) for his kindly help.

Funding
This work was supported by National Natural Science Foundation of China (Grant Nos. 31371996 and 31672081), Visiting Scholar Foundation of State Key Laboratory of Biocontrol at Sun Yat-sen University (Grant No. SKLBC12F01), Hangzhou Science and Technology Development Program of China (Grant No. 20140432B01) and the Program for Excellent Young Teachers in Hangzhou Normal University (Grant No. JTAS 2011-01-031).

Authors' contributions
BT and SW designed the study, PW and LZ developed the RNAi methodology, PW, LZ and ZS performed experiments. QS, MY and GX performed the statistical analysis, PW, BT and SW wrote the manuscript. All authors have read and approved the final manuscript.

Competing interests
The authors declared that they have no competing interests.

References

1. Fire A, Xu S, Montgomery MK, Kostas SA, Driver SE, Mello CC. Potent and specific genetic interference by double-stranded RNA in *Caenorhabditis elegans*. Nature. 1998;391:806–11.
2. Mao Y, Cai W, Wang J, Hong G, Tao X, Wang L, et al. Silencing a cotton bollworm P450 monooxygenase gene by plant-mediated RNAi impairs larval tolerance of gossypol. Nat Biotechnol. 2007;25:1307–13.
3. Baum JA, Bogaert T, Clinton W, Heck GR, Feldmann P, Ilagan O, et al. Control of coleopteran insect pests through RNA interference. Nat Biotechnol. 2007;25:1322–26.
4. Chen X, Li L, Hu Q, Zhang B, Wu W, Jin F, Jiang J. Expression of dsRNA in recombinant Isaria fumosorosea strain targets the TLR7 gene in *Bemisia tabaci*. BMC Biotechnol. 2015;15:64.
5. Belles X. Beyond *Drosophila*: RNAi *in vivo* and functional genomics in insects. Annu Rev Entomol. 2010;55:111–28.
6. Wang Y, Fan HW, Huang HJ, Xue J, Wu WJ, Bao YY, et al. Chitin synthase 1 gene and its two alternative splicing variants from two sap-sucking insects, *Nilaparvata lugens* and *Laodelphax striatellus* (Hemiptera: Delphacidae). Insect Biochem Mol Biol. 2012;42:637–46.
7. Wang JD, Wu M, Wang BJ, Han ZJ. Comparison of the RNA interference effects triggered by dsRNA and siRNA in *Tribolium castameum*. Pest Manag Sci. 2013;69:781–6.
8. Arakane Y, Hogenkamp DG, Zhu YC, Kramer KJ, Specht CA, Beeman RW, et al. Characterization of two chitin synthase genes of the red flour beetle, *Tribolium castaneum*, and alternate exon usage in one of the genes during development. Insect Biochem Mol Biol. 2004;34:291–304.

9. Arakane Y, Muthukrishnan S, Kramer KJ, Specht CA, Tomoyasu Y, Lorenzen MD, et al. The *Tribolium* chitin synthase genes TcCHS1 and TcCHS2 are specialized for synthesis of epidermal cuticle and midgut peritrophic matrix. Insect Mol Biol. 2005;14:453e463.
10. Arakane Y, Specht CA, Kramer KJ, Muthukrishnan S, Beeman RW. Chitin synthases are required for survival, fecundity and egg hatch in the red flour beetle, *Tribolium castaneum*. Insect Biochem Mol Biol. 2008;38:959e962.
11. Devine WP, Lubarsky B, Shaw K, Luschnig S, Messina L, Krasnow MA. Requirement for chitin biosynthesis in epithelial tube morphogenesis. Proc Natl Acad Sci U S A. 2005;102:17014e17019.
12. Chen XF, Tian HG, Zou LZ, Tang B, Hu J, Zhang WQ. Disruption of Spodoptera exigua larval development by silencing chitin synthase gene A with RNA interference. Bull Entomol Res. 2008;98:613e619.
13. Qu M, Yang Q. A novel alternative splicing site of class A chitin synthase from the insect *Ostrinia furnacalise* Gene organization, expression pattern and physiological significance. Insect Biochem Mol Biol. 2011;41:923e931.
14. Kramer KJ, Dziadik-Turner C, Koga D. Chitin metabolism in insects. Comp Insect Physiol Biochem Pharmacol. 1985;3:75–115.
15. Kramer KJ, Muthukrishnan S. Chitin metabolism in insects. In: Gilbert LI, Iatrou K, Gill SS, editors. Comprehensive molecular insect science, vol. 4. Oxford: Elsevier; 2005. p. 111–44.
16. Rockstein M. Biochemistry of insect. San Francisco: Academic Press New York; 1978.
17. Elbein AD, Pan YT, Pastuszak I, Carroll D. New insights on trehalose: a multifunctional molecule. Glycobiology. 2003;13:17R–27R.
18. Montooth KL, Marden JH, Clark AG. Mapping determinants of variation in energy metabolism, respiration and flight in Drosophila. Genetics. 2003;165:623–35.
19. Kunieda T, Fujiyuki T, Kucharski R, Foret S, Ament SA, Toth AL, et al. Carbohydrate metabolism genes and pathways in insects: insights from the honey bee genome. Insect Mol Biol. 2006;15:563–76.
20. Tang B, Xu Q, Zou Q, Fang Q, Wang SG, Ye GY. Sequencing and characterization of glycogen synthase and glycogen phosphorylase genes from *Spodoptera exigua* and analysis of their function in starvation and excessive sugar intake. Arch Insect Biochem Physiol. 2012;80:42–62.
21. Takiguchi M, Niimi T, Su ZH, Yaginuma T. Trehalase from male accessory gland of an insect, *Tenebrio molitor*. cDNA sequencing and developmental profile of the gene expression. Biochem J. 1992;288:19–22.
22. Parkinson NM, Conyers CM, Keen JN, MacNicoll AD, Smith I, Weaver RJ. cDNAs encoding large venom proteins from the parasitoid wasp *Pimpla hypochondriaca* identified by random sequence analysis. Comp Biochem Physiol. 2003;134C:513–20.
23. Mitsumasu K, Azuma M, Niimi T, Yamashita O, Yamashita T. Membrane-penetrating trehalase from silkworm *Bombyx mori*. Molecular cloning and localization in larval midgut. Insect Mol Biol. 2005;14:501–8.
24. Lee JH, Saito S, Mori H, Nishimoto M, Okuyama M, Kim D, et al. Molecular cloning of cDNA for trehalase from the European honeybee, *Apis mellifera L* and its heterologous expression in *Pichia pastoris*. Biosci Biotechnol Biochem. 2007;71:2256–65.
25. Tang B, Chen XF, Liu Y, Tian HG, Liu J, Hu J, et al. Characterization and expression patterns of a membrane-bound trehalase from *Spodoptera exigua*. BMC Mole Biol. 2008;9:51.
26. Tatun N, Singtripop T, Tungjitwitayakul J, Sakurai S. Regulation of soluble and membrane-bound trehalase activity and expression of the enzyme in the larval midgut of the bamboo borer *Omphisa fuscidentalis*. Insect Biochem Mol Biol. 2008;38:788–95.
27. Tatun N, Singtripop T, Sakurai S. Dual control of midgut trehalase activity by 20-hydroxyecdysone and an inhibitory factor in the bamboo borer *Omhisa fuscidentalis* Hampson. J Insect Physiol. 2008;54:351–57.
28. Chen J, Tang B, Chen HX, Yao Q, Huang XF, Chen J, et al. Different functions of the insect soluble and membrane-bound trehalase genes in chitin biosynthesis revealed by RNA Interference. PLoS One. 2010;5:e10133.
29. Zhang Q, Lu DH, Pu J, Wu M, Han ZJ. Cloning and RNA interference effects of trehalase genes in *Laodelphax striatellus* (Homoptera: Delphacidae). Acta Entomol Sin. 2012;55(8):911–20.
30. Tang B, Wei P, Chen J, Wang SG, Zhang WQ. Progress in gene features and functions of insect trehalases. Acta Entomol Sin. 2012;55(11):1315–21.
31. Richards S, Gibbs RA, Weinstock GM, Brown SJ, Denell R, Beeman RW, et al. The genome of the model beetle and pest *Tribolium castaneum*. Nature. 2008;452:949–55.
32. Noh MY, Beeman RW, Arakane Y. RNAi-based functional genomics in *Tribolium castaneum* and possible application for controlling insect pests. Entomol Res. 2012;42:1–10.
33. Yoshinori T, Denell RE. Larval RNAi in *Tribolium* (Coleoptera) for analyzing adult development. Dev Genes Evol. 2004;214:575–78.
34. Haliscak JP, Beeman RW. Status of malathion resistance in five genera of beetles infesting farm-stored corn, wheat, and oats in the United States. J Econ Entomol. 1983;76:717–22.
35. Jaubert-Possamai S, Le Trionnaire G, Bonhomme J, Christophides GK, Rispe C, Tagu D. Gene knockdown by RNAi in the pea aphid *Acyrthosiphon pisum*. BMC Biotechnol. 2007;7:63.
36. Tang T, Zhao CQ, Feng XY, Liu XY, Qiu LH. Knockdown of several components of cytochrome P450 enzyme systems by RNA interference enhances the susceptibility of *Helicoverpa armigera* to fenvalerate. Pest Manag Sci. 2012;68:1501–11.
37. Liu S, Ding Z, Zhang C, Yang B, Liu Z. Gene knockdown by intro-thoracic injection of double-stranded RNA in the brown planthopper, *Nilaparvata lugens*. Insect Biochem Mol Biol. 2010;40:666–71.
38. Bellés X. Beyond *Drosophila*: RNAi in vivo and functional genomics in insects. Ann Rev Entomol. 2010;55:111–28.
39. Scott JG, Michel K, Bartholomay LC, Siegfried BD, Hunter WB, Smagghe G, et al. Towards the elements of successful insect RNAi. J Insect Physiol. 2013;59(12):1212–21.
40. Dong Y, Friedrich M. Nymphal RNAi: systemic RNAi mediated gene knockdown in juvenile grasshopper. BMC Biotechnol. 2005;5:25.
41. Brown SJ, Mahaffey JP, Lorenzen MD, Denell RE, Mahaffey JW. Using RNAi to investigate orthologous homeotic gene function during development of distantly related insects. Evol Dev. 1999;1:11–5.
42. Bucher G, Scholten J, Klingler M. Parental RNAi in *Tribolium* (Coleoptera). Curr Biol. 2002;5:R85–6.
43. Yamashita O, Sumida M, Hasegawa K. Developmental changes in midgut trehalase activity and its localization in the silkworm, *Bombyx mori*. J Insect Physiol. 1974;20:1079–85.
44. Mitsumasu K, Azuma M, Niim T, Yamashita O, Yaginuma T. Changes in the expression of soluble and integral-membrane trehalases in the midgut during metamorphosis in *Bombyx mori*. Zool Sci. 2008;25:693–98.
45. Sumida M, Yamashita O. Trehalase transformation in silkworm midgut during metamorphosis. J Comp Physiol. 1977;115:241–53.
46. Wegener G, Tschiedel V, Schlöder P, Ando O. The toxic and lethal effects of the trehalase inhibitor trehazolin in locusts are caused by hypoglycemia. J Exp Biol. 2003;206:1233–40.
47. Kamei Y, Hasegawa Y, Niimi T, Yamashita O, Yaginuma T. Trehalase-2 protein contributes to trehalase activity enhanced by diapause hormone in developing ovaries of the silkworm, *Bombyx mori*. J Insect Physiol. 2011;57(5):608–13.
48. Zhang WQ, Chen XF, Tang B, Tian HG, Chen J, Yao Q. Insect chitin biosynthesis and its regulation. Chin J Appl Entomol. 2011;48(3):475–79.
49. Terra WR, Ferreira C. Insect digestive enzymes: properties, compartmentalization and function. Comp Biochem Physiol. 1994;109B:1–62.
50. Silva CPM, Terra RW, Ferreira C. The role of carboxyl, guanidine and imidazole groups in catalysis by a midgut trehalase purified from an insect larva. Insect Biochem Mol Biol. 2004;34:1089–99.
51. Thompson SN. Trehalose—the insect 'blood' sugar. Adv Insect Physiol. 2003;31:203–85.
52. Friedman S. Trehalose regulation, one aspect of metabolic homeostasis. Annu Rev Entomol. 1978;23:389–407.
53. Clegg JS, Evans DR. Blood trehalose and flight metabolism in the blowfly. Science. 1961;134:54–5.
54. Almeida FM, Bonini BM. Heterologous expression in *Escherichia coli* of *Neurospora crassa* neutral trehalase as an active enzyme. Protei Expre Purif. 2009;65:185–89.
55. Lipiski Z, Ótowska K, Wawrowska J, Zaleska M. The concentration of carbohydrates in the body of development all stages of *Apis mellifera* carnica drone brood. J Apicult Sci. 2008;52:5–11.
56. Yuhki K, Yuh H, Teruyuki N, Okitsugu Y, Toshinobu Y. Trehalase-2 protein contributes to trehalase activity enhanced by diapause hormone in developing ovaries of the silkworm, *Bombyx mori*. J Insect Physiol. 2011;57:608–13.
57. Su ZH, Ikeda M, Sato Y, Saito H, Imai K, Isobe M, et al. Molecular characterization of ovary trehalase of the silkworm, *Bombyx mori* and its

transcriptional activation by diapause hormone. Biochim Biophys Acta. 1994;1218:366–74.
58. Santos R, Alves-Bezerra M, Rosas-Oliveira R, Majerowicz D, Meyer-Fernandes JR, Gondim KC. Gene identification and enzymatic properties of a membrane-bound trehalase from the ovary of *Rhodnius prolixus*. Arch Insect Biochem. 2012;81:199–213.
59. Becker A, Schlder P, Steele JE, Wegener G. The regulation of trehalose metabolism in insects. Experientia. 1996;52:433–39.
60. Żółtowska K, Lipiński Z, Łopieńska-Biernat E, Farjan M, Dmitryjuk M. The activity of carbohydrate-degrading enzymes in the development of brood and newly emerged workers and drones of the carniolan honeybee, *Apis mellifera* carnica. J Insect Sci. 2012;12:1–11.
61. Arakane Y, Dixit R, Begum K, Park Y, Specht CA, Merzendorfer H, et al. Analysis of functions of the chitin deacetylase gene family in *Tribolium castaneum*. Insect Biochem Mol Biol. 2009;39:355e365.
62. Arakane Y, Baguinon MC, Jasrapuria S, Chaudhari S, Doyungan A, Kramer KJ, et al. Both UDP N-acetylglucosamine pyrophosphorylases of *Tribolium castaneum* are critical for molting, survival and fecundity. Insect Biochem Mol Biol. 2011;41:42e50.
63. Yang MM, Zhao LN, Shen QD, Xie GQ, Wang SG, Tang B. Knockdown of two trehalose-6-phosphate synthases severely affects chitin metabolism gene expression in the rice brown planthopper *Nilaparvata lugens*. Pest Manag Sci. 2016; doi: 10.1002/ps.4287.
64. Zhao LN, Yang MM, Shen QD, Liu XJ, Shi ZK, Wang SG, et al. Functional characterization of three trehalase genes regulating the chitin metabolism pathway in rice brown planthopper using RNA interference. Sci Repts. 2016;6:27841.
65. Villarreal AM, Adamson SW, Browning RE, Budachetri K, Sajid MS, Karim S. Molecular characterization and functional significance of the Vti family of SNARE proteins in tick salivary glands. Insect Biochem Mol Biol. 2013;43(5):483–93.
66. Ge LQ, Zhao KF, Huang LJ, Wu JC. The effects of triazophos on the trehalose content, trehalase activity and their gene expression in the brown planthopper *Nilaparvata lugens* (Stål) (Hemiptera: Delphacidae). Pestic Biochem Physiol. 2011;100:172–81.

4

Improvement of alkalophilicity of an alkaline xylanase Xyn11A-LC from *Bacillus* sp. SN5 by random mutation and Glu135 saturation mutagenesis

Wenqin Bai[1,2*], Yufan Cao[1,2], Jun Liu[1,2], Qinhong Wang[1] and Zhenhu Jia[2*]

Abstract

Background: Family 11 alkaline xylanases have great potential economic applications in the pulp and paper industry. In this study, we would improve the alkalophilicity of family 11 alkaline xylanase Xyn11A-LC from *Bacillus* sp. SN5, for the better application in this field.

Results: A random mutation library of Xyn11A-LC with about 10,000 clones was constructed by error-prone PCR. One mutant, M52-C10 (V116A and E135V), with improved alkalophilicity was obtained from the library. Site-directed mutation showed that the mutation E135V was responsible for the alkalophilicity of the mutant. The variant E135V shifted the optimum pH of the wild-type enzyme from 7.5 to 8.0. Compared to the relative activities of the wild type enzyme, those of the mutant E135V increased by 17.5, 18.9, 14.3 and 9.5 % at pH 8.5, 9.0, 9.5 and 10.0, respectively. Furthermore, Glu135 saturation mutagenesis showed that the only mutant to have better alkalophilicity than E135V was E135R. The optimal pH of the mutant E135R was 8.5, 1.0 pH units higher than that of the wild-type. In addition, compared to the wild-type enzyme, the mutations E135V and E135R increased the catalytic efficiency (k_{cat}/K_m) by 57 and 37 %, respectively. Structural analysis showed that the residue at position 135, located in the eight-residue loop on the protein surface, might improve the alkalophilicity and catalytic activity by the elimination of the negative charge and the formation of salt-bridge.

Conclusions: Mutants E135V and E135R with improved alkalophilicity were obtained by directed evolution and site saturation mutagenesis. The residue at position 135 in the eight-residue loop on the protein surface was found to play an important role in the pH activity profile of family 11 xylanases. This study provided alkalophilic mutants for application in bleaching process, and it was also helpful to understand the alkaline adaptation mechanism of family 11 xylanases.

Keywords: Xylanase, Alkalophilicity, Error-prone PCR, Site-directed mutation, Site-saturation mutagenesis, Catalytic efficiency

Background

Xylan is the second most abundant renewable resource in nature, following cellulose. It is composed of a backbone of β-1,4-linked D-xylopyranose residues, and its side chain is modified by different groups, such as acetate, D-glucuronate, 4-O-methyl-D-glucuronate, and α-L-arabinofuranose [1]. Because of its complex structure, the complete degradation of xylan requires the cooperative action of a variety of enzymes. Among them, endo-β-1, 4-xylanase (EC 3.2.1.8) is the crucial enzyme to randomly hydrolyze β-1,4-xylosidic linkages of xylan to xylooligosaccharides [2]. Based on hydrophobic cluster analysis and amino acid sequence similarities, xylanases are mainly classified into glycoside hydrolase (GH) families 10 and 11 [3]. Due to their ability to degrade xylan, xylanases have potential biotechnological applications in

* Correspondence: bai_wq@tib.cas.cn; jiaxueshi@163.com
[1]Department of Strategic and Integrative Research, Tianjin Institute of Industrial Biotechnology, Chinese Academy of Sciences, 300308 Tianjin, China
[2]College of Life Science, Shanxi Normal University, Linfen 041004, China

many fields [2]. A major application of xylanases is in the paper industry, in which the enzyme can reduce the amount of toxic chlorine-containing chemicals and improve the pulp brightness [4]. For example, pretreatment with Cartazyme HS-10 could reduce the chlorine consumption by 31 % at C-stage and 30 % reduction in total organochlorine content in the extraction stage effluent. The brightness was increased by 4.9 points in CEH sequence [5].

Family 11 xylanases can easily penetrate cellulose fiber networks without damaging the fibers because of their small molecular mass (about 20 kDa) and cellulase-free activity. Therefore, they are more suitable for the pulp bleaching process than family 10 enzymes [6]. Because the pulp bleaching process is usually in a high temperature (60–80 °C) and high pH (8–10) environment [2], the xylanases for this application are required to be thermophilic, thermostable, alkaliphilic, and alkali-stable [7, 8]. Although many xylanases have been cloned and characterized (http://www.cazy.org/GH11_characterized.html), few xylanases were found to be active and stable in the high temperature and alkaline condition. Therefore, engineering the native enzymes to achieve high activity and stability at high temperature and pH is in great demand.

Although many attempts on the modification of xylanases by rational design or directed evolution have been successful, such as increasing the optimal temperature or the catalytic activity of the enzyme [9–11], research on the improvement of alkalophilicity of the xylanase is limited [12]. Most of the mutations decreased the optimal pH of the enzyme [13, 14], or only shifted the optimum from acidity to neutrality rather than to the alkaline range [7, 15, 16]. Introducing excess arginine residues on the protein surface could increase the optimal pH of xylanase XynJ from aklaliphilic *Bacillus* sp. 41M-1 from 8.5 to 9.5 [17], but we could not replicate the result with the equivalent mutational change in the other xylanse in our previous study.

In the previous study, an alkaline xylanase Xyn11A-LC from *Bacillus* sp. SN5 was characterized [18]. It exhibited the highest catalytic activity at pH7.5, but little enzyme activity could be detected at pH 10.0. This property was not suitable for the pulp bleaching process. The three dimensional structure of xylanase Xyn11A-LC has been determined [19]. The molecular basis of alkaline adaptation of family 11 xylanase has been revealed [20]. These studies will contribute to engineer non-alkaline xylanase to function at a higher pH condition, but it is still hard to further improve the alkalophilicity of an alkaline xylanase by rational protein design. In this study, in order to obtain more alkaline mutants, a random mutation library of xylanase Xyn11A-LC was constructed by error-prone PCR. One mutant with improved alkalophilicity was obtained by a high-throughput screening system. Site-directed mutations and site-saturation mutagenesis were then carried out to validate the key role of the residue at position 135 on the pH activity profile of Xyn11A-LC. The mechanism of alkaline adaptation of the mutant was discussed by structural analysis.

Results

Construction and screening of random mutagenesis libraries

Over 10,000 colonies were obtained from the random mutagenesis libraries. About 6000 transformants showing clear halos on the medium with Remazol Brilliant Blue xylan (RBB-xylan) were picked into 96-well plates for screening for the mutants with improved alkalophilicity. The selection criterion was an increase of at least 10 % in the pH 10/pH 7.5 activity ratio of the mutants compared to that of the wild type. For the second screening, a total of 54 mutants showing higher ratio than that of the wild type were obtained. Then, according to the pH activity profiles of crude enzymes of the wild type and 54 mutants, 8 mutants with improved alkalophilicity were obtained in the third screening. For the fourth screening, one mutant M52-C10 with improved alkalophilicity was obtained from above mentioned 8 mutants by determining pH activity profiles of the purified enzymes (Additional file 1: Figure S1). M52-C10 exhibited an optimal activity at pH 8.0 and corresponded to a basic shift of 0.5 pH units compared to the wild-type enzyme. Furthermore, the relative activities of mutant M52-C10 increased by 10, 15, 8 % at pH 8.5, 9.0 and 9.5, respectively (Fig. 1).

Mutation site analysis

DNA sequencing showed that the mutant M52-C10 had two mutations, V116A and E135V. Site-directed mutagenesis revealed that only the E135V mutation showed the improvement of alkalophilicity (Fig. 1). The optimal pH of the mutant E135V increased to pH 8.0, which was 0.5 pH units higher than that of the wild-type enzyme. Compared to the relative activities of the wild type enzyme, those of the mutant E135V increased by 17.5, 18.9, 14.3 and 9.5 % at pH 8.5, 9.0, 9.5 and 10.0, respectively. Nevertheless, the pH activity profile of the mutant V116A was similar to that of the wild type enzyme. According to the above results, it could be concluded that Glu 135 might be one of the key residues involved in determining the pH activity profile of Xyn11A-LC.

In order to get more information about the influence of Glu 135 on the pH activity profile of Xyn11A-LC and to obtain more alkalophilic mutant, site-saturation mutagenesis was carried out by overlap extension PCR. The result showed that the only mutant to have better alkalophilicity than E135V was E135R. Its optimal pH was 8.5, 0.5 pH units higher than that of E135V. Mutations

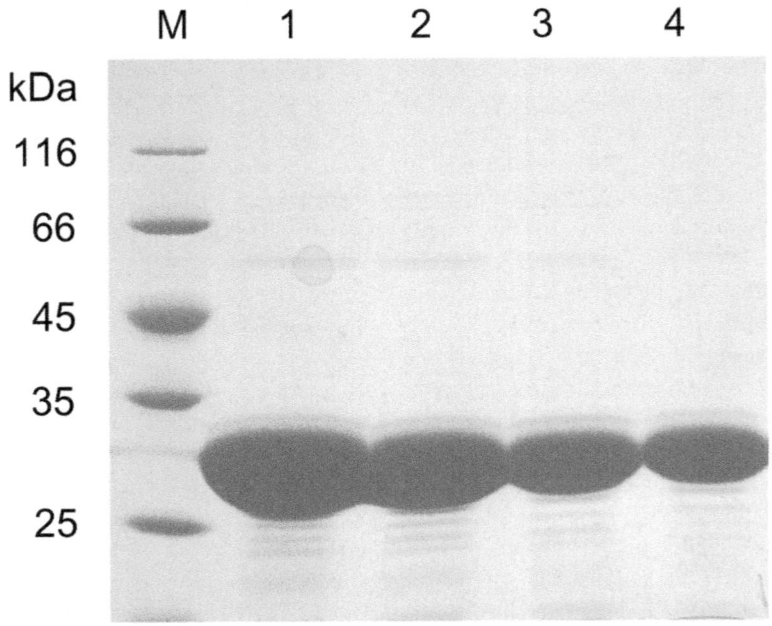

Fig. 1 SDS–PAGE analysis of the purified xylanase Xyn11A-LC and mutants. Lane *M*, the protein molecular mass markers; Lane *1–4*, the purified recombinant Xyn11A-LC, M52-C10, E135V and E135R, respectively

E135H, E135K, E135Q, E135M, E135Y and E135A exhibited maximal activity at pH 8.0, similar to the mutant E135V. The optimal pH of E135P was 6.0, 1.5 pH units lower than that of the wild-type. The ten other mutations exhibited maximal activity at pH 7.5, similar to the wild-type (Fig. 2).

Characterization of the mutants and the wild type enzyme

To further elucidate the effects of Glu 135 site substitutions on enzyme characterization, other characterizations of the wild type and the mutants were determined. The mutants E135V and E135R were both stable over a broad pH range, retained over 80 % of original activity after incubation at 37 °C for 1 h over a pH range from 4.5 to 10.0, while the wild-type enzyme was stable only at alkaline pH from 8.0 to 10.0 (Fig. 2b). The optimal temperatures of the mutants M52-C10, E135V and E135R were all 55 °C, which was similar to that of the wild type. However, the mutants exhibited higher relative activity at the temperature range of 30–70 °C (Fig. 2c). In addition, the thermostability of the mutants was also determined. After incubation at 60 °C for 30 min, the mutants M52-C10, E135V and E135R retained 57.4, 73.1 and 77.8 % of their maximal activities, respectively, while the wild type enzyme only retained 33.5 % of its activity under the same conditions (Fig. 2d). This result showed that the displacement of the negative charge on the surface increased the thermostability of xylanase Xyn11A-LC.

The kinetic parameters of the wild type enzyme and the mutants toward beechwood xylan were determined under their optimal conditions. The K_m, V_{max}, and k_{cat} values of the enzyme were shown in Table 1. Compared with the wild type, the mutants M52-C10, E135V and E135R had lower K_m values, which indicated that the Glu 135 mutation increased the substrate affinity. Additionally, the V_{max} values and k_{cat} values of the mutant E135V and E135R are all higher than those of the wild type. Compared with the wild-type enzyme, the mutations E135V and E135R increased the catalytic efficiency (k_{cat}/K_m) by 57 and 37 %, respectively.

Structural analyses of the mutants

Some amino acid residues close to the catalytic residues (Glu 93 and Glu 183) are presumed to be responsible for the pH activity profile of Xyn11A-LC by changing the electrostatic potential of the active site [20]. However, the residue Glu 135 is far away from the two catalytic residues (Glu 93 and Glu 183). The distance between Glu 135 Oε1/Oε2 and Glu 93 Oε1/Oε2 is 12.12, 12.30, 13.22 and 13.23 Å, respectively, whereas the distance between Glu 135 Oε1/Oε2 and Glu 183 Oε1/Oε2 is 19.12, 19.60, 19.90 and 20.53 Å, respectively. Xyn11A-LC has the β-jelly roll structure typical of family 11 xylanases (Fig. 3a), which is also likened to the shape of a 'right hand' [21]. The eight-residue loop (Gln 131, Pro 132, Ser 133, Ile 134, Glu 135, Gly 136, Thr 137, and Ala 138) forms a 'thumb' that partly encloses the catalytic cleft. Glu 135 is located at the edge of the 'thumb' (Fig. 3a). The sequence and structure of the

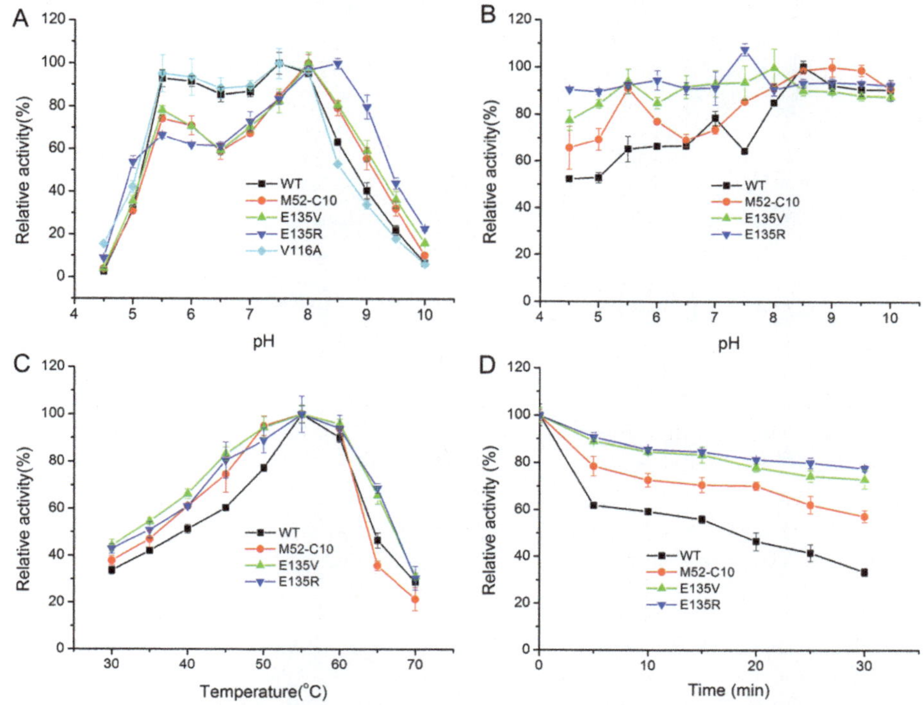

Fig. 2 Effects of pH and temperature on the activity and stability of the recombinant Xyn11A-LC and mutants. **a** Effect of pH on the activity of Xyn11A-LC and mutants. The assay was performed in different pH buffer ranging from pH 4.5 to pH 10.0 at optimal temperature for 10 min. **b** Effect of pH on the stability of Xyn11A-LC and mutants. The enzyme was diluted in different buffer (pH 4.5-pH 10.0) at 37 °C for an hour, and the residual activities were measured at the optimal condition for 10 min. The activity of the purified enzyme without pre-incubation was set as 100 %. **c** Effect of temperature on the activity of Xyn11A-LC and mutants. The assay was performed at different temperature ranging from 30 to 70 °C for 10 min. **d** Effect of temperature on the stability of Xyn11A-LC and mutants. The purified enzymes were incubated in 50 mM Tris–HCl buffer (pH 8.0) without substrate for 30 min at 60 °C, respectively and taken out every 5 min. The residual xylanase activities were measured under the optimal condition for 10 min. The 0 min value was set as 100 %

eight-residue loop of 13 family 11 xylanases, which were optimally active in acidic, neutral, or alkaline pH range, were compared, as shown in Table 2. The triad of residues that forms the tip of the thumb (Pro 132, Ser 133, and Ile 134) and Gly 136 are conserved in family 11 xylanase. The position 135 residue was a lysine residue in those xylanases with higher optimal pH, the corresponding residue is a glutamate or asparte residue in Xyn 11A-LC and some xylanases with lower optimal pH. In addition, the lysine residue in the eight-residue loop in higher alkaline xylanases was involved in the formation of a salt bridge (Table 2). It would be speculated that the elimination of the negative charge, the introduction of the positive charge, or the salt-bridge in the position 135 might be involved in the alkaline adaptation of xylanase.

The mutation E135V with improved alkalophilicity might prove that the elimination of the negative charge at position 135 could contribute to the alkaline adaptation of xylanase. In order to validate the positive charge at position 135, the pH activity profiles of the mutations E135H, E135K and E135R were compared with E135V. The result showed that the pH activity profiles of the mutations E135H and E135K were similar to that of the E135V mutation, but E135R had higher alkapholicity than E135V, its optimal pH was increased to 8.5 (Fig. 2a). The structural analysis showed that a putative salt-bridge could be established between the introduced Arg 135 and Asp 89 because of a large guanidinium group of arginine, but it didn't exist between Lys/His 135 and Asp 89 (Fig. 3b). The superior performance for Arg 135

Table 1 The kinetic parameters of the wild-type enzyme and the mutants

Parameter	WT	M52-C10	E135V	E135R
K_m (mg ml^{-1})	3.3 ± 0.2	3.0 ± 0.4	2.5 ± 0.6	2.6 ± 0.5
V_{max} (μmol min^{-1} mg^{-1})	7178 ± 186	7355 ± 324	8546 ± 654	7734 ± 602
k_{cat} (s^{-1})	3230 ± 84	3310 ± 146	3846 ± 294	3480 ± 271
k_{cat}/K_m (ml mg^{-1}s^{-1})	978 ± 25	1103 ± 48	1538 ± 118	1339 ± 104

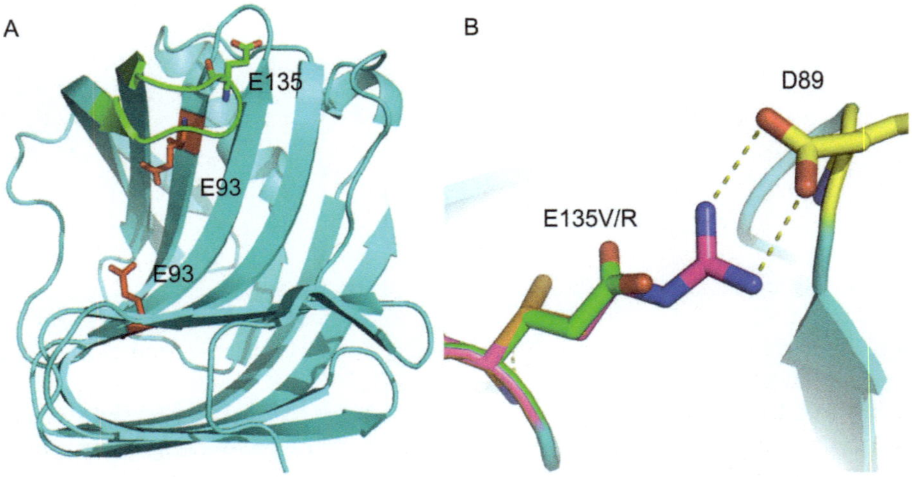

Fig. 3 Locations and structural analysis of three key mutation sites. **a** Mutation location in the crystal structure of Xyn11A-LC (PDB accession no. 4IXL). The side chains of the catalytic residues (E93 and E183) and mutation (E135) are shown in *red* and *green*, respectively. A 'thumb' formed by the eight-residue loop is shown in *green*. **b** Location and mutation of E135V/R. E135, V135 and R135 are shown in *green*, *orange*, and *purple*, respectively. Salt bridges are represented in *yellow* by *dashed lines*

relative to Lys 135 or His 135 suggests that the bidentate hydrogen bonding that is geometrically feasible for Arg 135 has more importance than simply the presence of a positive charge at this site.

Discussion

Research on xylanase used in the paper industry has attracted increasing attention because it can reduce the cost, lower environmental pollution and improve the pulp quality. It requires the xylanase to be stable and active at high temperature and alkaline pH [2]. However, most of the xylanases are reported to be mesophilic or acidophilic enzymes. In this study, directed evolution of the enzyme was used by error-prone PCR. Xyn11A-LC was successfully engineered to improve its alkaline adaptation for potential industrial application for the pulp bleaching process.

The mutations E135V, E135K, E135H, E135A, E135Q, E135M, and E135Y all increased the optimal pH from 7.5 to 8.0. Furthermore, E135R increased the optimal pH from 7.5 to 8.5 (Fig. 2a). However, the optimal pH of E135P was 6.0, 1.5 pH units lower than that of the wild-type. It is speculated that a proline at position 135 is incompatible with the phi/psi for this beta turn, or it is likely to cause disruption through steric incompatibility with neighboring residues. Other studies also proved

Table 2 Sequence and structure comparison of the eight-residue loops of family 11 xylanases

Enzyme abbreviation	Source	PDB code	Eight-residue loop sequence	Salt-bridge in the eight-residue loop	Optimum pH
XynJ (ΔXBD)	*Bacillus* sp. 41M-1	2DCK	QPSI**K**GTA	Lys135- Glu55	9 [17]
Xyl C	*alkalophilic Bacillus* (NCL 87-6-10)	2F6B	QPSI**K**GIA	Lys136-Glu56	8 [35]
BadX	*Bacillus agaradhaerens* AC13	1H4G	QPSI**K**GIA	Lys136- Glu56	8 [36]
Xyn11X	*Bacillus subtilis* B230	1IGO	QPSI**I**GIA	NA	8 [37]
Xyn11A-LC	*Bacillus* sp SN5	4IXL	QPSI**E**GTA	NA	7.5 [18]
XynA	*Bacillus subtilis* 168(1A1)	1XXN	APSI**D**GDR	NA	6–6.5 [38]
Xyl1	*Streptomyces* sp. S38	1HIX	APSV**E**GTK	NA	6 [39]
Bcx	*Bacillus circulans*	1XNB	APSI**D**GDR	NA	5.7 [40]
Xyn2	*Hypocrea jecorina* RUT-C30	1XYO	QPSI**I**GTA	NA	5.3 [21]
XYNI	*Trichoderma reesei*	1XYN	EPSI**Q**GTA	NA	3.5 [41]
XYL1	*Scytalidium acidophilum*	3M4F	EPSI**T**GTS	NA	3.2 [42]
XynA	*Aspergillus niger* CBS 513.88	1UKR	EPSI**T**GTS	NA	3 [43]
XynC	*Aspergillus kawachii*	1BK1	EPSI**T**GTS	NA	2 [44]

The key residues involved in the pH adaptation of enzymes are in bold and underlined

that the substitution of the residue in other xylanases corresponding to Glu 135 in xylanase Xyn11A-LC could also change the pH activity profile of enzymes. The E141A mutation in XYL1p from *Scytalidium acidophilum* increased the optimal pH from 3.2 to 4.0 [22]. The E141Q mutation in XynI from *Aureobasidium pullulans* changed the optimal pH from 2.5 to 3.5 [23]. The mutation E139K of Xyl1 of *Streptomyces* sp. S38 significantly raised the pH optimum from 6.0 to 7.5 [24]. Therefore, the position 135 site may play an important role in determining the pH activity profile of family 11 xylanase.

The residue Glu 135 located at the edge of the 'thumb' could also influence the catalytic activity of the enzyme. Two mutations (E135V and E135R) of xylanase Xyn11A-LC increased the catalytic activities to a different extent. The mutations E118A and E118Q of xylanase from *Aspergillus niger* for the same key position increased the activity by 50 and 16 %, respectively [25]. The same result was shown in XYL1p from *S. acidophilum* where E141A mutation elevated the specific activity by 50 % [22]. Furthermore, the importance of this residue in the catalytic activity of family 11 xylanases was confirmed by the inverse mutation, the mutation A139E in Xyl1 of *Streptomyces* sp. S38 decreased its specific activity by 75 % [26].

It was reported that the residues in the eight-residue loop were important in determining the pH optima of family 11 xylanase [7, 27]. The residue Glu 135 was located at the edge of the eight-residue loop on the protein surface (Fig. 3a). The substitution of Glu 135 residue with the valine residue might improve the alkalophilicity of the enzyme by eliminating the negative charge on the surface. The mutation E135R might further improve the alkalophilicity by introduction of arginines residues and a salt-bridge on the surface. The pH-dependent activity of the enzyme depends on the ionization of catalytic acid residues sensitive to the surface charge of the protein. Changing the surface charge of the enzyme could significantly shift the pH-activity profiles of the enzymes by changing the pK_a value of the active site, despite the two groups are far away from each other [28]. In our previous work, an increase in the number of the positive charge residues (Arg and Lys) and a decrease in the Glu residue on the protein surface were involved in the mechanism of alkaline adaptation of the family 11 xylanases, because the basic residues have high p*K*a values and can retain net positive charge even at high pH [20]. Several studies showed that increasing the number of arginines on the protein surface could improve the alkalophilicity of enzymes [17, 29]. In addition, there was a positive correlation between the number of salt bridges and alkaline adaptation of family 11 xylanases [20]. Other studies also indicated that the characteristic salt bridges in the catalytic cleft could contribute to the alkalophilicity of an alkaline xylanase XynJ from aklaliphilic *Bacillus* sp. 41M-1 [17, 30]. The salt bridge in the eight-residue loop might be beneficial to the stability of the enzyme in the alkaline condition and thus improve the alkalophilicity of the enzyme.

Research has indicated that the eight-residue loop possesses an opening and closing capability that modifies both the topology and the binding capacity of the active site, and the loop's precise position determines the width of the catalytic cleft [27, 31]. Although, the residue Glu135 is far from the active site, it is located at the edge of the substrate binding cleft. The displacement of Glu 135 residue might eliminate the electrostatic repulsions between the cleft and the xylan which has a certain degree of negative charge because of glucuronic acid substituents [23]. This could increase the binding ability of the substrate and the cleft [23]. An intimate interaction with the substrate might protect the catalytic residues within the active site from the solvent and therefore allow the enzyme to function at a higher pH condition. It is also the possible cause of the decrease of the K_m values in the two mutations E135V and E135R. The position of the loop may be changed when a larger substrate binds in a slightly different conformation, which changes the distances between the catalytic residues and the substrate and thus affects the catalytic activity of the enzyme [32].

Conclusion

Using directed evolution followed by site-directed mutagenesis, the residue at position 135 in Xyn11A-LC was speculated to be one of the key amino acids responsible for the pH activity profile of the enzyme. Furthermore, sequence alignment and structural analysis of family 11 xylanases with different optimum pH showed that the residue at position 135 located at the edge of the 'thumb' could affect the alkalophilicty by the elimination of the negative charge and the information of the salt-bridge. It will be useful to understand the alkaline adaptation of family 11 xylanases and engineer the enzyme to function at a higher pH condition. The mutant E135R is a promising and suitable candidate for paper pulp bio-bleaching.

Methods

Strains, plasmids, enzymes and reagents

The recombinant plasmid pET 28a-Xyn11A-LC was constructed by our research group [18]. *E.coli* BL21-Gold (DE3) strain was used as a host for the random mutagenesis library. *E.coli* BL21 (DE3) was used for protein production. GeneMorph II Random Mutagenesis Kit was purchased from Agilent technologies (USA). Beechwood xylan and RBB-xylan were purchased from Sigma (USA).

Construction of random mutagenesis library

The random mutagenesis library was constructed by error-prone PCR using GeneMorph II Random Mutagenesis Kit according to the manual. The recombinant plasmid pET 28a-Xyn11A-LC was used as a template. In order to achieve medium mutation rate, initial target amount of template DNA is about 100 ng. The mutant gene was amplified by PCR using the primers ep-F and ep-R listed in Additional file 1: Table S1. The PCR cycling conditions were as follows: 2 min at 95 °C, then 30 cycles of 30 s at 95 °C, 30 s at 57 °C, and 30 s at 72 °C, followed by 10 min at 72 °C. The PCR products were purified using the gel extraction kit (Omega), digested with *Bam*HI and *Xho*I, and then ligated into the pET28a between *Bam*HI and *Xho*I sites. The resulting plasmids were transformed into *E.coli* BL21-Gold (DE3) competent cells by electro transformation and plated on LKX medium (Luria-Bertani agar plate, 50 μg/ml kanamycin, 0.2 % RBB-xylan, pH 9.0) [18].

Screening for alkalophilic xylanase mutants

The colonies showing clear halos on above mentioned LKX medium were picked up and inoculated into 160 μl LBK medium (LB medium supplemented with 50 μg/ml kanamycin) in 96-well microplates for growth overnight at 37 °C with shaking at 900 rpm. Every mutant in the microplates was then replicated into fresh LBK medium containing 0.1 mM isopropyl β-D-1-thiogalactopyranoside (IPTG). After cultivation at 37 °C for 10 h, the cells were harvested by centrifugation at 4000 rpm at 4 °C for 10 min, re-suspended with lysis buffer (50 mM Tris–HCl buffer containing 10 mg/ml lysozyme, pH 8.0), and incubated at 37 °C for 60 min. The cell lysates were diluted 100-fold and used as the crude enzyme extract for downstream enzymatic reactions.

To screen the mutants with improved alkaline adaptation, xylanase activities were measured at pH 7.5 and pH 10.0, respectively. 10 μl of the crude enzyme extract was incubated with 100 μl 1 % beechwood xylan (*w/v*) dissolved in 50 mM Tris–HCl buffer (pH 7.5) and 50 mM Glycine–NaOH (pH 10.0), respectively. After incubation at 55 °C for 10 min, the released reducing sugars were assayed by the 3,5-dinitrosalicylic acid (DNS) method using xylose as the standard [33]. The variants with higher pH 10.0/pH 7.5 activity ratio than the wild type were considered as mutants with improved alkalophilicity. For the third screening, the mutants with significantly improved alkalophilicity were selected by detecting crude xylanase activity at pH from 7.0 to 10.0. For the fourth screening, the positive mutants were selected by determining the purified enzyme activity at pH from 4.5 to 10.0.

Site-directed mutagenesis and site-saturation mutagenesis

Site-directed mutagenesis and site-saturation mutagenesis were performed using the overlap-extension PCR method with Phusion Hot Start II High-Fidelity DNA Polymerase as described [19]. The plasmid pET28a-Xyn11A-LC was used as the template. The primers containing mutated codons were listed in Additional file 1: Table S1. The PCR cycling conditions were listed as follows: 30 s at 98 °C, then 20 cycles of 30 s at 98 °C, 30 s at 55 °C, and 3 min at 72 °C, followed by 10 min at 72 °C. The amplification products were digested by *Dpn*I overnight at 37 °C and then transformed into *E. coli* BL21 (DE3) competent cells by electro transformation. The mutant xylanase gene from the transformants was verified by DNA sequencing performed by Genewiz BioSciences Company (Beijing, China).

Expression and purification of enzymes

A single colony of strain BL21(DE3) harboring plasmid carrying gene of Xyn11A-LC or its mutant was grown in LBK medium at 37 °C until the OD600 reached 0.6–0.8. After a supplement with 1 mM IPTG, the culture was induced at 37 °C for 6 h. Cells were harvested by centrifugation at 8000 rpm for 10 min and disrupted by sonication in the lysis buffer (20 mM Tris–HCl, 500 mM NaCl, 10 mM imidazole, pH 8.0). Cell debris was removed by centrifugation at 16,000 rpm at 4 °C for 20 min. The supernatants were loaded on a nickel–nitrilotriacetic acid (Ni–NTA) column (Qiagen, Valencia, CA, USA) which was pre-equilibrated with the lysis buffer. Then the column was washed with washing buffer (20 mM Tris–HCl, 500 mM NaCl, 60 mM imidazole, pH8.0) and the bound protein was eluted with the elution buffer (50 mM Tris–HCl, 500 mM NaCl, 1 M imidazole, pH 8.0). Imidazole was removed by using a desalting column (GE Healthcare Bio-Sciences AB, Uppsala, Sweden) with 20 mM Tris–HCl buffer (pH 8.0). The purity of the proteins was analyzed by the sodium dodecyl sulfate polyacrylamide gel electrophoresis (SDS-PAGE). Protein concentrations were determined by a protein assay kit (Bio-Rad).

Enzyme activity

The enzyme activities of the wild type Xyn11A-LC and its mutants were assayed by measuring the amount of reducing sugars from beechwood xylan by DNS method [33]. The reaction mixture containing 20 μl of 1 μg/ml xylanase solution and 480 μl of 1 % beechwood xylan (*w/v*) in different buffer was incubated in 55 °C for 10 min. After supplemented with 500 μl of DNS reagent, the mixture was boiled for 5 min and cooled to room temperature. Then the absorbance at 540 nm was determined with Spectra Max 190 microplate reader

(Molecular Devices, Sunnyvale, CA, USA). One unit (U) of xylanase activity was defined as the amount of enzyme that released reducing sugar from xylan equivalent to 1 μmol xylose per minute under the assay condition. The assays were all performed in triplicate.

pH activity profile and pH stability of the wild type and mutants

The effects of pH on the enzyme activity of the wild type Xyn11A-LC and its mutants was measured at 55 °C at pH from 4.5 to 10.0. The maximum value was set as 100 %. For pH stability, the purified xylanases were diluted to 2 μg/ml in different buffers with pH ranging from 4.5 to 10.0 and pre-incubated without substrate at 37 °C for 1 h. The residual activities were measured under optimal conditions. The enzyme activity of the purified xylanases without pre-incubation was set as 100 %. The buffers used were citrate-Na_2HPO_4 for pH 4.5–8.0, 50 mM boric acid/borate for pH 8.0–9.0, and 50 mM borate/NaOH for pH 9.5–10.0, respectively.

Optimal temperature and thermostability

The effect of temperature on the xylanase activity was measured at optimal pH and a temperature range (30–70 °C). The maximum value was set as 100 %. For thermal stability, the purified xylanases were diluted to 2 μg/ml in 50 mM Tris–HCl buffer (pH 8.0), incubated at 60 °C and then taken out at an interval of 5 min. After incubation, the enzyme solution was immediately put in an ice bath for 10 min. The residual enzyme activities were assayed under optimal condition. The enzyme activity of purified protein without pre-incubation was set as 100 %.

The kinetic parameters assay

To assay the kinetic parameters K_m and V_{max}, reaction mixtures containing 1 μg/ml purified xylanases and different concentration of beechwood xylan (1–10 mg/ml) were incubated at optimum pH and temperature. The Michaelis-Menten parameters were calculated by nonlinear regression using Graphpad Prism 6.0 software (San Diego, CA, USA). The turn over number (k_{cat}) was calculated based on the V_{max} value and the molecular weight of the targeted enzyme.

Structural analysis

In our previous work, the structure of the wild type Xyn11A-LC was determined, and the atomic coordinates and structure factors have been submitted to the Protein Data Bank (PDB) under accession code 4IXL [19]. The coordinates of other twelve family 11 mesophilic xylanases with known pH activity profile were extracted from PDB. Multiple sequence alignment of the amino acid sequences was performed using the ClustalW program. The three dimensional structures of the mutants were modeled using Discovery Studio software. The surface residues of Xyn11A-LC with 30 % accessible surface areas were identified in Swiss-PDB Viewer using the default parameters [34]. Salt bridges were calculated online by submitting the coordinate files to website (http://bioinformatica.isa.cnr.it/ESBRI/); salt bridges were assigned when the distance between the two atoms of opposite charge was less than 4 Å. Structure visualization was performed using PyMOL software.

Abbreviations

DNS: 3,5-dinitrosalicylic acid; IPTG: Isopropyl β-D-1-thiogalactoside; Ni-NTA: Nickel-nitrilotriacetic acid; PDB: Protein Data Bank; RBB-xylan: Remazol Brilliant Blue xylan; SDS-PAGE: Sodium dodecyl sulfate polyacrylamide gel electrophoresis

Acknowledgement

We thank Maria McClintock and Kevin J. Fox in the University of Minnesota for editing the language.

Funding

This study was supported by the Natural Science Foundation of China (31400679), the Key Deployment Program of Chinese Academy of Sciences (KSZD-EW-Z-015), the National High Technology Research and Development Program of China (863 Program, 2012AA022203).

Authors' contributions

WQB: corresponding author, conceived this study, designed and supervised the experiments, performed the experiments, analyzed the data and structures, wrote the manuscript; YFC: performed construction and screening of mutant library, site-directed mutagenesis, and characterization of enzyme; JL: performed site-saturation mutagenesis and characterization of enzyme; QHW: conceived this study, supervised the experiments, revised the manuscript; ZHJ: corresponding author, conceived the study. All authors have read and approved the manuscript.

Competing interests

The authors declare that they have no competing interests.

References

1. Berrin JG, Juge N. Factors affecting xylanase functionality in the degradation of arabinoxylans. Biotechnol Lett. 2008;30(7):1139–50.
2. Collins T, Gerday C, Feller G. Xylanases, xylanase families and extremophilic xylanases. FEMS Microbiol Rev. 2005;29(1):3–23.
3. Henrissat B. A classification of glycosyl hydrolases based on amino acid sequence similarities. Biochem J. 1991;280(Pt 2):309–16.
4. Beg QK, Kapoor M, Mahajan L, Hoondal GS. Microbial xylanases and their industrial applications: a review. Appl Microbiol Biot. 2001;56(3–4):326–38.
5. Bajpai P, Bhardwaj NK, Bajpai PK, Jauhari MB. The impact of xylanases on bleaching of eucalyptus kraft pulp. J Biotechnol. 1994;38(1):1–6.
6. Bajpai P. Application of enzymes in the pulp and paper industry. Biotechnol Progr. 1999;15(2):147–57.
7. Li F, Xie J, Zhang X, Zhao L. Improvement of the optimum pH of *Aspergillus niger* xylanase towards an alkaline pH by site-directed mutagenesis. J Microbiol Biotech. 2015;25(1):11–7.
8. Shoham Y, Schwartz Z, Khasin A, Gat O, Zosim Z, Rosenberg E. Delignification of wood pulp by a thermostable xylanase from *Bacillus stearothermophilus* strain T-6. In: Microorganisms to combat pollution. Springer; 1992. p. 83–94

9. Dumon C, Varvak A, Wall MA, Flint JE, Lewis RJ, Lakey JH, Morland C, Luginbühl P, Healey S, Todaro T. Engineering hyperthermostability into a GH11 xylanase is mediated by subtle changes to protein structure. J Biol Chem. 2008;283(33):22557–64.
10. Zhang ZG, Yi ZL, Pei XQ, Wu ZL. Improving the thermostability of *Geobacillus stearothermophilus* xylanase XT6 by directed evolution and site-directed mutagenesis. Bioresource Technol. 2010;101(23):9272–8.
11. Joo JC, Pohkrel S, Pack SP, Yoo YJ. Thermostabilization of *Bacillus circulans* xylanase via computational design of a flexible surface cavity. J Biotechnol. 2010;146(1):31–9.
12. Zheng H, Liu Y, Sun M, Han Y, Wang J, Sun J, Lu F. Improvement of alkali stability and thermostability of *Paenibacillus campinasensis* Family-11 xylanase by directed evolution and site-directed mutagenesis. J Ind Microbiol Biot. 2014;41(1):153–62.
13. Gibbs MD, Reeves RA, Choudhary PR, Bergquist PL. Alteration of the pH optimum of a family 11 xylanase, XynB6 of *Dictyoglomus thermophilum*. N Biotechnol. 2010;27(6):803–9.
14. Liu X, Qu Y, You F, Liu Y. Studies on the key amino acid residues responsible for the alkali-tolerance of the xylanase by site-directed or random mutagenesis. J Mol Catal B Enzym. 2002;18(4):307–13.
15. Turunen O, Vuorio M, Fenel F, Leisola M. Engineering of multiple arginines into the Ser/Thr surface of *Trichoderma reesei* endo-1,4-beta-xylanase II increases the thermotolerance and shifts the pH optimum towards alkaline pH. Protein Eng. 2002;15(2):141–5.
16. Chen Y-L, Tang T-Y, Cheng K-J. Directed evolution to produce an alkalophilic variant from a *Neocallimastix patriciarum* xylanase. Can J Microbiol. 2001;47(12):1088–94.
17. Umemoto H, Ihsanawati, Inami M, Yatsunami R, Fukui T, Kumasaka T, Tanaka N, Nakamura S. Improvement of alkaliphily of *Bacillus* alkaline xylanase by introducing amino acid substitutions both on catalytic cleft and protein surface. Biosci Biotech Bioch. 2009;73(4):965–7.
18. Bai W, Xue Y, Zhou C, Ma Y. Cloning, expression, and characterization of a novel alkali-tolerant xylanase from alkaliphilic *Bacillus* sp. SN5. Biotechnol Appl Biochem. 2015;62(2):208–17.
19. Bai W, Zhou C, Xue Y, Huang CH, Guo RT, Ma Y. Three-dimensional structure of an alkaline xylanase Xyn11A-LC from alkalophilic *Bacillus* sp. SN5 and improvement of its thermal performance by introducing arginines substitutions. Biotechnol Lett. 2014;36(7):1495–501.
20. Bai W, Zhou C, Zhao Y, Wang Q, Ma Y. Structural insight into and mutational analysis of family 11 xylanases: implications for mechanisms of higher ph catalytic adaptation. PLoS One. 2015;10(7):e0132834.
21. Torronen A, Harkki A, Rouvinen J. Three-dimensional structure of endo-1,4-beta-xylanase II from *Trichoderma reesei*: two conformational states in the active site. EMBO J. 1994;13(11):2493–501.
22. Al Balaa B, Brijs K, Gebruers K, Vandenhaute J, Wouters J, Housen I. Xylanase XYL1p from *Scytalidium acidophilum*: site-directed mutagenesis and acidophilic adaptation. Bioresour Technol. 2009;100(24):6465–71.
23. Tanaka H, Okuno T, Moriyama S, Muguruma M, Ohta K. Acidophilic xylanase from *Aureobasidium pullulans*: efficient expression and secretion in *Pichia pastoris* and mutational analysis. J Biosci Bioeng. 2004;98(5):338–43.
24. De Lemos Esteves F, Gouders T, Lamotte-Brasseur J, Rigali S, Frere JM. Improving the alkalophilic performances of the Xyl1 xylanase from *Streptomyces* sp. S38: structural comparison and mutational analysis. Protein Sci. 2005;14(2):292–302.
25. Tahir TA, Berrin JG, Flatman R, Roussel A, Roepstorff P, Williamson G, Juge N. Specific characterization of substrate and inhibitor binding sites of a glycosyl hydrolase family 11 xylanase from *Aspergillus niger*. J Biol Chem. 2002;277(46):44035–43.
26. de Lemos Esteves F, Ruelle V, Lamotte-Brasseur J, Quinting B, Frere JM. Acidophilic adaptation of family 11 endo-beta-1,4-xylanases: modeling and mutational analysis. Protein Sci. 2004;13(5):1209–18.
27. Kim SH, Pokhrel S, Yoo YJ. Mutation of non-conserved amino acids surrounding catalytic site to shift pH optimum of *Bacillus circulans* xylanase. J Mol Catal B Enzym. 2008;55(3):130–6.
28. Russell AJ, Fersht AR. Rational modification of enzyme catalysis by engineering surface charge. Nature. 1986;328(6130):496–500.
29. Shirai T, Suzuki A, Yamane T, Ashida T, Kobayashi T, Hitomi J, Ito S. High-resolution crystal structure of M-protease: phylogeny aided analysis of the high-alkaline adaptation mechanism. Protein Eng. 1997;10(6):627–34.
30. Umemoto H, Ihsanawati, Inami M, Yatsunami R, Fukui T, Kumasaka T, Tanaka N, Nakamura S. Contribution of salt bridges to alkaliphily of Bacillus alkaline xylanase. Nucleic Acids Symp Ser (Oxf). 2007;51:2.
31. Muilu J, Torronen A, Perakyla M, Rouvinen J. Functional conformational changes of endo-1,4-xylanase II from *Trichoderma reesei*: a molecular dynamics study. Proteins. 1998;31(4):434–44.
32. Wakarchuk WW, Campbell RL, Sung WL, Davoodi J, Yaguchi M. Mutational and crystallographic analyses of the active site residues of the *Bacillus circulans* xylanase. Protein Sci. 1994;3(3):467.
33. Miller GL. Use of Dinitrosalicylic acid reagent for determination of reducing sugar. Anal Chem. 1959;31(3):426–8.
34. Guex N, Peitsch MC. SWISS-MODEL and the Swiss-PdbViewer: an environment for comparative protein modeling. Electrophoresis. 1997;18(15):2714–23.
35. Balakrishnan H, Kamal Kumar B, Dutta-Choudhury M, Rele MV. Characterization of alkaline thermoactive cellulase-free xylanases from alkalophilic *Bacillus* (NCL 87-6-10). J Biochem Mol Biol Biophys. 2002;6(5):325–34.
36. Poon DK, Webster P, Withers SG, McIntosh LP. Characterizing the pH-dependent stability and catalytic mechanism of the family 11 xylanase from the alkalophilic *Bacillus agaradhaerens*. Carbohyd Res. 2003;338(5):415–21.
37. Oakley AJ, Heinrich T, Thompson CA, Wilce MC. Characterization of a family 11 xylanase from *Bacillus subtillis* B230 used for paper bleaching. Acta Crystallogr D Biol Crystallogr. 2003;59(Pt 4):627–36.
38. Ruller R, Rosa JC, Faca VM, Greene LJ, Ward RJ. Efficient constitutive expression of *Bacillus subtilis* xylanase A in *Escherichia coli* DH5alpha under the control of the *Bacillus* BsXA promoter. Biotechnol Appl Biochem. 2006;43(Pt 1):9–15.
39. Georis J, Giannotta F, De Buyl E, Granier B, Frere J. Purification and properties of three endo-beta-1,4-xylanases produced by *Streptomyces* sp. strain S38 which differ in their ability to enhance the bleaching of kraft pulps*(2). Enzyme Microb Technol. 2000;26(2–4):178–86.
40. McIntosh LP, Hand G, Johnson PE, Joshi MD, Korner M, Plesniak LA, Ziser L, Wakarchuk WW, Withers SG. The pKa of the general acid/base carboxyl group of a glycosidase cycles during catalysis: a 13C-NMR study of *Bacillus circulans* xylanase. Biochemistry-US. 1996;35(31):9958–66.
41. Torronen A, Rouvinen J. Structural comparison of two major endo-1,4-xylanases from *Trichoderma reesei*. Biochemistry-US. 1995;34(3):847–56.
42. Al Balaa B, Wouters J, Dogne S, Rossini C, Schaus JM, Depiereux E, Vandenhaute J, Housen I. Identification, cloning, and expression of the *Scytalidium acidophilum* XYL1 gene encoding for an acidophilic xylanase. Biosci Biotechnol Biochem. 2006;70(1):269–72.
43. Krengel U, Dijkstra BW. Three-dimensional structure of endo-1,4-beta-xylanase I from *Aspergillus niger*: Molecular basis for its low pH optimum. J Mol Biol. 1996;263(1):70–8.
44. Ito K, Ogasawara H, Sugimoto T, Ishikawa T. Purification and properties of acid stable xylanases from *Aspergillus kawachii*. Biosci Biotech Bioch. 1992; 56(4):547–50.

5

Expression of a fungal manganese peroxidase in *Escherichia coli*: a comparison between the soluble and refolded enzymes

Nan Wang, Kai Ren, Rong Jia*, Wenting Chen and Ruirui Sun

Abstract

Background: Manganese peroxidase (MnP) from *Irpex lacteus* F17 has been shown to have a strong ability to degrade recalcitrant aromatic pollutants. In this study, a recombinant MnP from *I. lacteus* F17 was expressed in *Escherichia coli* Rosetta (DE3) in the form of inclusion bodies, which were refolded to achieve an active enzyme. Further, we optimized the in vitro refolding conditions to increase the recovery yield of the recombinant protein production. Additionally, we attempted to express recombinant MnP in soluble form in *E. coli*, and compared its activity with that of refolded MnP.

Results: Refolded MnP was obtained by optimizing the in vitro refolding conditions, and soluble MnP was produced in the presence of four additives, TritonX-100, Tween-80, ethanol, and glycerol, through incubation at 16 °C. Hemin and Ca^{2+} supplementation was crucial for the activity of the recombinant protein. Compared with refolded MnP, soluble MnP showed low catalytic efficiencies for Mn^{2+} and H_2O_2 substrates, but the two enzymes had an identical, broad range substrate specificity, and the ability to decolorize azo dyes. Furthermore, their enzymatic spectral characteristics were analysed by circular dichroism (CD), electronic absorption spectrum (UV-VIS), fluorescence and Raman spectra, indicating the differences in protein conformation between soluble and refolded MnP. Subsequently, size exclusion chromatography (SEC) and dynamic light scattering (DLS) analyses demonstrated that refolded MnP was a good monomer in solution, while soluble MnP predominantly existed in the oligomeric status.

Conclusions: Our results showed that two forms of recombinant MnP could be expressed in *E. coli* by varying the culture conditions during protein expression.

Keywords: *Escherichia coli*, Inclusion bodies, Kinetic properties, Manganese peroxidase, Protein expression, Protein refolding, Spectral characteristics

Background

Manganese peroxidase (MnP, E.C. 1.11.1.13) can be isolated from many species of white-rot fungi, and it catalyzes the oxidation of Mn^{2+} to Mn^{3+} in the presence of hydrogen peroxide (H_2O_2) [1]. The produced Mn^{3+} is stabilized by organic acid and forms a Mn^{3+}-organic acid complex that acts as a diffusible oxidizer, which in turn attacks a variety of aromatic compounds, including lignin, industrial dyes, and other organic contaminants [2, 3]. Crystal structures and spectroscopic studies demonstrated that MnP is a two-domain, globular protein that consists of 11 or 12 α-helices. A heme prosthetic group, most commonly Fe protoporphyrin IX, is located between the two domains [4]. Two Ca^{2+}-binding sites, which provide structural support, are located on both sides of the heme group, and a Mn-binding site, which is required for catalysis, is located near the internal heme propionate [4, 5]. In addition, MnP has four or five disulfide bridges that stabilize its protein structure [6].

Commercial applications of fungal MnP are mainly based on the high level of protein expression of some fungal strains. However, as described by Saroj et al. [7], the yield of native MnP in its native hosts was too low,

* Correspondence: ahdxjiarong@126.com
School of Life Science, Anhui University, 111 Jiulong Road, Economic and Technology Development Zone, Hefei, Anhui 230601, People's Republic of China

and approximately 1.0 to 5 mg of purified recombinant MnP was obtained per liter of extracellular culture fluid. And an appropriate medium composition and critical culture conditions are necessary to induce fungi to produce ligninolytic enzymes [8]. Therefore, improving the enzyme yield and reducing the cost of preparation are needed to meet the requirements of commercial use. In recent years, the production of recombinant proteins has been among the main options. Both homologous and heterologous expression in prokaryotic and eukaryotic systems have been used to achieve the overproduction of MnP. For instance, homologous expression systems in *Phanerochaete chrysosporium* [9] and *Pleurotus ostreatus* [10] used a primary metabolic promoter, but the resulting MnP yield was still low. *Pichia pastoris*, which is a common heterologous expression system, has been used to produce active MnP at a maximum yield of only 120 U/L [11]. In addition, numerous studies have used *Escherichia coli* to express MnP [12, 13]. In general, the recombinant protein accumulated in the form of in inclusion bodies is an inactive, insoluble form; functional, active MnP could be obtained by in vitro refolding based on the incorporation of exogenous hemin and Ca^{2+}. Nevertheless, an efficient process to produce high yields of the recombinant enzyme is still lacking, even after optimizing the refolding procedure [12, 14]. Additionally, refolding is a result of trial and error, and it requires much effort to obtain a renatured protein. Therefore, it is of great interest and importance to explore soluble protein production in *E. coli.* In this regard, alternative methods have been examined, such as choosing a suitable vector and host [15], changing the type of promoter [16], switching the location of an affinity tag [16], decreasing the growth temperature [17], optimizing codon usage [18], adding osmolytes to the expression medium [19], coexpressing chaperones [20], and site-directed mutagenesis [21]. However, thus far, no strategy has yielded soluble MnP in *E. coli.*

Irpex lacteus F17 is a local, MnP-producing, white-rot basidiomycete fungus, and an MnP-encoding gene, *imnp*, from this species has been cloned and expressed in *E. coli* by our laboratory [22]. As expected, the expressed protein accumulated in an insoluble form as inclusion bodies. Subsequent in vitro refolding and incorporating exogenous hemin into the recombinant protein led to the formation of biologically active MnP [22]. In the present study, we further optimized the in vitro refolding conditions to improve the efficiency of renaturation from inclusion bodies using a one-factor-at-a-time method and an orthogonal experimental design based on our previous results. In parallel, we used a strategy to express MnP in a soluble form in *E. coli* by adding chemicals to the cultures, and simultaneously lowering the growth temperature during protein expression. Additionally, the biochemical and spectroscopic characterizations of soluble and refolded MnP were analyzed, and their abilities to decolorize azo dyes were investigated.

Results

Optimal conditions for refolding MnP from inclusion bodies

In the single-factor experiment, the refolding efficiency was estimated from different shades of red in a 48 deep-well plate by changing various refolding parameters, as well as their concentrations or quantities, to preliminarily screen for the important factor(s). The results indicated that urea, GSSG, glycerol, and pH significantly affected the enzyme activity, while $MnSO_4$ and KCl had no obvious effects. No activity was detected if $CaCl_2$ and hemin were omitted from the refolding solution. To search for the optimal combination of significant factors, as well as their concentrations or quantities, an orthogonal experiment design, which consisted of six factors (urea, $CaCl_2$, hemin, glycerol, GSSG, and pH), was tested at three levels using the $L_{18}(3^6)$ orthogonal array. Details of the three levels for each factor were shown in Table 1. Eighteen experiments according to the $L_{18}(3^6)$ orthogonal array were performed. Results of the effects of the six factors on enzymatic specific activity were presented in Table 2. K_1 was the sum of the specific activities for each factor when the factor level is 1, and k_1 was the average number of K_1, and so on, for K_2 and K_3, as well as for k_2 and k_3. Meanwhile, R was the difference of the largest and smallest k values of each factor. A large value of R indicates that the effect of factor on the results is significant. As shown in Table 2, assay "6" produced the maximum MnP specific activity of 3.63 U/mg, which was almost twofold higher than that of the optimized condition obtained in the single-factor experiment (1.88 U/mg). In addition, according to the maximum of k_1, k_2 and k_3 of each factor, a combination (i.e., 1.0 M urea, 150 mM $CaCl_2$, 10 μM hemin, 10% glycerol, 0.5 mM GSSG, and pH 8.5) was further tested, and the specific activity of refolded MnP was 2.65 U/mg, indicating that it was inferior to experiment "6". Thus, the optimized refolding solution comprised 1.5 M urea, 150 mM $CaCl_2$, 25 μM hemin, 10% glycerol, 0.5 mM GSSG, 0.05 mM $MnSO_4$, and 20 mM KCl in 50 mM Tris-HCl (pH 8.5) at 4 °C for 24 h. The value of R in Table 2 showed that pH and GSSG had important effects on specific activity of refolding MnP.

Table 1 Contents of orthogonal factors and levels $L_{18}(3^6)$

Factor level	A Urea (M)	B $CaCl_2$ (mM)	C Hemin (μM)	D Glycerol (%)	E GSSG (mM)	F pH
1	1.0	75	5.0	10	0.5	8.0
2	1.5	100	10	15	0.7	8.5
3	2.0	150	25	20	1.0	9.0

Table 2 Effect of six factors on specific activity of refolded MnP

Experiment	A Urea (M)	B $CaCl_2$ (mM)	C Hemin (μM)	D Glycerol (%)	E GSSG (mM)	F pH	Specific activity (U/mg)
1	1	1	1	1	1	1	1.63 ± 0.23
2	1	2	2	2	2	2	1.89 ± 0.59
3	1	3	3	3	3	3	0.65 ± 0.02
4	2	1	1	2	2	3	0.86 ± 0.15
5	2	2	2	3	3	1	0.86 ± 0.04
6	2	3	3	1	1	2	3.63 ± 0.19
7	3	1	2	1	3	2	2.04 ± 0.24
8	3	2	3	2	1	3	1.10 ± 0.06
9	3	3	1	3	2	1	1.01 ± 0.25
10	1	1	3	3	2	2	2.23 ± 0.39
11	1	2	1	1	3	3	1.83 ± 0.25
12	1	3	2	2	1	1	3.38 ± 0.06
13	2	1	2	3	1	3	1.02 ± 0.23
14	2	2	3	1	2	1	0.86 ± 0.07
15	2	3	1	2	3	2	2.03 ± 0.10
16	3	1	3	2	3	1	0.96 ± 0.12
17	3	2	1	3	1	2	1.52 ± 0.36
18	3	3	2	1	2	3	1.16 ± 0.09
K_1	11.61	8.742	8.88	11.148	12.282	8.7	
K_2	9.258	8.058	10.35	10.218	8.01	13.338	
K_3	7.788	11.862	9.432	7.29	8.37	6.618	
k_1	1.935	1.457	1.48	1.858	2.047	1.45	
k_2	1.543	1.343	1.725	1.703	1.335	2.223	
k_3	1.298	1.977	1.572	1.215	1.395	1.103	
R	0.637	0.634	0.245	0.643	0.712	1.12	

Specific activity was obtained from the orthogonal design experiment. Each value is the average of three independent runs of experiment
K_1 represents the sum of the specific activity of each factor if the factor level is 1, K_2 represents the sum of the specific activity of each factor if the factor level is 2, and so K_3. k_1 represents the average number of K_1, and so on, for k_2 and k_3. R represents the difference of the largest and smallest k values of each factor

Optimal conditions for expressing soluble MnP

Similarly, the procedure searching for the optimal expression conditions for soluble MnP was similar to that used to optimize MnP refolding from inclusion bodies. According to the results of a single-factor experiment, Triton X-100 played a significant role in the expression of soluble MnP. Tween-80, glycerol, and ethanol improved the activity of soluble MnP, while sorbitol, L-proline, and glycine had no distinct impact on the expression of soluble MnP. Thus, an orthogonal experimental design comprising four factors (Triton X-100, Tween-80, glycerol, and ethanol), with each factor being tested at three levels, was performed using the $L_9(3^4)$ orthogonal array (Table 3). Results of the effects of the four factors on enzymatic specific activity were presented in Table 4. As shown, assay "1" produced a specific activity (0.411 U/mg), which was higher than that of the optimized condition obtained in the single-factor experiment (0.391 U/mg). In addition, according to the maximum of k_1, k_2 and k_3 of each factor in Table 4, a combination was further tested, which resulted in a maximum of 0.56 U/mg, higher than the specific activity of experiment "1". As a result, the production of soluble MnP could be induced by the addition of 0.3 mM IPTG, along with 0.25% Triton X-100, 0.25% Tween-80, 0.5% glycerol, and 1% ethanol, to the cell cultures, followed by cultivation for 16 h at 16 °C. The value of R in Table 4 indicated that

Table 3 Contents of orthogonal factors and levels $L_9(3^4)$

Factor level	A Triton X-100 (%)	B Tween-80 (%)	C Glycerol (%)	D Ethanol (%)
1	0.25	0.25	0.25	1.0
2	0.5	0.5	0.5	1.5
3	1.0	1.0	1.0	2.0

Table 4 Effect of four factors on specific activity of soluble MnP

Experiment	A Triton X-100 (%)	B Tween-80 (%)	C Glycerol (%)	D Ethanol (%)	Specific activity (U/mg)
1	1	1	1	1	0.411 ± 0.03
2	1	2	2	2	0.355 ± 0.02
3	1	3	3	3	0.361 ± 0.14
4	2	1	2	3	0.320 ± 0.03
5	2	2	3	1	0.185 ± 0.01
6	2	3	1	2	0.208 ± 0.07
7	3	1	3	2	0.167 ± 0.02
8	3	2	1	3	0.121 ± 0.03
9	3	3	2	1	0.252 ± 0.27
K_1	1.127	0.897	0.741	0.849	
K_2	0.714	0.66	0.927	0.729	
K_3	0.54	0.822	0.714	0.801	
k_1	0.376	0.299	0.247	0.283	
k_2	0.238	0.22	0.309	0.243	
k_3	0.18	0.274	0.238	0.267	
R	0.196	0.079	0.071	0.04	

Specific activity was obtained from the orthogonal design experiment. Each value is the average of three independent runs of experiment
K_1 represents the sum of the specific activity of each factor if the factor level is 1. K_2 represents the sum of the specific activity of each factor if the factor level is 2, and so K_3. k_1 represents the average number of K_1, and so on, for k_2 and k_3. R represents the difference of the largest and smallest k values of each factor

Triton X-100 was the most important factor for expressing soluble MnP in *E. coli*.

Purification of soluble and refolded MnP

Two forms of the recombinant MnP were obtained from *E. coli* under the optimal conditions for insoluble inclusion body and for soluble protein, respectively. Both soluble and refolded MnP, each with a C-terminal, six-histidine tag, were purified by Ni-NTA affinity chromatography. They were visualized as a single band on SDS-PAGE, and they exhibited a similar molecular weight of 43 kDa (Fig. 1). Under the optimal refolding conditions, the refolding yield of dialyzed MnP obtained from inclusion bodies was 13%, and the yield of purified MnP was 2.4% (Table 5). The final recovery, which was calculated from the total activity, was approximately 78%. For soluble MnP, approximately 1.5 mg of purified, active MnP was obtained from 2 L of *E. coli* fermentation broth.

Biochemical characterization of soluble and refolded MnP

Enzyme kinetic properties

Table 6 shows the kinetic constants of soluble and refolded MnP using Mn^{2+} and H_2O_2 as substrates. The K_m (336 μM) of soluble MnP for Mn^{2+} was higher than that (148 μM) of refolded MnP, and the K_m (23.5 μM) of soluble MnP for H_2O_2 was reduced 90% than that (234.5 μM) of refolded MnP. The k_{cat}/K_m values of refolded MnP for Mn^{2+} and H_2O_2 were 12.2-fold and 2.9-fold higher than those of soluble MnP, respectively. However, the k_{cat}/K_m values of the two forms of recombinant MnP were both lower than that of the native MnP for Mn^{2+} (9.62×10^3 (mmol)$^{-1}$ L s^{-1}) and H_2O_2 (5.22×10^4 (mmol)$^{-1}$ L s^{-1}), respectively.

Temperature stability and pH optimum

As illustrated in Fig. 2a, soluble MnP showed less than 65% relative activity (the ratio of actual to maximum activity) in a broad pH range (2.2–6.6). However, a rapid increase in activity occurred at pHs higher than 7.0, and the optimal pH was 8.0. Additionally, the enzyme was optimally stable at pH 5.0 during a 6-h incubation in citrate-phosphate buffer (pH 5.0). Unlike soluble MnP, refolded MnP showed maximum activity at pH 6.5, and retained more than 80% of its activity over a wide range of pH (3.0–7.0). In addition, the pH stability of refolded MnP was superior to that of soluble MnP (Fig. 2b). As shown in Fig. 2c and d, soluble MnP exhibited the highest activity at 45 °C, and retained 90% of its activity after a 6-h incubation at temperatures below 25 °C. Refolded MnP displayed optimum activity at 15 °C and was very stable below 35 °C; no activity was detected at 45 °C after incubation for 6 h.

Substrate specificity

To examine the substrate specificity of soluble and refolded MnP, the monocyclic phenolic compounds DMP and guaiacol, as well as the polycyclic compound ABTS, were selected as substrates. The results showed that the oxidation of these substrates by soluble MnP was similar to that for refolded MnP, and they were all dependent upon Mn^{2+} (Table 7). The results indicated that both of the soluble and refolded MnP are powerful oxidants that are capable of oxidizing aromatic compounds.

Decolorization of Acid Red 18 and Orange G

Acid Red 18 and Orange G, as important azo dyes, have been extensively used in the textile, printing, food, and drug industries, and they are usually released into industrial wastewater. Accordingly, we chose them to examine the degradation capability of soluble and refolded MnP. As shown in Fig. 3, the decolorization efficiency of Acid Red 18 was 63 ± 1.41% for the soluble MnP after a 60-min reaction at pH 4.0 and 35 °C, while the decolorization efficiency was 78.5 ± 0.71% for the refolded MnP. Moreover, for Orange G, 35 ± 0.01 and 56 ± 0.03% decolorization efficiencies were observed for soluble and refolded MnP, respectively. However, the decolorization efficiencies of the two forms of recombinant MnP were both lower than that of the native MnP for Acid Red 18 (84 ± 1.3%) and Orange G (68 ± 0.6%), respectively.

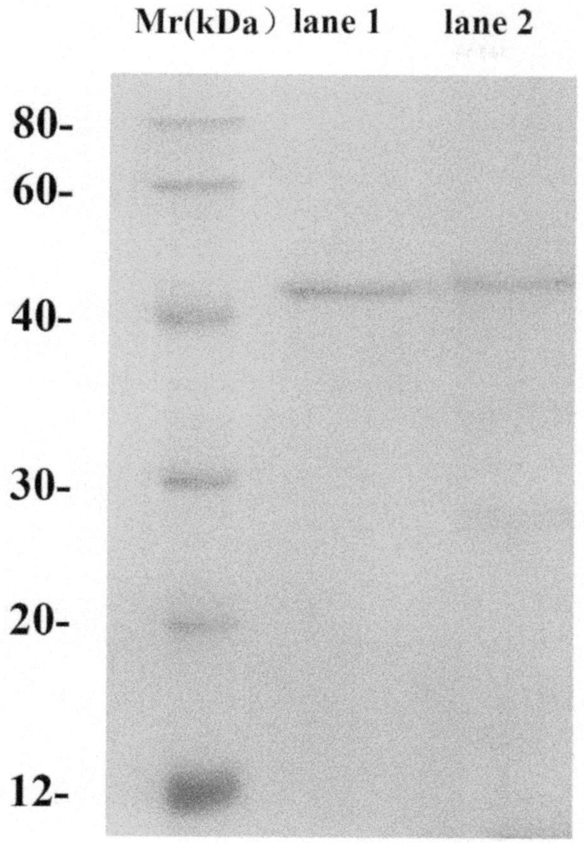

Fig. 1 SDS-PAGE of recombinant MnP purified by a Ni^{2+} His-tag column. Lane M: protein standards; Lane 1: purified refolded MnP; Lane 2: purified soluble MnP

Spectroscopic characterization of soluble and refolded MnP

CD spectra and UV-VIS spectra

It is well known that a circular dichroism (CD) spectrum in far-UV region provides information on the secondary structure of a protein. The far-UV CD spectra of soluble and refolded MnP exhibited two negative ellipticity bands at 208 nm and 222 nm (Fig. 4a), respectively, displaying characteristics of an α-helical structure, but they showed notable differences. Furthermore, the α-helix and the β-sheet contents of the two MnPs were calculated using the SELCON3 program, and the data showed that the helical content of soluble MnP (18.7%) was lower than that of refolded MnP (24.3%), whereas the β-sheet content of soluble MnP (27%) was higher than that of refolded MnP (18.2%).

To evaluate the secondary structure of soluble and refolded MnP, we investigated the far-UV CD spectrum of the purified, native MnP from fungus *Irpex lacteus* F17 using CD spectrometer under the same conditions, as shown in Fig. 4a. The native fungal MnP was obtained according to our previous experimental protocols described in the literature [23]. On the basis of the analysis of SELCON3 program, 39.7% α-helix and 13.7% β-sheet were involved in the structure of CD of native MnP produced by fungus *Irpex lacteus* F17.

The UV-VIS spectrum of soluble MnP was similar to that of refolded MnP (Fig. 4b); however, their absorption intensities obviously differed. Refolded MnP exhibited a Soret peak at 408 nm, Q band at 527 nm and a charge transfer (CT) band at 631 nm, while soluble MnP shifted the Soret peak to 410 nm, Q band to 536 nm with a tiny CT band at 631 nm, and the absorbance of soluble MnP was clearly lower than that of refolded MnP. The Reinheitszahl (R_Z) values (calculated from the A408/A280 and A410/A280 absorption ratios) for the purified, refolded MnP and soluble MnP were 2.5 and 1.1, respectively, indicating that the heme environment of the two enzymes differed. These results are in agreement with the lower specific activity of soluble MnP compared with refolded MnP, as shown in the inset of Fig. 4b.

Fluorescence spectra

Figure 5 showed the fluorescence spectra of soluble and refolded MnP at room temperature following excitation at 295 nm. As shown, refolded MnP had a maximum emission at 337 nm, whereas the maximum emission of soluble MnP occurred at 341 nm, which represented a 4-nm red shift and an increased quantum yield.

Raman spectra

Figure 6 represented the Raman spectra of soluble and refolded MnP at room temperature following excitation at 785 nm. The C-C stretching vibration near 890–945 cm^{-1} of Raman spectra indicated that soluble and refolded MnP were primarily α-helix structures. The relative weak amide I bands centered at 1598 cm^{-1} and 1675 cm^{-1} were both correlated with an increase in β-sheet contents for soluble MnP, while the β-sheet structure of refolded MnP was involved in an amide I band centered at 1680 cm^{-1}. The band at 1070 cm^{-1} exhibited in soluble MnP and the bands at 1101 and 1126 cm^{-1} presented in refolded MnP, originated from C-N stretching vibrations, revealed some

Table 5 Isolation and purification of MnP from inclusion bodies

Sample	Protein concentration (mg/ml)	Protein (mg)	Specific activity (U/mg)	Total activity (U)	Yield (%)
Inclusion body	18.7	93.5			
Dialyzed	0.43	12.35	3.04	37.5	100
Ni-NTA	0.076	2.26	12.9	29.15	78

Table 6 Kinetic parameters of refolded and soluble MnP

Enzyme	*Km* (Mn^{2+}) μmol L^{-1}	*Vmax* (Mn^{2+}) μmol L^{-1} min^{-1}	*Kcat* (Mn^{2+}) s^{-1}	*Kcat/Km* (Mn^{2+}) $(mmol)^{-1}$ L s^{-1}
Refolded MnP	148 ± 7.1	1276 ± 23.3	141.9 ± 2.62	930 ± 30
Soluble MnP	336 ± 8.0	236.6 ± 24.7	25.6 ± 2.7	76 ± 9.9
Native MnP	100.7 ± 0.87	1116.3 ± 18.3	969 ± 22.6	9620 ± 420
Enzyme	*Km* (H_2O_2) μmol L^{-1}	*Vmax* (H_2O_2) μmol L^{-1} min^{-1}	*Kcat* (H_2O_2) s^{-1}	*Kcat/Km* (H_2O_2) $(mmol)^{-1}$ L s^{-1}
Refolded MnP	234.5 ± 12	3326.5 ± 179	554.4 ± 29	2370 ± 7
Soluble MnP	23.5 ± 4.9	308.1 ± 26.9	19 ± 1.7	820 ± 100
Native MnP	30.7 ± 4.4	1844.5 ± 25	1601.1 ± 34.8	52200 ± 450

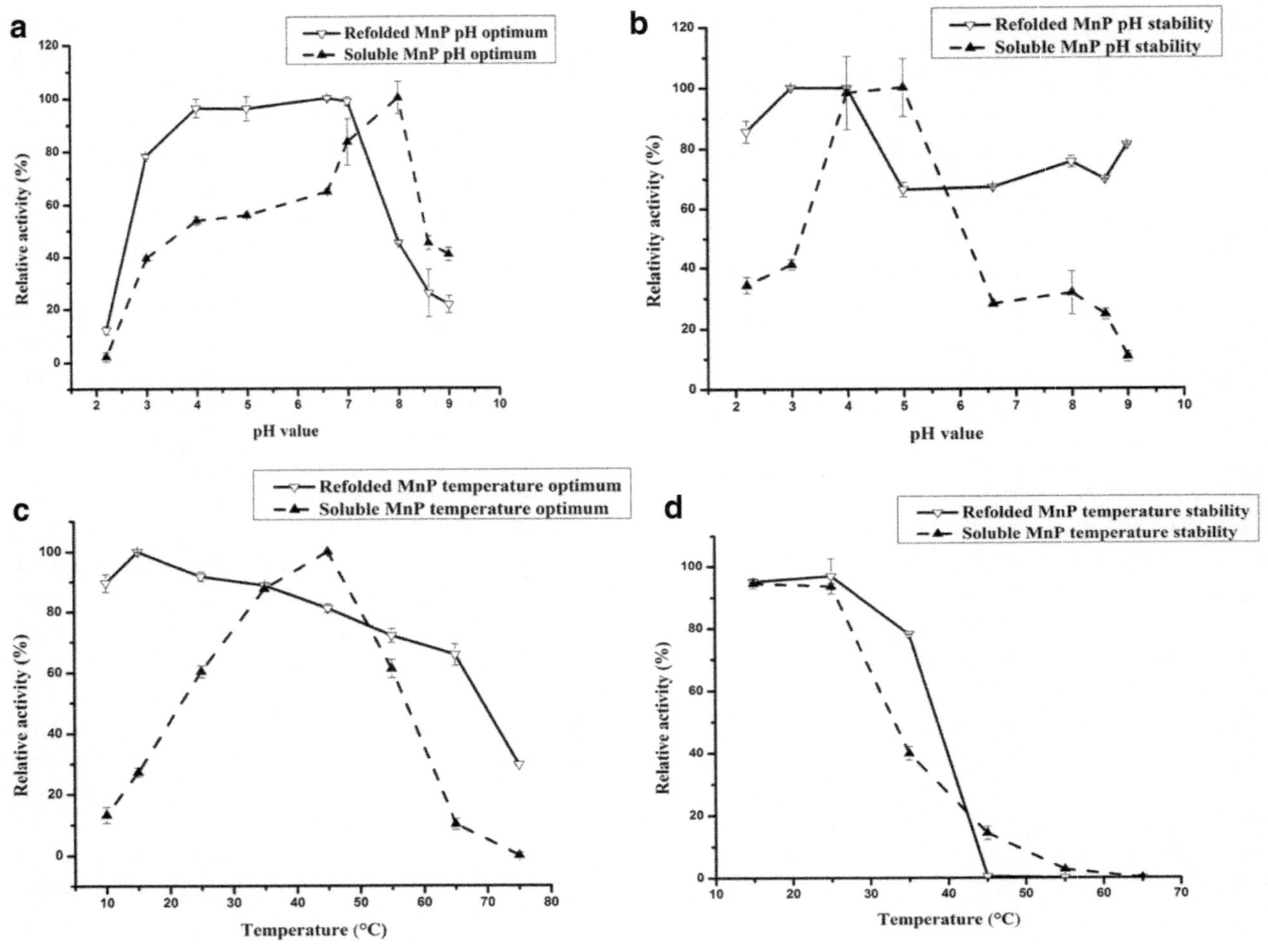

Fig. 2 Effect of pH and temperature on the activity and stability of refolded and soluble MnP. And MnP activity at the beginning was considered to be 100%. Deviation values are standard deviations based on triplicate determinations. **a** The pH optimum of refolded and soluble MnP. The MnP activity was determined in the citrate-phosphate buffer, pH 2.2-8.0 and Tris–HCl, pH 8.6-9.0, at 25 °C. **b** The pH stability of refolded and soluble MnP. Refolded and soluble MnP was incubated for 6 h at 25 °C in various pH vaules in citrate-phosphate (2.2-8.0) or Tris–HCl buffer (8.6-9.0). The residual MnP activity was measured according to the method of MnP activity assay. **c** The temperature optimum of refolded and soluble MnP. The enzyme reaction was performed in 0.11 M sodium lactate buffer, pH 4.5 at 10–75 °C. **d** The temperature stability of refolded and soluble MnP. Refolded and soluble MnP were incubated for 6 h at 15–65 °C. The residual MnP activity was measured according to the method of MnP activity assay

Table 7 Substrate specificities of refolded and soluble MnP

Substrate	Concentration (mM)	Emax (M^{-1} cm^{-1})	Wavelength (nm)	Refolded MnP (U L^{-1})		Soluble MnP (U L^{-1})	
				Mn^{2+} present	Mn^{2+} absent	Mn^{2+} present	Mn^{2+} absent
ABTS	1	36000	420	344.93 ± 0.03	37.73 ± 0.01	310.0 ± 0.046	24.4 ± 0.006
DMP	1	49600	469	105.42 ± 0.04	0.32 ± 0.001	104.35 ± 0.05	0
Guaiacol	1	12100	456	122.31 ± 0.003	2.27 ± 0.001	112.40 ± 0.008	0

conformation modifications between the two enzymes in the protein polypeptide backbone. A low intensity band at 540 cm^{-1} appeared in the spectrum of soluble MnP indicated the existence of the C-C-S-S-C-C stretching vibrations in trans-gauche-trans arrangements [24], whereas the structure of disulfide bonds was revealed by a weak band near 552 cm^{-1} in refolded MnP. Furthermore, a band at 759 cm^{-1} indicated the existence of tryptophan in refolded MnP, while a change occurred in soluble MnP by shifting the band to 763 cm^{-1}. Also, the phenylalanine bands of refolded MnP were observed at 621 cm^{-1} and 1203 cm^{-1}. Additionally, the bands at 1380 cm^{-1} and 1468 cm^{-1} for soluble MnP, and a band at 1451 cm^{-1} for refolded MnP were attributed to the CH_2 bending vibrations, showing some changes in protein side chains between the two enzymes.

Size exclusion chromatography and dynamic light scattering analysis

To further investigate the structural properties of soluble and refolded MnP, SEC and DLS experiments were used to analyze their protein behaviors and status in solution. The DLS distribution patterns of soluble and refolded MnP were given in Fig. 7 and the corresponding elution profiles were shown in the inset. The data from SEC and DLS were summarized in Table 8. The results revealed some notable differences between the two enzymes. In the case of refolded MnP (Fig. 7a, Table 8), the main peak in FPLC column showed a high specific activity (15.0 U/mg) and a high Rz (3.1), which corresponded to a monomodal distribution in DLS pattern (Fig. 7a). The hydrodynamic radius of the particle 1 which constituted 99.6% of the protein was 2.9 ± 0.5 nm, and represented an estimated molecular weight of 42.0 kDa. The data is corresponding with the determined MW (43.0 kDa) of recombinant MnP on SDS-PAGE, as shown in Fig. 1, and is also consistent with that of native MnP from *Irpex lacteus* F17, as reported in the literature [23], as well as the molecular range of the MnP family (38–50 kDa) [25, 26]. Thus, it could be concluded that refolded MnP was a monomer in solution. In the case of soluble MnP (Fig. 7, Table 8), two prominent peaks of I and II in FPLC column showed low MnP activities and low Rz values, which corresponded to peak I and peak II of soluble MnP in DLS pattern, respectively (Fig. 7b, c). The DLS results showed that the size distribution of soluble MnP peak I displayed a bimodal distribution with the mass percentage of 70.9% (particle 1) and 22.3% (particle 2), respectively (Fig. 7b); particle 1 centered at around 12.9 nm was dominant and indicated low aggregated oligomers, while particle 2 around 45.3 nm was comparatively weak and showed large aggregated status. Meanwhile, DLS of soluble MnP also showed a peak II with an average hydrodynamic radius of 4.5 nm (particle 3), which constituteed 99.3% of the protein and represented an estimated molecular weight of 111.0 kDa (Fig. 7c). This particle 3 probably was related to the formation of protein dimers. Therefore, it could be suggested that the overwhelming majority of soluble MnP was the oligomeric status in solution under the experimental conditions, different from a single monomeric form of refolded MnP.

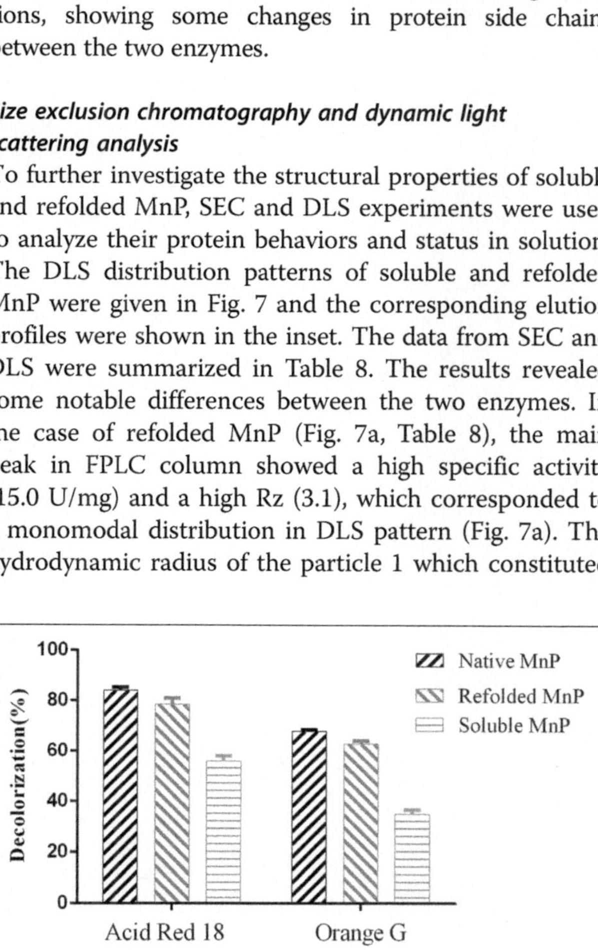

Fig. 3 Acid Red 18 and Orange G decolorization by refolded and soluble MnP

Discussion

Our previous studies indicated that the MnP preprotein from *I. lacteus* F17 is composed of 359 amino acid residues, including a 26-amino acid signal peptide and four disulfide bridges essential for the native conformation of the enzyme [22]. A cDNA encoding the mature 333-amino acid MnP protein without the signal sequence was subcloned into pET28a and expressed in *E. coli* Rosetta (DE3) cells. As expected, this protein formed inclusion bodies and required in vitro refolding to convert it into an active enzyme, as is true for recombinant MnP from *P. chrysosporium* [12] and *Phlebia radiata* [27].

One of the purposes of the present study was to enhance the MnP yield by optimizing the in vitro refolding conditions. Orthogonal design is a good mathematical

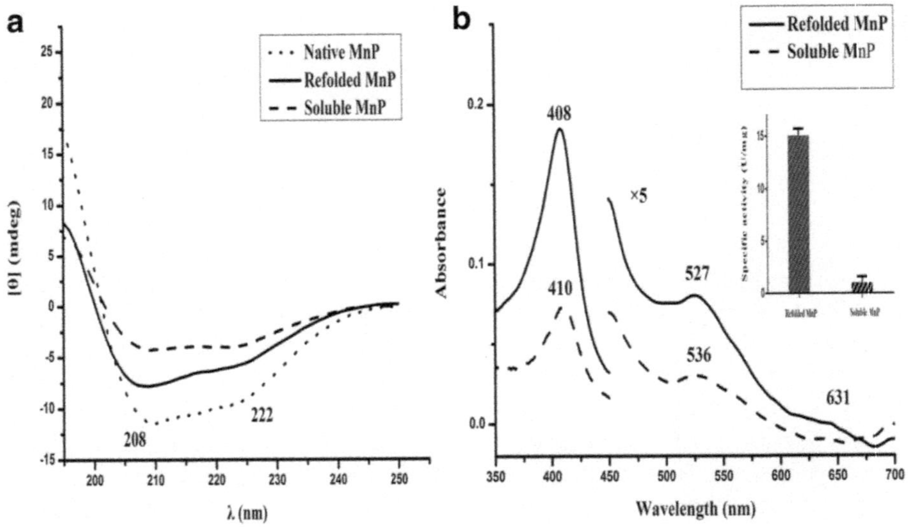

Fig. 4 The far-UV CD and absorbance spectra of refolded and soluble MnP. **a** The far UV-CD circular dichroism spectra of native, refolded and soluble MnP at room temperature and 0.1 mg/ml of protein concentration in 0.15 M phosphate buffer at pH 6.5. **b** Electronic absorption spectra of refolded and soluble MnP at room temperature. The region between 450 and 700 nm has been expanded (×5 absorbance). Enzymatic specific activities are represented in the inset

method used for planning multivariate tests [28]. Its greatest applications have been in healthcare evaluations or comparative effectiveness research [29]. Using an orthogonal design in the present study, the optimal refolding conditions were obtained, and the MnP yield attained was higher than that obtained in our previous studies, even surpassing those reported for recombinant MnP from *P. chrysosporium* [12], horseradish peroxidase [30], and lignin peroxidase [31].

Another purpose of the study was to determine whether MnP could be expressed in a soluble form in *E. coli* to avoid the tedious process of renaturing inclusion bodies. In this study, a biologically active form of soluble MnP was expressed by simultaneously decreasing the expression temperature and adding chemicals to the culture medium. Namely, the recombinant MnP could be expressed as a soluble protein in the presence of four chemicals in the medium, at 16 °C. At low temperature, the rate of protein synthesis slows; thus, proteins have sufficient time to fold, which increases their solubility [32]. Esposito et al. [33] also reported that a lower expression temperature improved the solubility of *E. coli*-expressed proteins, while a higher expression temperature was conducive to the production of insoluble protein in *E. coli* Rosetta-gami strains [34].

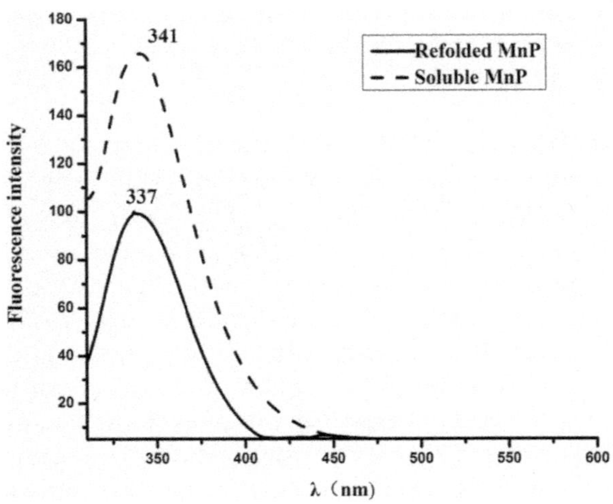

Fig. 5 The emission fluorescence spectra of refolded and soluble MnP. The emission fluorescence spectra of refolded and soluble MnP at room temperature and 0.1 mg/ml of protein concentration in the phosphate buffer at pH 6.5

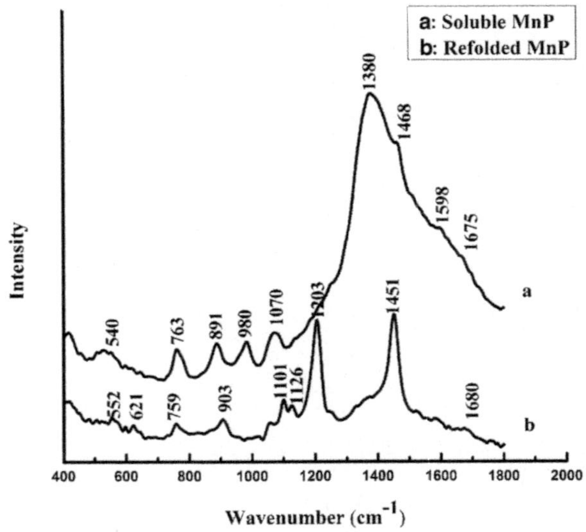

Fig. 6 The Raman spectra of refolded and soluble MnP

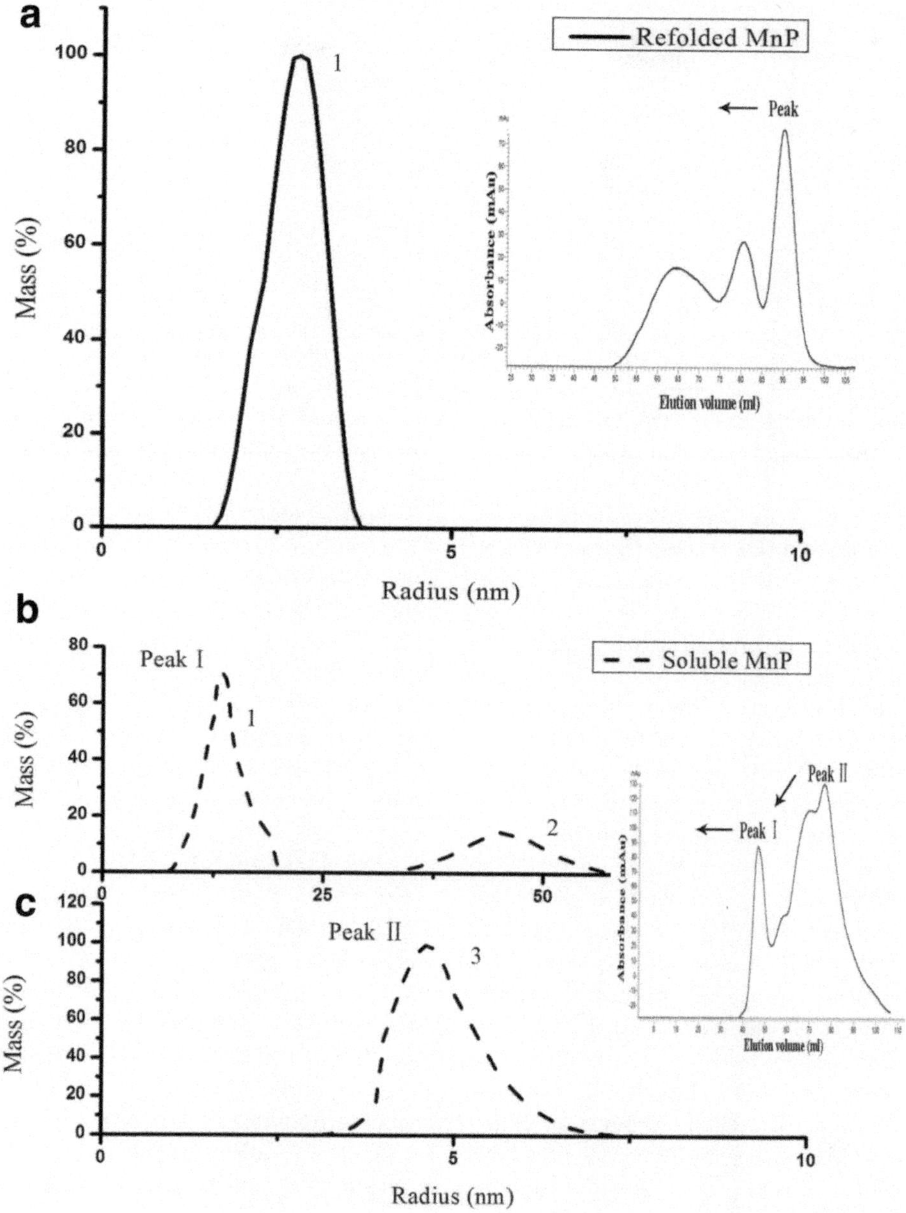

Fig. 7 Protein size analysis of refolded and soluble MnP. **a** Dynamic light scattering results of refolded MnP. The percentage of the protein mass (%) was plotted against the hydrodynamic radius (nm). Size exclusion chromatography of refolded MnP purified from Ni-NTA affinity chromatography was shown in the inset. Protein samples were collected after elution with 120 and 200 mM imidazole using 5-ml Ni Sepharose™ 6 Fast Flow. Elution profile of protein was plotted against the absorbance at 280 nm. **b** & **c** Dynamic light scattering results of soluble MnP. The percentage of the protein mass (%) was plotted against the hydrodynamic radius (nm). Size exclusion chromatography of soluble MnP from Ni-NTA affinity chromatography was shown in the inset

Our results also demonstrated that an expression temperature of 16 °C could increase the production of soluble MnP, but an expression temperature of 37 °C could not. In addition, some chemical additives play a useful role in increasing the solubility of recombinant proteins in *E. coli* [19]. Triton X-100 and Tween-80 are surfactants that can change the osmotic pressure in bacterial membranes [35]. Moreover, Triton X-100 supplementation in culture medium can hinder the formation of inclusion bodies in *E. coli*, and facilitate the secretion of recombinant proteins to the periplasmic space or culture medium [36]. Glycerol is commonly used as a protective agent to increase protein stability, as it is able to decrease the surface tension of water and increase viscosity [37]. An optimum glycerol concentration can stabilize protein structures, even at high protein concentrations. Ethanol has also been shown to increase the fraction of soluble protein [38]. Steczko et al. [39] reported that adding 3% ethanol dramatically

Table 8 Summary of the data from DLS and FPLC for refolded and soluble MnP

Size distribution	Radius (nm)	Molecular weight (kDa)	Mass (%)	Specific activity (U/mg)	Rz (Soret/280)
refolded MnP					
Particle 1	2.9 ± 0.5	42.0	99.6	15.0 ± 2.1	3.1 ± 0.6
soluble MnP peak I					
Particle 1	12.9 ± 0.7	1314.0	70.9	0.6 ± 1.5	1.0 ± 0.3
Particle 2	45.3 ± 1.4	25199.0	22.3		
soluble MnP peak II					
Particle 3	4.5 ± 0.8	111.0	99.3	0.2 ± 0.8	0.3 ± 1.1

"Particle 1" of refolded MnP was presented in Fig. 7a; "Particle 1", "Particle 2" and "Particle 3" of soluble MnP were presented in Fig. 7b and Fig. 7c. The results of hydrodynamic radius, specific activity and Rz were given as the mean ± standard deviation, calculated from at least three repetitions

increased the amount of soluble, active soybean seed lipoxygenase L-1. Using an orthogonal design, we determined the optimized combination of Triton X-100, Tween-80, glycerol, and ethanol, which resulted in the expression of soluble MnP in *E. coli* Rosetta (DE3) cells. Thus, it was also confirmed that these solutes did not impair cell metabolism and that they were compatible solutes [40]. MnP activity was detected in the cell lysates after hemin and Ca^{2+} supplementation at pH 8.0. However, recombinant MnP was not secreted into the culture medium, as determined by SDS-PAGE and enzyme activity assays, in our study.

As a result, soluble MnP was successfully induced and expressed in *E. coli* when grown in the presence of the four additives at 16 °C, as opposed to forming inclusion bodies. These additives could be used as agents to increase the yield of soluble proteins at lower temperatures. Although the concrete mechanism underlying this phenomenon is unclear, a case like this has been observed previously. Bacterial cytoplasm is a rather reducing environment, it generally inhibits the formation of disulfide bonds [41], while the periplasmic space provides an oxidative environment that facilitates proper protein folding [42]. Lee and Choi [36] described the secretory and extracellular production of recombinant proteins, which were exported to periplasm space or culture medium, using protein secretion systems in *E. coli*. Prasad et al. [19] suggested that chemical chaperones, such as sorbitol and arginine, increased the solubility of recombinant proteins. Yang et al. [43] reported that culturing conditions had effects on the production of heterologous proteins in *E. coli*. However, in the present study, the detailed information to reveal which pathways take part in the formation of soluble MnP have not been determined and need to be further investigated.

It is worth mentioning that hemin is crucial to activate soluble MnP. Heme is a non-covalently bound prosthetic group in MnP; hence, adding exogenous hemin to the cell lysates is necessary to yield an active holoenzyme. Additionally, Ca^{2+} supplementation is crucial to achieve the correct conformation, as well as the stability, of recombinant MnP [44, 45], because both Ca^{2+} ions are easily lost in a thermal or alkaline environment [6, 46].

Compared with refolded MnP, soluble MnP showed low catalytic efficiencies (k_{cat}/K_m) for Mn^{2+} and H_2O_2 substrates and was more susceptible to H_2O_2. The kinetic parameters are in agreement with those obtained from the spectroscopic characterizations. Additionally, both of the soluble and refolded MnP exhibited a broad range of substrate specificities in the presence of Mn^{2+}. Interestingly, soluble MnP had an optimum pH of 8.0 and an optimum temperature of 45 °C, which are obviously different from those of refolded MnP (pH 6.5 and 15 °C, respectively) and most of the reported fungal MnPs [47–49]. Furthermore, the two enzymes had the ability to decolorize the azo dyes Acid Red 18 and Orange G, but the decolorization efficiency of soluble MnP was lower than that of refolded MnP.

Experiments of CD and UV-VIS absorption spectra showed certain differences in the secondary structures, Soret peaks and other bands between soluble and refolded MnP (Fig. 4a and b). Compared with refolded MnP, soluble MnP showed the Soret peak height reduced by about 40% and red-shifted 2 nm, Q band reduced by about 37% and red-shifted 9 nm, as well as an inconspicuous CT band (631 nm). A characteristic CT band at 631 nm was observed in the spectrum of refolded MnP, indicating a high-spin configuration of the protein, because the wavelength of the region between 600 and 650 nm is sensitive to changes in the heme microenvironment [50, 51]. This band, assigned to a charge transfer transition from the porphyrin to the iron, was typical of a high-spin heme Fe^{3+} states [51]. The specific activities and R_Z values demonstrated that soluble MnP was less active than refolded MnP. The CD spectra analyses indicated that refolded MnP was more closer to native MnP, in terms of its α-helix and β-sheet contents, than soluble MnP. Thus, it appears that the optimized in vitro refolding conditions helped MnP to refold into its native conformation. Nevertheless, soluble MnP exhibited a lower α-helix content than refolded MnP, while its β-sheet content was higher, indicating that the two structures differed. Hence, the lower specific activity of soluble MnP could be ascribed to a change in protein structure, and the lower R_Z value could result from an insufficient incorporation of exogenous hemin. Fluorescence spectra also showed the differences in protein conformation. Compared with refolded MnP, soluble MnP exhibited a 4-nm red shift, implying that the tryptophan residue in the protein was more exposed to a hydrophilic environment (MnP contains one tryptophan residue, GenBank accession no. AGO86670.2). Raman spectroscopy is also an efficient technique that is capable of investigating protein molecular structures, good for solid samples as well as aqueous solutions [24]. Recently, Raman spectra have been used for the analysis of different structures of various proteins [24, 52, 53]. Therefore, in the present study, comparisons of major Raman scattering bands were made between soluble and refolded MnP. The results indicated that both enzymes have different Raman characteristic bands, which reflected their structural differences in protein backbone conformations and protein side chains. Moreover, the results from SEC and DLS also supported these findings. The elution profile of refolded MnP produced a main fraction, corresponding to monomeric protein with an Rz value greater than 3. This data (Rz > 3) is a significant characteristic of highly purified MnP and almost identical to that of the native fungal MnP [12]. Further, we confirmed the monomeric nature of the native MnP from *Irpex lacteus* F17 in solution by DLS studies (Additional file 1: Figure S1). Therefore, it was obvious that the quality of refolded MnP was good and the optimized refolding conditions in vitro were achieved. However, we noticed that soluble MnP showed aggregate behavior through dimerization and polymerization in solution. As such, it can be considered that a decrease in enzymatic activity and Rz value could be ascribed to the aggregate formation of soluble MnP. In addition, the DLS results provided reasonable explanation for their differences of CD, fluorescence and Raman spectra between soluble and refolded MnP. Nevertheless, the mechanism of such aggregate formation for soluble MnP still needs further study. Chihi et al. [54] demonstrated that the hydrophobic interactions and the formation of new disulfide bridges resulted in soluble protein aggregates; while, Wang et al. [55] emphasized the importance of the hydrophobic interaction in promoting the protein aggregate, as compared with disulfide bonds. Therefore, we will shed light on the intriguing mechanism of polymerization for soluble MnP in the future.

Although secretory production of soluble MnP in *E. coli* did not result in an ideal enzyme activity, it has some advantages over inclusion bodies [36]. Furthermore, several strategies for the efficient secretory production of recombinant proteins in *E. coli* have been explored [16, 19, 34]. To our knowledge, ours is the first report of soluble MnP expression in *E. coli*, and it provides an optional solution for obtaining biologically active forms of recombinant proteins.

Conclusions

In summary, using a one-factor-at-a-time method and an orthogonal design, the optimal conditions for the production of refolded and soluble MnP were obtained. The refolded protein yield was improved, and the specific activity of refolded MnP increased by approximately twofold. Soluble MnP was also successfully induced and expressed in *E. coli*, but it was less active than refolded MnP. This phenomenon was confirmed by biochemical and spectroscopic characterizations, as well as their different abilities to decolorize azo dyes. Based on SEC and DLS analyses, refolded MnP was a monomer in solution, while soluble MnP behaved as the protein aggregates. Further studies to improve the activity of soluble MnP and clarify its aggregation mechanism are required and are currently in progress in our laboratory.

Methods

Chemicals

Lysozyme, glutathione (GSSG), dithiothreitol (DTT), H_2O_2, 2,6-dimethoxyphenol (DMP), 2,2'-azino-bis(3-ethylbenzothiazoline-6-sulfonic acid) (ABTS), phenylmethanesulfonyl fluoride (PMSF), and isopropyl-β-D-thiogalactopyranoside (IPTG) were purchased from Sigma-Aldrich (St. Louis, MO, USA). Triton X-100, Tween-80, ethanol, and other chemicals were obtained from Macklin (Shanghai, China) and were of analytical grade.

Host strain and expression vector

The *E. coli* strain Rosetta (DE3) was used as a host for recombinant MnP production. The mature MnP cDNA sequence, named *imnp*, was amplified by polymerase chain reaction using the genomic DNA of *I. lacteus* F17 as a template [22]. The *imnp* cDNA was excised with *Xho*I and *Nco*I, and inserted into the expression vector pET28a(+). The resulting construct (named pET28a-*imnp*) was used to transform *E. coli*.

MnP activity assay

MnP activity was determined by the oxidation of Mn^{2+} to Mn^{3+} at 25 °C and 240 nm ($\varepsilon = 6500\ M^{-1}cm^{-1}$). Incubation mixtures (1 ml) contained 0.11 M sodium lactate buffer (pH 4.5), 1 mM $MnSO_4$, 0.1 mM H_2O_2, and 25 μl of enzyme sample. One unit (U) of MnP activity was defined as the amount of enzyme needed to oxidize 1 μmol of Mn^{2+} per min.

Additionally, when screening parameters that influence enzyme activity, MnP activity was also estimated using the naked eye to detect a color change based on the oxidation of 2,6-dimethoxyphenol (2,6-DMP) by

MnP, which forms a quinone dimer at 469 nm. The DMP oxidation assay was conducted at 30 °C in 0.1 M sodium lactate buffer (pH 4.5) containing 1 mM $MnSO_4$, 0.1 mM H_2O_2, 1 mM DMP, and 25 µl of MnP. MnP activity assay was performed in triplicate. All the data are the mean values of triplicates.

Expression of MnP as inclusion bodies

E. coli Rosetta (DE3) harboring the recombinant vector pET28a-*imnp* was grown at 37 °C, and recombinant MnP was expressed in the form of inclusion bodies. In vitro refolding was performed according to the method of Chen et al. [22]. The optimal refolding conditions were further studied in small-scale experiments. To begin with, a one-factor-at-a-time method was used to screen influencing factors and corresponding levels on MnP refolding. In this set of experiments, parameter ranges in the refolding solutions were set at: urea (0.1–4.0 M), GSSG (0–5.0 mM), glycerol (0–25%), $MnSO_4$ (0–5.0 mM), KCl (0–100 mM), $CaCl_2$ (0–300 mM), and hemin (0–500 µM), as well as pH (7.0–9.5) and incubation time (0–110 h). Factors exhibiting significant effects were selected out by observing DMP oxidation in 48-deep well plates. To achieve a higher refolding efficiency and test mutual interactions of those factors obtained by the single-factor experiment, an orthogonal experimental design was used to determine the optimum combination of significant factors and their concentrations. The refolding efficiency was estimated by analyzing the specific Mn^{2+} oxidizing activity of the dialyzed refolded MnP.

Expression of soluble MnP

E. coli cells harboring pET28a-*imnp* were grown at 37 °C in Luria–Bertani broth containing 50 µg/ml kanamycin sulfate and 34 µg/ml chloramphenicol, as described by Chen et al. [22]. When the optical density at 600 nm reached 0.5–0.6, 0.3 mM IPTG was added to the cultures. Simultaneously, different concentrations of chemicals were also added to the culture medium. Then, the growth temperature was lowered from 37 °C to 16 °C, and the cultures were incubated for 16 h, with shaking at 120 rpm, to induce the expression of soluble MnP. Bacterial pellets were obtained by centrifugation at 4 °C at 7800 *g* for 10 min, and they were resuspended in lysis buffer (containing 5 mM imidazole, 0.02 mM PMSF, 0.5 M NaCl, 20 mM Tris, and pH 8.0) in 1/10 of the original culture volume. Then bacteria were lysed by sonication for 30 min. After that, the lysed cells were centrifuged at 4 °C at 10,000 *g* for 30 min, and then, 10 µM hemin and 25 mM $CaCl_2$ were added to the supernatant, which was stored at 4 °C for 6 h. Finally, the solution containing soluble MnP was centrifuged at 4 °C for 30 min to remove any excess hemin.

Like the optimization design for the refolding of inclusion bodies, several chemical additives (Triton X-100, Tween-80, ethanol, glycerol, sorbitol, L-proline, and glycine), as well as their concentrations, were tested using the one-factor-at-a-time method to achieve the soluble expression of MnP. Factors exhibiting significant effects were selected by observing DMP oxidation in 48-deep well plates. Subsequently, the optimum combination of significant factors, as well as their concentrations, was further optimized by an orthogonal experimental design to obtain greater soluble MnP expression. The optimum factor combination was estimated by analyzing the specific Mn^{2+} oxidizing activity of MnP after adding exogenous hemin and Ca^{2+} to the cell lysates.

Purification of recombinant MnP

Recombinant MnP was purified using a Ni^{2+}-affinity column (Sangon Biotech, Shanghai, China) according to the manufacturer's instructions. Sodium dodecyl sulfate-polyacrylamide gel electrophoresis (SDS-PAGE) was conducted using a 12% Tris–HCl separation gel and a 5% Tris–HCl staking gel, followed by staining with Coomassie Brilliant Blue R-250.

Biochemical characterization of MnP

Determination of kinetic parameters of soluble and refolded MnP

The K_m and k_{cat} for Mn^{2+} were determined by the hyperbolic, non-linear least squares method at varying $MnSO_4$ concentrations (0.025–1.0 mM) in the presence of 0.1 mM H_2O_2. The K_m and k_{cat} for H_2O_2 were determined at varying H_2O_2 concentrations (0.01–0.15 mM) in the presence of 1 mM $MnSO_4$. The reactions were conducted at 25 °C in 0.11 M of sodium lactate buffer (pH 4.5).

Effect of pH and temperature on the activity and stability of MnP

The optimum pH for activites of soluble and refolded MnP was evaluated at 25 °C in 0.11 M citrate-phosphate buffer (pH 2.2–8.0) or Tris–HCl buffer (pH 8.6–9.0). pH stability was determined at 25 °C by incubating the enzyme at different pH values (2.2–9.0) for 6 h. The residual MnP activity was measured according to the method used for the MnP activity assay.

The optimum temperature for MnP activity was measured from 10 to 75 °C in appropriate increments in 0.11 M sodium lactate buffer (pH 4.5). To evaluate the thermal stability of MnP, the enzyme was incubated at temperatures ranging from 15 to 65 °C in appropriate increments for 6 h. The residual MnP activity was measured according to the method used for the MnP activity assay.

Substrate specificity

The oxidation of guaiacol ($\varepsilon = 12{,}100\ M^{-1}\ cm^{-1}$), DMP ($\varepsilon = 49{,}600\ M^{-1}\ cm^{-1}$) and ABTS ($\varepsilon = 36{,}000\ M^{-1}\ cm^{-1}$) was measured at 456, 469, and 420 nm, respectively, to estimate the substrate specificities of soluble and refolded MnP. The reaction mixtures contained 0.11 M sodium tartrate (pH 4.5), 0.1 mM H_2O_2, 25 µl of MnP, with or without 1 mM Mn^{2+}, in a total volume of 1 ml. The Mn^{2+} oxidizing activity of both soluble and refolded MnP was 530 U/L.

Decolorization of Acid Red 18 and Orange G by recombinant MnP

Dye decolorization was measured spectrophotometrically at 506 and 478 nm (WFV UV-2100 spectrophotometer, UNICO, Shanghai, China), which are the maximum visible absorbances of Acid Red 18 and Orange G, respectively.

The decolorization reaction system contained sodium tartrate (0.11 M, pH 4.0), $MnSO_4$ (1.25 mM), H_2O_2 (0.15 mM)), dye (50 mg/L), and MnP (100 U/L) in a total volume of 1 ml. The reaction was initiated by the addition of H_2O_2. Control samples, without enzyme, were performed in parallel under identical conditions.

The decolorization efficiency was obtained from the following equation:

$$\text{Decolorization}(\%) = \frac{A_0 - A_t}{A_0} \times 100\%$$

where A_0 is the initial absorbance at 506 nm for Acid Red 18 or at 478 nm for Orange G, and A_t refers to the absorbance at 506 nm or 478 nm at reaction time t. The data are presented as the mean values of triplicate experiments.

Spectroscopic characterization of MnP

Far-UV circular dichroism (CD) spectra and UV–VIS spectra

The CD measurements were implemented with a MOS-500 CD spectrometer (Bio-Logic, Grenoble, France), with a 1-mm light path cell, at room temperature. The CD spectra were recorded using a 2-mm bandwidth in the far-UV region (190–250 nm). The protein concentration was 0.1 mg/ml in 0.15 M phosphate buffer, pH 6.5. The UV–VIS spectra were obtained in the 200–700 nm region at room temperature using a DU 730 UV spectrophotometer (Beckman Coulter, Brea, CA, USA), and the protein concentration was 0.5 mg/ml in 0.15 M phosphate buffer, pH 6.5.

Fluorescence spectroscopy

All fluorescence spectra of enzymes were obtained at room temperature using an F-4500 FL spectrophotometer (Hitachi, Tokyo, Japan) using a cuvette with a 10-mm path length. Excitation and emission bandwidths were set at 5 nm. Fluorescence measurements were performed at a protein concentration of 0.1 mg/ml.

Raman spectra

After protein purification, enzyme solution was dried under vacuum freezing condition for 24 h to produce the solid powder samples. Protein samples were prepared and stored at −20 °C. The Raman spectra of soluble and refolded MnP were obtained in the 400 cm^{-1} to 1800 cm^{-1} at room temperature using a InVla-Reflex instrument (Renishaw, London, England) equipped with excitation from the 785 nm line of a semiconductor laser and a charge-coupled-device (CCD) array detector. The laser was focused through an Olympus 50× objective lens on the solid protein samples. The laser power was 8.4 mW, and the resolution was 0.48 cm^{-1}. An acquisition time of 30s was used.

Size exclusion chromatography and dynamic light scattering

After being purified using Ni-NTA affinity chromatography, protein samples were further purified by a HiLoad ™ 16/60 superdex 200 gel filtration column (10–600 kDa, GE health care, USA) equilibrated with protein buffer (20 mM Tris–HCl (pH 8.0), 200 mM NaCl). Elution was performed with the same buffer at a flow rate of 1 ml/min. Absorbance was monitored at 280 nm. Fractions containing MnP activity peak were pooled and kept at 4 °C to be used for dynamic light scattering studies. Dynamic light scattering (DLS) spectra were obtained at room temperature using an Dynapro-MS800 instrument (Wyatt, Santa Barbara, CA, USA) and high precision cell made of quartz suprasil with 1.5 mm path length. The protein concentration was diluted to a final concentration of 0.1 mg/ml with 20 mM Tris–HCl (pH 8.0) and 200 mM NaCl. Before measurement, all samples were centrifuged at 4 °C at 10,000 *g* for 10 min. The radius and estimated molecular weight were provided by the Dynamics software.

Abbreviations

ABTS: 2,2′-azino-bis(3-ethylbenzothiazoline-6-sulfonic acid); CD: Circular dichroism; DLS: Dynamic light scattering; DMP: 2,6-dimethoxyphenol; DTT: Dithiothreitol; *E. coli*: *Escherichia coli*; FPLC: Fast protein liquid chromatography; GSSG: Glutathione; H_2O_2: Hydrogen peroxide; *I. lacteus* F17: *Irpex lacteus* F17; IPTG: Isopropyl-β-D-thiogalactopyranoside; MnP: Manganese peroxidase; PMSF: Phenylmethanesulfonyl fluoride; Rz: Reinheitszahl; SDS-PAGE: Sodium dodecyl sulfate-polyacrylamide gel electrophoresis; SEC: Size exclusion chromatography

Acknowledgements

We thank Chao He for her help to the dynamic light scattering experiment.

Funding
This work was supported by the National Natural Science Foundation of China (31570102, 31070109) and the College Students' Innovation and Entrepreneurship Training Project of Anhui University.

Authors' contributions
WN carried out all expression, purification and characterization analyses of recombinant MnP, participated in the design of the experiments and drafted the manuscript. RK participated in the optimal experiment for the expression of soluble MnP and performed the data analysis. JR conceived of the study, and participated in its design and helped to revise the manuscript. CWT carried out the cloning and transforming experiments. SRR participated in enzymatic assays. All authors read and approved the final manuscript.

Competing interests
I confirm that I have read BioMed Central's guidance on competing interests and we declare that we have no competing interests.

References
1. Hammel KE, Kalyanaraman B, Kirk TK. Substrate free radicals are intermediates in ligninase catalysis. Proc Natl Acad Sci. 1986;83(11):3708–12.
2. Michel FC, Dass SB, Grulke EA, Reddy CA. Role of manganese peroxidases and lignin peroxidases of *Phanerochaete chrysosporium* in the decolorization of kraft bleach plant effluent. Appl Environ Microbiol. 1991;57(8):2368–75.
3. Maijala P, Kleen M, Westin C, Poppius-Levlin K, Herranen K, Lehto JH, Reponen P, Mäentausta O, Mettälä A, Hatakka A. Biomechanical pulping of softwood with enzymes and white-rot fungus *Physisporinus rivulosus*. Enzyme Microb Technol. 2008;43:169–77.
4. Sundaramoorthy M, Kishi K, Gold MH, Poulos TL. Preliminary crystallographic analysis of manganese peroxidase from *Phanerochaete chrysosporium*. J Mol Biol. 1994;238:845–8.
5. Fernández-Fueyo E, Ruiz-Dueñas FJ, Martínez MJ, Romero A, Hammel KE, Medrano FJ, Martínez AT. Ligninolytic peroxidase genes in the *oyster mushroom* genome: heterologous expression, molecular structure, catalytic and stability properties, and lignin-degrading ability. Biotechnol Biofuels. 2014;7:2.
6. Martínez AT. Molecular biology and structure-function of lignin-degrading heme peroxidases. Enzyme Microb Tech. 2002;30:425–44.
7. Saroj S, Agarwal P, Swati D, Singh RP. Manganese peroxidases: molecular diversity, heterologous expression, and applications. Advances in Enzyme Biotechnology. 2013:67–87.
8. Singh D, Chen S. The white-rot fungus *Phanerochaete chrysosporium*: conditions for production of lignin-degrading enzymes. App Microbiol Biotechnol. 2008;81:399–417.
9. Mayfield MB, Kishi K, Alic M, Gold MH. Homologous expression of recombinant manganese peroxidase in *Phanerochaete chrysosporium*. Appl Environ Microb. 1994;60:4303–9.
10. Irie T, Honda Y, Watanabe T, Kuwahara M. Homologous expression of recombinant manganese peroxidase genes in ligninolytic fungus *Pleurotus ostreatus*. Appl Microbiol Biotechnol. 2001;55:566–70.
11. Gu L, Lajoie C, Kelly C. Expression of a *Phanerochaete chrysosporium* manganese peroxidase gene in the yeast *Pichia pastoris*. Biotechnol Progr. 2003;19:1403–9.
12. Whitwam R, Ming T. Heterologous expression and reconstitution of fungal Mn peroxidase. Arch Biochem Biophys. 1996;333:439–46.
13. Ambert-Balay K, Dougherty M, Tien M. Reactivity of manganese peroxidase: site-directed mutagenesis of residues in proximity to the porphyrin ring. Arch Biochem Biophys. 2000;382:89–94.
14. Ignatova Z, Mahsunah A, Georgieva M, Kasche V. Improvement of posttranslational bottlenecks in the production of penicillin amidase in recombinant *Escherichia coli* strains. App Environ Microb. 2003;69:1237–45.
15. Järvinen J, Taskila S, Isomäki R, Ojamo H. Screening of white-rot fungi manganese peroxidases: a comparison between the specific activities of the enzyme from different native producers. AMB Express. 2012;2:62.
16. Pacheco B, Crombet L, Loppnau P, Cossar D. A screening strategy for heterologous protein expression in *Escherichia coli* with the highest return of investment. Protein Expres Purif. 2012;81:33–41.
17. Shaw MK, Ingraham JL. Synthesis of macromolecules by *Escherichia coli* near the minimal temperature for growth. J Bacteriol. 1967;94:157–64.
18. Basu A, Li X, Leong SSJ. Refolding of proteins from inclusion bodies: rational design and recipes. Appl Microbiol Biotechnol. 2011;92:241–51.
19. Prasad S, Khadatare PB, Roy I. Effect of chemical chaperones in improving the solubility of recombinant proteins in *Escherichia coli*. Appl Environ Microb. 2011;77:4603–9.
20. Schlieker C, Bukau B, Mogk A. Prevention and reversion of protein aggregation by molecular chaperones in the *E. coli* cytosol: implications for their applicability in biotechnology. J Biotehnol. 2002;96:13–21.
21. Jenkins TM, Engelman A, Ghirlando R, Craigie R. A soluble active mutant of HIV-1 integrase. J Biol Chem. 1996;271:7712–8.
22. Chen WT, Zheng LL, Jia R, Wang N. Cloning and expression of a new manganese peroxidase from *Irpex lacteus* F17 and its application in decolorization of reactive black 5. Process Biochem. 2015;50:1748–59.
23. Zhao XS, Huang XJ, Yao JT, Zhou Y, Jia R. Fungal growth and manganese peroxidase production in a deep tray solid-state bioreactor, and in vitro decolorization of poly R-478 by MnP. J Microbiol Biotech. 2014;25:803–13.
24. Herrero AM. Raman spectroscopy for monitoring protein structure in muscle food systems. Crit Rev Food Sci. 2008;48:512–23.
25. Fakoussa RM, Hofrichter M. Biotechnology and microbiology of coal degradation. Appl Microbiol Biotechnol. 1999;52:25–40.
26. Perie FH, Sheng D, Gold MH. Purification and characterization of two manganese peroxidase isozymes from the white-rot basidiomycete *Dichomitus squalens*. Biochim Biophys Acta. 1996;1297:139–48.
27. Ufot UF, Akpanabiatu MI. An engineered *Phlebia radiata* manganese peroxidase: expression, refolding, purification and preliminary characterization. Am J Resp Cell Mol. 2012;2:359–70.
28. Liu Y, Deng RJ, Hao MF, Yan H, Yang WM. Orthogonal design study on factors effecting on fibers diameter of melt electrospinning. Polym Eng Sci. 2010;50(10):2074–8.
29. Jiao Y, Han Y, Li X, Fang YG, Liu ZH, Zhou WN, Zhou JC, Wu ZC, Yang JH, Li SY, Meng FY, Xu WW. Comparison of body, auricular, and abdominal acupuncture treatments for insomnia differentiated as internal harassment of phlegm-heat syndrome: an orthogonal design. Evid-Based Compl Alt. 2015;5:1–9.
30. Smith AT, Santama N, Dacey S, Edwards M, Bray RC, Thorneley RNF, Burke JF. Expression of a synthetic gene for horseradish peroxidase C in *Escherichia coli* and folding and activation of the recombinant enzyme with Ca^{2+} and heme. J Biol Chem. 1990;265:13335–43.
31. Miki Y, Morales M, Ruiz-Dueñas FJ, Martínez MJ, Wariishi H, Martínez AT. *Escherichia coli* expression and in vitro activation of a unique ligninolytic peroxidase that has a catalytic tyrosine residue. Protein Expres Purif. 2009;68:208–14.
32. Schein CH. Production of soluble recombinant proteins in bacteria. Bio Technology. 1989;7:1141–9.
33. Esposito D, Chatterjee DK. Enhancement of soluble protein expression through the use of fusion tags. Curr Opin Biotech. 2006;17:353–8.
34. Mohorčič M, Benčina M, Friedrich J, Jerala R. Expression of soluble versatile peroxidase of *Bjerkandera adusta* in *Escherichia coli*. Bioresource Technol. 2009;100:851–8.
35. Tang JB, Yang HM, Song SL, Zhu P, Ji AG. Effect of glycine and Triton X-100 on secretion and expression of ZZ-EGFP fusion protein. Food Chem. 2008;108:657–62.
36. Choi JH, Lee SY. Secretory and extracellular production of recombinant proteins using *Escherichia coli*. Appl Microbiol Biotechnol. 2004;64:625–35.
37. Alibolandi M, Mirzahoseini H. Chemical assistance in refolding of bacterial inclusion bodies. Biochemistry Research International. 2011; doi:10.1155/2011/631607.
38. Thomas JG, Baneyx FD. Divergent effects of chaperone overexpression and ethanol supplem entation on in clusion body formation in recombinant *Escherichia coli*. Protein Expr Purif. 1997;11(3):289–96.
39. Steczko J, Donoho GA, Dixon JE, Sugimoto T, Axelrod B. Effect of ethanol and low-mperature culture on expression of soybean lipoxygenase L-1 in *Escherichia coli*. Protein Expr Purif. 1991;2:221–7.
40. Samuel D, Kumar TKS, Ganesh G, Jayaraman G, Yang PW, Chang MM, Trivedi VD, Wang SL, Hwang KC, Chang DK, Yu C. Proline inhibits aggregation during protein refolding. Protein Sci. 2000;9:344–52.
41. Pollitt S, Zalkin H. Role of primary structure and disulfide bond formation in b-lactamase secretion. J Bacteriol. 1983;153:27–32.

42. Makrides SC. Strategies for achieving high-level expression of genes in *Escherichia coli*. Microbiol Rev. 1996;60:512–38.
43. Yang JB, Moyana T, Mackenzie S, Xia Q, Xiang J. One hundred seventy-fold increase in excretion of an FV fragment-tumor necrosis factor alpha fusion protein (sFV/TNF-alpha) from *Escherichia coli* caused by the synergistic effects of glycine and triton X. App Environ Microb. 1998;64(8):2869–74.
44. Pérez-Boada M, Doyle WA, Ruiz-Dueñas FJ, Martínez MJ, Martínez AT. Expression of *Pleurotus eryngii* versatile peroxidase in *Escherichia coli* and optimisation of in vitro folding. Enzyme Microb Tech. 2002;30:518–24.
45. Ninomiya R, Zhu B, Kojima T, Iwasaki Y, Nakano H. Role of disulfide bond isomerase DsbC, calcium ions, and hemin in cell-free protein synthesis of active manganese peroxidase isolated from *Phanerochaete chrysosporium*. J Biosci Bioeng. 2014;117:652–7.
46. Sutherland GRJ, Aust SD. The effects of calcium on the thermal stability and activity of manganese peroxidase. Arch Biochem Biophys. 1996;332:128–34.
47. Sklenar J, Niku-Paavola ML, Santos S, Man P, Kruus K, Novotny C. Isolation and characterization of novel pI 4.8 MnP isoenzyme from white-rot fungus *Irpex lacteus*. Enzyme Microb Tech. 2010;46:550–6.
48. Boer CG, Obici L, Souza CGM, Peralta RM. Purification and some properties of Mn peroxidase from Lentinula edodes. Process Biochem. 2006;41:1203–7.
49. Knop D, Yaeden O, Hasar Y. The ligninolytic peroxidases in the genus *pleurotus*: divergence in activities, expression, and potential applications. Appl Microbiol Biotechnol. 2015;99:1025–38.
50. Kishi K, Hildebrand DP, Someren MK, Gettemy J, Mauk AJ, Gold MH. Site-directed mutations at phenylalanine-190 of manganese peroxidase: effects on stability, function, and coordination. Biochem. 1997;36:4266–77.
51. Aceves MA, Baratto MC, Basosi R, Vazquez-Duhalt R, Pogni R. Spectroscopic characterization of a manganese–lignin peroxidase hybrid isozyme produced by *Bjerkandera adusta*, in the absence of manganese: evidence of a protein centred radical by hydrogen peroxide. J Mol Catal B-Enzym. 2001; 16:159–67.
52. Mendes S, Brissos V, Gabriel A, Catarino T, Turner DL. An integrated view of redox and catalytic properties of B-type PpDyP from *Pseudomonas putida*, MET94 and its distal variants. Arch Biochem Biophy. 2015;574:99–107.
53. Rygula A, Majzner K, Marzec KM, Kaczor A, Pilarczyk M. Raman spectroscopy of proteins: a review. J Raman Spectrosc. 2013;44:1061–76.
54. Chihi ML, Mession JL, Sok N, Saurel R. Heat-induced soluble protein aggregates from mixed pea globulins and β-lactoglobulin. J Agr Food Chem. 2016;64:2780–91.
55. Wang JJ, Liu GY, Liu G, Zeng QH, Shen X, Hou Y, Li L, Hu SQ. The soluble recombinant N-terminal domain of HMW 1Dx5 and its aggregation behavior. Food Res Int. 2015;78:201–8.

6

Targeted optimization of central carbon metabolism for engineering succinate production in *Escherichia coli*

Ying Zhao[1,2,3], Chang-Song Wang[1,2,3,4], Fei-Fei Li[1,2,3], Zhen-Ning Liu[1,2,3] and Guang-Rong Zhao[1,2,3*]

Abstract

Background: Succinate is a kind of industrially important C4 platform chemical for synthesis of high value added products. Due to the economical and environmental advantages, considerable efforts on metabolic engineering and synthetic biology have been invested for bio-based production of succinate. Precursor phosphoenolpyruvate (PEP) is consumed for transport and phosphorylation of glucose, and large amounts of byproducts are produced, which are the crucial obstacles preventing the improvement of succinate production. In this study, instead of deleting genes involved in the formation of lactate, acetate and formate, we optimized the central carbon metabolism by targeting at metabolic node PEP to improve succinate production and decrease accumulation of byproducts in engineered *E. coli*.

Results: By deleting *ptsG*, *ppc*, *pykA*, *maeA* and *maeB*, we constructed the initial succinate-producing strain to achieve succinate yield of 0.22 mol/mol glucose, which was 2.1-fold higher than that of the parent strain. Then, by targeting at both reductive TCA arm and PEP carboxylation, we deleted *sdh* and co-overexpressed *pck* and *ecaA*, which led to a significant improvement in succinate yield of 1.13 mol/mol glucose. After fine-tuning of *pykF* expression by anti-*pykF* sRNA, yields of lactate and acetate were decreased by 43.48 and 38.09 %, respectively. The anaerobic stoichiometric model on metabolic network showed that the carbon fraction to succinate of engineered strains was significantly increased at the expense of decreased fluxes to lactate and acetate. In batch fermentation, the optimized strain BKS15 produced succinate with specific productivity of 5.89 mmol $gDCW^{-1}$ h^{-1}.

Conclusions: This report successfully optimizes succinate production by targeting at PEP of the central carbon metabolism. Co-overexpressing *pck-ecaA*, deleting *sdh* and finely tuning *pykF* expression are efficient strategies for improving succinate production and minimizing accumulation of lactate and acetate in metabolically engineered *E. coli*.

Keywords: Succinate, *Escherichia coli*, sRNA, Metabolic engineering, Synthetic biology

Background

Succinate, an important member of C4-dicarboxylic acid family, has been widely used in agricultural, food, pharmaceutical, cosmetic, textile and fine chemicals industries [1, 2]. Meanwhile, succinate has received considerable attention to synthesize various valuable molecules such as 1,4-butanediol, tetrahydrofuran, γ-butyrolactone and adipic acid [3]. Petrochemistry-based succinate production requires various metal catalysts and discharges organic wastes, which make petrochemical processes costly and not environmental friendly. Bio-based succinate production is a promising and green process as it uses renewable bioresources as substrates and fixes greenhouse gas CO_2 [4]. Therefore, the concomitant economical and environmental advantages stimulate the efforts to engineer microorganisms for efficient succinate production.

Succinate can be naturally produced by many strict anaerobic bacteria and facultative anaerobes. *Escherichia coli* is most widely studied for succinate production due to its convenience for genetic manipulation and fast growth with flexible nutrient requirements [5]. However, the wild *E. coli* strain prefers to produce lactate and acetate as major products with a small amount of succinate in mixed-acid fermentation under anaerobic conditions [6].

* Correspondence: grzhao@tju.edu.cn
[1]Department of Pharmaceutical Engineering, School of Chemical Engineering and Technology, Tianjin University, Tianjin 300072, China
[2]Key Laboratory of Systems Bioengineering, Ministry of Education, Tianjin 300072, China
Full list of author information is available at the end of the article

Efforts of metabolic engineering and adaptive evolution have been made to obtain succinate-producing *E. coli*. Inactivation of genes accounting for biosyntheses of those byproducts was first pursued to produce succinate as the predominant fermentation product. However, the mutant *E. coli* strains deficient in *ldhA* (coding lactate dehydrogenase) and *pflB* (coding pyruvate-formate lyase), *adhE* (coding alcohol dehydrogenase) and *pta* (coding phosphotransacetylase) or their combinations were unable to anaerobically grow on glucose media and the titer and yield of succinate were relatively low. For example, the mutant *E.coli* strain NZN111 deficient in *ldhA* and *pflB* only produced minor amount of succinate [7]. Evolutionary engineering of strain NZN111 led to spontaneous chromosomal mutant strain AFP111, which was able to ferment glucose anaerobically and produced higher succinate yield, as well as higher acetate [8]. Similarly, by combining metabolic engineering and evolution of over 2000 generations screened on glucose minimal medium, *E. coli* strain KJ073 with deletions of *ldhA*, *adhE*, *ackA* (coding acetate kinase), *focA* (coding formate channel), *pflB*, *mgsA* (coding methylglyoxal synthase) and *poxB* (coding pyruvate oxidase) was capable of producing high succinate yield, but significant amounts of acetate and malate were also produced [9].

Metabolic targets of the central carbon metabolism have been used to improve succinate production in *E. coli*. In order to enhance carbon flux to succinate, formation of oxaloacetate (OAA) from pyruvate or phosphoenolpyruvate (PEP) was chosen as metabolic target. Heterologous expressions of *pyc* (coding pyruvate carboyxlase, PYC) from *Rhizobium etli* [10] or from *Lactococcus lactis* [11, 12], *pck* (coding PEP carboxykinase, PCK) from *Actinobacillus succinogenes* [13, 14] and overexpression of native *ppc* (coding PEP carboxylase, PPC) [15] were shown to increase succinate production in recombinant *E. coli* strains. Subtle co-overexpression of both *ppc* and *pck* genes regulated by promoters with different strengths improved succinate production [16]. To increase NADH availability in succinate-producing *E. coli*, several genes involved in redox reactions were identified to improve cell growth impairment under microaerobic conditions [17]. Heterologous NAD^+-dependent formate dehydrogenase gene *fdh* of *Candida boidinii* or native nicotinate phosphoribosyltransferase gene *pncB* were co-overexpressed with *Lactococcus lactis pyc* gene to achieve the redox and ATP balance [18, 19]. Activation of pentose phosphate pathway, transhydrogenase and pyruvate dehydrogenase were identified for improved succinate production by increasing reducing power supplement [20]. To enhance glucose utilization in *E. coli* strain deficient in PEP carbohydrate phosphotransferase system (PTS), native *galP* (coding D-galactose transporter) and *glk* (coding glucokinase) were co-overexpressed or modulated to facilitate succinate production [21]. *Zymomonas mobilis glf* gene (coding glucose facilitator, Glf) was more efficient than *E. coli galP* gene due to the higher transport velocity and lower energetic cost of Glf [22]. In addition, C4-dicarboxylic acid transporter genes were also activated to decrease the feedback effects through accelerating succinate export [23, 24].

Although considerable metabolic targets are available to improve succinate production, genes involved in competing pathways such as the formation of lactate, acetate, formate and ethanol were inactivated in previous works. In this study, targeted engineering strategy was employed to optimize metabolic pathway of succinate production from glucose without deletions of *ldhA*, *pflB*, *pta-ackA*, and *adhE* (Fig. 1). Focusing on PEP node as the engineering target, metabolic flux from PEP was enhanced to OAA and minimized to lactate and acetate. By pentuple deletions of genes *ptsG* (coding glucose phosphotransferase), *pykA* (coding pyruvate kinase II), *ppc*, *maeA* and *maeB* (coding malic enzymes) of the central carbon metabolism, we reconstructed initial *E. coli* strain to increase PEP pool for succinate production. Then we optimized metabolic flux to succinate from PEP by deletion of *sdh* (coding succinate dehydrogenase) and *iclR* (coding transcriptional repressor IclR) as well as co-overexpression of *pck-ecaA* (coding carbonic anhydrase). We further attenuated the accumulation of lactate and acetate by fine tuning of *pykF* (coding pyruvate kinase I) expression via antisense sRNA strategy to prevent metabolic flux to pyruvate from PEP. Finally, the fermentation process was carried out with optimized succinate-producing strains.

Results and discussion

Initial construction for succinate production

The wildtype *E. coli* BW25113 (DE3) produced a small amount of succinate in the acid mixture (Fig. 2) from glucose under anaerobic fermentation conditions, which was consistent with the previous report [6]. Glucose uptake through PTS system consumes almost half of the available PEP that is the precursor of succinate, which leads to the significantly decreased amounts of PEP for succinate production. In *E. coli*, the inactivation and mutation of genes involved in the PTS system was beneficial for succinate production [25, 26]. Thus, to save PEP from consumption of PTS system, we deleted *ptsG* gene in strain BW25113 (DE3) and constructed strain BKS4. Succinate production of strain BKS4 was significantly increased with 2.0-fold higher yield than that of strain BW25113 (DE3) ($p < 0.01$) (Fig. 2). Meanwhile, the yields of lactate and acetate in strain BKS4 were decreased by 17.65 % ($p < 0.05$) and 19.83 % ($p < 0.01$), respectively. The results indicated that the inactivation of PTS system played an essential role in the availability of PEP to support succinate production.

In succinate metabolic pathway, the carboxylation of PEP catalyzed by PPC or PCK is a rate-limiting step committed to succinate production. ATP is essentially consumed for

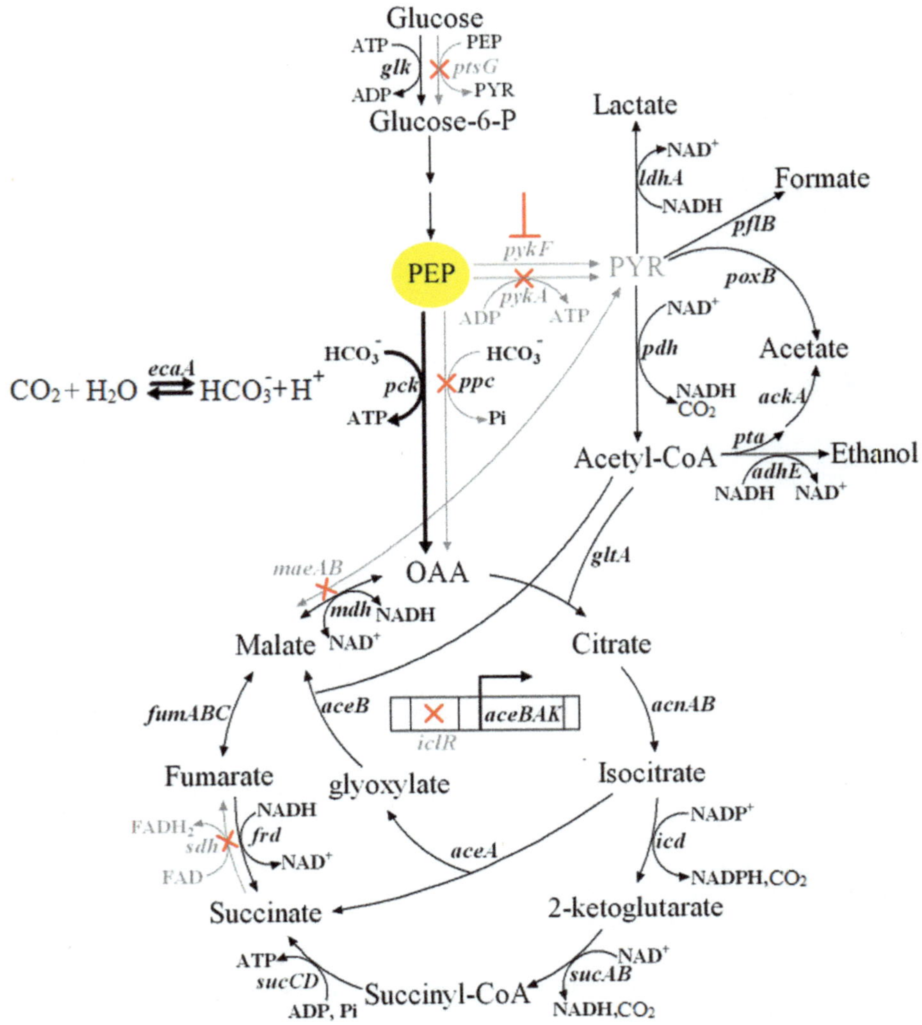

Fig. 1 Targeted engineering of the central carbon metabolism for succinate production in *E. coli*. Red crosses represent deletion of gene and the reactions affected by the deletion are indicated with grey. The black arrows of the reactions involved in the overexpressed genes are thickened. Red ⊣ represents inhibition of *pykF* expression by anti-*pykF* sRNA either on high-copy-number plasmid (pRSFDuet-1) (H) or low-copy-number plasmid (pBldgbrick2) (L). Genes coding the corresponding enzymes in the pathways: *ptsG*, glucose phosphotransferase; *pykF*, pyruvate kinase I; *pykA*, pyruvate kinase II; *ppc*, PEP carboxylase; *pck*, PEP carboxykinase; *ecaA*, carbonic anhydrase; *iclR*, transcriptional repressor IclR; *aceA*, isocitrate lyase; *aceB*, malate synthase; *aceK*, isocitrate dehydrogenase kinase/phosphatase; *ldhA*, lactate dehydrogenase; *pflB*, pyruvate formate lyase; *pdh*, pyruvate dehydrogenase; *poxB*, pyruvate oxidase; *pta*, phosphotransacetylase; *ackA*, acetate kinase; *adhE*, alcohol/acetaldehyde dehydrogenase; *maeAB*, malic enzyme; *mdh*, malate dehydrogenase; *fumABC*, fumaraseABC; *frd*, fumarate reductase; *sdh*, succinate dehydrogenase; *sucABCD*, succinyl CoA synthase; *icd*, isocitrate dehydrogenase; *acnAB*, aconitate hydratase; *gltA*, citrate synthase

PPC catalyzing the formation of OAA from PEP [27]. On the contrary, one molecule ATP is generated from carboxylation of one molecule PEP catalyzed by PCK. The deletion of *pck* gene in *E. coli* remarkably inhibited succinate production as well as the cell growth [27], indicating that PCK might be more efficient than PPC. In addition, the function of PCK was partially inhibited by PPC under anaerobic fermentation [13, 14]. Thus, we deleted *ppc* gene to enhance energy supplement and activate PCK. Furthermore, both PEP and malate would convert to pyruvate, which is smoothly turned into byproducts lactate, acetate and formate via the decarboxylation, dehydrogenation, and pyruvate-formate lyase, respectively. Formate is further split into carbon dioxide and water by formate dehydrogenase, while lactate and acetate accumulate in fermentation broth. Since the substrate specificity of malic enzymes for malate is 6-fold higher than that for pyruvate, malic enzymes encoded by *maeA* and *maeB* tend to catalyze the decarboxylation of malate to pyruvate [28]. The formation of pyruvate and its derivative byproducts strongly compete with succinate production for PEP and malate. Inactivation of *pykA* and *pykF* has been shown to be effective in inhibiting the conversion of PEP to pyruvate [29]. Consequently, in order to inhibit the formation of pyruvate from PEP and

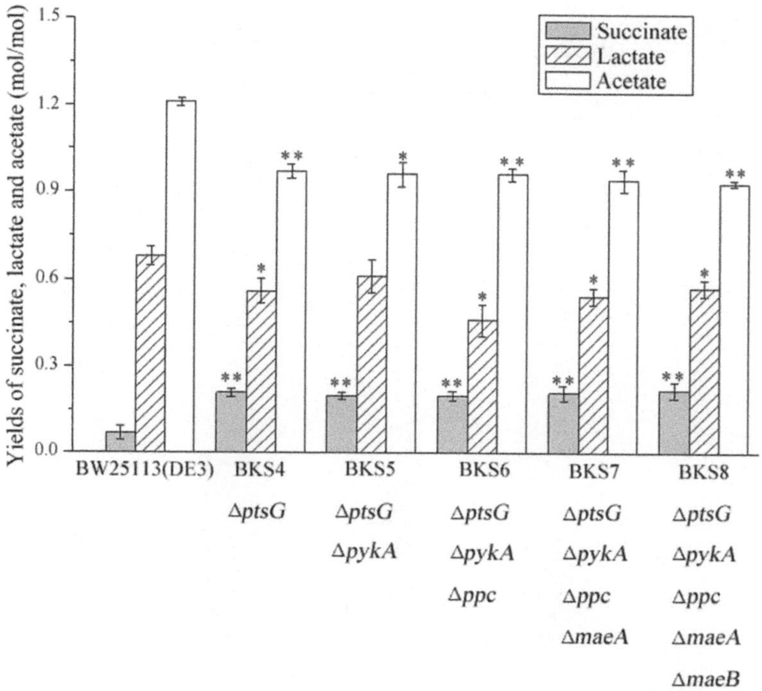

Fig. 2 Yields of succinate, lactate and acetate of initial succinate-producing strains. Error bars represent SD for three replicates. Asterisks indicate *p*-values (**$p < 0.01$, *$p < 0.05$) compared to BW25113 (DE3)

malate, we deleted *pykA*, *maeA* and *maeB* genes. Unfortunately, compared to strain BKS4, strain BKS8 with deletion of *pykA*, *ppc*, *maeA* and *maeB* did neither significantly attenuate the accumulation of lactate and acetate, nor increase the succinate yield (Fig. 2). The low expression level of *pck* gene in wild-type *E. coli* could result in the insufficient metabolic flux to OAA [27], and *pykF* might be more active than *pykA* in the formation of pyruvate from PEP. It suggested that *pck* and *pykF* genes could be the potential targets. Therefore, using initial strain BKS8, we further optimize these two targets of succinate metabolic pathway to improve succinate production.

Combined optimization of targeting at TCA cycle and carboxylation of PEP to increase succinate production

Succinate, an essential intermediate of TCA cycle, cannot be efficiently accumulated in *E. coli* fermentation. In order to increase succinate production, we optimized succinate metabolic pathway by preventing the backflow of succinate to fumarate, activating glyoxylate shunt bypass to decrease the requirement of reducing power, and co-overexpressing *pck-ecaA* to fix CO_2 more efficiently.

Succinate dehydrogenase (SHD) encoded by *sdh* gene catalyzes the dehydrogenation of succinate to fumarate. The *sdh* expression was not totally inhibited under anaerobic conditions [30]. Herein, we deleted *sdh* gene to enhance the reductive TCA arm and block the conversion of succinate to fumarate in strain BKS8 background. As expected, the titer and yield of succinate in strain BKS9 were increased by 55.24 % (7.11 mM) ($p < 0.05$) and 50.00 % (0.33 mol/mol glucose) ($p < 0.05$), respectively (Fig. 3). The inactivation of *sdh* gene showed to increase succinate production in *E. coli* and *Corynebacterium glutamicum* under aerobic conditions [31–33]. To the best of our knowledge, *sdh* gene was first deleted to improve anaerobic succinate production in our study.

Glyoxylate shunt bypass could recover the metabolic flux of the oxidative TCA arm and acetyl-CoA of pyruvate metabolism with less reducing power used, and might contribute to succinate production. The *aceBAK* operon coding isocitrate lyase, malate synthase and isocitrate dehydrogenase kinase is responsible for the glyoxylate shunt bypass. The transcription of the *aceBAK* operon is tightly repressed by transcription factor IclR, but induced by inactivating *iclR* gene [34]. Thus, the deletion of *iclR* gene resulted in strain BKS10. As shown in Fig. 3, the titer and yield of succinate in strain BKS10 was not apparently increased. It was likely that the gene expression involved in glyoxylate bypass are complex and regulated by multiple factors [35] and deletion of *iclR* was not sufficient for activating glyoxylate shunt bypass [36]. Conversion of PEP to OAA in succinate metabolic pathway is net carbon integrated via CO_2 fixation catalyzed by PCK. In fact, the active substrate for PCK is not CO_2, but the chemically less reactive bicarbonate anion (HCO_3^-) [37]. Thus, $CaCO_3$, $MgCO_3$ or $NaHCO_3$ were often added to the culture media. CO_2 is more permeable

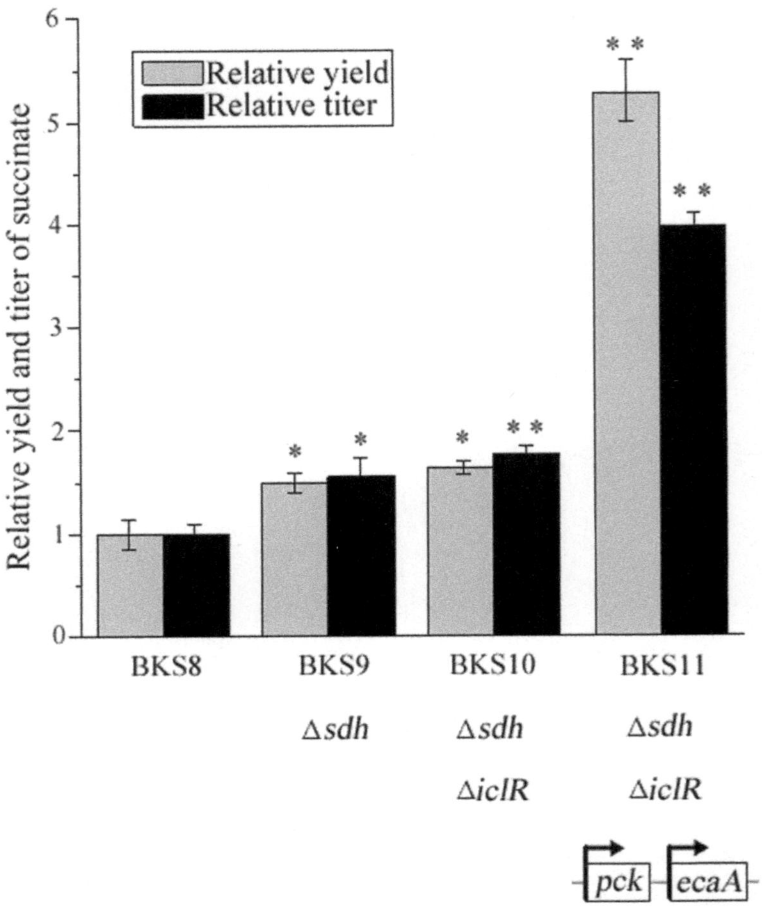

Fig. 3 Deletion of *sdh* and *iclR*, and co-overexpression of *pck-ecaA* increased succinate production. Error bars represent SD for three replicates. Asterisks indicate *p*-values (**$p < 0.01$, *$p < 0.05$) in which BKS9 and BKS10 were compared to BKS8 and BKS11 was compared to BKS10

across cell membrane than HCO_3^-, but the hydration reaction rate of CO_2 to HCO_3^- is relatively slow. There might not be enough HCO_3^- spontaneously made in vivo to access succinate production. Carbonic anhydrase encoded by *ecaA* gene catalyzes the hydration of intracellular CO_2 to HCO_3^-. Expression of *ecaA* gene of cyanobacterium *Anabaena* in *E. coli* led to an obvious increase in succinate production [38, 39]. Thus, the *ecaA* gene was co-expressed with *pck* in strain BKS10, generating strain BKS11. Compared to strain BKS10, combinatorial expression of *pck-ecaA* in strain BKS11 resulted in a 2.2-fold increase in succinate yield (1.16 mol/mol glucose) ($p < 0.01$) and a 1.2-fold increase in succinate titer (18.17 mM) ($p < 0.01$) (Fig. 3).

Fine tuning of *pykF* expression to improve succinate production

Although succinate production was increased remarkably in engineered strains, the yields and titers of lactate and acetate remained high by using the strategies aforementioned in the text (Fig. 4b, c, d), which suggested that metabolic flux from PEP to pyruvate was relatively strong. Deletion of *maeA* and *maeB* and *pykA* did not significantly attenuated the accumulation of lactate and acetate (Fig. 2), suggesting that *pykF* gene might dominate the formation of pyruvate. Thus the strategy of synthetic small RNA (sRNA) engineering [40] was used to finely tune the expression of *pykF* to attenuate the accumulation of lactate and acetate.

Using AUG to nucleotide +24 of the *pykF* mRNA as the binding sequence and selecting *E. coli micC* as the scaffold, anti-*pykF* sRNA working sequence was designed (Fig. 4a). We used two kinds of plasmids with different copy number and tested the inhibitory effects of anti-*pykF* sRNA on the accumulation of lactate and acetate in strain BKS12 with overexpression of *pck* gene. When anti-*pykF* sRNA was expressed on the high-copy-number plasmid pRSF and under the control of T7 promoter, no obvious changes were observed in the yields of succinate, lactate and acetate (Fig. 4b). Then, we constructed the low-copy-number plasmid pBldg-anti-pykF with a pY15A origin of replication, and expression of anti-*pykF* was controlled under *lacUV5* promoter. The metabolite analysis of engineered strain BKS14 showed that the yields of lactate and acetate were decreased by 55.77 % ($p < 0.01$) and 47.73 % ($p < 0.01$), respectively,

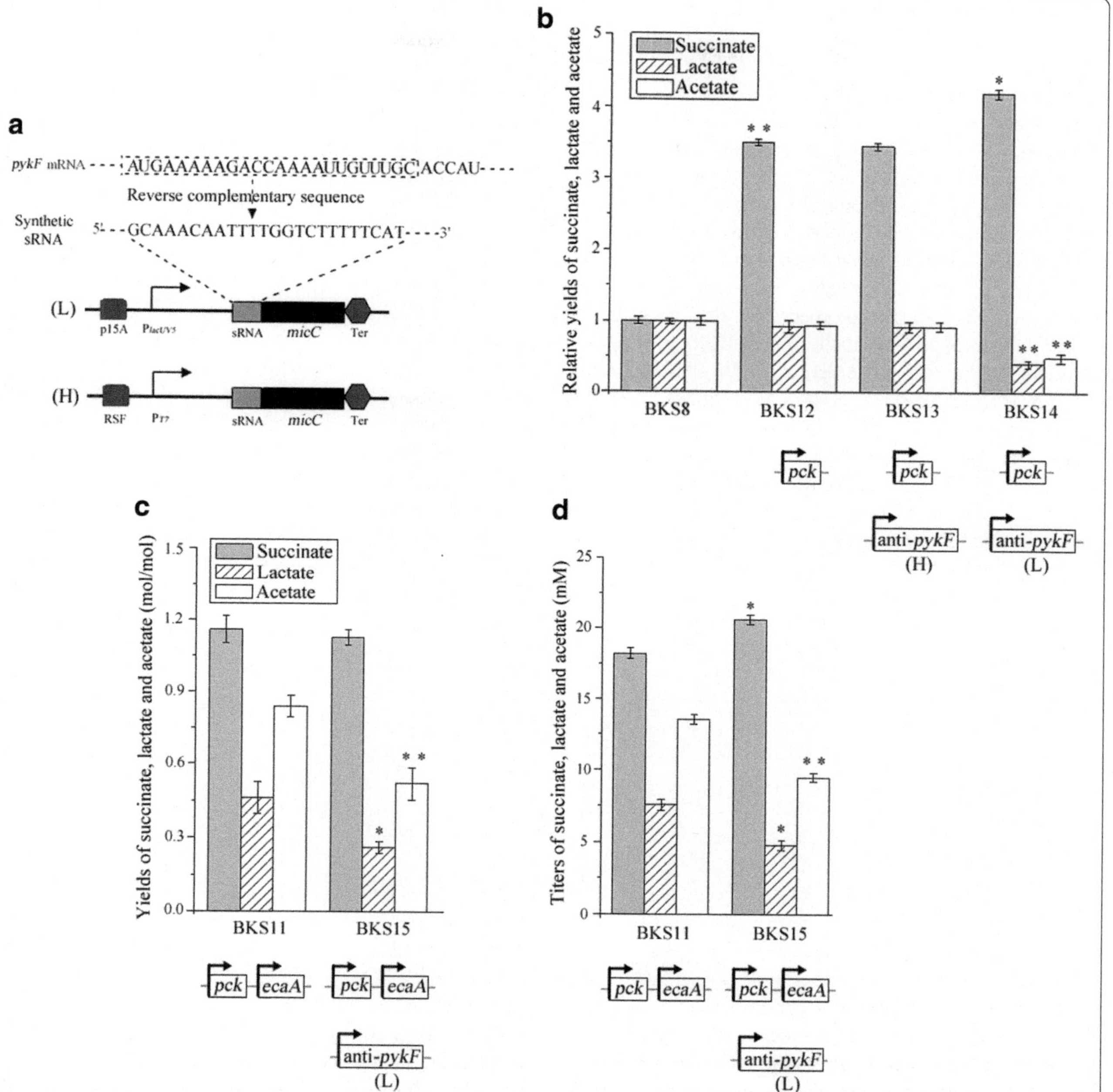

Fig. 4 Fine tuning of *pykF* expression strength to improve succinate production and attenuate accumulation of lactate and acetate. **a** Two anti-*pykF* sRNA plasmids were designed and constructed at different expression levels by combinations of promoters and plasmid copy number. (H) and (L) represented high-copy-number plasmid (pRSF) and low-copy-number plasmid (pBldgbrick2), respectively. **b** Relative yields of succinate, lactate and acetate. BKS12 was compared to BKS8 and BKS13 and BKS14 were compared to BKS12. **c** Yields of succinate, lactate and acetate. The significance was compared to BKS11. **d** Titers of succinate, lactate and acetate. The significance was compared to BKS11. Error bars represent SD for three replicates. Asterisks indicate *p*-values (**$p < 0.01$, *$p < 0.05$)

and the yield of succinate was increased by 23.38 % ($p < 0.05$) compared to BKS12(Fig 4b).

We further tested whether the expression of anti-*pykF* under the control of *lacUV5* promoter in strain BKS11 would improve succinate production and attenuate accumulation of byproducts. pBldg-anti-pykF was transformed into strain BKS11, generating strain BKS15. Compared to strain BKS11, the low expression of anti-*pykF* in strain BKS15 led to the decrease of 43.48 % ($p < 0.05$) and 38.09 % ($p < 0.01$) in the yields of lactate and acetate, respectively (Fig 4c). Although succinate yield of strain BKS15 was not improved, succinate titer was increased by 13.43 % ($p < 0.05$) (Fig. 4d). The results showed that the down-regulated formation of pyruvate by expressing anti-

pykF would enhance the metabolic flux from PEP to succinate.

Distribution of intracellular metabolic flux

Genetic and metabolic modification used in this study remarkably increased succinate production and attenuated the accumulation of lactate and acetate. However, the intracellular metabolic flux distribution of the metabolic network was still unclear. In order to demonstrate in detail how previous efforts changed the metabolic flux directions and optimized the performance of succinate-producing strains step by step, global metabolic flux analysis was made. The simplified metabolic model that described the metabolic relationship in anaerobically fermentative *E. coli* was shown in Fig. 5. This model was comprised of fifteen intermediates and sixteen metabolic reactions designated by V1-V16 (Additional file 1: Table S1). Among these sixteen reactions, the measurable quantities V1, V6, V16 and (V7 + V10) were used to calculate the metabolic fluxes of other intermediates according to relationships shown in Additional file 1: Table S2. The estimated metabolic fluxes in mM gDCW^{-1} h^{-1} of *E. coli* stains BW25113(DE3), BKS8, BKS9, BKS10, BKS11 and BKS15 under anaerobic fermentation were presented in Additional file 1: Table S3.

As shown in Fig. 5, metabolic modifications led to the fact that fluxes to OAA (V5), malate (V12), fumarate (V13), succinate (V15 and V16) were significantly increased and that fluxes to pyruvate (V4), lactate (V6), and acetate (V7 + V10) were remarkably decreased from strains BW25113(DE3) to BKS15. The results indicated that our strategies favored the improvement of succinate production and the decrease of byproduct accumulation.

The split ratios of fluxes to OAA, PYR, lactate, acetate and succinate were obtained by analyzing the PEP, PYR, acetyl-CoA and succinate nodes. As shown in Table 1, compared to strain BW25113 (DE3), the fraction of the metabolic flux diverted to OAA from PEP node (V5/V3) in strain BKS8 increased by 1.8-fold ($p < 0.01$), corresponding 2.2-fold fraction increase of the metabolic flux to succinate (V16/V3) ($p < 0.01$). Pentuple deletions of *ptsG*, *ppc*, *pykA*, *maeA* and *maeB* could significantly streamline PEP pool for succinate production. Strain BKS9 showed the increase of the metabolic flux to succinate (V16/V3), indicating the deletion of *sdh* gene resulted in more metabolic flux to

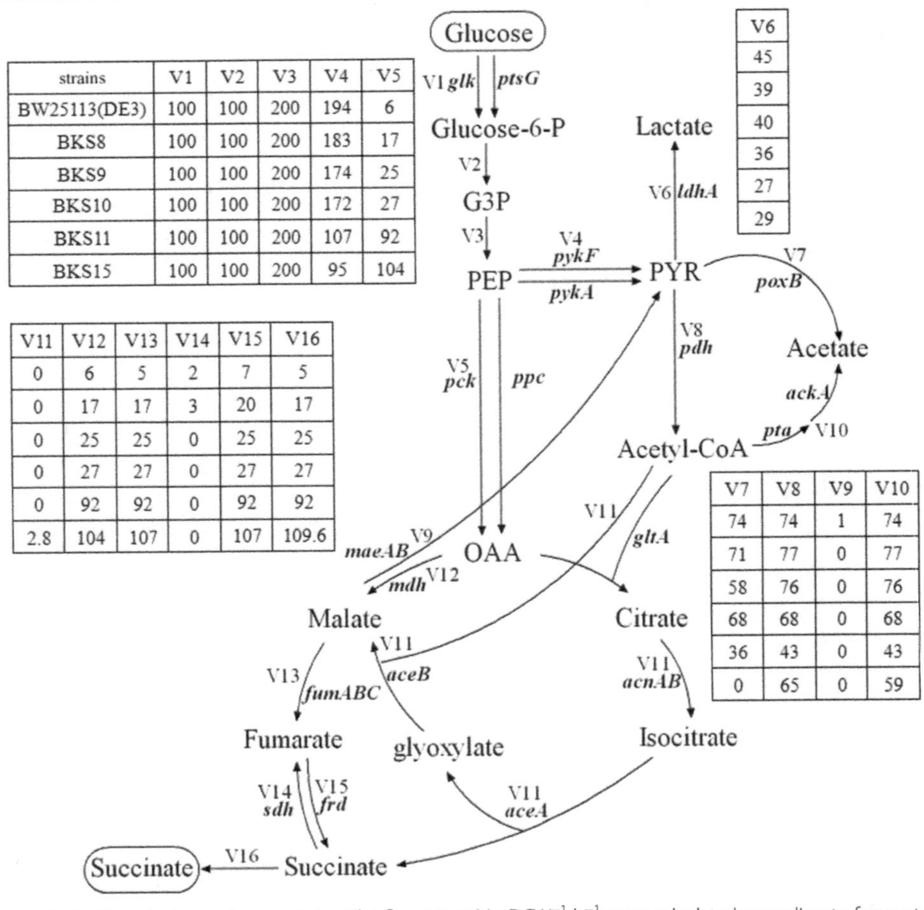

strains	V1	V2	V3	V4	V5
BW25113(DE3)	100	100	200	194	6
BKS8	100	100	200	183	17
BKS9	100	100	200	174	25
BKS10	100	100	200	172	27
BKS11	100	100	200	107	92
BKS15	100	100	200	95	104

V6
45
39
40
36
27
29

V11	V12	V13	V14	V15	V16
0	6	5	2	7	5
0	17	17	3	20	17
0	25	25	0	25	25
0	27	27	0	27	27
0	92	92	0	92	92
2.8	104	107	0	107	109.6

V7	V8	V9	V10
74	74	1	74
71	77	0	77
58	76	0	76
68	68	0	68
36	43	0	43
0	65	0	59

Fig. 5 Metabolic flux analysis of succinate-producing strains. The fluxes in mM gDCW^{-1} h^{-1} were calculated according to fermentation data at 40 h and normalized by glucose consumption rate as well as expressed in a basis of 100

Table 1 Split ratios of fluxes to OAA, PYR, lactate, acetate and succinate

Strains	Fraction of PEP to OAA (V5/V3)	Fraction of PYR production (V4/V3)	Fraction of lactate production (V6/V3)	Fraction of acetate production (V7 + V10)/V3	Fraction of succinate production (V16/V3)
BW25113(DE3)	3.08 ± 0.02 %	96.92 ± 0.02 %	23.46 ± 0.71 %	77.25 ± 2.66 %	2.61 ± 0.02 %
BKS8	8.53 ± 0.03 %	91.47 ± 0.02 %	19.43 ± 0.48 %	73.70 ± 1.56 %	8.29 ± 0.01 %
BKS9	12.39 ± 0.45 %	86.93 ± 1.43 %	19.95 ± 0.14 %	66.74 ± 0.08 %	12.61 ± 0.47 %
BKS10	13.66 ± 0.22 %	86.34 ± 1.10 %	18.28 ± 0.44 %	67.84 ± 2.89 %	13.66 ± 0.22 %
BKS11	45.94 ± 0.73 %	53.87 ± 0.72 %	16.97 ± 0.59 %	36.90 ± 0.32 %	55.54 ± 0.98 %
BKS15	52.31 ± 0.83 %	47.69 ± 0.67 %	14.88 ± 0.76 %	29.91 ± 0.70 %	67.20 ± 0.78 %

OAA from PEP node (V5/V3). Strain BKS10 did not show carbon flux through glyoxylate shunt bypass (V11 = 0) in the stoichiometric model, indicating that deletion of *iclR* gene did not activate glyoxylate shunt bypass.

In strain BKS11, 45.94 % of PEP was converted to OAA (V5/V3), 2.4-fold higher than that of strain BKS10 ($p < 0.01$). As a result, the fraction of the metabolic flux to succinate (V16/V3) increased from 13.66 % in strain BKS10 to 55.54 % in strain BKS11 (Table 1) ($p < 0.01$). Meanwhile, strain BKS11 showed lower acetic fluxes ((V7 + V10)/V3). This indicated that co-overexpression of *pck-ecaA* could significantly enhanced the metabolic flux of PEP to OAA, and simultaneously inhibit other metabolic branches. Compared to strain BKS11, the fractions of the metabolic flux to lactate (V6/V3) and acetate ((V7 + V10)/V3) of strain BKS15 decreased by 12.32 % ($p < 0.05$) and 18.94 % ($p < 0.01$), respectively (Table 1), indicating that expression of anti-*pykF* attenuated the accumulation of lactate and acetate. At last, with a series of metabolic modifications, compared to strain BW25113(DE3), the final fraction of the metabolic flux to succinate in BKS15 was increased by 24.8 fold ($p < 0.01$) and those to lactate and acetate were decreased by 36.57 % ($p < 0.01$) and 61.28 % ($p < 0.01$), respectively.

Anaerobic batch fermentation for succinate production

To estimate the fermentation behaviors of engineered succinate-producing strains, anaerobic batch experiments were conducted. The titers, yields , specific productivities and productivities of succinate, lactate and acetate in 70 h fermentation were summarized in Table 2. As shown in Fig. 6, the distribution pattern of glucose metabolism and the production of succinate, lactate and acetate were remarkably changed. Strain BKS10 exhausted almost glucose, and accumulated large amounts of lactate and acetate, and a small amount of succinate in 70 h fermentation. Compared to strian BKS10, co-overexpression of *pck-ecaA* in strain BKS11 retarded glucose consumption, but achieved higher succinate production (25.51 mM), higher succinate yield (0.92 mol/mol glucose) and higher succinate specific productivity (3.96 mmol $gDCW^{-1}$ h^{-1}), increased by 1.9- ($p < 0.01$), 1.9- ($p < 0.01$) and 2.6-fold ($p < 0.01$), respectively. Moreover, the accumulation of lactate and acetate was significantly attenuated. When anti-*pykF* was further expressed in strain BKS15, glucose was completely consumed and largely distributed to succinate. Production of succinate in strain BKS15 was increased at a linear manner during the fermentation, and the specific productivity of succinate increased by 48.74 % ($p < 0.01$); the accumulation of acetate was greatly decreased, and the specific productivity of acetate decreased by 31.64 % ($p < 0.01$). Engineered strain BKS15 showed the optimal fermentation performance of higher productivity, titer and yield of succinate with the lower accumulation of lactate and acetate.

Conclusion

In this paper, PEP was selected as optimized target for increased succinate production and attenuated accumulation of byproducts in engineered *E. coli* under anaerobic conditions. By deleting *ptsG*, *pykA*, *ppc* and *maeAB* genes, we have designed and constructed initial succinate-producing *E. coli* strain. The succinate metabolic pathway was then enhanced with deletion of *sdh* and co-overexpression of *pck-ecaA*, resulting in succinate production of 25.51 mM. By introducing artificial sRNA of anti-*pykF*, the titer of succinate in the final optimized strain BKS15 was 30.12 mM with remarkable decrease in lactate and acetate. Metabolic flux analysis and fermentation kinetics showed that our optimization strategy could efficiently enhance the central carbon flux to succinate and decrease to byproducts. Recently, the progress in metabolic engineering suggested that limitation of cellular ATP supply and redox unbalance can be alleviated for improving succinate production in *E. coli* [41]. Combination of our strategies with those targets would further develop high succinate-producing microorganisms.

Methods

Bacterial strains and plasmids

E. coli DH5α was used for plasmids cloning and BW25113 was used as the wildtype strain for the construction of all engineered strains described in this study and succinate production. The *ecaA* gene was kindly donated by professor Jian-Min Xing, Chinese Academy of Sciences. Bacterial

Table 2 Parameters of succinate production by engineered *E. coli* strians during anaerobic fermentation

Strains	Growth rate (h^{-1})	Titer (mM)			Yield (mol/mol of glucose)			Specific productivity (mmol $gDCW^{-1}$ h^{-1})			Productivity (mmol L^{-1} h^{-1})		
		Succinate	Lactate	Acetate	Succinate	Lactate	Acetate	Succinate	Lactate	Acetate	Succinate	Lactate	Acetate
BKS10	0.071 ± 0.002	8.65 ± 0.73	12.06 ± 0.70	27.13 ± 2.56	0.31 ± 0.02	0.43 ± 0.02	0.98 ± 0.09	1.09 ± 0.06	1.47 ± 0.20	3.31 ± 0.13	0.12 ± 0.01	0.17 ± 0.01	0.39 ± 0.04
BKS11	0.052 ± 0.003**	25.51 ± 1.79**	7.82 ± 0.63**	23.52 ± 1.53	0.92 ± 0.06**	0.28 ± 0.02**	0.85 ± 0.05	3.96 ± 0.13**	1.18 ± 0.04*	3.54 ± 0.08	0.36 ± 0.03**	0.11 ± 0.01	0.34 ± 0.02
BKS15	0.043 ± 0.002*	30.12 ± 3.31	6.55 ± 0.33*	13.22 ± 1.64**	1.08 ± 0.11	0.24 ± 0.01*	0.48 ± 0.06**	5.89 ± 0.41**	1.20 ± 0.07	2.42 ± 0.19**	0.43 ± 0.05	0.09 ± 0.01	0.19 ± 0.02**

The data are shown as mean values ± standard deviation (SD) of three replicates. Asterisks indicate *p*-values (**$p < 0.01$, *$p < 0.05$) in which BKS11 was compared to BKS10 and BKS15 was compared to BKS11

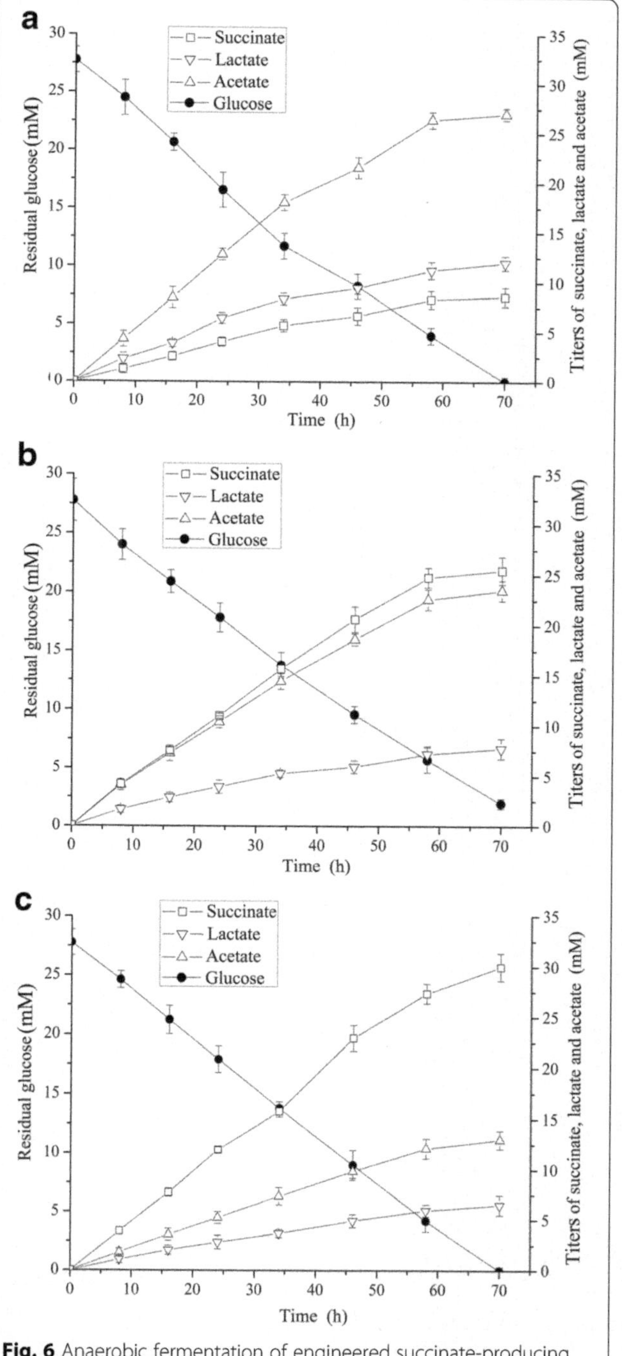

Fig. 6 Anaerobic fermentation of engineered succinate-producing strains. **a** BKS10, **b** BKS11, **c** BKS15

strains and plasmids used in this study were listed in Table 3.

Construction of engineered strains and plasmids

Restriction endonucleases and T4 DNA ligase were purchased from Thermo Scientific (USA), High-Fidelity DNA polymerase used for PCR amplification was purchased from Transgene Biotech (Beijing, China). Appropriate restriction sites were added to 5′and 3′ ends of the primers and all primers used in this study were listed in Additional file 1: Table S4. All plasmids was constructed through the enzymatic digestion of PCR products and plasmids with appropriate restriction sites, followed by the ligation of the appropriate fragments. Clones bearing inserted gene were screened by PCR and recombinant plasmids were confirmed by DNA sequencing.

By using the lambda Red recombinase system [42], the gene coding for T7 RNA polymerase was inserted into the genome of *E. coli* BW25113. The DNA fragment containing 500 bp upstream of the *ybhB* gene, T7 RNA polymerase gene, chloramphenicol resistance cassette and 500 bp downstream of the *ybhC* gene was constructed. The detailed procedure was shown in Additional file 1: Figure S1 and the primers used were shown in Additional file 1: Table S4. This DNA fragment was then electrotransformed into *E. coli* BW25113 which expressed lambda Red system for homologous recombination. The positive clones were confirmed with primers F-ybhB and R-ybhC. Next, the chloramphenicol resistance cassette was removed with the help of pCP20 and its removal was confirmed with primers F-ybhB and R-ybhC. The function of T7 RNA polymerase in BW25113 (DE3) was verified by SDS-PAGE of BW25113 (DE3) carrying pCDF-pck, using BL21 (DE3) harboring pCDF-pck and BW25113 harboring pCDF-pck as positive and negative controls, respectively (Additional file 1: Figure S2).

All in-frame gene deletion strains were constructed in *E. coli* BW25113 (DE3) according to the procedure described previously [42] and confirmed by PCR. Briefly, for deleting *ptsG* as example, the DNA fragment containing the chloramphenicol resistance cassette for homologous recombination was amplified by PCR using F-ptsG-Q and R-ptsG-Q as primers and the plasmid pKD3 as the template. The DNA fragment was then electrotransformed into *E. coli* BW25113 (DE3) which expressed lambda Red system for homologous recombination. The replacement of *ptsG* gene was confirmed by PCR using the primers F-ptsG and R-ptsG and the removal of chloramphenicol resistance was confirmed with primers F-ptsG and R-ptsG listed in Additional file 1: Table S4. The same procedure was performed for deletions of *pykA*, *ppc*, *maeA*, *maeB*, *sdh*, and *iclR*.

For construction of pRSF-anti-pykF and pBldg-anti-pykF, the complementary sequence that spans to + 24 nucleotides of *pykF* coding mRNA was used as the binding sequence and was designed in the primer. In order to construct pRSF-anti-pykF, the sequence between RBS and terminator was removed from pRSFDuet-1 using primers F-RSF and R-RSF, followed by the ligation, resulting in pRSFM1. The scaffold *micC* with 24 bp binging sequence at the 5′ end [40] was amplified with primers F-RSF-anti-pykF and R-RSF-anti-pykF and cloned into the *Spe*I site of

Table 3 *E. coli* strains and plasmids used in this study

Name	Characteristics	Source
Strains		
BW25113	$lacI^q rrnB_{T14}\Delta lacZ_{WJ16}hsdR514\Delta araBAD_{AH33}$	NBRP-*E. coli* at NIG
BW25113(DE3)	$lacI^q rrnB_{T14}\Delta lacZ_{WJ16}hsdR514\Delta araBAD_{AH33}$dcm (DE3)	This study
BKS1	BW25113(DE3) harboring pCDF-pck	This study
BKS2	BW25113 harboring pCDF-pck	This study
BKS3	BL21(DE3) harboring pCDF-pck	This study
BKS4	BW25113(DE3) Δ*ptsG*::FRT	This study
BKS5	BKS4 Δ*pykA*::FRT	This study
BKS6	BKS5 Δ*ppc*::FRT	This study
BKS7	BKS6 Δ*maeA*::FRT	This study
BKS8	BKS7 Δ*maeB*::FRT	This study
BKS9	BKS8 Δ*sdh*::FRT	This study
BKS10	BKS9 Δ*iclR*::FRT	This study
BKS11	BKS10 harboring pCDF-pck-ecaA	This study
BKS12	BKS8 harboring pCDF-pck	This study
BKS13	BKS8 harboring pCDF-pck and pRSF-anti-pykF	This study
BKS14	BKS8 harboring pCDF-pck and pBldg-anti-pykF	This study
BKS15	BKS11 harboring pBldg-anti-pykF	This study
Plasmids		
pKD3	FRT(FLP recognition target) sites; Cm^R	(Datsenko and Wanner 2000)
pKD46	Red recombinase expression vector; Amp^R	(Datsenko and Wanner 2000)
pCP20	FLP expression vector; Amp^R,Cm^R	(Datsenko and Wanner 2000)
pETDuet-1	pBR322 ori with P_{T7}; Amp^R	Novagen
pCDFDuet-1	CDF ori with P_{T7}; Str^R	Novagen
pRSFDuet-1	RSF ori with P_{T7}; Kan^R	Novagen
pBldgbrick2	p15A ori with P_{lacUV5}; Cm^R	(Yao et al, 2013)
pCDF-pck	pCDFDute-1 with *pck*	This study
pCDF-pck-ecaA	pCDFDuet-1 with *pck* and *ecaA*	This study
pRSFM1	pRSF without RBS sequence	This study
pRSF-anti-pykF	pRSFM1 with anti-*pykF*	This study
pBldg-anti-pykF	pBldgbrick2 with anti-*pykF*	This study

pRSFM1 (high-copy-number plasmid), and resulting in plasmid pRSF-anti-pykF. The correct construct pRSF-anti-pykF was screened by PCR using primers ACYC-DuetUP1 and R-RSF-anti-pykF, and confirmed DNA sequencing. The DNA fragment containing 24 bp binding sequence and *micC* was amplified by PCR with primers F-Bldg-anti-pykF and R-Bldg-anti-pykF listed in Additional file 1: Table S4 using *E. coli* BW25113 genome as template. Then, PCR product was cloned into vector pBldgbrick2 (low-copy-number plasmid) [43] between *Hind*III and *Nco*I, resulting plasmid pBldg-anti-pykF. The plasmids with anti-*pykF* sequence were used to silence the expression of *pykF* gene.

Fermentation conditions

Dual phase fermentation mode was employed [38]. For all engineered *E. coli* strains, a seed inoculum of 500 μL from an overnight 3 mL of LB culture was first inoculated at 37 °C in 250 mL shake flask containing 100 mL of liquid LB medium for aerobic growth. When the optical density (OD) reached 1.0, cells were induced with a final concentration of 0.1 mM isopropyl-β-D-thiogalactopyranoside (IPTG) and grown for another 3 h for recombinant protein expression. Then, bacterial cells were collected by centrifugation and resuspended in 150 mL shake flask containing 100 mL of fresh YM9 medium (1*M9 salts, 1 g/L yeast extract) at an initial OD of 1.0 for anaerobic fermentation. At that point, 5 g/L $CaCO_3$, 2 g/L $NaHCO_3$, 0.1 mM IPTG were added. Flasks were sealed with non-ventilated plugs. The cells were incubated at 37 °C on a shaker (150 rpm) and sample were collected at 40 h for analysis. For kinetic study, samples were collected at 0, 8, 16, 24, 34, 46, 58 and 70 h. Appropriate amounts of antibiotics (50 mg/L ampicillin, 30 mg/L streptomycin, 30 mg/L kanamycin) were added to media when needed.

Analytical techniques

Cell growth was monitored by measuring the optical density (OD) at 600 nm (UV-vis spectrophotometer) and was transformed into dry cell weight using the coefficient as: dry cell mass (g L^{-1}) = 0.48*OD_{600} [44]. The concentration of glucose was measured using SBA-90B biosensor (Biology Institute of ShanDong Academy of Science, China). The sample was centrifuged and the supernatant of fermentation sample was filtered through 0.2 μm syringe filter and metabolites were analyzed using an Waters 1515 differential HPLC system equipped with a Bio-Red HPX-87H HPLC column. 10 μL of sample was injected into the HPLC at column temperature of 65 °C and ran isocratically with 5 mM H_2SO_4 as mobile phase sat on a flow rate of 0.6 ml/min.

Metabolic flux analysis

The metabolic network was constructed based on engineered pathways in anaerobically grown *E. coli* (Fig. 1). This network included glycolysis, TCA cycle and glyoxylate bypass (Fig. 5). As an attempt to analyze the distribution of carbon source, the fluxes through each pathway in the metabolic network were designated by V1-V16. The simplified central metabolic reactions were described in detail in Additional file 1: Table S1. According to the law of mass conversation and the quasi-steady-state assumption, these metabolic flux relationships were constructed to simplify the computational process, and shown in Additional file 1: Table S2, in which V1, V6, V16, and V7 + V10 were measurable quantities while the others were the metabolic fluxes of the corresponding intermediates. In this study, Lingo software [45] was used to obtain the solutions to distribution of metabolic fluxes that were limited by the formulas in Additional file 1: Table S2.

Statistical analysis

The data are shown as mean values ± standard deviation (SD) of three replicates. The Student's *t* test was used for all statistical analysis using SPSS 17.0. The *p*-value of < 0.05 and < 0.01 was considered statistically significant, more significant, respectively.

Additional file

Additional file 1: Figure S1. Assembly of DNA fragment for inserting the T7 RNA polymerase into BW25113 genome. **Figure S2.** SDS-PAGE of protein expression of strain BKS1, BKS2 and BKS3. **Table S1.** The simplified central metabolic reaction of engineered *E. coli* anaerobically grown in glucose. **Table S2.** Stoichiometric relationships for fluxes of metabolic reactions in anaerobic growth of *E. coli*. **Table S3.** Metabolic fluxes (mM $gDCW^{-1}$ h^{-1}) of engineered strains based on the anaerobic fermentation results at 40 h. **Table S4.** Primers used in this study.

Abbreviations

ATP, adenosine triphosphate; G3P, Glyceraldehyde 3-P; Glf, glucose facilitator; IPTG, isopropyl-β-D-thiogalactopyranoside; NADH, Nicotinamide adenine dinucleotide; OAA, oxaloacetate; PCK, PEP carboxykinase; PEP, phosphoenolpyruvate; PPC, PEP carboxylase; PTS, PEP carbohydrate phosphotransferase system; PYC, pyruvate carboxylase; PYR, pyruvate; SDH, succinate dehydrogenase

Acknowledgements

We acknowledge Professor Jian-Min Xing (Chinese Academy of Sciences) for providing *ecaA* gene. We also thank Associate Professor Tao Chen and Dr. Zhiwen Wang (Tianjin University) for their kind suggestions to this manuscript.

Funding

This work was supported by the National Basic Research Program of China (2011CBA00800), the National High-Tech R&D Program of China (2012AA02A701), the National Natural Science Foundation of China (31570087), and the Natural Science Foundation of Tianjin (13JCZDJC27600).

Authors' contributions

GRZ, YZ and CSW conceived method and designed experiment; YZ, CSW, FFL and ZNL performed experiment; GRZ, YZ and CSW analyzed the data; YZ analyzed metabolic fluxes; YZ wrote the manuscript with help by GRZ. All authors have read and approved the final manuscript.

Competing interests

The authors declare that they have no competing interests.

Author details

[1]Department of Pharmaceutical Engineering, School of Chemical Engineering and Technology, Tianjin University, Tianjin 300072, China. [2]Key Laboratory of Systems Bioengineering, Ministry of Education, Tianjin 300072, China. [3]SynBio Research Platform, Collaborative Innovation Center of Chemical Science and Engineering, Tianjin 300072, China. [4]Present address: PPG Coating (Tianjin) Co., Ltd. Tianjin Economic Technological Development Area (TEDA), 192 Huanghai Road, Tianjin 300457, China.

References

1. Jang YS, Kim B, Shin JH, Choi YJ, Choi S, Song CW, et al. Bio-based production of C2–C6 platform chemicals. Biotechnol Bioeng. 2012;109(10):2437–59.
2. Becker J, Lange A, Fabarius J, Wittmann C. Top value platform chemicals: bio-based production of organic acids. Curr Opin Biotechnol. 2015;36:168–75.
3. Cheng KK, Zhao XB, Zeng J, Zhang JA. Biotechnological production of succinic acid: current state and perspectives. Biofuels Bioprod Bioref. 2012;6(3):302–18.

4. Vuoristo KS, Mars AE, Sanders JPM, Eggink G, Weusthuis RA. Metabolic engineering of TCA cycle for production of chemicals. Trends Biotechnol. 2015;34(3):191–7.
5. Chen X, Zhou L, Tian K, Kumar A, Singh S, Prior BA, et al. Metabolic engineering of *Escherichia coli*: a sustainable industrial platform for bio-based chemical production. Biotechnol Adv. 2013;31(8):1200–23.
6. Clark DP. The fermentation pathways of *Escherichia coli*. FEMS Microbiol Rev. 1989;5(3):223–34.
7. Bunch PK, Mat-Jan F, Lee N, Clark DP. The *ldhA* gene encoding the fermentative lactate dehydrogenase of *Escherichia coli*. Microbiology. 1997; 143(1):187–95.
8. Donnelly MI, Millard CS, Clark DP, Chen MJ, Rathke JW. A novel fermentation pathway in an *Escherichia coli* mutant producing succinic acid, acetic acid, and ethanol. Appl Biochem Biotechnol. 1998;70-72:187–98.
9. Jantama K, Haupt MJ, Svoronos SA, Zhang XL, Moore JC, Shanmugam KT, et al. Combining metabolic engineering and metabolic evolution to develop nonrecombinant strains of *Escherichia coli* C that produce succinate and malate. Biotechnol Bioeng. 2008;99(5):1140–53.
10. Vemuri GN, Eiteman MA, Altman E. Effects of growth mode and pyruvate carboxylase on succinic acid production by metabolically engineered strains of *Escherichia coli*. Appl Environ Microbiol. 2002;68(4):1715–27.
11. Sanchez AM, Bennett GN, San KY. Novel pathway engineering design of the anaerobic central metabolic pathway in *Escherichia coli* to increase succinate yield and productivity. Metab Eng. 2005;7:229–39.
12. Thakker C, Zhu JF, San KY, Bennett G. Heterologous *pyc* gene expression under various natural and engineered promoters in *Escherichia coli* for improved succinate production. J Biotechnol. 2011;155(2):236–43.
13. Singh A, Soh KC, Hatzimanikatis V, Gill RT. Manipulating redox and ATP balancing for improved production of succinate in *E. coli*. Metab Eng. 2011; 13(1):76–81.
14. Kim P, Laivenieks M, Vieille C, Zeikus JG. Effect of overexpression of *Actinobacillus succinogenes* phosphoenolpyruvate carboxykinase on succinate production in *Escherichia coli*. Appl Environ Microbiol. 2004;70(2):1238–41.
15. Wang D, Li Q, Mao Y, Xing JM, Su ZG. High-level succinic acid production and yield by lactose-induced expression of phosphoenolpyruvate carboxylase in *ptsG* mutant *Escherichia coli*. Appl Microbiol Biotechnol. 2010;87(6):2025–35.
16. Tan Z, Zhu X, Chen J, Li Q, Zhang X. Activating phosphoenolpyruvate carboxylase and phosphoenolpyruvate carboxykinase in combination for improvement of succinate production. Appl Environ Microbiol. 2013;79(16):4838–44.
17. Singh A, Lynch MD, Gill RT. Genes restoring redox balance in fermentation-deficient *E. coli* NZN111. Metab Eng. 2009;11:347–54.
18. Balzer GJ, Thakker C, Bennett GN, San KY. Metabolic engineering of *Escherichia coli* to minimize byproduct formate and improving succinate productivity through increasing NADH availability by heterologous expression of NAD $^{+}$-dependent formate dehydrogenase. Metab Eng. 2013;20(5):1–8.
19. Ma JF, Gou DM, Liang LY, Liu RM, Chen X, Zhang CQ, et al. Enhancement of succinate production by metabolically engineered *Escherichia coli* with co-expression of nicotinic acid phosphoribosyltransferase and pyruvate carboxylase. Appl Microbiol Biotechnol. 2013;97(15):6739–47.
20. Zhu X, Tan Z, Xu H, Chen J, Tang J, Zhang X. Metabolic evolution of two reducing equivalent-conserving pathways for high-yield succinate production in *Escherichia coli*. Metab Eng. 2014;24:87–96.
21. Lu J, Tang J, Liu Y, Zhu X, Zhang T, Zhang X. Combinatorial modulation of *galP* and *glk* gene expression for improved alternative glucose utilization. Appl Microbiol Biotechnol. 2012;93(6):2455–62.
22. Tang J, Zhu X, Lu J, Liu P, Xu H, Tan Z, et al. Recruiting alternative glucose utilization pathways for improving succinate production. Appl Microbiol Biotechnol. 2013;97(6):2513–20.
23. Beauprez JJ, Foulquié-Moreno MR, Maertens J, Horen EV, Bekers K, Baart GJE, et al. Influence of C4-dicarboxylic acid transporters on succinate production. Green Chem. 2011;13(8):2179–86.
24. Chen J, Zhu XN, Tan ZG, Xu HT, Tang JL, Xiao DG, et al. Activating C4-dicarboxylate transporters DcuB and DcuC for improving succinate production. Appl Microbiol Biotechnol. 2014;98(5):1–9.
25. Chatterjee R, Millard CS, Champion K, Clark DP, Donnelly MI. Mutation of the *ptsG* gene results in increased production of succinate in fermentation of glucose by *Escherichia coli*. Appl Environ Microbiol. 2001;67(1):148–54.
26. Wang Q, Wu C, Chen T, Chen X, Zhao XM. Expression of galactose permease and pyruvate carboxylase in *Escherichia coli ptsG* mutant increases the growth rate and succinate yield under anaerobic conditions. Biotechnol Lett. 2006;28(2):89–93.
27. Zhang X, Jantama K, Moore JC, Jarboe LR, Shanmugam KT, Ingram LO. Metabolic evolution of energy-conserving pathways for succinate production in *Escherichia coli*. Proc Natl Acad Sci U S A. 2009;106(48):20180–5.
28. Stols L, Donnelly MI. Production of succinic acid through overexpression of NAD^{+}-dependent malic enzyme in an *Escherichia coli* mutant. Appl Environ Microbiol. 1997;63(7):2695–701.
29. Noda S, Shirai T, Oyama S, Kondo A. Metabolic design of a platform *Escherichia coli* strain producing various chorismate derivatives. Metab Eng. 2016;33:119–29.
30. Park SJ, Tseng CP, Gunsalus RP. Regulation of succinate dehydrogenase (*sdhCDAB*) operon expression in *Escherichia coli* in response to carbon supply and anaerobiosis: role of ArcA and Fnr. Mol Microbiol. 1995;15(3):473–82.
31. Li N, Zhang B, Chen T, Wang ZW, Tang YJ, Zhao XM. Directed pathway evolution of the glyoxylate shunt in *Escherichia coli* for improved aerobic succinate production from glycerol. J Ind Microbiol Biotechnol. 2013;40(12):1461–75.
32. Lin H, Bennett GN, San KY. Genetic reconstruction of the aerobic central metabolism in *Escherichia coli* for the absolute aerobic production of succinate. Biotechnol Bioeng. 2005;89(2):148–56.
33. Litsanov B, Kabus A, Brocker M, Bott M. Efficient aerobic succinate production from glucose in minimal medium with *Corynebacterium glutamicum*. Microb Biotechnol. 2012;5(1):116–28.
34. Waegeman H, Beauprez J, Moens H, Maertens J, Mey MD, Foulquié-Moreno WR, et al. Effect of *iclR* and *arcA* knockouts on biomass formation and metabolic fluxes in *Escherichia coli* K12 and its implications on understanding the metabolism of *Escherichia coli* BL21 (DE3). BMC Microbiol. 2011;11(15):70.
35. Shimizu K. Metabolic regulation of a bacterial cell system with emphasis on *Escherichia coli* metabolism. ISRN Biochem. 2013;6.
36. Skorokhodova AY, Gulevich AY, Morzhakova AA, Shakulov RS, Debabov VG. Comparison of different approaches to activate the glyoxylate bypass in *Escherichia coli* K-12 for succinate biosynthesis during dual-phase fermentation in minimal glucose media. Biotechnol Lett. 2013;35(4):577–83.
37. Kai Y, Matsumura H, Izui K. Phosphoenolpyruvate carboxylase: three-dimensional structure and molecular mechanisms. Arch Biochem Biophys. 2003;414(2):170–9.
38. Wang D, Li Q, Li WL, Xing JM, Su ZG. Improvement of succinate production by overexpression of a cyanobacterial carbonic anhydrase in *Escherichia coli*. Enzyme Microb Technol. 2009;45(6):491–7.
39. Wang J, Qin DD, Zhang BY, Li Q, Li S, Zhou XH, et al. Fine-tuning of *ecaA* and *pepc* gene expression increases succinic acid production in *Escherichia coli*. Appl Microbiol Biotechnol. 2015;99(20):8575–86.
40. Na D, Yoo SM, Chung H, Park H, Park JH, Lee SY. Metabolic engineering of *E. coli* using synthetic small regulatory RNAs. Nat Biotechnol. 2013;31(2):170–4.
41. Thakker C, Martínez I, San KY, Bennett GN. Succinate production in *Escherichia coli*. Biotechnol J. 2012;7(2):213–24.
42. Datsenko KA, Wanner BL. One-step inactivation of chromosomal genes in *Escherichia coli* K-12 using PCR products. Proc Natl Acad Sci U S A. 2000;97:6640–5.
43. Yao YF, Wang CS, Qiao J, Zhao GR. Metabolic engineering of *Escherichia coli* for production of salvianic acid A via an artificial biosynthetic pathway. Metab Eng. 2013;19(5):79–87.
44. Gokarn RR, Eiteman MA, Altman E. Expression of pyruvate carboxylase enhances succinate production in *Escherichia coli* without affecting glucose uptake. Biotechnol Lett. 1998;20(8):795–8.
45. Li XJ, Chen T, Chen X, Zhao XM. Redirection electron flow to high coupling efficiency of terminal oxidase to enhance riboflavin biosynthesis. Appl Microbiol Biotechnol. 2006;73(2):374–83.

7

Highly efficient mesophyll protoplast isolation and PEG-mediated transient gene expression for rapid and large-scale gene characterization in cassava (*Manihot esculenta* Crantz)

Jun-Zheng Wu[1], Qin Liu[1], Xiao-Shan Geng[1], Kai-Mian Li[2], Li-Juan Luo[1*] and Jin-Ping Liu[1*]

Abstract

Background: Cassava (*Manihot esculenta* Crantz) is a major crop extensively cultivated in the tropics as both an important source of calories and a promising source for biofuel production. Although stable gene expression have been used for transgenic breeding and gene function study, a quick, easy and large-scale transformation platform has been in urgent need for gene functional characterization, especially after the cassava full genome was sequenced.

Methods: Fully expanded leaves from in vitro plantlets of *Manihot esculenta* were used to optimize the concentrations of cellulase R-10 and macerozyme R-10 for obtaining protoplasts with the highest yield and viability. Then, the optimum conditions (PEG4000 concentration and transfection time) were determined for cassava protoplast transient gene expression. In addition, the reliability of the established protocol was confirmed for subcellular protein localization.

Results: In this work we optimized the main influencing factors and developed an efficient mesophyll protoplast isolation and PEG-mediated transient gene expression in cassava. The suitable enzyme digestion system was established with the combination of 1.6% cellulase R-10 and 0.8% macerozyme R-10 for 16 h of digestion in the dark at 25 °C, resulting in the high yield (4.4×10^7 protoplasts/g FW) and vitality (92.6%) of mesophyll protoplasts. The maximum transfection efficiency (70.8%) was obtained with the incubation of the protoplasts/vector DNA mixture with 25% PEG4000 for 10 min. We validated the applicability of the system for studying the subcellular localization of MeSTP7 (an H^+/monosaccharide cotransporter) with our transient expression protocol and a heterologous *Arabidopsis* transient gene expression system.

Conclusion: We optimized the main influencing factors and developed an efficient mesophyll protoplast isolation and transient gene expression in cassava, which will facilitate large-scale characterization of genes and pathways in cassava.

Keywords: Cassava, *Manihot esculenta*, Protoplast isolation, Transient expression system, Subcellular localization

Background

Transient gene expression in plant protoplasts is a powerful technique for studying subcellular localization of proteins, gene and promoter activities, protein-protein interactions, and signal transduction [1–7]. Compared with the stable gene expression in transgenic plants, transient gene expression represents a fast, convenient and efficient alternative system with higher expression levels [8–10]. Moreover, the transient gene expression assay enable the high-throughput analysis of gene functions while the stable gene expression is relatively expensive and time-consuming thus limiting the utilization of this technique for large-scale analyses of plant genes [8–10]. Although different transfection techniques such as polyethylene glycol (PEG)-mediated, electroporation and microinjection have been developed to deliver recombinant DNA plasmids into protoplasts, utilization of the PEG-mediated approach has high transformation efficiency, so it is widely applied in

* Correspondence: Luoljd@126.com; liu3305602@163.com
[1]Hainan Key Laboratory for Sustainable Utilization of Tropical Bioresources, College of Agriculture, Hainan University, Haikou, Hainan Province 570228, China
Full list of author information is available at the end of the article

molecular and cellular studies for both model plants and non-model plants [4, 7, 8, 11–20].

Cassava (*Manihot esculenta* Crantz) is a perennial and woody shrub of the Euphorbiaceae cultivated in tropical and subtropical regions for its starchy storage roots [21, 22]. These roots have been used as the sources for dietary carbohydrate, starch processing and potential biofuel production. Since that cassava is a staple crop for approximately 800 million people in developing regions of the tropics [23], many groups have been intensively performing research on cassava molecular breeding which relies on the identification of agronomically important genes and pathways. Recently, the cassava sequencing data set has been publicly released [24] and the determination of gene function has been a major goal of cassava molecular biology in genomic and post-genomic era, thus it will necessitate efficient high-throughput transient gene expression for gene function analysis.

Stable gene transfer can be developed for gene function studies at whole plant level. Much progress has been made in cassava genetic transformation of both *Agrobacterium tumefaciens*-mediated and biolistic-mediated system [23, 25–27]. However, the transformation efficiency is relatively low and genotypic dependent [28], and the process requires well-trained tissue culture specialists, is lengthy and difficult to repeat [29, 30]. Hence, the development of an efficient transient gene expression system that could be exploited to characterize gene functions and investigate molecular processes in cassava is required.

Due to the fact that cassava is recalcitrant for plant regeneration from protoplasts, only two reports about the protoplast culture and regeneration of cassava have been published. Shahin and Shephard reported shoot regeneration from the isolated mesophyll protoplasts of cassava [31] and Sofiari et al. reported the plant regeneration from protoplasts isolated from friable embryogenic callus of cassava [32]. Anthony et al. developed a protocol for the culture of cassava leaf protoplasts but plant regeneration from protoplasts had not achieved [33]. To date, transient gene expression in cassava protoplasts has not been developed. Here we report an effective and reliable mesophyll protoplast isolation and PEG-mediated transient gene expression in cassava. The protocol has been successfully used for subcellular localization of a sugar transporter, MeSTP7 protein.

Methods

Materials

In vitro plantlets of *Manihot esculenta* cv. South China 8 (SC8) were used in this study. Cultures were kept on 1/2MS (Murashige and Skoog) medium (supplemented with 2% sucrose and 0.8% agarose, pH5.8) at 28 °C, under lighting with a cycle of 12 h/8 h (light/darkness) for 6-10 weeks to obtain fully expanded leaves.

Protoplast isolation

Protoplast isolation was conducted using the protocols described by Yoo et al. [3] and Anthony et al. [33] with slight modification. The fully expanded green leaves were cut into about 0.5-1.0 mm strips with sharp razors. The strips of 0.3 ± 0.03 g (for each treatment) were transferred quickly into 10 mL of enzyme solution (0.1% BSA, 9% mannitol, 20 mM KCl, 10 mM $CaCl_2$, 20 mM MES, pH 5.8) with different concentrations of cellulase R-10 (0.8%, 1.6% or 2.4%) (Yakult, Japan) and macerozyme R-10 (0.4%, 0.8% or 1.2%) (Yakult, Japan) for cell wall hydrolysis by shaking at 45 rpm for 16 h in the dark (25 °C) for plasmolysis.

The digested tissues were filtered through a 0.75-mm nylon sieve, collected by centrifugation (80 *g* for 3 min) and suspended in 20 mL pre-cooled W5 solution (154 mM NaCl, 125 mM $CaCl_2$, 5 mM KCl and 2 mM MES, pH 5.8). The protoplast pellet was purified twice in W5 solution by repeated resuspension and centrifugation (80 *g* for 3 min).

The resulting protoplasts were resuspended in 2 mL W5 solution, placed on the ice for 30 min and counted under a microscope equipped with a hemocytometer. The viability of protoplasts was measured with 0.2% fluorescein diacetate (FDA) staining and determined as follows: protoplast viability (%) = (fluorescent protoplast number in view/protoplast total number in view) × 100%.

Cassava mesophyll protoplast transformation

The protoplasts were collected by centrifugation (80 *g* for 2 min) and resuspended in MMg solution (9% mannitol, 15 mM $MgCl_2$, 4 mM MES) to a density of 1.0×10^7 protoplasts/ml. 15 μg of pA7-GFP plasmid [34] were mixed gently with 200 μl protoplasts. Then different concentrations of freshly prepared PEG4000 solutions (9% mannitol, 100 mM $CaCl_2$) were immediately added and mixed by gentle inversion to obtain a 15%, 20%, 25% or 30% final PEG4000 concentration. The mixture was incubated at room temperature for 5, 10, 15 and 20 min, respectively. After incubation, the transfection mixture was gently diluted with three volumes of W5 solution.

The transfected protoplasts were centrifuged (80 *g*, 2 min) twice, then resuspended with 300-500 μl WI solution (4 mM MES, 9% mannitol, 20 mM KCl, pH 5.8) and incubated overnight at room temperature in dark for the induction of protein expression. Transformation efficiency was detected under fluorescence microscopy and the expression of GFP tag was observed by confocal laser scanning microscope. Transformation efficiency was calculated as follows: transformation efficiency (%) = (bright green fluorescent protoplast number in view/ total protoplast number in view).

Plasmid construction and subcellular localization of MeSTP7 protein

Total RNA was extracted from cassava leaves with RNAplant Plus (Tiangen biotech, Beijing, China) and cDNA was synthesized using the PrimeScript RT reagent Kit with gDNA Eraser (TaKaRa, Dalian, China) following the manufacturer's instructions.

The specific primers with *Xho* I and *Spe* I restriction enzyme sites (bold) introduced to their ends respectively, STP7-F (5'-CCG**CTCGAG**ATGCCTGCAGGAGGTTT-3') and STP7-R (5'-GG**ACTAGT**GAAAACTGGGTAACAGGATCAA-3'), were designed according to the conserved region of *MeSTP7* (GenBank accession number: Manes.03G180400). The gene of interest was amplified using Phusion DNA Polymerase under the following conditions: 98 °C for 30 s followed by 35 cycles of amplification (98 °C for 10 s, 57 °C for 30 s, 72 °C for 45 s) and a 10-min final extension at 72 °C.

The PCR products were confirmed by sequencing and inserted between the 35S promoter and the *GFP* of the expression vector *pA7-GFP*, designated as *pA7-MeSTP7-GFP* (Fig. 1). The plasmid construct was propagated in competent *Escherichia coli* Trans-T1 (TransGen Biotech, Beijing, China). Subcellular localizations of MeSTP7 protein in cassava and *Arabidopsis* mesophyll protoplasts were carried out according to our transformation procedure described above and the method established by Yoo et al [3], respectively.

Results

Isolation of protoplasts from cassava leaves

Different enzyme combinations significantly influenced the yield and viability of cassava protoplasts. Protoplast yield increased with increasing concentrations of cellulase R-10 and macerozyme R-10, but protoplast viability decreased with increasing enzyme concentrations. It was found that the highest production of protoplasts (5.1×10^7 protoplasts/ g FW) was obtained with 2.4% cellulose and 1.2% macrerozyme, with the lowest viability of protoplasts (83.8%). For obtaining the highest number viable protoplasts for the subsequent transfection experiments, the optimum combination of enzymes for cassava protoplast isolation was determined as 1.6% cellulase + 0.8% macerozyme. This combination gave the yield of 4.4×10^7 protoplasts/g FW and the viability of 92.6%. The enzyme solution turned green, indicating the release of mesophyll protoplasts. Leaf mesophyll protoplasts were uniformly spherical, and containing single intact nucleus and green chloroplasts. (Table 1, Fig. 2).

Establishment of cassava mesophyll protoplast expression system

In this study, the effects of PEG4000 concentration and transfection duration on cassava protoplast transformation efficiency was investigated with the transient expression vector *pA7-GFP*. The results showed that when the transfection time was 10 min, the transfection efficiency increased along the increasing concentration of PEG4000 and reached the maximum level (70.8%) at the concentration of 25% (Figs. 3 and 4a). When the concentration of PEG4000 was 25%, transfection efficiency increased with increases in the transfection time until it peaked at 10 min (Figs. 3 and 4b). Therefore, factors affecting transfection efficiency were optimized: 25% of PEG4000 concentration and 10 min of transfection time. In addition, stability of transient expression of GFP has been verified as GFP signal could be detected 8-30 h after transformation.

Subcellular localization of GFP-fused MeSTP1 protein in cassava mesophyll protoplasts

To test the feasibility of the protoplast transient expression system for the subcellular localization of protein in cassava cells, the transient expression of GFP-fused MeSTP7 protein in cassava mesophyll protoplasts was performed. Twelve-eighteen hours after transfection, brilliant green fluorescence distributed in intracellular compartments was observed. MeSTP7:GFP was specifically detected in cell membrane, but empty vector control (GFP) was distributed throughout the nucleus and cytoplasm (Fig. 5). Similar pattern of MeSTP7 subcellular localization in *Arabidopsis* mesophyll protoplasts validated the authenticity of the results described above (Fig. 6).

Discussion

A reliable protoplast isolation with high yield and viability is prerequisite for the success of transient gene

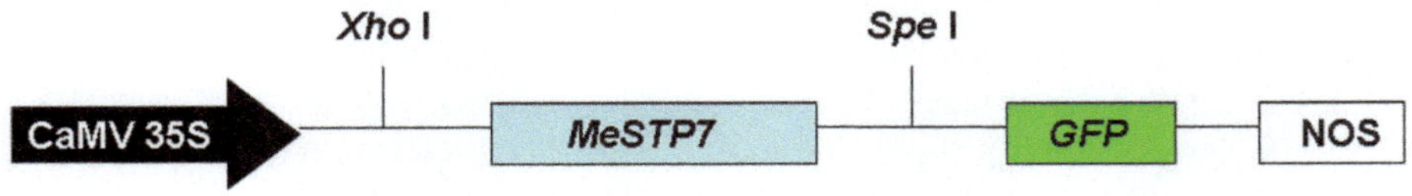

Fig. 1 Schematic representation of the *pA7-MeSTP7-GFP* construct driven by the constitutive CaMV 35S promoter and terminated by nopaline synthase terminator (NOS)

Table 1 Effects of enzyme combinations on the yield and viability of protoplasts isolated from leaves of cassava

Cellulase R-10 (*W/V*)	Macerozyme R-10 (*W/V*)	Protoplast yield (10^7#/g FW)	Protoplast viability (%)
0.8	0.4	1.08 ± 0.14f	93.64 ± 1.00a
0.8	0.8	1.45 ± 0.09ef	93.05 ± 3.00a
0.8	1.2	1.89 ± 0.14e	91.32 ± 1.00a
1.6	0.4	3.08 ± 0.14d	93.43 ± 2.00a
1.6	0.8	4.42 ± 0.29b	92.60 ± 2.00a
1.6	1.2	3.25 ± 0.39 cd	90.14 ± 3.00a
2.4	0.4	1.80 ± 0.15e	88.38 ± 4.00ab
2.4	0.8	3.66 ± 0.18c	87.67 ± 1.00ab
2.4	1.2	5.06 ± 0.25a	83.82 ± 5.00b

Note: The averages of three technical replicates ± standard error of the mean (SEM) are shown. The different letters indicate significant differences ($P \leq 0.05$) according to the Duncan test

expression. Source tissue, the type and concentration of enzymes are critical factors affecting the release of protoplasts [3, 35, 36]. Leaves from field or greenhouse-grown plants usually have a thick cuticle and recalcitrant cell walls, which are likely to be the major limiting factors for protoplast isolation [12]. In cassava, Shahin et al. [31] used well expanded leaves which were collected from preconditioned plants (in growth chamber) and then incubated on a solution supplemented with naphthaleneacetic acid (NAA) and benzylaminopurine (BAP) as protoplast source. Sofiari et al. [32] isolated cassava protoplasts from friable embryogenic callus (FEC) and from suspensions derived from FEC. Anthony et al. [33] took juvenile leaves from axenic shoot cultures of cassava as protoplast source. For convenience sake, we chose the leaves from *in vitro* propagated cassava plants as the protoplast source.

For a given species and source tissue, suitable enzymolysis time and enzymolysis composition are vital for protoplast isolation. Anthony et al. [33] had established an improved protocol for cassava protoplast isolation with an enzyme mixture for 16 h. Referring to the method described by Anthony et al. [33], optimization of the enzymolysis composition was conducted in this study. The combinations of cellulase and macerozyme which have been commonly used for protoplast isolation, were evaluated to obtain protoplasts with the highest yield and viability. As the concentration of total enzyme in the mixture increased, protoplast yield seemed to increase but protoplast viability to decrease. High levels of cellulose R-10 and macrerozyme R-10 increased protoplast yield but decreased the viability of protoplasts, possibly due to the influence of enzymes on the integrity of membrane and the physiological activities of protoplasts [37, 38]. The highest yield (5.1×10^7 protoplasts/ g FW) and the lowest viability of protoplasts (83.8%) were recorded with the 2.4% cellulose and 1.2% macrerozyme, which was the most concentrated enzyme mixture in our study. This lowest viability of protoplasts may not be suitable for subsequent experiments. However, the lowest concentrations of enzymes in the isolation mixture gave a lower yield (1.8×10^7 protoplasts/g FW) but with a slightly higher viability of protoplasts (93.6%). The enzyme mixture containing 1.6% Cellulase R10 and 0.8% macerozyme R10 was the best mixture in terms of both yield (4.4×10^7 protoplasts/g FW) and viability (92.6%) of protoplasts. The maximum yield and viability of leaf mesophyll protoplasts achieved in our study were substantially higher than the results previously reported [28, 30]. The reason possibly attributed to the differences in tissue source used [31, 32], and/or the type and concentration of enzymes [32, 33]. In addition, the high yield (4.4×10^7 protoplasts/g FW) and vitality (92.6%) of cassava mesophyll protoplasts obtained in our study were comparable to the values reported for *Populus* (1×10^7 protoplasts/g FW, more than 90%) [13], *Brachypodium distachyon* (1.7×10^7 protoplasts/g FW, more than 90%) [19], cucumber ($6\text{-}7 \times 10^7$ protoplasts/g FW, about 90%) [15]. The high yield and viability allowed the application of this system for the subsequent transfection experiment.

Although heterologous transient gene expression systems with *Arabidopsis* and tobacco protoplasts are commonly adopted for studying gene function in non-model plants such as cassava, these heterologous systems may display aberrant results due to a foreign genetic background. In this study, we established an efficient cassava protoplast transfection system by optimizing PEG4000 concentration and transfection time. This protocol will be an extraordinarily valuable tool for cassava gene functional study. In addition, transfection efficiency over 50% is generally required for obtaining reliable and reproducible results with the protoplast system [3]. The highest transfection efficiency (70.84%) with strong GFP signal was obtained with 25% of PEG4000 concentration for 10 min of transfection time. This result demonstrated that the transfection efficiency of cassava protoplasts was

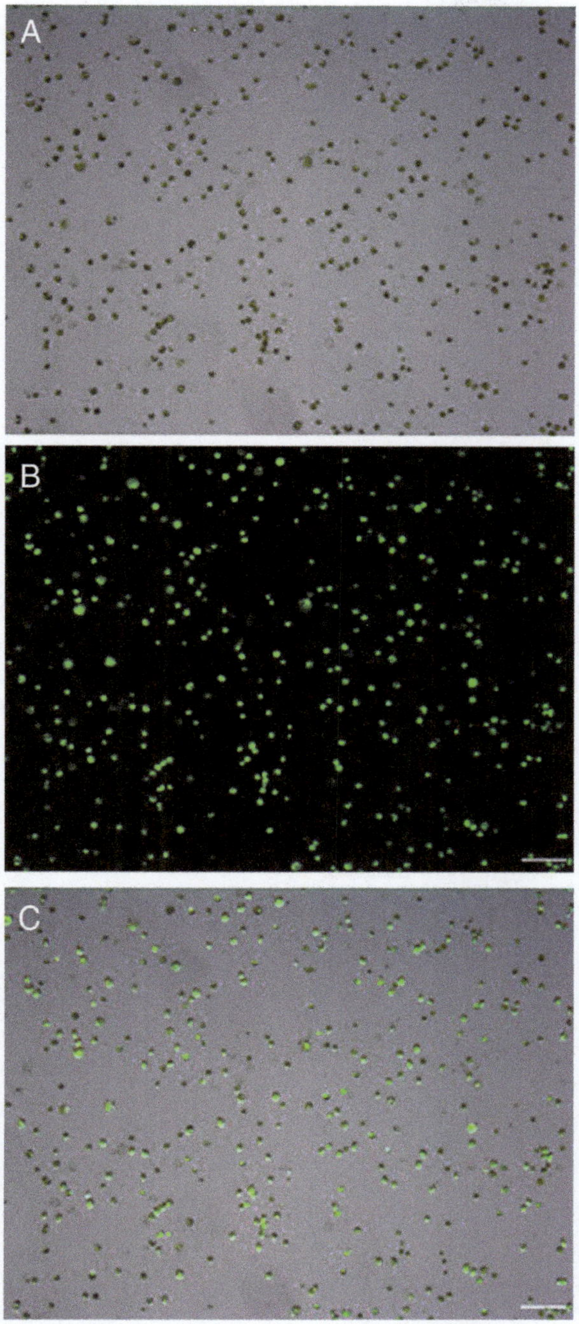

Fig. 2 The viability assays of cassava mesophyll protoplasts using the optimized protocol, i.e. preparation of mesophyll protoplasts in an enzyme solution (1.6% cellulase + 0.8% macerozyme) incubated at 25 °C (45 rpm) for 16 h in darkness. **a** Bright field image of protoplasts; (**b**) Protoplasts stained with FDA; (**c**) Merged image of protoplasts (Scale bar = 100 μm).

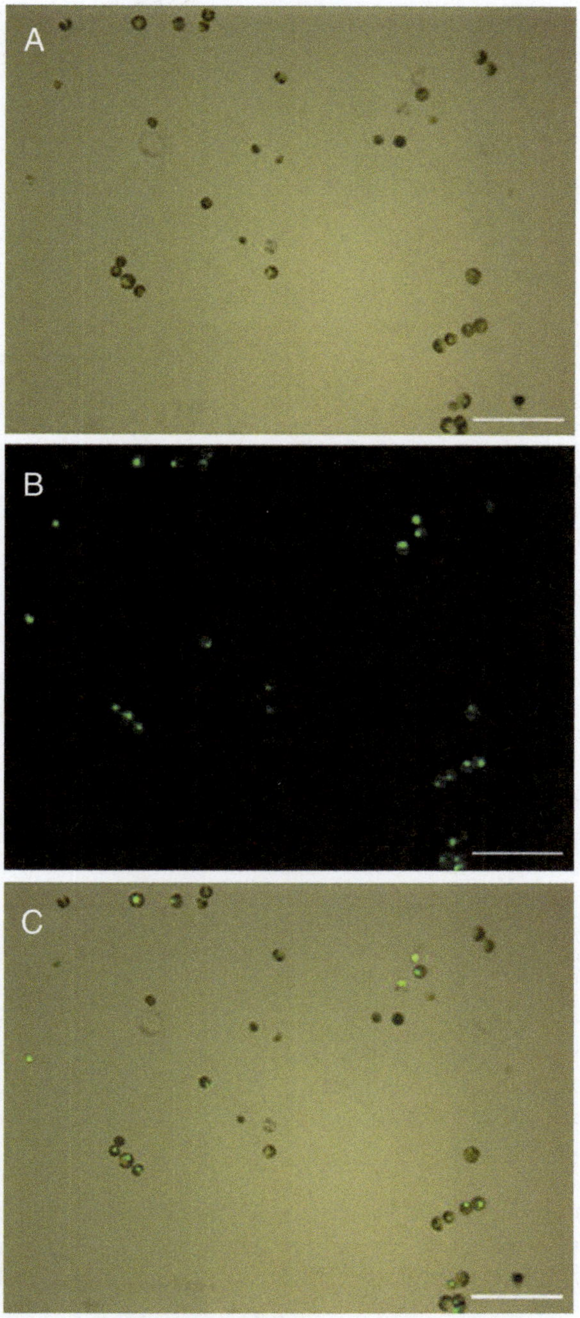

Fig. 3 High-efficiency transformation of cassava mesophyll protoplasts with pA7-GFP plasmid. **a** Bright field image of protoplasts; (**b**) Image of GFP; (**c**) Image of GFP merged with chlorophyll autofluorescence (Scale bar = 100 μm)

sufficiently high to be used for gene functional analysis in the future.

Using the optimized protocol that we developed, subcellular localization of MeSTP7 was investigated. Cassava *MeSTP7* is a homologous gene of *STP1* (*SUGAR TRANSPORTER PROTEIN 1*) in *Arabidopsis thaliana* and *Nicotiana tabacum*, which encodes an H^+/monosaccharide cotransporter containing 12 putative transmembrane domains and is a putative plasma membrane protein [39–41]. Our results showed that MeSTP7 have been likely localized to the cassava plasma membrane. Similar to the localization in

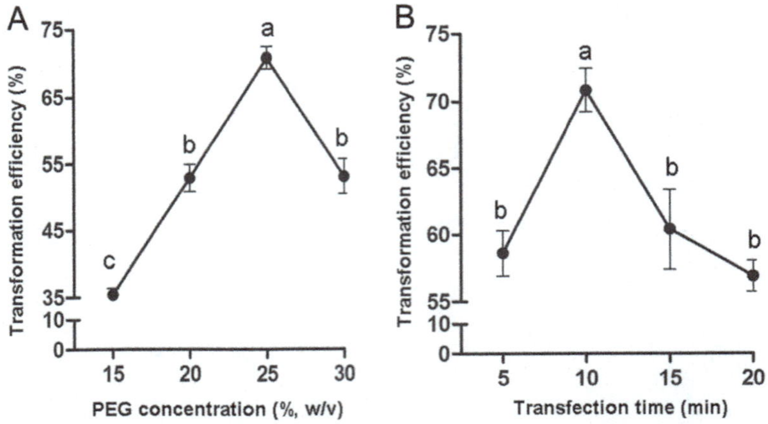

Fig. 4 Effects of PEG4000 concentration (**a**) and transfection time (**b**) on cassava protoplast transformation efficiency. Values are expressed as mean ± SEM, the error bars represent at least five independent replicates. The different letters indicate significant differences ($P \leq 0.05$) according to the Duncan test

cassava protoplasts, the MeSTP7 protein was distributed in *Arabidopsis* plasma membrane using a heterologous transient gene expression system. This confirmed that the cassava mesophyll protoplast expression is a reliable procedure for subcellular protein localization.

Conclusions

In summary, a highly efficient protocol for cassava mesophyll protoplast isolation and PEG-mediated transient expression was obtained through optimization of the type and concentration of enzymes for protoplast preparation, and PEG4000 concentration and treatment time for transfection. The method reported here can thus be very useful for molecular and cellular studies in cassava, especially combined with genetics, genomics, transcriptomics and proteomics. It is noteworthy that the simple incubation buffers described in this article do not support division of mesophyll protoplasts. Therefore, the system will not

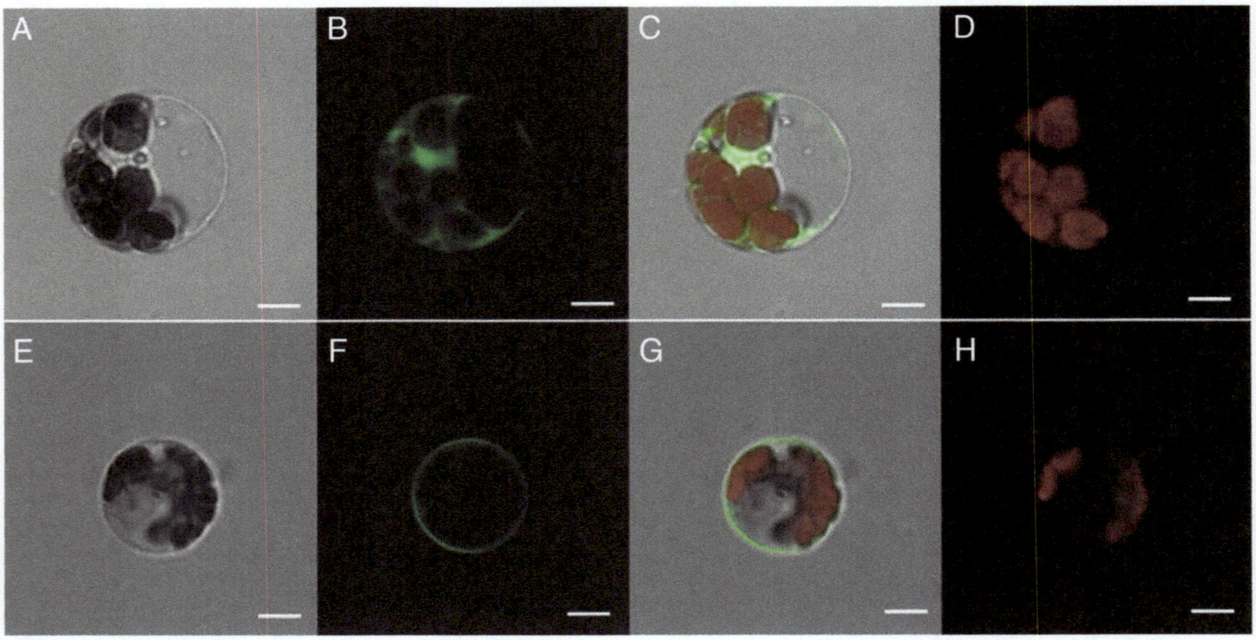

Fig. 5 Subcellular localization of MeSTP7 in cassava mesophyll protoplasts. Transient expression of GFP, showing that the GFP is distributed throughout the nucleus and cytoplasm (**a-d**). The laser-scanning confocal microscopy images are the bright field image (**a**), fluorescence image (**b**), merged image (**c**) and autofluorescence image (**d**), respectively. The transient expression of GFP-fused MeSTP7 protein, showing that the MeSTP7-GFP fusion protein is likely localized to plasma membrane (**e-h**). The laser-scanning confocal microscopy images are the bright field image (**e**), fluorescence image (**f**), merged image (**g**) and chloroplast fluorescence (H), respectively. The bars = 5 μm.)

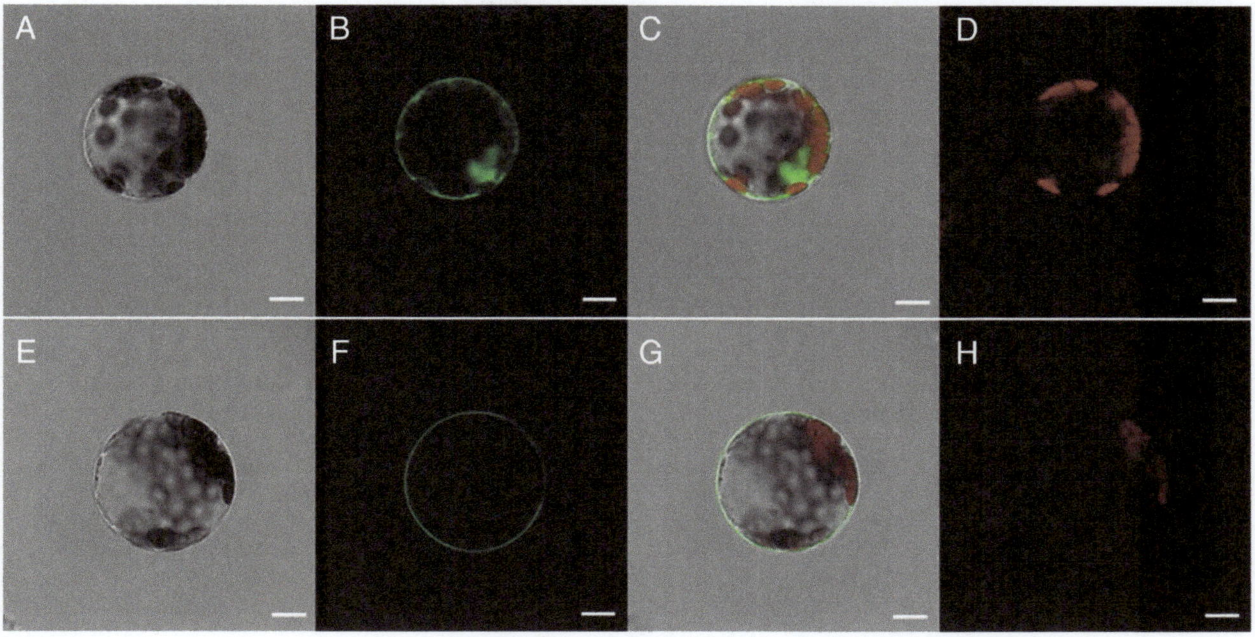

Fig. 6 Subcellular localization analysis of MeSTP7 in *Arabidopsis* mesophyll protoplasts. Transient expression of GFP, showing that the GFP is distributed throughout the nucleus and cytoplasm (**a-d**). The laser-scanning confocal microscopy images are the bright field image (**a**), fluorescence image (**b**), merged image (**c**) and chloroplast fluorescence image (**d**), respectively. The transient expression of GFP-fused MeSTP7 protein, showing that the MeSTP7-GFP fusion protein is likely localized to cytoplasm membrane (**e-h**). The laser-scanning confocal microscopy images are the bright field image (**e**), fluorescence image (**f**), merged image (**g**) and chloroplast fluorescence image (**h**), respectively. The bars = 10 μm

appropriate to evaluate the expression of cassava genes involved in, e.g., cell cycle, plant dedifferentiation, or genes that require long-term follow-up (several cycles of cell division). The protoplast regeneration protocols developed by Shahin and Shephard [31], Sofiari et al [32] and Anthony et al [33] can be used for this purpose.

Acknowledgements
We would like to thank the anonymous reviewers for their helpful comments on the manuscript.

Funding
This work was supported by the National Natural Science Foundation of China (31660412), Cassava Molecular Breeding Technology Integration and Germplasm Innovation (2010CB126606-001) and Mechanisms of Sugar Metabolism, Barren-resistance and Disease Resistance in Cassava and Rubber Tree (ZXBJH-XK001-1630032016008).

Authors' contributions
L-JL, J-PL and K-ML designed and supervised the study. J-ZW carried out the experiments, analyzed the data and wrote the preliminary masuscript. QL and X-SG participated in experimental procedures. L-JL and J-PL prepared the final manuscript. The manuscript was read and approved by all the authors.

Competing interests
The authors declare that they have no competing interests.

Author details
[1]Hainan Key Laboratory for Sustainable Utilization of Tropical Bioresources, College of Agriculture, Hainan University, Haikou, Hainan Province 570228, China. [2]The Institute of Tropical Bioscience and Biotechnology, Chinese Academy of Tropical Agricultural Sciences, Haikou, Hainan Province 571101, China.

References
1. Abel S, Theologis A. Transient transformation of *Arabidopsis* leaf protoplasts: a versatile experimental system to study gene expression. Plant J. 1994;5: 421–7.
2. Faraco M, Di Sansebastiano GP, Spelt K, Koes RE, Quattrocchio FM. One protoplast is not the other! Plant Physiol. 2011;156:474–8.
3. Yoo SD, Cho YH, Sheen J. *Arabidopsis* mesophyll protoplasts: a versatile cell system for transient gene expression analysis. Nat Protoc. 2007;2:1565–72.
4. Sheen J. Signal transduction in maize and *Arabidopsis* mesophyll protoplasts. Plant Physiol. 2001;127:1466–75.
5. Siritunga D, Arias-Garzon D, White W, Sayre R. Over-expression of hydroxynitrile lyase in transgenic cassava (*Manihot esculenta* Crantz) roots accelerates cyanogenesis. Plant Biotech J. 2004;2:37–43.
6. Miao YS, Jiang LW. Transient expression of fluorescent fusion proteins in protoplasts of suspension cultured cells. Nat Protoc. 2007;2:2348–53.
7. De Caroli M, Lenucci MS, Manualdi F, Dalessandro G, De Lorenzo G, Piro G. Molecular dissection of *Phaseolus vulgaris* polygalacturonase-inhibiting protein 2 reveals the presence of hold/release domains affecting protein trafficking toward the cell wall. Front Plant Sci. 2015;6:660.
8. Nanjareddy K, Arthikala MK, Blanco L, Arellano ES, Lara M. Protoplast isolation, transient transformation of leaf mesophyll protoplasts and improved Agrobacterium-mediated leaf disc infiltration of *Phaseolus vulgaris* tools for rapid gene expression analysis. BMC Biotechnol. 2016;16:53.
9. Chen SB, Tao LZ, Zeng LR, Vega-Sanchez M, Umemura KJ, Wang GL. A highly efficient transient protoplast system for analyzing defence gene expression and protein-protein interactions in rice. Mol Plant Pathol. 2006;7:417–27.
10. Kim MJ, Baek K, Park CM. Optimization of conditions for transient *Agrobacterium*-mediated gene expression assays in *Arabidopsis*. Plant Cell Rep. 2009;28:1159–67.

11. Walter M, Chaban C, Schütze K, Batistic O, Weckermann K, Nake C, Blazevic D, Grefen C, Schumacher K, Oecking C, Harter K, Kudla J. Visualization of protein interactions in living plant cells using bimolecular fluorescence complementation. Plant J. 2004;40:428–38.
12. Guo J, Morrell-Falvey JL, Labbe JL, Muchero W, Kalluri UC, Tuskan GA, Chen J-G. Highly Efficient isolation of *Populus* mesophyll protoplasts and its application in transient expression assays. PLoS One. 2012;7, e44908.
13. Guo Y, Song X, Zhao S, Lv J, Lu M. A transient gene expression system in Populus euphratica Oliv. protoplasts prepared from suspension cultured cells. Acta Physiol Plant. 2015;37:160.
14. Tan BY, Xu M, Chen Y, Huang MR. Transient expression for functional gene analysis using *Populus* protoplasts. Plant Cell Tissue Org Cult. 2013;114:11–8.
15. Huang H, Wang Z, Cheng J, Zhao W, Li X, Wang H, Zhang Z, Sui X. An efficient cucumber (*Cucumis sativus* L.) protoplast isolation and transient expression system. Sci Hortic. 2013;150:206–12.
16. Bai Y, Han N, Wu J, Yang Y, Wang J, Zhu M, Bian H. A transient gene expression system using barley protoplasts to evaluate microRNAs for post-transcriptional regulation of their target genes. Plant Cell Tiss Organ Cult. 2014;119:211–9.
17. Wang H, Wang W, Zhan J, Huang W, Xu H. An efficient PEG-mediated transient gene expression system in grape protoplasts and its application in subcellular localization studies of flavonoids biosynthesis enzymes. Sci Hortic. 2015;191:82–9.
18. Burris KP, Dlugosz EM, Collins AG, Stewart Jr CN, Lenaghan SC. Development of a rapid, low-cost protoplast transfection system for switchgrass (*Panicum virgatum* L.). Plant Cell Rep. 2016;35:693–704.
19. Hong SY, Seo PJ, Cho SH, Park CM. Preparation of leaf mesophyll protoplasts for transient gene expression in *Brachypodium distachyon*. J Plant Biol. 2012;55:390–7.
20. Shen J, Fu J, Ma J, Wang X, Gao C, Zhuang C, Wan J, Jiang L. Isolation, culture, and transient transformation of plant protoplasts. Curr Protoc Cell Biol. 2014;63:2.8.1–17.
21. Alves AGC. Cassava botany and physiology. In: Hillocks RJ, Thresh JM, Bellotti AC, editors. Cassava biology, production and utilization. Wallingford: CABI; 2001. p. 67–90.
22. Jennings DL, Iglesias C. Breeding for crop improvement. In: Hillocks RJ, Thresh JM, Bellotti AC, editors. Cassava biology, production and utilization. Wallingford, CABI/Embrapa, Colombia. 2001. p. 149-65.
23. Liu J, Zheng Q, Ma Q, Gadidasu KK, Zhang P. Cassava genetic transformation and its application in breeding. J Integr Plant Biol. 2011;53: 552–69.
24. Prochnik S, Marri PR, Desany B, Rabinowicz PD, Kodira C, Mohiuddin M, Rodriguez F, Fauquet C, Tohme J, Harkins T, Rokhsar DS, Rounsley S. The cassava genome: current progress, future directions. Trop Plant Biol. 2012;5:88–94.
25. Zhang P, Legris G, Coulin P, Puonti-Kaerlas J. Production of stably transformed cassava plants via particle bombardment. Plant Cell Rep. 2000; 19:939–45.
26. Siritunga D, Sayre R. Generation of cyanogen-free transgenic cassava. Planta. 2003;217:367–73.
27. Siritunga D, Sayre R. Engineering cyanogens in cassava. Plant Mol Biol. 2004; 56:661–9.
28. Zhang P, Puonti-Kaerlas J. PEG-mediated cassava transformation using positive and negative selection. Plant Cell Rep. 2000;19:1041–48.
29. Nyaboga EN, Njiru JM, Tripathi L. Factors influencing somatic embryogenesis, regeneration, and *Agrobacterium*-mediated transformation of cassava (*Manihot esculenta* Crantz) cultivar TME14. Front Plant Sci. 2015;6:411.
30. Zainuddin IM, Schlegel K, Gruissem W, Vanderschuren H. Robust transformation procedure for the production of transgenic farmer-preferred cassava landraces. Plant Methods. 2012;8:24.
31. Shahin EA, Shephard JF. Cassava mesophyll protoplasts: Isolation, proliferation, and shoot formation. Plant Sci Lett. 1980;17:459–65.
32. Sofiari E, Raemakers CJJM, Bergervoet JEM, Jacobsen E, Visser RGF. Plant regeneration from protoplasts isolated from friable embryogenic callus of cassava. Plant Cell Rep. 1998;18:159–65.
33. Anthony P, Davey MR, Power JB, Lowe KC. An improved protocol for the culture of cassava leaf protoplasts. Plant Cell Tiss Organ Cult. 1995;42:299–302.
34. De Angeli A, Monachello D, Ephritikhine G, Frachisse JM, Thomine S, Gambale F, Barbier-Brygoo H. The nitrate/proton antiporter AtCLCa mediates nitrate accumulation in plant vacuoles. Nature. 2006;442:939–42.
35. Kantharajah A, Dodd W. Factors that influence the yield and viability of cucumber (*Cucumis sativus* L.) cotyledon protoplasts. Aust J Bot. 1990;38: 169–75.
36. Davey MR, Anthony P, Power JB, Lowe KC. Plant protoplasts: status and biotechnological perspectives. Biotechnol Adv. 2005;23:131–71.
37. Zhu L, Wang B, Zhou J, Chen L, Dai C, Duan C. Protoplast isolation of callus in *Echinacea augustifolia*. Colloid Surface B. 2005;44:1–5.
38. Raikar SV, Braun RH, Bryant C, Conner AJ, Christey MC. Efficient isolation, culture and regeneration of *Lotus corniculatus* protoplasts. Plant Biotechnol Rep. 2008;2:171–7.
39. Cordoba E, Aceves-Zamudio DL, Hernández-Bernal AF, Ramos-Vega M, León P. Sugar regulation of *SUGAR TRANSPORTER PROTEIN 1* (*STP1*) expression in *Arabidopsis thaliana*. J Exp Bot. 2015;66:147--59.
40. Sauer N, Friedländer K, Gräml-Wicke U. Primary structure, genomic organization and heterologous expression of a glucose transporter from *Arabidopsis thaliana*. EMBO J. 1990;9:3045–50.
41. Sauer N, Stadler R. A sink-specific H+/monosaccharide co-transporter from *Nicotiana tabacum*: cloning and heterologous expression in baker's yeast. Plant J. 1993;4:601–10.

8

A novel pyruvate kinase and its application in lactic acid production under oxygen deprivation in *Corynebacterium glutamicum*

Xin Chai[1,2], Xiuling Shang[1], Yu Zhang[1,2], Shuwen Liu[1], Yong Liang[1], Yun Zhang[1*] and Tingyi Wen[1,3*]

Abstract

Background: Pyruvate kinase (Pyk) catalyzes the generation of pyruvate and ATP in glycolysis and functions as a key switch in the regulation of carbon flux distribution. Both the substrates and products of Pyk are involved in the tricarboxylic acid cycle, anaplerosis and energy anabolism, which places Pyk at a primary metabolic intersection. Pyks are highly conserved in most bacteria and lower eukaryotes. *Corynebacterium glutamicum* is an industrial workhorse for the production of various amino acids and organic acids. Although *C. glutamicum* was assumed to possess only one Pyk (*pyk1*, NCgl2008), NCgl2809 was annotated as a pyruvate kinase with an unknown role.

Results: Here, we identified that NCgl2809 was a novel pyruvate kinase (*pyk2*) in *C. glutamicum*. Complementation of the WTΔ*pyk1*Δ*pyk2* strain with the *pyk2* gene restored its growth on D-ribose, which demonstrated that Pyk2 could substitute for Pyk1 *in vivo*. Pyk2 was co-dependent on Mn^{2+} and K^+ and had a higher affinity for ADP than phosphoenolpyruvate (PEP). The catalytic activity of Pyk2 was allosterically regulated by fructose 1,6-bisphosphate (FBP) activation and ATP inhibition. Furthermore, *pyk2* and *ldhA*, which encodes L-lactate dehydrogenase, were co-transcribed as a bicistronic mRNA under aerobic conditions and *pyk2* deficiency had a slight effect on the intracellular activity of Pyk. However, the mRNA level of *pyk2* in the wild-type strain under oxygen deprivation was 14.24-fold higher than that under aerobic conditions. Under oxygen deprivation, *pyk1* or *pyk2* deficiency decreased the generation of lactic acid, and the overexpression of either *pyk1* or *pyk2* increased the production of lactic acid as the activity of Pyk increased. Fed-batch fermentation of the *pyk2*-overexpressing WTΔ*pyk1* strain produced 60.27 ± 1.40 g/L of lactic acid, which was a 47% increase compared to the parent strain under oxygen deprivation.

Conclusions: Pyk2 functioned as a pyruvate kinase and contributed to the increased level of Pyk activity under oxygen deprivation.

Keywords: *Corynebacterium glutamicum*, Pyruvate kinase, Oxygen deprivation, Lactate dehydrogenase, Fructose 1,6-bisphosphate

Background

Pyruvate kinase catalyzes the transphosphorylation from PEP to ADP to generate pyruvate and ATP in glycolysis [1]. Pyks are highly conserved in both eukaryotes and prokaryotes [2, 3]. In mammals, four types of Pyk isoenzymes (M1, M2, L and R) with different kinetic properties are distributed in various tissues to satisfy the diverse metabolic requirements. Most bacteria and lower eukaryotes have only one form of Pyk. However, there are two identified Pyk isoenzymes in *Escherichia coli*, *Salmonella typhimurium* and *Saccharomyces cerevisiae*. The activities of Pyks are dependent on a bivalent and/or a univalent cation (Mn^{2+}/Mg^{2+} and/or K^+) [1]. In the majority of organisms, Pyk is a homotetramer that exhibits inactive/active reversible allosteric mechanisms [4]. The functional forms of Pyks are divided into the following two broad types based on their allosteric effectors: type I is allosterically activated by FBP, whereas type II is allosterically activated by AMP and monophosphorylated sugars [1]. The allosteric behavior of Pyk allows it to be an important switch point for the regulation of carbon flux distribution.

* Correspondence: zhangyun@im.ac.cn; wenty@im.ac.cn
[1]CAS Key Laboratory of Pathogenic Microbiology and Immunology, Institute of Microbiology, Chinese Academy of Sciences, Beijing 100101, People's Republic of China
Full list of author information is available at the end of the article

The two most studied allosteric Pyk isoenzymes are those of *E. coli* and *S. cerevisiae*. In *E. coli*, PykF exhibited higher activity than PykA *in vitro*, and PykF and PykA played different roles in pyruvate synthesis and cell growth [5–7]. In *S. cerevisiae*, FBP-activated Pyk1p was the major enzyme that catalyzed the conversion of PEP to pyruvate, whereas Pyk2p with relatively low enzymatic activity was insensitive to FBP and subject to glucose repression, which suggested that Pyk2p played a major role under conditions of low glycolytic flux [8]. To date, little is known about other Pyk isoenzymes and their physiological function in other bacteria.

Corynebacterium glutamicum is widely used for the industrial production of various amino acids and vitamins during aerobic cultivation due to the robustness of its genetic modifications and fermentation [9]. Under oxygen deprivation, *C. glutamicum* can primarily convert glucose to pyruvate, which is an intermediate for the production of organic acids and amino acids despite the cessation of growth [10–13]. Two genes [*NCgl2008* (*pyk1*) and *NCgl2809* (*pyk2*)] were annotated as pyruvate kinases in the genome of *C. glutamicum* ATCC 13032 [9]. The deletion of *pyk1* resulted in marginal Pyk activity that was below the detection limit, which supported the assertion that *C. glutamicum* possessed only one pyruvate kinase [14, 15]. Pyk1 exhibited high catalytic activity using Mn^{2+} or Co^{2+} as a cation and subjected to AMP activation and ATP inhibition [16]. The expression of *pyk1* was repressed by the global transcriptional regulator SugR, which binds to the promoters of many target genes involved in glycolysis (i.e., *gapA*, *pfkA*, *fba*, *eno*, *pyk1* and *ldhA*) to regulate sugar metabolism [17, 18]. The deletion of *pyk1* led to decrease biomass and increase glutamic acid production [14]. Therefore, Pyk1 functioned as an effective metabolic engineering node for the production of various metabolites [19]. The genetic organization of the homologous *pyk2*, which is adjacent to *ldhA*, exists in many *Corynebacterium* species. However, the function and biological significance of Pyk2 has not been understood.

In this study, we identified NCgl2809 (Pyk2) as a functional pyruvate kinase in *C. glutamicum*. Pyk2 exhibited different enzymatic properties compared to Pyk1 and could substitute for Pyk1 *in vivo*. Under oxygen deprivation, *pyk1* or *pyk2* deficiency decreased the generation of lactic acid, and the overexpression of either *pyk1* or *pyk2* increased the production of lactic acid.

Results

Pyk2 is a pyruvate kinase in *C. glutamicum*

NCgl2809 was annotated as a pyruvate kinase in the genome of *C. glutamicum* ATCC 13032 (WT). This gene is adjacent to *ldhA* and distant from *pyk1* (Additional file 1: Figure S1a). A protein homology search and alignment indicated that Pyk2 had moderate sequence identity to the functionally identified Pyk1 of *C. glutamicum* (32%), PykF of *E. coli* (30%) and PykA of *E. coli* (27%), but showed high sequence identity to the annotated homologs from other species of *Corynebacterium*. Using the PykF crystal structure from *E. coli* (PDB code 1PKY) as a template, the secondary structure of Pyk2, which ranged from 128 to 213 residues and 346 to 606 residues, consisted of 12 α-helices and 16 β-strands structures. This structure was similar to the catalytic domain A of PykF (Additional file 1: Figure S1b and c), in which the GDLGVE motif served as the PEP binding site and the conserved sequences assisted the FBP allosteric regulation. A long extra N-terminal sequence of 120 amino acids made Pyk2 (67.6 kDa) larger than other typical bacterial Pyks (51 kDa).

To demonstrate the function of Pyk2 *in vivo*, the *pyk2* and/or *pyk1* genes were deleted in the chromosome. The WTΔ*pyk1*Δ*pyk2* strain grown on glucose under aerobic conditions exhibited only a slightly lower growth rate than the WT strain ($P < 0.05$), which indicated that the conversion of PEP to pyruvate during glucose transport by the phosphotransferase transport system (PTS) could supply pyruvate for cell growth (Additional file 1: Table S1). To eliminate the effect of the pyruvate generated from the glucose transport on growth, D-ribose, which was imported by the ribose-specific ATP-binding cassette transport system, was employed as the sole carbon source to evaluate the role of Pyk isoenzymes under aerobic conditions [20]. As shown in Fig. 1a, the WTΔ*pyk2* strain showed similar growth compared to the WT strain. However, the WTΔ*pyk1* strain showed a 52% decrease in the final biomass and the WTΔ*pyk1*Δ*pyk2* strain failed to grow when D-ribose was the sole carbon source, which suggested that the *pyk2* gene maintained cell growth on D-ribose to some extent. Moreover, complementation of the WTΔ*pyk1* strain with the *pyk2* gene restored its growth in the same manner as the *pyk1* gene (Fig. 1b), which indicated that Pyk2 could substitute for Pyk1 in *C. glutamicum*. Furthermore, complementation of the WTΔ*pyk1*Δ*pyk2* strain with *pyk1* and *pyk2* completely restored its growth (Fig. 1c). In addition, the overexpression of *pyk2* in the WTΔ*pyk1*Δ*pyk2* strain made the specific activities of Pyk (448.81 ± 19.25 mU/mg) comparable to that of overexpression of *pyk1* (458.76 ± 7.99 mU/mg) in the WTΔ*pyk1*Δ*pyk2* strain. Altogether, our results strongly supported that Pyk2 functioned as a pyruvate kinase *in vivo*.

Enzymatic properties of Pyk2

To detect the activity of Pyk2 *in vitro*, the *pyk2* gene of *C. glutamicum* was expressed in *E. coli* BL21 (DE3) using the pET-28a expression system. N-terminal His-tagged Pyk2 exhibited similar activity to Pyk2 with the His-tag removed by thrombin cleavage. After further purification using ion exchange chromatography, the recombinant Pyk2 showed

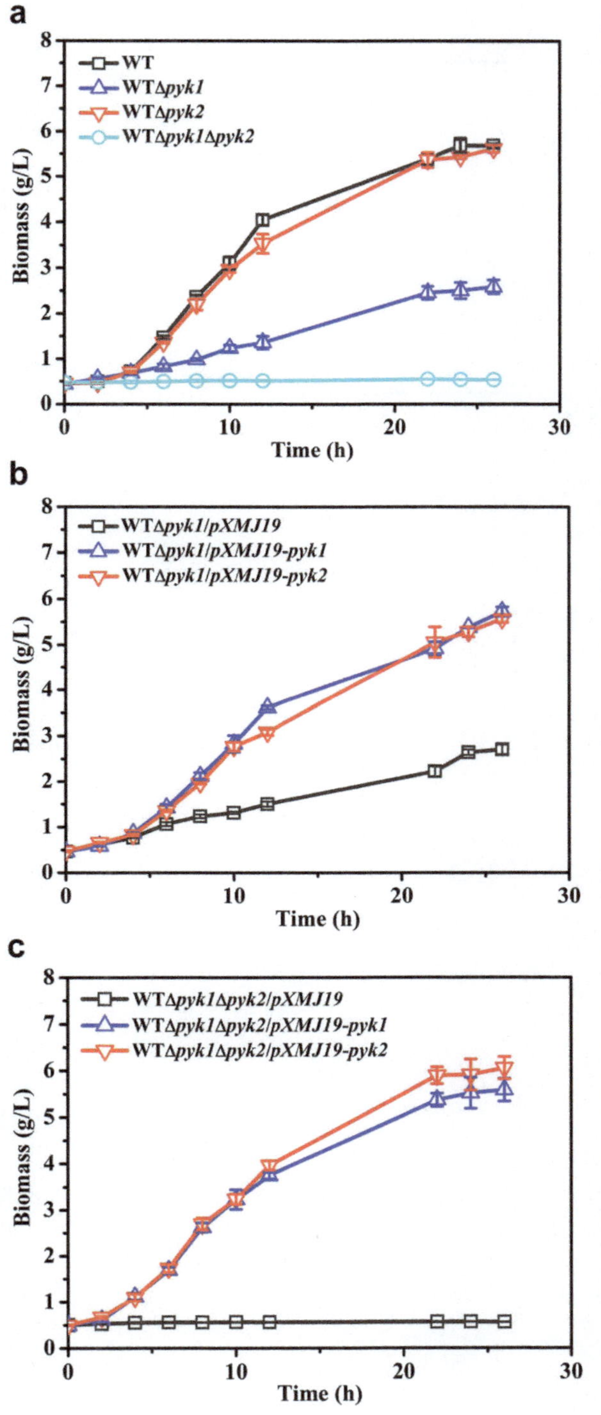

Fig. 1 Growth of *C. glutamicum* strains on CGXII medium with 100 mM D-ribose under aerobic conditions. *C. glutamicum* strains were aerobically pre-cultivated in 50 mL of CGIII medium. After washing the cells in CGXII minimal medium, the cultures were inoculated to obtain optical densities at 600 nm of approximately 1.5 (dry cell weight of 0.45 g/L). When necessary, 0.5 mM IPTG was supplemented after inoculation. **a** WT, WTΔ*pyk1*, WTΔ*pyk2* and WTΔ*pyk1*Δ*pyk2*. **b** WTΔ*pyk1*/pXMJ19, WTΔ*pyk1*/pXMJ19-*pyk1* and WTΔ*pyk1*/pXMJ19-*pyk2*. **c** WTΔ*pyk1*Δ*pyk2*/pXMJ19, WTΔ*pyk1*Δ*pyk2*/pXMJ19-*pyk1* and WTΔ*pyk1*Δ*pyk2*/pXMJ19-*pyk2*. The data were derived from experiments performed at least in three times, and the *error bars* represent the standard deviations

an approximately 12.81-fold purification with a yield of 37% and a specific activity of 1.64 U/mg. The molecular weight of the recombinant Pyk2 subunit was 67,572 Da (Fig. 2a). Gel filtration chromatography yielded a 430,881 Da fraction that corresponded to the predicted size of a homohexamer, which was different from the homotetramer observed for other Pyks (Fig. 2b).

Pyk2 was active at temperatures between 20 and 50 °C, with a maximal activity at 35 °C (Fig. 2c). The enzyme maintained greater than 80% of its relative activity at a pH range between 6 and 9, with a maximal activity at pH 7.0 (Fig. 2d). Pyk2 exhibited an absolute dependence on Mn^{2+} together with a monovalent cation. This finding was quite different from Pyk1 in *C. glutamicum*, which only used Mn^{2+} or Co^{2+} as cations [16]. Among all the tested monovalent cations, the maximum activity of Pyk2 was detected in the presence of 5 mM Mn^{2+} and 100 mM K^+. When NH_4^+, Na^+, or Li^+ was substituted for K^+, the relative activities of Pyk2 were maintained at 73, 51, and 38%, respectively (Fig. 2e).

Kinetic parameters of Pyk2

The kinetics of Pyk2 were determined with respect to PEP and ADP as described in the Methods section (Fig. 3 and Additional file 1: Table S2). As shown in Table S2, the $S_{0.5}$ value of Pyk2 for PEP was 4.58 ± 0.74 mM, and the k_{cat} value was 5.58 ± 0.53 s^{-1}, which was lower than that of Pyk1 [16]. The k_{cat} value of Pyk2 for ADP was 4.36 ± 0.22 s^{-1}, and the $S_{0.5}$ value was 0.33 ± 0.03 mM (Fig. 3b); this latter value was 4.71-fold higher than that of *C. glutamicum* Pyk1 (0.07 mM), which indicated that Pyk2 had a relatively lower affinity for ADP compared to Pyk1.

The kinetic properties of Pyk2 in response to different effectors were determined to understand its allosteric regulation. In the presence of 5 mM FBP (Fig. 3c), the $S_{0.5}$ of Pyk2 for PEP was dramatically decreased to 2.17 ± 0.12 mM and the k_{cat} value was increased to 7.67 ± 0.26 s^{-1}, which demonstrated that Pyk2 was dependent on FBP activation. However, glucose-6-phosphate (G6P) had no significant effect on the $S_{0.5}$ and k_{cat} of Pyk2 (Additional file 1: Figure S2a). In addition, there was no significant effect of AMP on the $S_{0.5}$ and k_{cat} of Pyk2 (Additional file 1: Figure S2b). The kinetic properties of Pyk2 were similar to *E. coli* PykF and different from *C. glutamicum* Pyk1, which was activated by AMP [1]. Moreover, ATP had an inhibitory effect on Pyk2 because the k_{cat} for PEP decreased to 2.09 ± 0.10 s^{-1} (Fig. 3d). In addition, in the presence of citrate (CIT), the sigmoidal nature of the activity curve increased

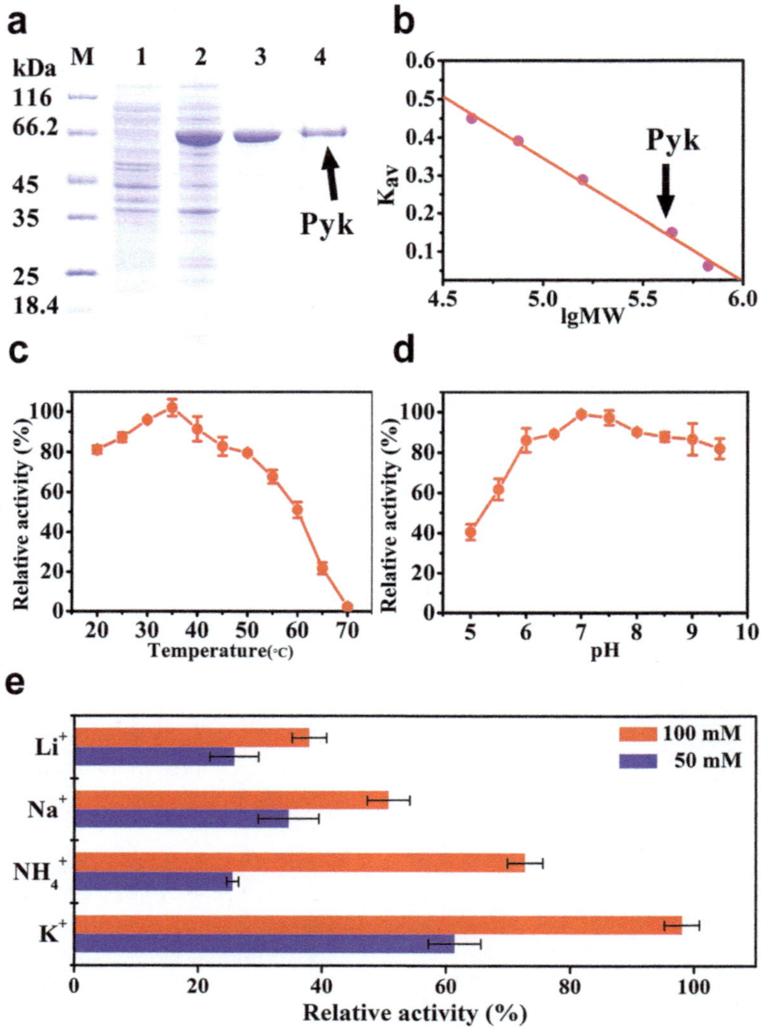

Fig. 2 Enzymatic properties of Pyk2 from *C. glutamicum*. **a** SDS-PAGE of purified Pyk2. Lane 1, supernatant of crude extract from *E. coli* BL21(DE3)/pET-28a; Lane 2, supernatant of crude extract from *E. coli* BL21(DE3)/pET-28a-*pyk2*; Lane 3, Pyk2 purified by His-tagged affinity chromatography; and Lane 4, Pyk2 purified by ion exchange chromatography. **b** Molecular weight determination by gel filtration chromatography. The molecular weight of a protein could be determined from the calibration curve (plot of K_{av} versus the logarithm of molecular weight). The K_{av} value was calculated from the measured elution volume using the following equation: $K_{av} = (V_e - V_o)/(V_c - V_o)$, where V_e is the elution volume, V_o is the column void volume, and V_c is the geometric column volume. The logarithm of the molecular weight was represented as lgMW. The standard proteins were shown as follows (●): 1, ovalbumin (44 kDa); 2, conalbumin (75 kDa); 3, aldolase (158 kDa); 4, ferritin (440 kDa); and 5, thyroglobulin (669 kDa). The native molecular weight of Pyk2 was determined to be a homohexamer of 430,881 (*arrow*). **c** The enzyme activity of Pyk2 at different temperatures. The maximum value of 2.19 ± 0.09 U/mg was set as relative 100%. **d** Enzyme activities of Pyk2 at various pHs (The maximum value of 2.43 ± 0.02 U/mg was set as relative 100%). **e** Enzyme activities of Pyk2 in the presence of various monovalent ions (The maximum value of 1.77 ± 0.05 U/mg was set as relative 100%). The enzyme was incubated in 100 mM Tris-HCl buffer (pH 7.4) and 5 mM Mn^{2+} with 50 mM or 100 mM monovalent ions. The data were derived from at least three replicate experiments, and the *error bars* represent the standard deviations

to a Hill coefficient of 3.66 that was indicative of increased cooperativity (Fig. 3e). Altogether, the FBP-activated Pyk2 was different from the AMP-activated *C. glutamicum* Pyk1.

The expression of *pyk2* is induced under oxygen deprivation

Considering that the transcriptional direction of *pyk2* and the neighboring *ldhA* genes in the chromosome are similar, RT-PCR was performed using two primers designed against the overlapping adjacent region of the two genes to determine whether *pyk2* and *ldhA* were co-transcribed. Under aerobic conditions, a single 491 bp PCR product was obtained, which indicated that *pyk2* and *ldhA* were co-transcribed as a bicistronic mRNA (Fig. 4a). The expression levels of *pyk1*, *pyk2* and *ldhA* in the WT strain were then further investigated using qRT-PCR under aerobic and oxygen-deprived conditions. The mRNA level of *pyk2* under oxygen deprivation was 14.24-fold higher than

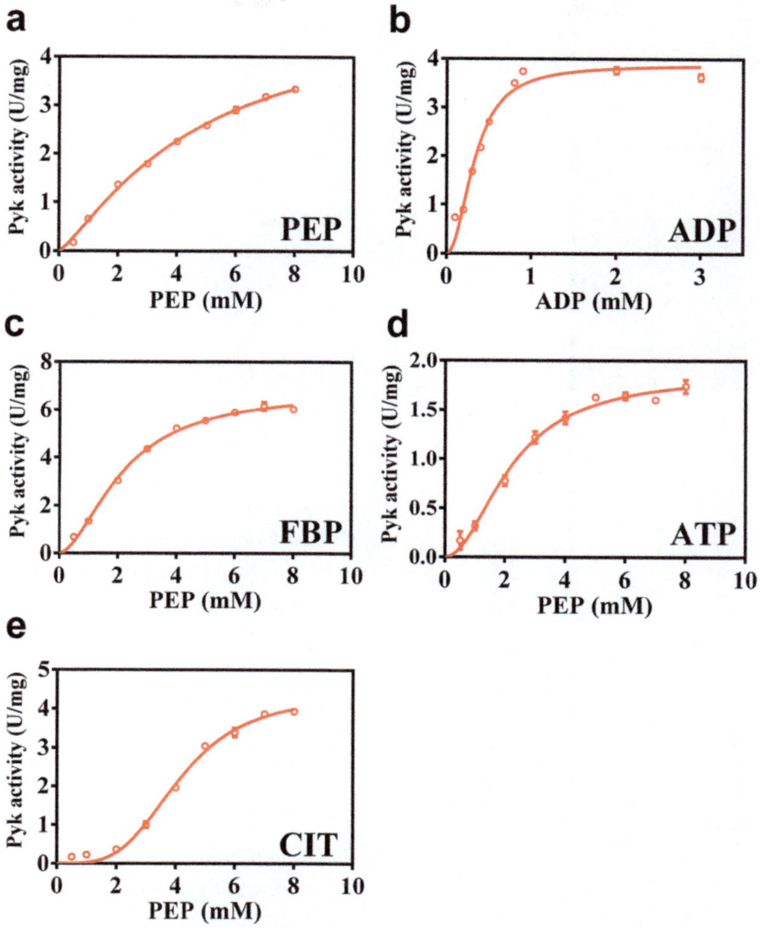

Fig. 3 Kinetic parameters of Pyk2. The kinetic parameters of Pyk2 with respect to the substrates are shown. **a** PEP. **b** ADP. The kinetic parameters of Pyk2 with respect to PEP at a concentration of 5 mM different effectors are shown. **c** FBP. **d** ATP. **e** CIT. The data points were fitted to the allosteric-sigmoidal equation using Prism software (GraphPad, USA). The $S_{0.5}$ (the substrate concentration giving one-half of the V_{max}) and k_{cat} values were calculated. The experiments were performed at least in three times, and the *error bars* represent the standard error of the regression

that under aerobic conditions (Fig. 4b), which indicated that *pyk2* was induced at the transcription level under oxygen deprivation. The mRNA level of *ldhA* under oxygen deprivation was 2.61-fold higher than under aerobic conditions, which was consistent with the observed increase of *ldhA* expression in the WT strain under low oxygen tension [21]. In contrast, the mRNA level of *pyk1* under oxygen deprivation exhibited a 1.65-fold increase compared to the mRNA level obtained under aerobic conditions.

The Pyk activities in the WT, WTΔ*pyk1*, WTΔ*pyk2* and WTΔ*pyk1*Δ*pyk2* strains were investigated under aerobic and oxygen-deprived conditions. The Pyk activity in the WT strain under oxygen-deprived conditions increased by 12% compared to the activity under aerobic conditions, which was mainly attributed to the increased expression of *pyk2* (Table 1). The deficiency of *pyk1* resulted in a 97% decrease in Pyk activity; however, the Pyk activity in the WTΔ*pyk2* strain was comparable to that in the WT strain under aerobic conditions. In addition, there was no significant difference in the specific glucose consumption rate and the specific growth rate of the WTΔ*pyk2* strain compared with the WT strain (Additional file 1: Table S1), which indicated that the *pyk2* gene might maintain a relatively low expression level under aerobic conditions. In contrast, *pyk2* deficiency led to a 15% decrease in Pyk activity under oxygen-deprived conditions, and Pyk1 could only maintain 85% of Pyk activity in the WTΔ*pyk2* strain compared to the WT strain. Therefore, these results illustrated that the expression of *pyk2* was induced and contributed to increased whole cell Pyk activity under oxygen deprivation.

The effect of *pyk1* or *pyk2* expression on the production of lactic acid under oxygen deprivation

Pyk2 had a slight effect on Pyk activity under aerobic conditions, and therefore, the role of Pyk2 was further investigated in the WT, WTΔ*pyk1*, WTΔ*pyk2* and

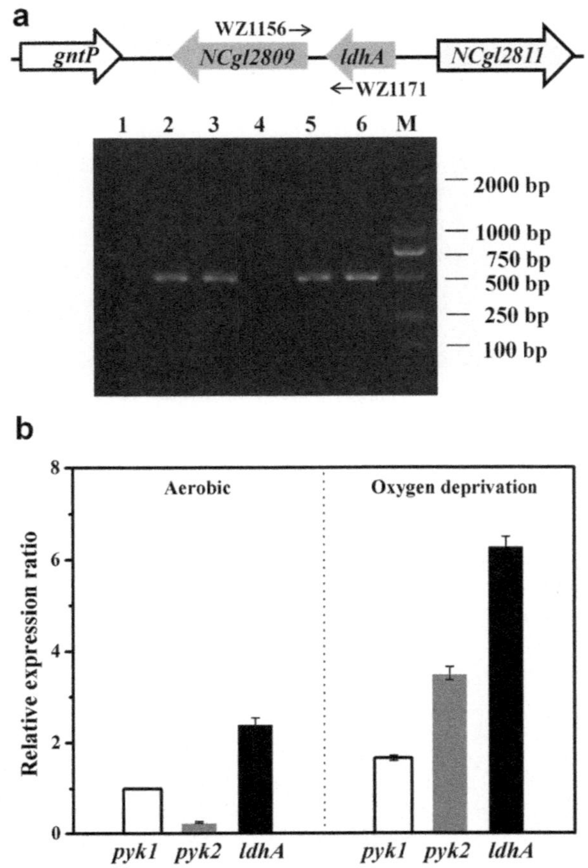

Fig. 4 Expression of the *ldhA* and *pyk2* genes in *C. glutamicum*. **a** Identification of the co-transcription of *ldhA* and *pyk2* in the *ldhA-pyk2* cluster using RT-PCR. The *C. glutamicum* WT strain was cultured in minimal medium with glucose under aerobic conditions. The templates used for the PCR were as follows: lanes 1 and 4, total RNA reverse transcribed without reverse transcriptase; lanes 2 and 5, genomic DNA; and lanes 3 and 6, cDNA. The fragments in lanes 1, 2 and 3 were amplified using the primers WZ1181/WZ868 for *rpoB*. In addition, the primers WZ1171 and WZ1156 were used for the *ldhA-pyk2* region in lanes 4, 5 and 6. **b** The relative transcription levels of the *pyk1*, *pyk2* and *ldhA* genes were analyzed by qRT-PCR. Total RNA was isolated from WT cells harvested at the exponential phase under aerobic conditions and at 3 h cultivation under oxygen-deprived conditions. The expression levels of *pyk1*, *pyk2* and *ldhA* under different conditions were compared against the expression of *pyk1* under aerobic conditions (=1). The mean values from at least three independent cultures are shown with the standard deviations

WTΔ*pyk1*Δ*pyk2* strains under oxygen-deprived conditions. There was no obvious difference in the glucose consumption rate; however, a significant change was observed in the production of organic acids among these strains (Table 1). The yield of lactic acid from glucose in the *pyk1*- or *pyk2*-deficient strains was similar, but there was a 13 and 7% decrease in the production of lactic acid in the WTΔ*pyk1* and WTΔ*pyk2* strains, respectively, compared to the WT strain. The production of lactic acid in the double mutant decreased by 22%, and the yield further decreased by 23%. In contrast, the variation tendency of production of succinic acid was opposite to that of lactic acid. Compared to the WT strain, the production of succinic acid was 1.68 and 1.97-fold higher in the WTΔ*pyk2* and WTΔ*pyk1* strains, respectively. In addition, the titers of acetic acid were lower in the three mutants. These results indicated that *pyk1* or *pyk2* deficiency decreased the generation of lactic acid and increased the formation of succinic acid.

Pyk1 or *pyk2* was overexpressed in the WTΔ*pyk1*Δ*pyk2* and WT strains to determine their effects on the production of organic acids. Complementation of the WTΔ*pyk1*Δ*pyk2* strain with *pyk1* and *pyk2* resulted in a lactic acid production that was comparable to that of the WT strain (Table 1). Pyk activity was restored to a similar level as the WT strain. Furthermore, the titers of acetic acid in the three mutants were not significantly different from those of the WT strain. However, the overexpression of the *pyk1* or *pyk2* gene produced a 20 and 17% increase in the production of lactic acid in the WT strain, with 1.90 and 1.87-fold higher Pyk activities, respectively. In addition, the production of succinic acid was significantly decreased, and the production of acetic acid was increased in the mutants. Therefore, the overexpression of either *pyk1* or *pyk2* increased the production of lactic acid under oxygen deprivation.

Fed-batch lactic acid production of the *pyk2*-overexpressing strain

Because both Mn^{2+} and K^+ are necessary for Pyk2 activity, the effects of the concentrations of $MnSO_4$ and KCl in the medium on the production of lactic acid were investigated in the WTΔ*pyk1*/pXMJ19-*pyk2* strain under oxygen deprivation. The production of lactic acid was enhanced by increasing the concentration of Mn^{2+} and reached the highest level of 160.97 ± 1.11 mM with a yield of 82.95 ± 0.91% at 1 g/L $MnSO_4$ (Fig. 5a). The production of lactic acid was improved and then gradually decreased with increasing concentrations of KCl (Fig. 5a). The highest concentration of 158.38 ± 3.57 mM with a yield of 81.76 ± 1.31% was obtained in the presence of 3 g/L KCl. These results demonstrated that the activation of Pyk2 activity by Mn^{2+} and K^+ contributed to the increased production of lactic acid.

Fed-batch fermentation was performed in a 7.5 L fermenter to estimate the production of lactic acid at the optimal cation concentrations. The WTΔ*pyk1* strain produced 454.04 ± 5.55 mM of lactic acid with the productivity of 18.86 ± 0.21 mM/h (Fig. 5b and Table 2). To determine the distribution of carbon flux towards organic acids, the yields were expressed as a percentage of the theoretical yield (100% represented 2 mol of lactic acid, succinic acid or acetic acid per 1 mol of glucose consumed) [22, 23]. In the WTΔ*pyk1* strain, 65.60 ± 1.76% of the carbon flux was

Table 1 The production of organic acids by *C. glutamicum* strains under oxygen deprivation[a]

Strains	Glucose (mM/h)	Lactic acid (mM)	Succinic acid (mM)	Acetic acid (mM)	Lactic acid Yield (%)[b]	Pyk activity (mU/mg)[c]
WT	28.46 ± 0.56	126.18 ± 1.70	22.59 ± 1.65	5.17 ± 0.05	73.90 ± 0.99	471.06 ± 6.61 (418.56 ± 6.54)*
WTΔ*pyk1*	27.72 ± 0.28	109.90 ± 4.00*	44.49 ± 0.88*	5.89 ± 0.91	66.09 ± 2.41*	69.98 ± 5.00* (11.34 ± 0.42)*
WTΔ*pyk2*	29.75 ± 0.11	117.30 ± 2.31*	38.02 ± 0.82*	5.35 ± 0.76	65.71 ± 1.29*	402.29 ± 8.98* (406.87 ± 14.69)
WTΔ*pyk1*Δ*pyk2*	28.83 ± 0.38	98.80 ± 1.11*	47.83 ± 0.86*	6.06 ± 0.64	57.12 ± 0.64*	ND (ND)
WTΔ*pyk1*Δ*pyk2*/pXMJ19	29.51 ± 0.43	99.17 ± 2.31*	49.45 ± 1.69*	4.40 ± 1.80	56.02 ± 1.30*	ND
WTΔ*pyk1*Δ*pyk2*/pXMJ19-*pyk1*	29.81 ± 0.85	127.29 ± 3.57	28.42 ± 3.13*	6.30 ± 2.19	71.16 ± 1.99	472.94 ± 21.29
WTΔ*pyk1*Δ*pyk2*/pXMJ19-*pyk2*	29.69 ± 1.20	128.77 ± 2.22	29.11 ± 5.13*	6.23 ± 2.07	72.28 ± 1.25	442.72 ± 10.64
WT/pXMJ19	29.63 ± 0.56	125.44 ± 4.00	27.59 ± 0.76*	6.72 ± 1.16	70.56 ± 2.25	486.38 ± 31.68
WT/pXMJ19-*pyk1*	32.78 ± 0.74*	150.98 ± 2.94*	13.37 ± 0.14*	12.78 ± 1.29*	76.77 ± 1.49*	926.55 ± 63.96*
WT/pXMJ19-*pyk2*	33.58 ± 0.28*	146.17 ± 3.20*	19.20 ± 1.27*	15.38 ± 0.89*	72.54 ± 1.59	911.45 ± 31.86*

[a] The cells were suspended to a final dry cell weight of 10 g/L in 100 mL of CGXII medium with 100 mM bicarbonate. The values were analyzed after 3 h under oxygen deprivation. The data were derived from experiments performed at least in three times, and the *error bars* represent the standard deviations

[b] The yields were calculated as a percentage of the theoretical yield (100% represented 2 mol of lactic acid per 1 mol of glucose consumed)

[c] Pyk activities were measured under oxygen deprivation as described above. And the values in parentheses show the Pyk activities at the exponential growth phase under aerobic conditions. ND, not detected

*,significant differences between WT and mutants under oxygen deprivation analyzed by *t*-test ($P < 0.05$) were marked. And the differences of Pyk activities of strains between aerobic and oxygen-deprived conditions are also shown ($P < 0.05$)

directed to lactic acid, and 25.10 ± 0.40% was directed to succinic acid. When the *pyk2* gene was overexpressed, the specific glucose uptake rate of the WTΔ*pyk1*/pXMJ19-*pyk2* strain increased by 25%, and the lactic acid production reached 669.03 ± 15.59 mM (60.27 ± 1.40 g/L) (Fig. 5c and Table 2). The lactic acid production rate of the WTΔ*pyk1*/pXMJ19-*pyk2* strain increased by 48% compared to the WTΔ*pyk1* strain. Furthermore, 77.82 ± 2.24% of the carbon flux was directed to lactic acid, and 10.57 ± 0.24% was directed to succinic acid. Taken together, these results demonstrated that the overexpression of either *pyk1* or *pyk2* increased the carbon flux into lactic acid formation under oxygen deprivation.

Discussion

Pyk is an important control node in response to changes in glycolysis and energy charges. Two isoenzymes of Pyk have been found to be subjected to allosteric regulation by different effectors in *E. coli* and *S. cerevisiae* [1, 8]. *C. glutamicum* was assumed to possess only one Pyk with higher activity compared to other bacterial Pyks [14, 16]. In this study, Pyk2 was identified as a functional isoenzyme in *C. glutamicum*. Because Pyk2 was Mn^{2+} and K^{+} co-dependent, marginal Pyk2 activity was detected in the *pyk1*-deficient mutant in the absence of the K^{+} cation. This indicated that the binding of K^{+} might induce the conversion of Pyk2 from the inactive conformation into the active conformation [24]. Our results confirmed that Pyk2 was activated by FBP despite the deficiency in the C-terminal domain containing the binding sites for the allosteric effector FBP in other Pyks [2, 25]. Consequently, we inferred that *C. glutamicum* Pyk2 might utilize a distinct allosteric mechanism that differed from other known FBP-activated Pyks. In *C. glutamicum*, the different allosteric activators of Pyk1 and Pyk2 made them subject to distinct metabolic regulation mechanisms similar to the regulation of the two isoenzymes in *E. coli* and *S. cerevisiae* [1, 8]. Under oxygen deprivation, the restriction of glyceraldehyde-3-phosphate dehydrogenase (GAPDH), a glycolysis rate-limiting enzyme, might result in the accumulation of FBP [12], which activated Pyk2 to direct the metabolic flux of pyruvate towards the formation of lactic acid. The allosteric effect of Pyk2 resulted in a regulatory role in pyruvate metabolism under oxygen deprivation.

The *pyk2* and *ldhA* in *C. glutamicum* were co-transcribed under aerobic conditions; therefore, *pyk2* followed the complicated regulatory mode of *ldhA*, which was subjected to dual repression under aerobic conditions mediated by the global regulator SugR and the special regulator LldR [17, 18]. SugR-mediated primary repression of *ldhA* and *pyk2* transcription was relieved in the presence of sugar to induce lactic acid formation. In turn, the generated lactic acid alleviated the LldR-mediated repression to enhance the transcription of *ldhA* and *pyk2* [26]. However, the mRNA level of *pyk2* was significantly higher than the mRNA level of *ldhA* under oxygen deprivation, which indicated that *pyk2* had a different transcription mode.

Under oxygen deprivation, *C. glutamicum* produces organic acids such as lactic acid, succinic acid and acetic acid from glucose with cell growth arrested [27]. The deletion or overexpression of two *pyk* genes had a significant influence on the production of organic acids. The deficiency of *pyk2* led to a decrease in the production of lactic acid and an increase in the synthesis of succinic acid. Moreover, the overexpression of *pyk2* promoted the formation of pyruvate and the production of lactic

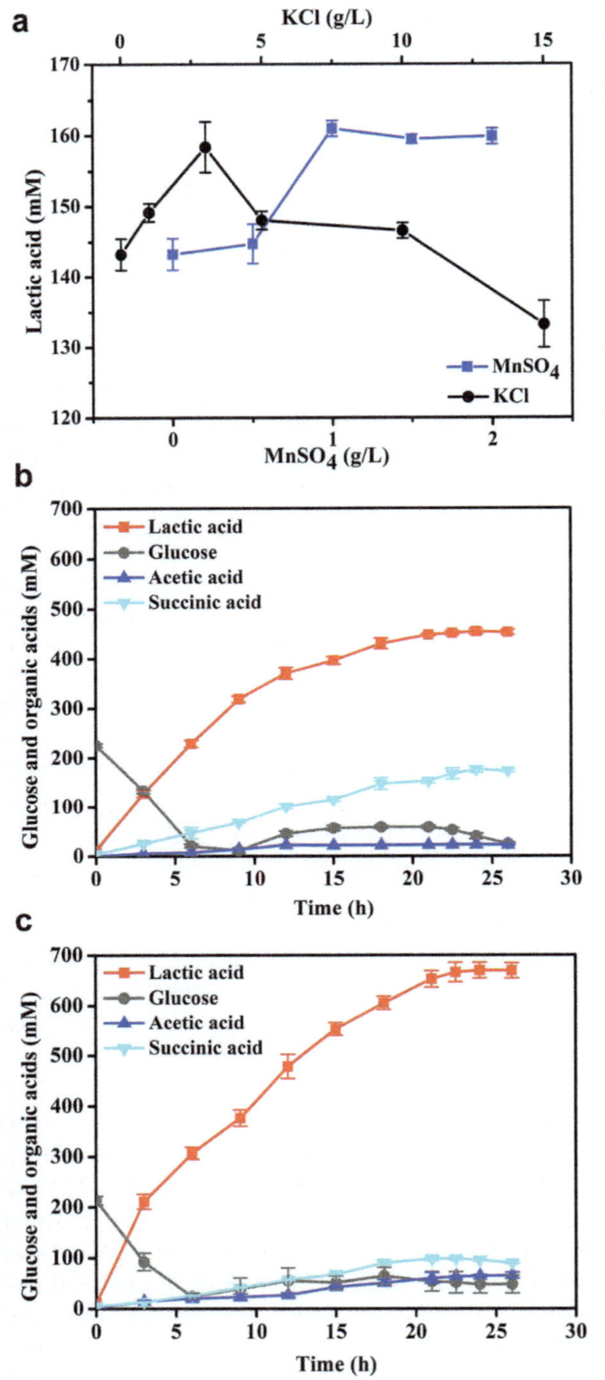

Fig. 5 Lactic acid production by *C. glutamicum* strains under oxygen deprivation. **a** The effects of $MnSO_4$ and KCl on lactic acid production by the WTΔ*pyk1*/pXMJ19-*pyk2* strain. The cells were suspended to a final dry cell weight of 10 g/L in 100 mL of CGXII medium with 100 mM bicarbonate. **b** Fed-batch production by the WTΔ*pyk1* strain under oxygen deprivation. **c** Fed-batch production by the WTΔ*pyk1*/pXMJ19-*pyk2* strain under oxygen deprivation. For the fed-batch fermentation experiments, the cells were suspended to a final dry cell weight of 20 g/L in 3 L of CGXII medium with 100 mM bicarbonate. The data were derived from experiments performed at least in three times, and the *error bars* represent the standard deviations

acid, whereas the metabolic flux from PEP towards oxaloacetate might decrease and led to decreased synthesis of succinic acid. Under oxygen deprivation, when the availability of pyruvate was decreased, the redundant PEP was redirected to synthesize succinic acid by PEP carboxylase, which was the predominant anaplerotic enzyme [10, 28, 29]. Our findings confirmed that *C. glutamicum* regulated the synthesis of lactic acid and succinic acid by Pyk and PEP carboxylase to maintain the intracellular $NADH/NAD^+$ equilibrium under oxygen deprivation [27]. Therefore, our results suggested that Pyk is a key factor that influences the formation of lactic acid in *C. glutamicum* under oxygen deprivation.

Furthermore, Pyk2 homologs had been annotated in the genomes of *Corynebacterium*, *Mycobacterium*, *Geobacillus*, etc. The *ldhA-pyk2* cluster has been found in many representative *Corynebacterium* species (Additional file 1: Figure S3). A transcriptional unit composed of several genes encoding enzymes involved in the catalysis of related metabolic reactions has been proposed to be an effective means to rapidly adjust bacterial metabolism under different conditions [30, 31]. These results support the hypothesis that lactate dehydrogenase and Pyk2 might be interdependent and play an important role in diverting the metabolic flux into lactic acid formation in *C. glutamicum* under low oxygen tensions. Therefore, the evolution of the *ldhA-pyk2* cluster for lactic acid synthesis grants *C. glutamicum* more metabolic flexibility in response to environmental changes.

Conclusions

In the present study, we identified that Pyk2 is a novel pyruvate kinase in *C. glutamicum* and has significantly different biochemical properties than Pyk1. The catalytic activity of Pyk2 was allosterically regulated by FBP activation. The allosteric behavior of Pyk made it an important switch point for the regulation of the glycolytic carbon flux distribution. *Pyk2* and *ldhA* were co-transcribed as a bicistronic mRNA under aerobic conditions. However, the mRNA level of *pyk2* in the WT strain was increased under oxygen deprivation and *pyk1* or *pyk2* deficiency decreased the production of lactic acid. The overexpression of either *pyk1* or *pyk2* increased the activity of Pyk and the production of lactic acid. The fed-batch fermentation of the *pyk2*-overexpressing strain resulted in increased lactic acid production than the parent strain, which indicated that overexpression of Pyk redirected more carbon flux into lactic acid production under oxygen deprivation.

Methods

Bacterial strains, plasmids, and media

The strains and plasmids used in this study are listed in Table S3 (Additional file 1: Table S3). *E. coli* EC135 [32] and BL21 (DE3) (Novagen, Germany) were used for the gene cloning and the recombinant enzyme expression

Table 2 The production of organic acids by *C. glutamicum* strains in fed-batch fermentation[a]

Strains	Glucose (mM/h)	Lactic acid (mM)	Lactic acid productivity (mM/h)	Yield[b] (%)			
				Lactic acid	Succinic acid	Acetic acid	Total
WTΔ*pyk1*	14.38 ± 0.27	454.04 ± 5.55	18.86 ± 0.21	65.60 ± 1.76	25.10 ± 0.40	3.20 ± 0.18	93.90 ± 2.34
WTΔ*pyk1*/pXMJ19-*pyk2*	17.92 ± 0.50	669.03 ± 15.59	27.88 ± 0.65	77.82 ± 2.24	10.57 ± 0.24	7.70 ± 0.68	96.10 ± 3.16

[a]The cells were suspended to a final dry cell weight of 20 g/L in 3 L of CGXII medium with 100 mM bicarbonate. The data were derived from experiments performed at least in three times, and the *error bars* represent the standard deviations
[b]The yields were calculated as a percentage of the theoretical yield (100% represented 2 mol of lactic acid, succinic acid or acetic acid per 1 mol of glucose consumed)

experiments, respectively. The *E. coli* strains were aerobically grown at 37°C in Luria-Bertani medium [33]. The *C. glutamicum* wild-type strain ATCC 13032 (WT) was routinely cultivated in brain heart infusion (BHI) medium (37 g/L brain heart infusion with 91 g/L sorbitol) at 30°C for the genetic disruption and complementation experiments [9]. The antibiotic concentrations were 50 μg/mL kanamycin and 20 μg/mL chloramphenicol for *E. coli*, and 25 μg/mL kanamycin and 10 μg/mL chloramphenicol for *C. glutamicum*.

Culture conditions

For the aerobic growth experiments, pre-cultures inoculated from agar plates were aerobically grown in 50 mL of CGIII medium with 4% glucose in 500 mL shake flasks on a shaker at 220 rpm for 12 h [34]. After washing the cells with CGXII minimal medium without carbon and nitrogen sources, the *C. glutamicum* strains were inoculated in 50 mL of CGXII minimal medium in 500 mL shake flasks with an initial optical density of 600 nm of approximately 1.5 (dry cell weight of 0.45 g/L) and incubated at 220 rpm [35]. A total of 100 mM glucose or 100 mM D-ribose was added as the carbon sources. The growth of the strains was assessed by measuring the OD_{600}. When necessary, 0.5 mM isopropylthio-β-D-galactopyranoside (IPTG) was supplemented after inoculation.

For the oxygen-deprived growth experiments, the *C. glutamicum* strains were pre-cultured as described above for the aerobic growth experiments [34]. The cell suspension was harvested by centrifugation at 8000 × *g* and 4°C and washed with a CGXII minimal medium without carbon and nitrogen sources. The washed cells were suspended to a final dry cell weight of 10 g/L in 100 mL of CGXII minimal medium with 100 mM bicarbonate in 150 mL lidded bottles and incubated by preventing aeration and gentle agitation [10]. Oxygen deprivation was obtained using N_2 flux. The pH was maintained at pH 7.0 by adding ammonia. Under oxygen deprivation, the growth of *C. glutamicum* was arrested. The effects of Mn^{2+} or K^+ on fermentation were studied using $MnSO_4$ (0–2 g/L) or KCl (0–15 g/L) supplementation. For the fed-batch fermentation experiments, pre-cultures inoculated from agar plates were aerobically grown in 1 L shake flasks on a shaker at 220 rpm for 12 h. Afterward, the cells were harvested by centrifugation at 8000 × *g* and 4°C, and washed with CGXII minimal medium without carbon and nitrogen sources. The washed cells were resuspended at a final dry cell weight of 20 g/L in 3 L of CGXII minimal medium with 100 mM bicarbonate and 222 mM glucose in a 7.5 L fermenter (New Brunswick BioFlo 115, Germany). N_2 flux was flushed to maintain oxygen deprivation. When needed, 50% glucose was continuously fed to maintain the residue glucose between 0 and 50 mM. The strains were incubated at 30°C and 200 rpm, and the pH was maintained at 7.0 using a pH controller with automatic addition of an ammonia solution.

Construction of plasmids and strains

Total genomic DNA of the WT strain was extracted according to previously described methods [36]. The oligonucleotides used in this study are shown in Table S4 (Additional file 1: Table S4). For the gene disruption experiments, the upstream and downstream homologous fragments were amplified using PCR with the primers WZ539/WZ540 and WZ541/WZ542 for the *pyk1* gene, and WZ545/WZ546 and WZ547/WZ548 for the *pyk2* gene. The amplified DNA fragments were spliced using overlap extension PCR and ligated into the suicide vector pK18*mobsacB* [37]. The plasmids were verified by DNA sequencing and then transformed into the WT strain through electroporation to generate the WTΔ*pyk1* strain, which was further verified by sequencing. Similarly, the *pyk2* gene was disrupted in the WT and WTΔ*pyk1* strains. The *C. glutamicum* and *E. coli* shuttle vector pXMJ19 was used for the gene overexpression studies. The replication origin of this vector was derived from that of pBL1 with a cryptic low copy number [38]. For the gene complementation experiments, the *C. glutamicum* strain was transformed by the plasmid pXMJ19, which contained the intact *pyk1* and *pyk2* genes. To induce the expression of the *pyk1* and *pyk2* genes, 0.5 mM IPTG was supplemented.

Expression and purification of Pyk2

The *pyk2* gene was cloned into the vector pET-28a and expressed in the BL21 (DE3) strain. When the OD_{600} reached 0.6, the cells were induced with 0.1 mM IPTG at 16°C and harvested by centrifugation after the OD_{600} reaching 4.0. The cells were resuspended in buffer (20 mM Tris-

HCl (pH 7.5), 150 mM NaCl, and 10% glycerol) and then disrupted by sonication. The supernatants were collected by centrifugation at 12,000 × *g* for 30 min at 4°C. The recombinant Pyk2 was purified using Ni^{2+}-affinity chromatography (GE Healthcare, USA). The eluted protein solutions were concentrated by ultrafiltration (Merck Millipore, Germany). Subsequently, the protein solution was applied to a Resource Q column (GE Healthcare, USA), which was equilibrated with a standard buffer (20 mM Tris-HCl (pH 7.5), 1 M NaCl, and 10% glycerol). The fraction was pooled and concentrated for further analysis. The eluted protein solution was incubated with thrombin (Merck Millipore, Germany) in 1× PBS (pH 7.3) at 22 °C for 16 h, which efficiently cleaved the N-terminal His-tag. After the cleavage reaction, the reaction mixture was eluted through a Ni^{2+}-affinity column to remove the His-tagged thrombin. The molecular mass of Pyk2 was analyzed using gel filtration chromatography on a Superdex 200 10/300 GL column (GE Healthcare, USA), which was equilibrated with 50 mM potassium phosphate buffer (pH 7.0) containing 150 mM NaCl. A standard curve was determined based on the elution volumes of the protein standards (GE Healthcare, USA). The purity of the purified Pyk2 was determined by SDS-PAGE. The protein concentrations were determined by the Bradford method using the Bio-Rad Protein Assay Reagent (Bio-Rad, USA).

Enzymatic activity and property determination

The assay for Pyk activity was based on changes in the absorbance of NADH at 340 nm coupled with the L-lactate dehydrogenase [39]. The reaction mixture contained 100 mM Tris-HCl (pH 7.4), 5 mM $MnSO_4 \cdot H_2O$, 5 mM PEP, 2 mM ADP, 100 mM KCl, 0.2 mM NADH, 6 units L-lactate dehydrogenase, and the enzyme in a total volume of 0.5 mL. To avoid a decrease in L-lactate dehydrogenase activity at extreme conditions (pH and temperature), the amount of L-lactate dehydrogenase was increased to 15 units. The reaction was initiated by the addition of PEP, and the decrease in the absorbance at 340 nm was monitored by a UV-visible spectrophotometer (Unico, USA). One unit of activity was defined as the amount of enzyme that converted 1 μmol of PEP to pyruvate per min at 25°C. To determine the effect of the pH on the enzyme activity, various buffer systems (pH 5.0–9.5) were used at 100 mM. The optimum temperature of Pyk2 was determined at temperatures ranging from 20 to 70°C. To exclude PEP hydrolysis with increasing temperatures, the relative Pyk2 activity was calculated by subtracting the background. The effects of different cations on Pyk activity were assayed by replacing Mn^{2+} with 5 mM or 10 mM of other cations (Mg^{2+}, Cu^{2+}, Co^{2+}, Ca^{2+}, Ni^{2+}, Zn^{2+}, Ba^{2+}, Fe^{2+}, Fe^{3+} or Al^{3+}) or replacing K^+ with 50 mM or 100 mM of monovalent cations (NH_4^+, Li^+, or Na^+). The enzyme activity was also measured without any cations. The data were obtained from three independent experiments.

Analysis of kinetic properties

The enzymatic activities were assayed using various concentrations of PEP and ADP under conditions identical to those described above. To determine the kinetics, the PEP concentration was varied with the concentration of ADP at 2 mM, or the ADP concentration was varied with the concentration of PEP at 5 mM. The effects of the various effectors tested on PEP were assessed with 2 mM ADP. The formation of the product was linear throughout the period. The data were fitted to an allosteric sigmoidal kinetic model using Prism software (GraphPad, USA). The $S_{0.5}$ (the substrate concentration giving one-half of V_{max}) and n_H (Hill coefficient) values were calculated. The k_{cat} value was calculated according to the V_{max} and the subunit molecular weight of 67,572 Da. All data were obtained from at least three independent experiments.

Analytical methods

Glucose was measured with an enzyme electrode glucose sensor (SBA-40D, Shandong Province Academy of Sciences, China). The dry cell weights of the *C. glutamicum* strains were calculated as one optical density unit at 600 nm to 0.3 g dry cell weight/L [40]. The specific growth rate μ (1/h) was estimated using a linear regression. The biomass yield ($Y_{X/S}$) was deduced from the individual biomass and glucose balancing, and the specific glucose uptake rate (q_S) was calculated from the specific growth rate and the biomass yield [41]. Organic acids were determined by high performance liquid chromatography equipped with an SB-Aq column (4.6 × 250 mm; Agilent Technologies, USA) at 210 nm. Mobile phase A (20 mM KH_2PO_4, pH 2.3) and mobile phase B (acetonitrile) were at a ratio of 95:5 [42]. To measure the enzyme activity, the cells were resuspended in the indicated buffer (100 mM Tris-HCl pH 7.5 and 10% glycerol) and disrupted by ultrasonic treatment at 4°C. The enzyme activity was monitored by measuring the changes in the NADH concentration at 340 nm. For Pyk, the reaction mixture was used as described above with 5 mM FBP. The Student's *t*-test was performed with SPSS software to determine whether the difference was statistically significant.

RNA preparation and RT-PCR analysis

Total RNA was isolated from the cells with the RNAprep Pure Cell/Bacteria Kit (Tiangen, China). The cDNA from approximately 300 ng of RNA was prepared using the specific primers listed in Table S4 (Additional file 1: Table S4) and the FastQuant RT Kit (Tiangen, China). The *rpoB* gene, which encodes RNA polymerase β subunit, was used as the reference gene [43]. The primer pair WZ1171 and WZ1156 was used to amplify the *ldhA-pyk2* region in the

RT-PCR. Genomic DNA was used as a positive control in the PCR reactions. RNA reverse transcribed without reverse transcriptase was used as a negative control to exclude DNA contaminants in the RNA preparation.

The qRT-PCR (quantitative reverse transcription-PCR) was performed with the GoTaq qPCR master mix (Promega, USA) in a 20 μL mixture using the LightCycler® 96 Real-Time PCR System (Roche, Switzerland). The data analysis was performed using the LightCycler® 96 software (Roche, Switzerland).

Sequence alignment

Database searches and sequence comparisons were performed with the BLAST programs at the BLAST tool on the National Center for Biotechnology Information website (https://blast.ncbi.nlm.nih.gov/Blast.cgi). Multiple sequence alignments were generated with the CLUSTAL X program [44]. Alignment of the secondary structures was produced using the high resolution X-ray crystallographic structure of *E. coli* PykF (PDB code 1PKY) as the template and ESPript 3.0 (http://espript.ibcp.fr/ESPript/ESPript/) with default settings.

Additional file

Additional file 1: Table S1. Growth kinetics of the *pyk* mutants on glucose under aerobic conditions. **Table S2.** Kinetic parameters of the Pyk2 for substrates. **Table S3.** Strains and plasmids used in this study. **Table S4.** Primers used in this study. **Figure S1.** The map positions of the *pyk* genes and the multiple alignment of the Pyk sequences. **Figure S2.** Kinetic parameters of Pyk2 with respect to PEP at a concentration of 5 mM G6P (**a**) or AMP (**b**). **Figure S3.** Conservation of *pyk2* gene and adjacent genes in species of the *Corynebacterium* and other species.

Abbreviations

BHI: Brain heart infusion; CIT: Citrate; FBP: Fructose 1,6-bisphosphate; G6P: Glucose-6-phosphate; GAPDH: Glyceraldehyde-3-phosphate dehydrogenase; IPTG: Isopropylthio-β-D-galactopyranoside; *ldhA*: L-lactate dehydrogenase encoding gene; PEP: Phosphoenolpyruvate; PTS: Phosphotransferase transport system; Pyk: Pyruvate kinase; qRT-PCR: quantitative reverse transcription-PCR; *rpoB*: RNA polymerase β subunit encoding gene; SugR: DeoR-type transcriptional regulator; WT: Wild-type

Acknowledgments

We thank Dr. Aihua Deng and Qian Liu for assistance with critical reading of the manuscript.

Funding

This work was supported by grants from National Hi-Tech Research and Development Program of China (2014AA021203), the Science and Technology Service Network Initiative of the Chinese Academy of Sciences (KFJ-EW-STS-078), National Natural Science Foundation of China (31100074) and the Ministry of Science and Technology of China (2008BAI63B01 and 2008ZX09401-05).

Authors' contributions

TW, YZ and XC designed the project. XC, XS and YZ performed experiments. XC, SL, YL and YZ analyzed the data. XC, XS, YZ and TW mainly wrote the manuscript. All authors read and approved the final manuscript.

Competing interests

The authors declare that they have no competing interests.

Author details

[1]CAS Key Laboratory of Pathogenic Microbiology and Immunology, Institute of Microbiology, Chinese Academy of Sciences, Beijing 100101, People's Republic of China. [2]University of Chinese Academy of Sciences, Beijing 100049, People's Republic of China. [3]Savaid Medical School, University of Chinese Academy of Sciences, Beijing 100049, People's Republic of China.

References

1. Muñoz ME, Ponce E. Pyruvate kinase: current status of regulatory and functional properties. Comp Biochem Physiol B Biochem Mol Biol. 2003;135:197–218.
2. Mattevi A, Valentini G, Rizzi M, Speranza ML, Bolognesi M, Coda A. Crystal structure of *Escherichia coli* pyruvate kinase type I: molecular basis of the allosteric transition. Structure. 1995;3:729–41.
3. Steiner P, Fussenegger M, Bailey JE, Sauer U. Cloning and expression of the *Zymomonas mobilis* pyruvate kinase gene in *Escherichia coli*. Gene. 1998;220:31–8.
4. Mattevi A, Bolognesi M, Valentini G. The allosteric regulation of pyruvate kinase. FEBS Lett. 1996;389:15–9.
5. Ponce E. Effect of growth rate reduction and genetic modifications on acetate accumulation and biomass yields in *Escherichia coli*. J Biosci Bioeng. 1999;87:775–80.
6. Al Zaid Siddiquee K, Arauzo-Bravo MJ, Shimizu K. Effect of a pyruvate kinase (*pykF*-gene) knockout mutation on the control of gene expression and metabolic fluxes in *Escherichia coli*. FEMS Microbiol Lett. 2004;235:25–33.
7. Ponce E, Martínez A, Bolívar F, Valle F. Stimulation of glucose catabolism through the pentose pathway by the absence of the two pyruvate kinase isoenzymes in *Escherichia coli*. Biotechnol Bioeng. 1998;58:292–5.
8. Boles E, Schulte F, Miosga T, Freidel K, Schlüter E, Zimmermann FK, Hollenberg CP, Heinisch JJ. Characterization of a glucose-repressed pyruvate kinase (Pyk2p) in *Saccharomyces cerevisiae* that is catalytically insensitive to fructose-1,6-bisphosphate. J Bacteriol. 1997;179:2987–93.
9. Kalinowski J, Bathe B, Bartels D, Bischoff N, Bott M, Burkovski A, Dusch N, Eggeling L, Eikmanns BJ, Gaigalat L, Goesmann A, Hartmann M, Huthmacher K, Krämer R, Linke B, McHardy AC, Meyer F, Möckel B, Pfefferle W, Pühler A, Rey DA, Rückert C, Rupp O, Sahm H, Wendisch VF, Wiegräbe I, Tauch A. The complete *Corynebacterium glutamicum* ATCC 13032 genome sequence and its impact on the production of L-aspartate-derived amino acids and vitamins. J Biotechnol. 2003;104:5–25.
10. Okino S, Inui M, Yukawa H. Production of organic acids by *Corynebacterium glutamicum* under oxygen deprivation. Appl Microbiol Biotechnol. 2005;68:475–80.
11. Jojima T, Fujii M, Mori E, Inui M, Yukawa H. Engineering of sugar metabolism of *Corynebacterium glutamicum* for production of amino acid L-alanine under oxygen deprivation. Appl Microbiol Biotechnol. 2010;87:159–65.
12. Yamamoto S, Gunji W, Suzuki H, Toda H, Suda M, Jojima T, Inui M, Yukawa H. Overexpression of genes encoding glycolytic enzymes in *Corynebacterium glutamicum* enhances glucose metabolism and alanine production under oxygen deprivation conditions. Appl Environ Microbiol. 2012;78:4447–57.
13. Hasegawa S, Suda M, Uematsu K, Natsuma Y, Hiraga K, Jojima T, Inui M, Yukawa H. Engineering of *Corynebacterium glutamicum* for high-yield L-valine production under oxygen deprivation conditions. Appl Environ Microbiol. 2013;79:1250–7.
14. Sawada K, Zen-in S, Wada M, Yokota A. Metabolic changes in a pyruvate kinase gene deletion mutant of *Corynebacterium glutamicum* ATCC 13032. Metab Eng. 2010;12:401–7.
15. Netzer R, Krause M, Rittmann D, Peters-Wendisch PG, Eggeling L, Wendisch VF, Sahm H. Roles of pyruvate kinase and malic enzyme in *Corynebacterium*

glutamicum for growth on carbon sources requiring gluconeogenesis. Arch Microbiol. 2004;182:354–63.

16. Jetten MSM, Gubler ME, Lee SH, Sinskey AJ. Structural and functional analysis of pyruvate kinase from *Corynebacterium glutamicum*. Appl Environ Microbiol. 1994;60:2501–7.
17. Engels V, Lindner SN, Wendisch VF. The global repressor SugR controls expression of genes of glycolysis and of the L-lactate dehydrogenase LdhA in *Corynebacterium glutamicum*. J Bacteriol. 2008;190:8033–44.
18. Toyoda K, Teramoto H, Inui M, Yukawa H. Molecular mechanism of SugR-mediated sugar-dependent expression of the *ldhA* gene encoding L-lactate dehydrogenase in *Corynebacterium glutamicum*. Appl Microbiol Biotechnol. 2009;83:315–27.
19. Becker J, Klopprogge C, Wittmann C. Metabolic responses to pyruvate kinase deletion in lysine producing *Corynebacterium glutamicum*. Microb Cell Fact. 2008;7:8.
20. Nentwich SS, Brinkrolf K, Gaigalat L, Hüser AT, Rey DA, Mohrbach T, Marin K, Pühler A, Tauch A, Kalinowski J. Characterization of the LacI-type transcriptional repressor RbsR controlling ribose transport in *Corynebacterium glutamicum* ATCC 13032. Microbiology. 2009;155:150–64.
21. Inui M, Suda M, Okino S, Nonaka H, Puskás LG, Vertès AA, Yukawa H. Transcriptional profiling of *Corynebacterium glutamicum* metabolism during organic acid production under oxygen deprivation conditions. Microbiology. 2007;153:2491–504.
22. Guillouet S, Rodal AA, An G-H, Lessard PA, Sinskey AJ. Expression of the *Escherichia coli* catabolic threonine dehydratase in *Corynebacterium glutamicum* and its effect on isoleucine production. Appl Environ Microbiol. 1999;65:3100–7.
23. Hasegawa S, Uematsu K, Natsuma Y, Suda M, Hiraga K, Jojima T, Inui M, Yukawa H. Improvement of the redox balance increases L-valine production by *Corynebacterium glutamicum* under oxygen deprivation conditions. Appl Environ Microbiol. 2012;78:865–75.
24. Oria-Hernández J, Riveros-Rosas H, Ramírez-Sílva L. Dichotomic phylogenetic tree of the pyruvate kinase family: K^+-dependent and -independent enzymes. J Biol Chem. 2006;281:30717–24.
25. Jurica MS, Mesecar A, Heath PJ, Shi W, Nowak T, Stoddard BL. The allosteric regulation of pyruvate kinase by fructose-1,6-bisphosphate. Structure. 1998; 6:195–210.
26. Georgi T, Engels V, Wendisch VF. Regulation of L-lactate utilization by the FadR-type regulator LldR of *Corynebacterium glutamicum*. J Bacteriol. 2008; 190:963–71.
27. Inui M, Murakami S, Okino S, Kawaguchi H, Vertès AA, Yukawa H. Metabolic analysis of *Corynebacterium glutamicum* during lactate and succinate productions under oxygen deprivation conditions. J Mol Microbiol Biotechnol. 2004;7:182–96.
28. Okino S, Noburyu R, Suda M, Jojima T, Inui M, Yukawa H. An efficient succinic acid production process in a metabolically engineered *Corynebacterium glutamicum* strain. Appl Microbiol Biotechnol. 2008;81:459–64.
29. Dietrich C, Nato A, Bost B, Le Marechal P, Guyonvarch A. Regulation of *ldh* expression during biotin-limited growth of *Corynebacterium glutamicum*. Microbiology. 2009;155:1360–75.
30. Han SO, Inui M, Yukawa H. Expression of *Corynebacterium glutamicum* glycolytic genes varies with carbon source and growth phase. Microbiology. 2007;153:2190–202.
31. Llanos RM, Harris CJ, Hillier AJ, Davidson BE. Identification of a novel operon in *Lactococcus lactis* encoding three enzymes for lactic acid synthesis: phosphofructokinase, pyruvate kinase, and lactate dehydrogenase. J Bacteriol. 1993;175:2541–51.
32. Zhang G, Wang W, Deng A, Sun Z, Zhang Y, Liang Y, Che Y, Wen T. A mimicking-of-DNA-methylation-patterns pipeline for overcoming the restriction barrier of bacteria. PLoS Genet. 2012;8:e1002987.
33. Sezonov G, Joseleau-Petit D, D'Ari R. *Escherichia coli* physiology in Luria-Bertani broth. J Bacteriol. 2007;189:8746–9.
34. Menkel E, Thierbach G, Eggeling L, Sahm H. Influence of increased aspartate availability on lysine formation by a recombinant strain of *Corynebacterium glutamicum* and utilization of fumarate. Appl Environ Microbiol. 1989;55:684–8.
35. Gaigalat L, Schlüter JP, Hartmann M, Mormann S, Tauch A, Pühler A, Kalinowski J. The DeoR-type transcriptional regulator SugR acts as a repressor for genes encoding the phosphoenolpyruvate:sugar phosphotransferase system (PTS) in *Corynebacterium glutamicum*. BMC Mol Biol. 2007;8:104.
36. Zhang Y, Shang X, Lai S, Zhang G, Liang Y, Wen T. Development and application of an arabinose-inducible expression system by facilitating inducer uptake in *Corynebacterium glutamicum*. Appl Environ Microbiol. 2012;78:5831–8.
37. Schäfer A, Tauch A, Jäger W, Kalinowski J, Thierbach G, Pühler A. Small mobilizable multipurpose cloning vectors derived from the *Escherichia coli* plasmids pK18 and pK19: selection of defined deletions in the chromosome of *Corynebacterium glutamicum*. Gene. 1994;145:69–73.
38. Jakoby M, Ngouoto-Nkili C-E, Burkovski A. Construction and application of new *Corynebacterium glutamicum* vectors. Biotechnol Tech. 1999;13:437–41.
39. Gubler M, Jetten M, Lee SH, Sinskey AJ. Cloning of the pyruvate kinase gene (*pyk*) of *Corynebacterium glutamicum* and site-specific inactivation of *pyk* in a lysine-producing *Corynebacterium lactofermentum* strain. Appl Environ Microbiol. 1994;60:2494–500.
40. Buchholz J, Schwentner A, Brunnenkan B, Gabris C, Grimm S, Gerstmeir R, Takors R, Eikmanns BJ, Blombach B. Platform engineering of *Corynebacterium glutamicum* with reduced pyruvate dehydrogenase complex activity for improved production of L-lysine, L-valine, and 2-ketoisovalerate. Appl Environ Microbiol. 2013;79:5566–75.
41. Blombach B, Buchholz J, Busche T, Kalinowski J, Takors R. Impact of different CO_2/HCO_3^- levels on metabolism and regulation in *Corynebacterium glutamicum*. J Biotechnol. 2013;168:331–40.
42. Lai SJ, Zhang Y, Liu SW, Liang Y, Shang XL, Chai X, Wen TY. Metabolic engineering and flux analysis of *Corynebacterium glutamicum* for L-serine production. Sci China Life Sci. 2012;55:283–90.
43. Shang X, Zhang Y, Zhang G, Chai X, Deng A, Liang Y, Wen T. Characterization and molecular mechanism of AroP as an aromatic amino acid and histidine transporter in *Corynebacterium glutamicum*. J Bacteriol. 2013;195:5334–42.
44. Jeanmougin F, Thompson JD, Gouy M, Higgins DG, Gibson TJ. Multiple sequence alignment with Clustal X. Trends Biochem Sci. 1998;23:403–5.

9

Microbial detoxification of eleven food and feed contaminating trichothecene mycotoxins

Rafiq Ahad[1,2], Ting Zhou[1*], Dion Lepp[1] and K. Peter Pauls[2]

Abstract

Background: Contamination of agricultural commodities with multiple trichothecene mycotoxins, produced by toxigenic *Fusarium* species, is a food safety issue, which greatly affects grain production and marketing worldwide. Importantly, exposure to multiple trichothecenes may increase toxicity in animals due to their synergistic and/or additive effects. To address the problem this study aimed to achieve a novel biological trait capable of detoxifying various food and feed contaminating trichothecenes under aerobic and anaerobic conditions and wide range of temperatures.

Results: A highly enriched microbial consortium (called DX100) capable of transforming eleven trichothecenes to significantly less toxic de-epoxy forms was achieved after prolonged incubation of soil microbial culture with 200 μg/mL deoxynivalenol (DON). DX100 demonstrated de-epoxidation activity under aerobic and anaerobic conditions, a greater range of temperatures and around neutral pH. The consortium contains 70% known and 30% unknown bacterial species, dominated by *Stenotrophomonas* species. Probably novel bacteria including strains of *Stenotrophomonas* and *Alkaliphilus-Blautia* species complex could be involved in aerobic and anaerobic de-epoxidation of trichothecenes, respectively. DX100 showed rapid and stable activity by de-epoxidizing 100% of 50 μg/mL deoxynivalenol at 48 h of incubation and retaining de-epoxidation ability after 100 subcultures in mineral salts broth (MSB). It was able to de-epoxidize high concentration of DON (500 μg/mL), and transformed ten more food contaminating trichothecenes into de-epoxy forms and/or other known/unknown compounds. Microbial de-epoxidation rate increased with increasing trichothecene concentrations in the broth media, suggesting that DX100 maintains a robust trichothecene detoxifying mechanism. Furthermore, the nature of microbial de-epoxidation reaction and inhibition of the reaction by sodium azide and the finding that bacterial cell culture lysate retained activity suggests that certain cytoplasmic reductases may be responsible for the de-epoxidation activity.

Conclusions: This study reports the enrichment procedure for obtaining an effective and stable microbial consortium DX100 capable of de-epoxidizing several food contaminating trichothecene mycotoxins. DX100, which has de-epoxidation ability under wide range of conditions, represents a unique enzymatic source which has great industrial potential for reducing contamination of foods/feeds with multiple trichothecenes, and minimizing their synergistic/additive cytotoxic effects on consumer health.

Keywords: Biodetoxification, Food contamination, *Fusarium* mycotoxins, Microbial de-epoxidation, Trichothecenes

* Correspondence: Ting.Zhou@agr.gc.ca
[1]Guelph Food Research Centre, Agriculture and Agri-Food Canada, 93 Stone Road West, Guelph, Ontario N1G 5C9, Canada
Full list of author information is available at the end of the article

Background

The toxigenic *Fusarium* (mold) and other fungal pathogens contaminate cereal-based foods and feeds by producing a variety of health injurious trichothecene mycotoxins [1]. The pathogens deposit multiple mycotoxins in grain during the course of disease development in the crop field and during post-harvest storage [2, 3]. Importantly, the fungal toxins often enter the food chain [4], and exposures to them through consumption may cause serious mycotoxicosis in humans and animals [5–8].

Trichothecenes are chemically tricyclic sesquiterpenes, characterized by a double bond at the C9, 10 position and a C12, 13 epoxy functional group. The latter is the principal moiety responsible for the toxicity of these fungal secondary metabolites in eukaryotic organisms [9]. Type-A and B are the two types of fungal trichothecenes that often co-contaminate cereals [1, 10]. The two types of trichothecenes are separated by the presence of a ketone at position C8 in type-B [(e.g. DON (deoxynivalenol), also known as vomitoxin], which is absent in type-A toxins (e.g. T2-toxin, HT-2 toxin) [11]. Although the toxicities of the members of two trichothecene types vary, exposure to multiple mycotoxins may increase their cytotoxicity because of their synergistic and/or additive cytotoxic effects on eukaryotic organisms [12, 13]. Common toxicity effects of the trichothecenes in humans and animals include diarrhea, vomiting, feed refusal, growth retardation, immunosuppression, reduced ovarian functions/reproductive disorders and even death [14, 15].

The magnitude of food contamination with trichothecenes is influenced by various biotic and abiotic factors, and the risk can be further increased due to climate change [16]. Therefore, reduction of the toxins levels in contaminated foods remains a very difficult task [17]. Physical and chemical processes to remove trichothecenes from foods and feeds are either detrimental, ineffective or prohibitively expensive [18–20]. However, the use of detoxifying microorganisms to biodegrade mycotoxins, through de-epoxidation or other mechanisms, could be an effective alternative [21, 22]. Previous studies suggested that the toxicity of DON is mainly due to the epoxide moiety, and there is a significant reduction of DON toxicity (by 54%) when it is reduced to produce de-epoxy DON [23, 24]. These studies further revealed reductions of nivalenol (NIV) and T-2 toxin toxicities by 55 and 400%, respectively, when their de-epoxidized derivatives were tested.

The potential value of microorganisms with trichothecene de-epoxidation activity for improving food/feed quality and safety has led to the identification of bacteria isolated from chicken intestine and rumen fluid, that are able to cleave the epoxide ring of food contaminating trichothecenes under anaerobic conditions and relatively high temperature (~37 °C) [25, 26]. Furthermore, the anaerobic bacterial strain BBSH 797, which can de-epoxidize DON, scirpentriol and T-2 triol trichothecene mycotoxins, has been used as a feed additive for poultry and swine diets [27, 28]. Earlier attempts were made to obtain de-epoxidizing aerobic bacteria, but the culture did not retain DON/NIV de-epoxidation activity after subculturing [29, 30].

In the process of developing effective biotechnological strategies for eliminating trichothecenes from contaminated agricultural commodities, we previously reported an aerobic microbial consortium with a low rate of DON de-epoxidation activity, most likely due to the existence of undesired fast growing microorganisms in the culture [31]. The present study aimed to increase its de-epoxidation activity through selective enrichment of that culture, and to characterize the enriched culture including its ability to de-epoxidize a variety of *Fusarium* produced trichothecenes under aerobic and anaerobic conditions and wide range of temperatures.

Methods

Media and chemicals used

The media used in this study were: minimal medium (MM) = six mineral salts and 10.0 g/L sucrose [31], MSB = six mineral salts containing 0.5% Bacto Peptone [31], MSC = MSB containing reduced quantity (0.3%) Bacto Peptone, MSY = six mineral salts + 0.5% yeast extract, MS-DON = six mineral salts + 200 μg/mL DON, SE = sterile water-soluble soil extract [31], PDB = potato dextrose broth, LB = Luria Bertani, NB = nutrient broth and 1/10 NB = one-tenth strength NB. The pH of the media was adjusted as needed before autoclaving. Solid media were prepared by adding 1.5% agar. The media ingredients and prepared media were purchased from Sigma-Aldrich (Oakville, ON, Canada), Fisher Scientific (Fair Lawn, New Jersey, USA), and/or Difco (Sparks, Maryland, USA). All trichothecene mycotoxins were obtained from Biomin (Romer Lab, Union, MO, USA).

Enrichment and functional stability of trichothecene de-epoxidizing microorganisms

A soil microbial culture having low level of DON de-epoxidation activity originating from southern Ontario, Canada was used for enrichment [31]. An aliquot (20 μL) of the culture was transferred to a 1.5 mL eppendorf tube containing 0.5 mL MSB (pH 7.0). The culture was spiked with 200 μg/mL pure DON and incubated with continuous shaking at 220 revolutions per minute (rpm) at room temperature. After 7–10 days of incubation, an aliquot (10 μL) of the culture was transferred to fresh MSB containing 200 μg/mL DON. In negative controls no DX100 was added. This procedure was repeated twelve times and the microbial consortium was designated DX100. Before every transfer to fresh MSB the de-epoxidation activity of the culture was examined by using a standard method, liquid chromatography-ultraviolet-mass spectrometry (LC-UV-MS) [32]. Briefly, equal volumes of the bacterial culture and methanol were mixed and the samples were incubated for 30 min at room temperature to lyse the bacterial cells. Insoluble cell materials and media components in the mixture were eliminated by passing it through a 0.45 μm polyvinylidene difluoride membrane filter (Whatman®, Florham Park, New Jersey, USA). The percentage of DON to dE-DON (de-epoxy deoxynivalenol), also known as DOM-1, transformation in the filtrate was determined by LC-UV-MS technique [32].

To determine the stability of the microbial de-epoxidation activity an active culture was subcultured for one hundred generations in MSB containing 50 μg/mL DON. The final subculture was tested for its de-epoxidation activity using LC-UV-MS technique.

16S rRNA metagenomic library preparation and sequencing

Genomic DNA was extracted from the highly enriched bacterial culture using Gentra Puregene Yeast/Bacteria Kit (Qiagen Inc., Mississauga, Ontario, Canada). Amplicon libraries of the V3-4 region of the 16S rRNA gene were prepared following the procedure of Illumina 16S Metagenomic Sequencing Library Preparation Guide (Rev. B). Briefly, a ~ 550 bp fragment of the 16S rRNA V3-4 region was amplified using the primers Bakt 341F (5-cctacgg gnggcwgcag) and Bakt 805R (5-gactachvgggtatctaatcc) [33]. The Bakt 341F and Bakt 805R primers contained 5-tcgtcggcagcgtcagatgtgtataagagacag and 5-gtctcgtgggctcgga gatgtgtataagagacag 5' Illumina overhang adapter sequences, respectively. Each 25 μl volume reaction mixture contained 12.5 ng of template DNA, 200 nM of each primer and 1x KAPA HiFi HotStart ReadyMix (VWR, Mississauga, ON, Canada). PCR was performed following the conditions: 95 °C (3 min), 25 cycles of 95 °C (30 s), 55 °C (30 s) and 72 °C (30 s), followed by 72 °C (5 min). The PCR amplicons were purified using Ampure XP beads (Beckman Coulter, Mississauga, ON, Canada). Sequencing adapters containing 8 bp indices were incorporated into the 3' and 5' ends of the purified amplicons by PCR using the Nextera XT Index kit (Illumina, San Diego, CA, USA). The 50 μl PCR reaction contained 5 μl PCR amplicon, 5 μl each indexing primer, and 25 μl 2x KAPA HiFi HotStart ReadyMix. PCR was performed under the following conditions: 95 °C (3 min), 8 cycles of 95 °C (30 s), 55 °C (30 s) and 72 °C (30 s), followed by 72 °C (5 min). The PCR amplicons were purified with Ampure XP bead and quantified using the Quant-iT PicoGreen double-stranded DNA assay (Invitrogen/Life Technologies Inc., Burlington, ON, Canada). After pooling equimolar amounts of amplicons, 5% equimolar of PhiX DNA (Illumina) was added, and sequencing was performed on an Illumina MiSeq instrument, using the Illumina MiSeq 600-cycle v3 kit. Taxonomic composition of the sequences was determined with Illumina MiSeq Reporter v.2.5.1 using the metagenomics workflow.

Transformation of A- and B type trichothecene mycotoxins

DX100 was examined for its ability to degrade five type-A trichothecene mycotoxins (diacetoxyscirpenol, HT-2 toxin, neosolaniol, T-2 toxin and T2-triol) and five type-B trichothecene mycotoxins (3-acetyl-deoxynivalenol, 15-acetyl-deoxynivalenol, fusarenonX, nivalenol and verrucarol). One hundred and fifty micrograms of each mycotoxin were added to 1.5 mL NB in 5 mL microcentrifuge tubes (Sigma Aldrich), resulting in a 100 μg/mL mycotoxin solutions. Each tube was inoculated with 100 μL of an overnight DX100 culture, and the cultures in NB + mycotoxin were incubated at 28 °C with continuous shaking at 180 rpm, while the negative controls lacked DX100. The percentage of trichothecene transformation to different compounds (including de-epoxy and de-acetylated forms) by the culture was analyzed [32]. Five replicate cultures were evaluated for each trichothecene mycotoxin.

Effects of DON concentration, substrate, temperature, pH, and aeration on microbial de-epoxidation activity

To determine the effects of DON concentration on the growth and de-epoxidation activity of DX100, tubes containing 3 mL of NB or MSB spiked with low (20 μg/mL), high (200 μg/mL) or very high (500 μg/mL) levels of DON were inoculated with DX100. The cultures were grown for 16 h with continuous shaking at 220 rpm and then the cell density was adjusted to 0.25 OD_{600nm} (10^7 cells/mL). While incubating the culture at 28 °C with shaking at 180 rpm, 1.5 ml was removed every 12 h and the OD_{600nm} was measured. The same volume of microbial culture (bacteria and broth) was analyzed for de-epoxidation activity by LC-UV-MS technique [32]. Three replicate cultures for each mycotoxin were tested.

The effect of oxygen was evaluated by inoculating 5 mL of NB or MSB in 15 mL falcon tubes with an overnight culture of DX100. A moderate amount of DON (50 μg/mL) was added to the cultures and they were incubated at 28 °C with shaking at 180 rpm. Bacterial growth (OD_{600nm}) and de-epoxidation activity were determined every 6 h [32]. The assay was repeated with MSB medium under anaerobic conditions [31]. Three replicate cultures were considered for each experiment.

To examine the effects of substrate on microbial de-epoxidation activity DX100 was grown at 28 °C for 72 h in NB, 1/10 NB, MM, MS-DON, MSB, MSC, MSY, SE, PDB or LB containing 50 μg/mL DON. The media pH values for this assay were maintained at pH 7.0 ± 0.2. To evaluate the effect of temperature, MSB containing 50 μg/mL DON was inoculated with DX100 and incubated at 10, 15, 20, 25, 30, 35, 40, 45 or 50 °C. The effect of pH on growth and de-epoxidation of DX100 was studied with MSB + 50 μg/mL DON adjusted to pH 5.5, 6.0, 6.5, 7.0, 7.5 or 8.0, and the cultures were grown at 28 °C for 72 h. Three replicate cultures were tested for each experiment.

Effects of bacterial energy inhibitor, and localization of microbial de-epoxidation enzyme

In order to determine if the trichothecene de-epoxidation is bacterial and enzymatic, overnight cultures of DX100 grown in MSB containing 50 μg/mL DON were treated with 0.01% and 0.001% (w/v) of the bacterial energy inhibitor, sodium azide (NaN_3) (Sigma-Aldrich), which blocks gram-negative bacterial electron transport [34]. Controls were prepared without SA, and three replicate cultures were maintained for each treatment. The de-epoxidation activity of the DX100 culture was analyzed after incubation at 28 °C for 72 h with shaking at 180 rpm [32].

In a separate assay, bacterial cell lysates were prepared using B-PER II Bacterial Protein Extraction Reagent following the manufacturer's instruction (PI 78260, Thermo Scientific, Rockford, IL, USA). DX100 was cultured for 48 h in 1.5 mL MSB under the conditions described above. After collecting the cell pellet by centrifuging one-third (0.5 mL) of DX100 culture at 5000 × g for 10 min 2 mL B-PER II reagent was added to it. The cells were disrupted by pipetting the pellet suspension up and down several times. Immediately 20 μL Halt Protease Inhibitor Cocktail (Thermo Scientific, Rockford, IL, USA) was added to the cell lysate. To prepare culture filtrate the DX100 culture (0.5 mL) was passed through a 0.22 μm EMD filter Millex™ Sterile Syringe Filters: Durapore™ PVFD Membrane (Fisher Scientific). The remaining 0.5 mL of the DX100 culture was used as positive control. Cell lysate, culture filtrate and control were spiked with 50 μg/mL DON and incubated at 28 °C. The de-epoxidation activity of the cell lysate, culture filtrate and control were analyzed after a 72 h incubation [32]. Three replicate cultures were tested.

Data analysis

Data were organized using Microsoft Excel Program 2007, and the statistical analyses were carried out using SAS procedures of General Linear Model for Windows version 9.1 (SAS Institute, Cary, NC, USA, 2002–03). Average values of the replications were expressed as means ± standard errors. The mean differences among treatments were considered significant at $p < 0.05$ according to Tukey's-HSD test.

Results

Microbial enrichment, functional stability and identification of bacteria in trichothecene transforming bacterial consortium

Three months of co-incubating the soil microbes with 200 μg/mL DON in MSB led to an enriched bacterial consortium DX100 that efficiently de-epoxidized trichothecenes, after 100 subcultures in broth. The next generation metagenomic sequencing technology determined an altered composition of the bacterial consortium in DX100 compared to the original culture that was used for enrichment. The enriched DX100 consortium is comprised of different known and unknown bacterial genera, dominated by aerobic *Stenotrophomonas* (80.03%), followed by anaerobic *Blautia* (11.19%) and unknown microbes (3.38%) (Fig. 1a). In contrast, the initial culture contained a heterogenous community comprised of species of aerobic and anaerobic bacterial genera (namely, *Serratia, Clostridium, Citrobacter, Enterococcus, Stenotrophomonas and Streptomyces*), and dominated by *Serratia* [35]. Further bioinformatics analyses elucidated the percentage occurrences of known and unknown bacterial species in DX100 (Fig. 1b). The predominant species was identified to be *Stenotrophomonas geniculata* (38.55%), followed by *S. pavanii* (21.72%), *Blautia coccoides* (4.81%), *S. retroflexus* (1.23%), *S. terrae* (0.56%), *S. maltophilia* (0.44%) and *Alkaliphilus crotonatoxidans* (0.38%). Noticeably, a significant percentage (30.68%) of species in DX100 was identified to be unknown.

Microbial transformation of A- and B type trichothecene mycotoxins

Aerobic cultures of DX100 grown for 72 h in NB + 100 μg/mL HT-2 transformed 74% of the mycotoxin to de-epoxy HT-2 toxin, and 26% to a de-acetylated derivative (Fig. 2, Table 1). The culture incubated with T2-toxin transformed 46% of the mycotoxin to the de-epoxy T-2 product, 16% to an unknown compound and left 38% of the toxin untransformed. In the DX100 culture containing diacetoxyscirpenol (DAS), 61% and 39% was transformed into de-epoxy DAS and de-acetylated DAS, respectively. DX100 transformed 55% neosolaniol (NEO) to de-epoxy NEO and 45% to de-acetylated NEO derivatives. The bacterial culture showed capacity to transform T2-triol into two derivatives, namely: 47% de-epoxy T2-triol and 53% de-acetylated T2-triol.

DX100 was also grown under aerobic conditions for 72 h in NB + 100 μg/mL of five type-B trichothecenes, separately. NIV was completely transformed into 92% de-

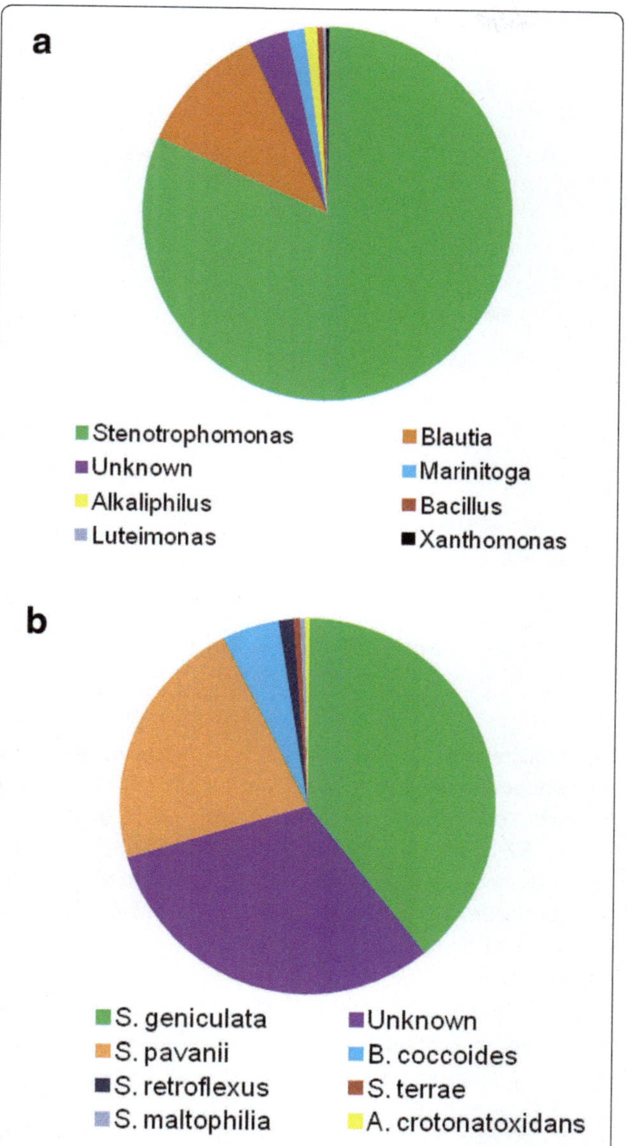

Fig. 1 Identification and abundance of bacterial genera and species in trichothecene de-epoxidizing DX100 microbial consortium. **a**: Relative abundance of known and unknown bacterial genera in DX100. **b**: Relative abundance of known and unknown bacterial species in DX100. The areas of different color represent abundance of different bacterial genera or species present in DX100. The initials of bacterial species A, B and S indicate genera *Alkaliphilus*, *Blautia* and *Stenotrophomonas*, respectively

epoxy and 8% DON products. The culture transformed 3-acetyldeoxynivalenol (3-ADON) into three products, namely: de-epoxy NIV (5%), de-epoxy DON (54%) and DON (41%). DX100 was able to transform acetyldeoxyni valenol (15-ADON) into de-epoxy 15-ADON (8%) and de-epoxy deoxynivalenol (dE-DON) (92%). Most (97%) of verrucarol (VER) was converted to de-epoxy VER, and 95% FusarenonX (FUS) was transformed into de-epoxy FUS. The remaining 3% VER and 5% FUS were not transformed. No spontaneous transformation of trichothecenes in the negative controls was observed.

Effects of different growth factors on bacterial de-epoxidation activity

To determine if trichothecene concentration can affect the growth and de-epoxidation activity of DX100, the most common contaminant in food, namely DON, was tested. Growth of DX100 bacteria, aerobically, in MSB with 20 and 200 μg/mL DON was not significantly different from that observed in control cultures without the trichothecene (Fig. 3a). However, a reduced rate of bacterial growth was determined in MSB containing 500 μg/mL DON. Similarly, increasing the mycotoxin concentration up to 200 μg/mL did not affect bacterial growth in NB medium (Fig. 3b). Bacterial de-epoxidation activity [determined as μg/mL DON converted/cell density (OD_{600})] was lowest in MSB + 20 μg/mL DON, and highest in MSB + 500 μg/mL DON containing cultures (Fig. 3c). The highest rate of dE-DON formation was 11.5 μg/h in DX100 cultures in MSB with 500 μg/mL DON between 48 to 72 h. All of the mycotoxin was converted in the cultures containing 20 μg/mL between 48–60 h and in the cultures containing 200 μg/mL between 60–72 h. The consortium showed complete de-epoxidation of a high quantity of DON (500 μg/mL) within 96 h. In 200 μg/mL and 500 μg/mL of DON, the rate of de-epoxidation increased significantly after 36 h.

DX100 grew more quickly in aerobic conditions in MSB + 50 μg/mL DON than in MSB + 50 μg/mL under anaerobic conditions (Fig. 4a). Aerobic growth was faster in broth containing NB + 50 μg/mL trichothecene compared to MSB + 50 μg/mL. However, the de-epoxidation activities, on a cell density basis, were very similar in NB + 50 μg/mL DON compared to MSB + 50 μg/mL (Fig. 4b). Also, de-epoxidation activity of DX100 under aerobic conditions in MSB + 50 μg/mL trichothecene was higher than in the same medium under anaerobic conditions.

Growth of DX100 in media containing 50 μg/mL DON under aerobic conditions at 72 h was three-fold higher in NB and LB than in SE and 1/10 NB and did not occur in MS-DON (Fig. 5a). Growth in MM, MSB and MSY was intermediate. Over the 72 h, de-epoxidation per cell density was highest in MSB and MSY and approximately 25% lower in NB and LB and lower by 75% and 87% in 1/10 NB and MSC, respectively (Fig. 5b). No de-epoxidation was observed in MM, MS-DON, SE and PDB in 72 h. Growth of DX100 under aerobic conditions for 72 h in MSB + 50 μg/mL DON was greatest between 25-40 °C, but no growth occurred at 50 °C (Fig. 5c). De-epoxidation activity was greatest and equivalent between 20-40 °C and lower at 10, 15 and 45 °C ($p < 0.05$) (Fig. 5d). Growth of the consortium

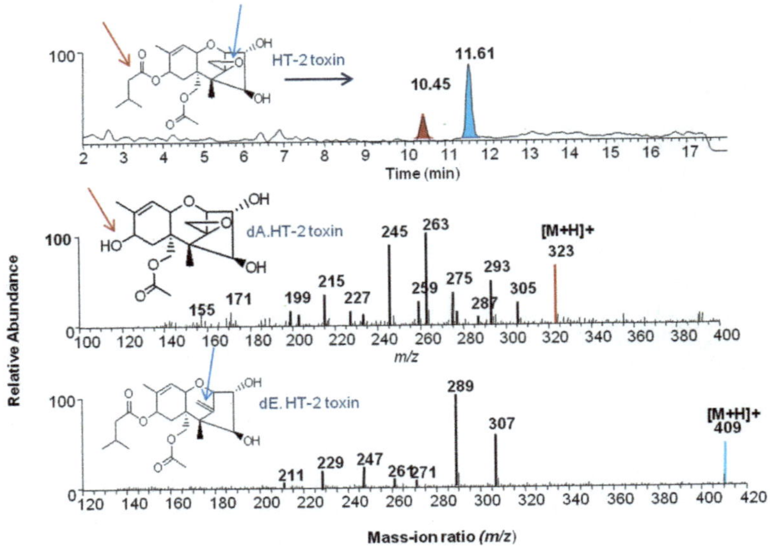

Fig. 2 Mass spectrometric evidence of microbial de-epoxidation and de-acetylation of HT-2 trichothecene mycotoxin. The upper panel illustrates the mass spectrometer chromatograph of the de-acetylated and de-epoxidized products. The *middle* and *lower panels* show the mass spectra of the transformed compounds. *Arrows* on the *upper panel* indicate the targeted acetyl (*red*) and epoxy (*blue*) functional groups of HT-2 toxin. *Red* and *blue arrows* on the *middle* and *lower panels* identify the degradation sites of HT-2 toxin. dA.HT-2 = de-acetylated HT-2, and dE.HT-2 = de-epoxidized HT-2 toxin

under aerobic conditions for 72 h in MSB + 50 μg/mL DON was limited to pHs near neutral (i.e. 6.0 - 7.5) and the activity was not significantly different between pH 6.5 to 7.5 ($p < 0.05$) (Fig. 5e, f). No de-epoxidation occurred below 6.0 or above 7.5 this pH.

Energy requirements and localization of bacterial de-epoxidation enzyme

Treatment of DX100 microbial consortium with the 0.001% and 0.01% NaN_3 for 72 h reduced de-epoxidation activity to 5% (2.5 μg/mL) and 0%, respectively. In contrast, the control with no NaN_3 added was unable to inhibit de-epoxidation activity of DX100. A cell lysate of DX100 transformed 20% (10 μg/mL) of 50 μg/mL DON to dE-DON within 72 h. In contrast, a cell-free culture filtrate did not show de-epoxidation activity, although the positive control having bacterial cells showed 100% de-epoxidation activity.

Discussion

This study reports a highly enriched, stable and effective microbial consortium DX100 capable of de-epoxidizing eleven trichothecene mycotoxins under aerobic and

Table 1 Transformation of type-A and B trichothecene mycotoxins by the DX100 microbial consortium. DAS = diacetoxyscirpenol, DON = deoxynivalenol, FUS = FusarenonX, 3-ADON = 3-acetyl-deoxynivalenol and 15-ADON = 15-acetyl-deoxynivalenol

Trichothecene mycotoxins		Untransformed trichothecenes (%)	Transformed products of trichothecenes (%)		
			De-epoxy	De-acetylate	Other[a]
Type-A	HT-2 toxin	0	74 ± 2.08	26 ± 2.08	-
	T-2 toxin	38	46 ± 0.58	-	16 ± 0.58 (unknown)
	DAS	0	61 ± 2.33	39 ± 2.33	-
	Neosolaniol	0	55 ± 2.51	45 ± 2.51	-
	T2-triol	0	47 ± 1.15	53 ± 1.15	-
Type-B	DON	0	100 ± 0.33	-	-
	Nivalenol	0	92 ± 3.38	-	8 ± 3.38 (DON)
	3-ADON	0	59 ± 1.76		41 ± 1.76 (DON)
	15-ADON	0	100 ± 0.66	-	-
	FUS	5	95 ± 2.90	-	-
	Varrucarol	3	97 ± 1.45	-	-

[a]Transformation of trichothecene mycotoxins to either DON or an unknown compound. The mean values and average ± standard errors of five replicate observations are presented

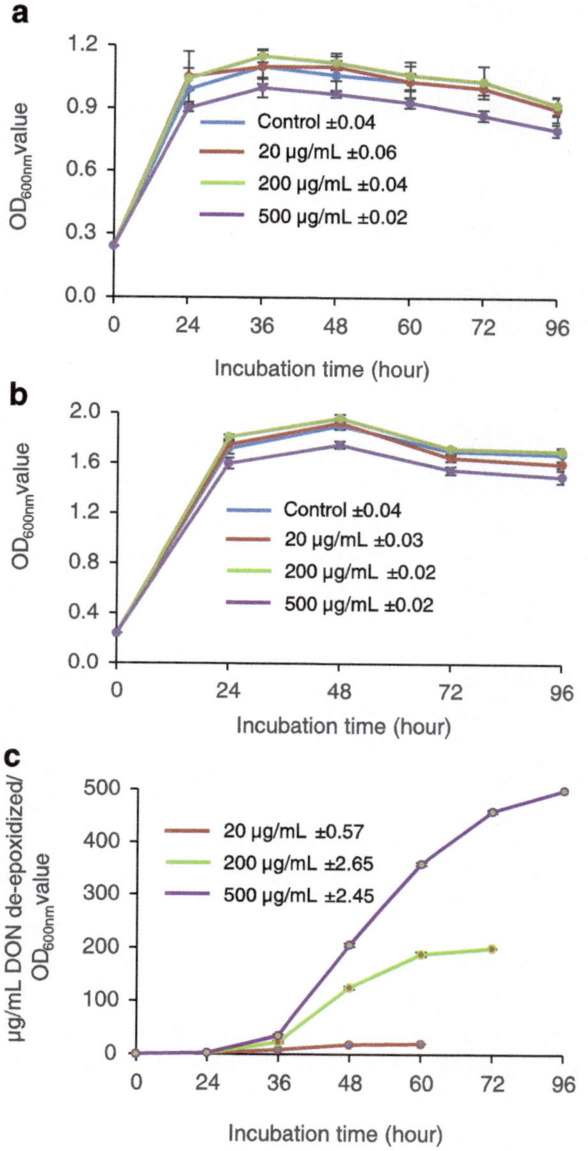

Fig. 3 Effect of deoxynivalenol concentrations on the growth and de-epoxidation activity of DX100 microbial consortium. **a**: Microbial cell density (OD$_{600nm}$) in MSB with different DON concentrations measured at 24, 36, 48, 60, 72 and 96 h. **b**: Microbial cell growth (OD$_{600nm}$) in NB with different DON concentrations, measured at 24, 48, 72 and 96 h. **c**: Temporal DON de-epoxidation activity of DX100 in MSB containing 20, 200 and 500 μg/mL DON. The relative activity was measured by quantifying de-epoxidation activity divided by microbial growth (OD$_{600nm}$ value) at different time points. Results are the mean of three replicate observations, and bars shown are ± standard errors of the means. The averages ± standard errors for three treatments and the control are presented in the graph. MSB = broth containing six mineral salts + 0.5% Bacto Peptone, DON = deoxynivalenol, and NB = nutrient broth

anaerobic conditions and broad range of temperatures. Since co-applications of multiple microbial strains often have greater beneficial activity [22, 36, 37], the DX100 microbial consortium has a great potential to be used as an efficient biodetoxifier of animal feedstuffs contaminated with multiple co-contaminating trichothecene mycotoxins; therefore, improving food/feed quality and safety.

The microbial consortium was found to be comprised of over 30% of unknown bacterial species. This indicates the possibility of novel bacterial involvement in the trichothecenes transformation, and such bacteria appeared to be difficult to isolate using the currently available methods. Our attempts, over a period of a few years, failed to produce a stable pure strain with trichothecene de-epoxidation activity. Although a DON de-epoxidizing aerobic bacterium (*Citrobacter freundii* A47) was isolated from the same source as the DX100 [35]. Unfortunately, the strain did not retain DON de-epoxidizing activity after subculturing on agar media. Also the bacterium could not be detected in DX100, suggesting that the strain died during enrichment process. Isolation of trichothecene de-epoxidizing bacteria through selective enrichment procedures is also challenging because the de-epoxidation reaction involves removal of an oxygen molecule [31, 38], which does not supply any nutrients to bacteria [39, 40] preventing the utilization of trichothecenes as sole carbon or nitrogen sources for selective isolation of de-epoxidizing bacteria from soil samples. This is supported by the present work, which demonstrated that DX100 was unable to grow in media containing DON as a sole carbon source. Previous attempts to obtain aerobic bacteria with de-epoxidation activity were also successful, but the cultures lost their ability to detoxify trichothecenes after subculturing [29, 30]. Remarkably, DX100 consortium which was achieved through prolonged incubation of soil culture with high concentrations of DON demonstrated stable activity under aerobic conditions by de-epoxidizing trichothecenes after 100 generations in broth medium, and retained its activity when it was tested after two years of preservation at −80 °C.

Although trichothecenes are toxic to eukaryotes, prokaryotes appear to be resistant to these fungal secondary metabolites [30, 41]. In agreement with those findings the present study demonstrated that DX100 grew well in cultures with high concentration of DON (500 μg/mL) and metabolized it completely. Taking advantage of this capacity, we were able to increase the population of trichothecene de-epoxidizing bacteria in a culture by prolonged subculturing of soil microbes in media with 200 μg/mL DON. This is supported by the observation that a 50-fold dilution of DX100 showed de-epoxidation activity, while before the enrichment up to 20-fold dilution of the original culture (which used for enrichment in this study) was able to de-epoxidize trichothecene. This suggests that during each of twelve seven-day long enrichment steps the de-epoxidizing bacterial cells survived by gaining energy from co-metabolically cleaving

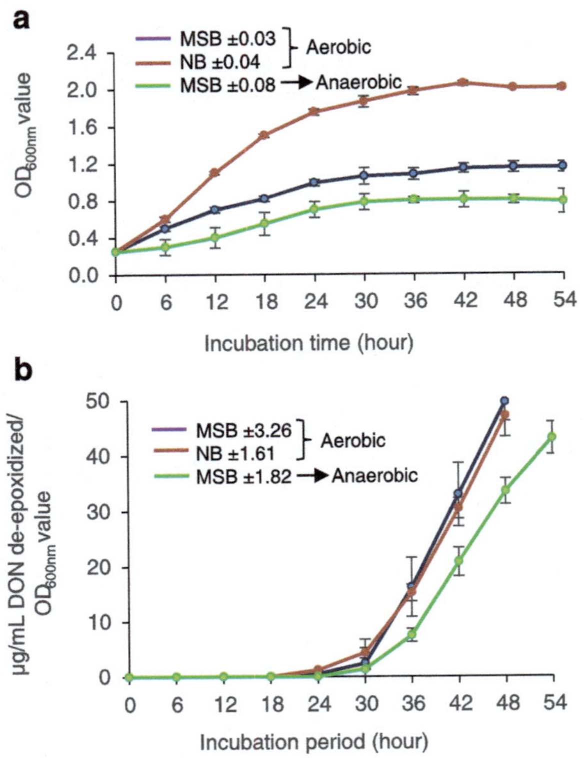

Fig. 4 Relative growth and de-epoxidation capabilities of DX100 microbial consortium under aerobic and anaerobic conditions at moderate temperature. **a**: Microbial growth (determined by OD_{600} value) under aerobic conditions in MSB and NB medium and anaerobic conditions in MSB medium. **b**: Microbial DON de-epoxidation ability under aerobic conditions in MSB and NB media and anaerobic conditions in MSB medium. Tubes containing MSB or NB were supplemented with 50 µg/mL DON. The activity was measured by quantifying de-epoxidation activity divided by microbial growth (OD_{600} value) at different time points. Values are means of triplicate experiments, and bars shown are ± standard errors of the means. The average ± standard errors for the treatments are indicated in the graph. MSB = broth containing six mineral salts + 0.5% Bacto Peptone, NB = nutrient broth, DON = deoxynivalenol

the epoxy ring of the trichothecene, while the non-trichothecene metabolizing microbial populations declined by starvation. This explanation is supported by our observation that microbial de-epoxidation in broth mainly occurred at the stationary growth phase, when the bacteria had fewer nutrients, since at this growth stage bacteria increased their de-epoxidation rate when grown with higher concentrations of DON (500 µg/mL). Such microbial energy gain from co-metabolic reactions has been reported by Chiu and colleagues [42].

Treating the DX100 culture with 0.01% NaN_3 completely blocked de-epoxidation activity. NaN_3 inhibits cellular energy production, specifically in gram-negative bacteria, by interfering with catalase/cytochrome c oxidase activity, while gram-positive bacteria are mostly resistant to this energy inhibitor [34, 43]. This suggests a possible role for certain gram-negative bacteria in the observed trichothecene de-epoxidation activity [30]. Although the next generation metagenomics technology identified a low percentage of gram-positive anaerobic bacterial species, namely *Alkaliphilus crotonatoxidans*, *Bacillus* spp., and *Blautia coccoides*, the aerobic species of *Stenotrophomonas* (gram-negative) were found to be the predominant ones present in DX100. It remains unknown if the de-epoxidation of the variety of trichothecenes observed in the present study is the additive activities of a number of individual bacteria and/or the consequences of intra- and interspecific microbial interactions. However, interpretation of the data predicts that certain strains of *Stenotrophomonas* could be the major agents for aerobic de-epoxidation of trichothecenes, and the role of gram-positive bacteria (which are strictly anaerobic and endospore forming) could attributed to anaerobic de-epoxidation. Strains belonging to *Stenotrophomonas* demonstrated efficient biodegradations of fungal toxins and other environmental pollutants [44]. These bacteria were not identified as the most abundant ones in our initial soil-derived microbial culture that was enriched to obtain DX100. Therefore, the increase of de-epoxidation activity of DX100 could be further attributed to the elevation of de-epoxidizing bacterial population during the three-month enrichment process with high concentration DON.

Among 11 trichothecenes tested, DX100 efficiently transformed 10 to de-epoxy forms and/or DON and/or unknown compounds. Probably, because of the activity of specific microbial dehydrooxylase(s) NIV and 3-ADON were transformed to DON via removal of a hydroxyl group at the C4 position and the acetyl moiety, respectively. However, the reasons for DX100's inability to subsequently de-epoxidize DON in those two assays are obscure.

Microbial de-epoxidation in this study occurred in intact and lysed bacterial cells. This implies that the small molecule DON entered into bacterial cells and became metabolized there by bacterial cytoplasmic enzymes. This agrees with previous studies that suggested that absorption of trichothecenes by human cells occurred via simple diffusion, and that bacterial cytoplasmic flavoproteins were involved in de-epoxidation of epoxyalkanes [45, 46]. The gene(s) and enzymes involved in trichothecene de-epoxidation have not been identified yet, however, the reductive nature of the bacterial trichothecene de-epoxidation reaction and inhibition of de-epoxidation activity by NaN_3 suggests that an active electron transport process is essential and an epoxy reductase is involved in this biological reaction [30].

The functionality of such reductases appeared to be greatly affected by media composition [47, 48]. For example, this study showed that decreasing the

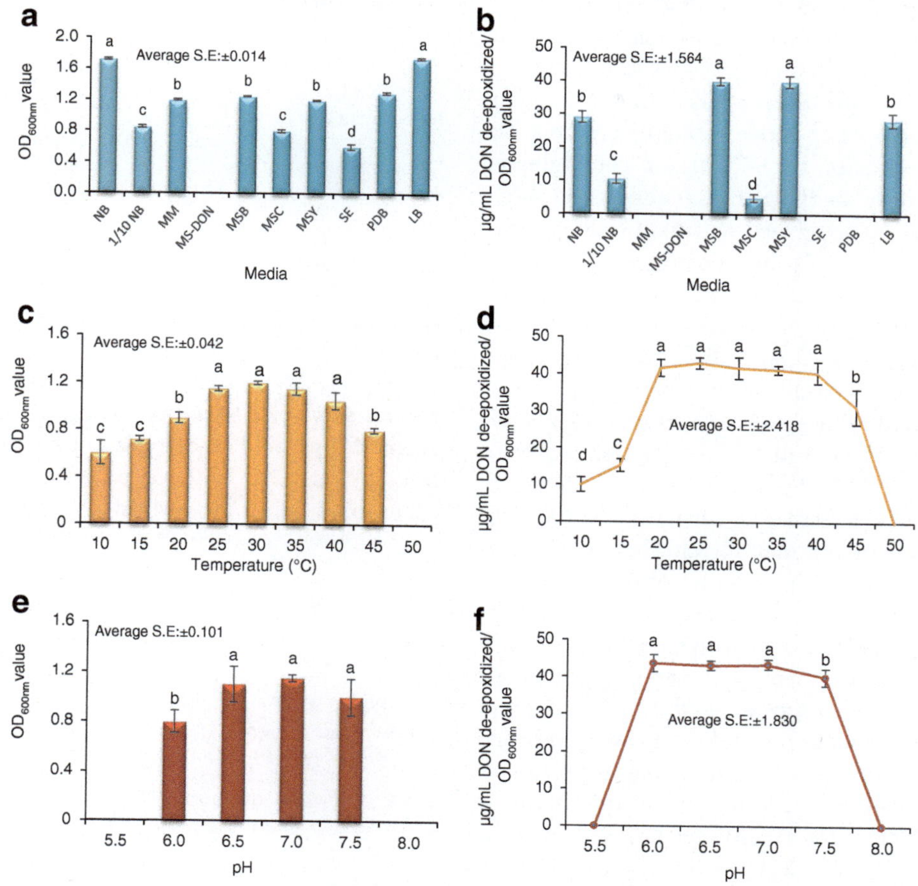

Fig. 5 Effects of temperature, substrate and pH on growth and de-epoxidation activities of DX100 under aerobic conditions. **a**: OD_{600nm} values show relative growth performance of DX100 in different broth media. **b**: Bacterial de-epoxidation activity (DON to dE-DON transformation) in different broth media. **c**: OD_{600nm} values demonstrate the effects of temperatures on the growth capacity of DX100 in MSB. **d**: Bacterial DON de-epoxidation in µg/mL under different temperatures in MSB. **e**: OD_{600nm} values show growth behavior of DX100 at different pHs in MSB. **f**: Bacterial de-epoxidation (DON to dE-DON transformation) in µg/mL in different pHs. The activity was measured by quantifying de-epoxidation activity divided by microbial growth (OD_{600} value) under different growth conditions. Results are the mean of five replicate observations, and bars shown are ± standard errors of the means. The averages of ± standard errors for different growth conditions are presented in the graph. Treatments labeled with the same letter are not significantly different, according to Tukey's HSD test ($p < 0.05$). Media: NB = nutrient broth, 1/NB = 1/10 strength NB, MM = Minimal medium, MS-DON = six mineral salts having 200 µg/mL DON. dE-DON = de-epoxy deoxynivalenol, MSB = six mineral salts + 0.5% Bacto Peptone, MSC = six mineral salts broth + 0.3% Bacto Peptone, MSY = six mineral salts with 0.5% yeast extract, S.E = standard error, SE = soil extract, PDB = potato dextrose broth and LB = Luria Bertani

concentration of Bacto Peptone in MS (from 0.5% to 0.3%) dramatically lowered the bacterial de-epoxidation performance, suggesting that the DON de-epoxidation activity of DX100 relies on the availability of certain proteins and amino acids. The inhibition of microbial de-epoxidation performance in high sugar containing MM and PDB media suggests the repressive effects of high sugar contents on the de-epoxidation enzymatic activity of DX100 [49]. Therefore, a screening strategy for microbial trichothecene de-epoxidation should use different media and enrichment procedures in parallel, as was done in the initial enrichment process of DX100.

All together, the achievement and characterization of DX100 microbial consortium may facilitate the development of an effective technology for biodetoxification of trichothecene contaminated foods and animal feeds. Furthermore, the knowledge we have gained from this study would assist in isolating the trichothecene de-epoxidizing genes and enzymes, which could, potentially, also be used to detoxify foods and feeds.

Conclusions

The capacity of DX100 to de-epoxidize several mycotoxins could be an advantage to develop effective strategies for minimizing the synergistic/additive toxic effects of these toxic compounds on consumer health. The de-epoxidation activity of DX100 under aerobic and anaerobic conditions and wide range of temperatures as well as its ability to retain activity after many generations and years makes it a unique microbial source. Following similar formulation method to Mycofix® (encapsulation of bacteria) DX100 can be used as a feed additive [27] as well as the consortium has a potential to be utilized as a biodetoxifier of feedstuffs. Such applications of the consortium are pending on positive evaluation of the de-epoxidation performance of DX100 on contaminated feedstuffs followed by studying the suitability and safety of the treated materials for feed and food uses. Further improvement of de-epoxidation activity under nutrient rich conditions via increasing inoculum size and/or optimizing microbial growth factors could make the consortium an effective feed detoxifier. As an alternative, the trichothecene de-epoxidizing enzymes that might be obtained from DX100 through fermentation or a suitable aerobic process could be utilized for treatments of contaminated foods and feeds. Furthermore, analyses of meta-transcriptomes of active and inactive DX100 consortium could lead to the identification of novel genes responsible for aerobic de-epoxidation of trichothecenes. Since the fungal toxins enhance disease severity in grain [50], transfer and constitutive expression of the identified novel gene(s) into cereal genomes could generate cultivars resistant to *Fusarium* pathogens and their toxins. Such achievements not only improve yields and quality of grain but also minimize the exposure of humans and animals to these health hazardous fungal toxins while protecting the environment by reducing the applications of toxic chemicals in the environment for controlling trichothecene producing fungal pathogens.

Abbreviations

ADON: 3-acetyldeoxynivalenol; DAS: Diacetoxyscirpenol; dE-DON: De-epoxy DON; DON: Deoxynivalenol; FUS: FusarenonX; LB: Luria bertani; LC-UV-MS: Liquid chromatography-ultraviolet-mass spectrometry; MM: Minimal medium; MSB: Mineral salts broth containing 0.5% Bacto Peptone; MSC: Six mineral salts broth containing 0.3% Bacto Peptone; MS-DON: Mineral salts broth containing DON; MSY: Six mineral salts broth containing 0.5% yeast extract; NaN_3: Sodium azide; NB: Nutrient broth; NEO: Neosolaniol; NIV: Nivalenol; OD: Optical density; PDB: Potato dextrose broth; rpm: Revolution per minute; SE: Soil extract; VER: Verrucarol

Acknowledgements

The authors thankfully acknowledge the AAFC/BARD (Canada-Israel) Binational Program for funding this project. We are grateful to the Natural Sciences and Engineering Research Council of Canada (NSERC) for awarding a scholarship to Rafiq Ahad for Ph.D. program at the University of Guelph, Canada. The authors appreciate the contributions of Dr. Chris Young, Dr. Xiu-Zhen Li, Dr. Jianwei He and Honghui Zhu to this work.

Funding

The work was financed by the AAFC/BARD (Canada-Israel) Binational Program to Dr. Ting Zhou.

Authors' contribution

RA, TZ and KPP were involved in developing the project. RA and DL conducted the experiments and analyzed data. RA drafted the manuscript, and TZ, KPP and DL reviewed and approved the final manuscript.

Competing interests

The authors declare that they have no competing interests.

Author details

[1]Guelph Food Research Centre, Agriculture and Agri-Food Canada, 93 Stone Road West, Guelph, Ontario N1G 5C9, Canada. [2]Department of Plant Agriculture, University of Guelph, Guelph, Ontario N1G 2W1, Canada.

References

1. Schollenberger M, Müller HM, Rüfle M, Suchy S, Plank S, Drochner W. Natural occurrence of 16 Fusarium toxins in grains and feedstuffs of plant origin from Germany. Mycopathologia. 2006;161:43–52.
2. Homdork S, Beck R. Influence of different storage conditions on the mycotoxin production and quality of fusarium-infected wheat grain. J Phytopathol. 2000;148:7–15.
3. Bianchini A, Horsley R, Jack MM, Kobielush B, Ryu D, Tittlemier S, Wilson WW, Abbas HK, Abel S, Harrison G, Miller JD. DON occurrence in grains: a North American perspective. Cereal Foods World. 2015;60:32–56.
4. Bryden WL. Mycotoxins in the food chain: human health implications. Asia Pac J Clin Nutr. 2007;16:95–101.
5. Pestka JJ, Smolinski AT. Deoxynivalenol: toxicology and potential effects on humans. J Toxicol Environ Health. 2005;14(8):39–69.
6. Sherif SO, Salama EE, Abdel-Wahhab MA. Mycotoxins and child health: The need for health risk assessment. Int J Hyg Environ Health. 2009;212:347–68.
7. Miller JD, Schaafsma AW, Bhatnagar D, Bondy G, Carbone I, Harris LJ, Harrison G, Munkvold GP, Oswald IP, Pestka JJ, Sharpe L. Mycotoxins that affect the North American agri-food sector: state of the art and directions for the future. World Mycotoxin J. 2013;7:63–82.
8. Streit E, Naehrer K, Rodrigues I, Schatzmayr G. Mycotoxin occurrence in feed and feed raw materials worldwide: long-term analysis with special focus on Europe and Asia. J Sci Food Agric. 2013;93:2892–99.
9. Desjardins AE. Fusarium mycotoxins: chemistry, genetics, and biology. Minnesota: American Phytopathological Society; 2006.
10. Vanhoutte I, Audenaert K, De Gelder L. Biodegradation of mycotoxins: Tales from known and unexplored worlds. Front Microbiol. 2016;7.
11. Omurtag GZ. Fumonisins, trichothecenes and zearalenone in cereals. Int J Mol Sci. 2008;9:2062–90.
12. Binder EM, Tan LM, Chin LJ, Handl J, Richard J. Worldwide occurrence of mycotoxins in commodities, feeds and feed ingredients. Anim Feed Sci Technol. 2007;137:265–82.
13. Alassane-Kpembi I, Kolf-Clauw M, Gauthier T, Abrami R, Abiola FA, Oswald IP, Puel O. New insights into mycotoxin mixtures: The toxicity of low doses of Type B trichothecenes on intestinal epithelial cells is synergistic. Toxicol Appl Pharmacol. 2013;272:191–8.
14. Pestka JJ. Deoxynivalenol: mechanisms of action, human exposure, and toxicological relevance. Arch Toxicol. 2010;84:663–79.
15. Alizadeh A, Braber S, Akbari P, Garssen J, Fink-Gremmels J. Deoxynivalenol impairs weight gain and affects markers of gut health after low-dose, short-term exposure of growing pigs. Toxins. 2015;7:2071–95.
16. Magan N, Aldred D, Medina A. Food security, climate change and mycotoxins. Qual Assur Saf Crop. 2012;4:145.
17. Miller JD. Mycotoxins in food and feed: a challenge for the 21st century. In Biology of Microfungi; Li, D.-W., Ed. Cham: Springer International Publishing. 2016; p 469–493.
18. Kushiro M. Effects of milling and cooking processes on the deoxynivalenol content in wheat. Int J Mol Sci. 2008;9:2127–45.
19. Numanoglu E, Gökmen V, Uygun U, Koksel H. Thermal degradation of deoxynivalenol during maize bread baking. Food Addit Contam Part A. 2012;29:423–30.
20. Cheli F, Pinotti L, Rossi L, Dell'Orto V. Effect of milling procedures on mycotoxin distribution in wheat fractions: A review. LWT-Food Sci Technol. 2013;54:307–14.
21. Awad WA, Ghareeb K, Böhm J, Zentek J. Decontamination and detoxification strategies for the Fusarium mycotoxin deoxynivalenol in animal feed and the effectiveness of microbial biodegradation. Food Addit Contam. 2010;27:510–20.

22. Murugesan GR, Ledoux DR, Naehrer K, Berthiller F, Applegate TJ, Grenier B, Phillips TD, Schatzmayr G. Prevalence and effects of mycotoxins on poultry health and performance, and recent development in mycotoxin counteracting strategies. Poultry Sci. 2015;94:1298–15.
23. Swanson SP, Helaszek C, Buck WB, Rood HD, Haschek WM. The role of intestinal microflora in the metabolism of trichothecene mycotoxins. Food Chem Toxicol. 1988;26:823–9.
24. Eriksen GS, Pettersson H, Lundh T. Comparative cytotoxicity of deoxynivalenol, nivalenol, their acetylated derivatives and de-epoxy metabolites. Food Chem Toxicol. 2004;42:619–24.
25. Fuchs E, Binder EM, Heidler D, Krska R. Structural characterization of metabolites after the microbial degradation of type A trichothecenes by the bacterial strain BBSH 797. Food Addit Contam. 2002;19:379–86.
26. Yu H, Zhou T, Gong J, Young C, Su X, Li XZ, Zhu H, Tsao R, Yang R. Isolation of deoxynivalenol-transforming bacteria from the chicken intestines using the approach of PCR-DGGE guided microbial selection. BMC Microbiol. 2010;10:182.
27. He J, Zhou T, Young JC, Boland GJ, Scott PM. Chemical and biological transformations for detoxification of trichothecene mycotoxins in human and animal food chains: a review. Trends Food Sci Technol. 2010;28(21):67–76.
28. Karlovsky P. Biological detoxification of the mycotoxin deoxynivalenol and its use in genetically engineered crops and feed additives. Appl Microbiol Biotechnol. 2011;91:491–504.
29. Ueno YO, Nakayama K, Ishii K, Tashiro F, Minoda Y, Omori T, Komagata K. Metabolism of T-2 toxin in *Curtobacterium* sp. strain 114–2. Appl Environ Microbiol. 1983;46:120–27.
30. He PI, Young LG, Forsberg C. Microbial transformation of deoxynivalenol (vomitoxin). Appl Environ Microbiol. 1992;58:3857–63.
31. Islam R, Zhou T, Young JC, Goodwin PH, Pauls KP. Aerobic and anaerobic de-epoxydation of mycotoxin deoxynivalenol by bacteria originating from agricultural soil. World J Microbiol Biotechnol. 2012;28:7–13.
32. Guan S, He J, Young JC, Zhu H, Li XZ, Ji C, Zhou T. Transformation of trichothecene mycotoxins by microorganisms from fish digesta. Aquaculture. 2009;290:290–95.
33. Herlemann DP, Labrenz M, Jurgens K, Bertilsson S, Waniek JJ, Andersson AF. Transitions in bacterial communities along the 2000 km salinity gradient of the Baltic Sea. ISME J. 2011;5:1571–9.
34. Lichstein HC, Soule MH. Studies of the effect of sodium azide on microbic growth and respiration: I. The action of sodium azide on microbic growth. J Bacteriol. 1944;47:221.
35. Islam R. Isolation, Characterization and Genome Sequencing of a Soil-Borne *Citrobacter Freundii* Strain Capable of Detoxifying Trichothecene Mycotoxins (Doctoral dissertation, University of Guelph, Guelph, ON, Canada). 2012. https://atrium.lib.uoguelph.ca/xmlui/bitstream/handle/10214/3580/rafiqul_islam_PhD%20thesis_May3.pdf?sequence=1.
36. Jetiyanon K, Kloepper JW. Mixtures of plant growth-promoting rhizobacteria for induction of systemic resistance against multiple plant diseases. Biol Control. 2002;24:285–91.
37. Siddiqui IA, Shaukat SS. Mixtures of plant disease suppressive bacteria enhance biological control of multiple tomato pathogens. Biol Fert Soils. 2002;36:260–68.
38. Yoshizawa T, Takeda H, Ohi T. Structure of a Novel Metabolite from Deoxynivalenol, a Tricho-thecene Mycotoxin, in Animals. Agric Biol Chem. 1983;47:2133–35.
39. Horvath RS. Microbial co-metabolism and the degradation of organic compounds in nature. Bacteriol Rev. 1972;36:146–55.
40. Fedorak PM, Grbić-Galić D. Aerobic microbial cometabolism of benzothiophene and 3-methylbenzothiophene. Appl Environ Microbiol. 1991;57:932–40.
41. Rocha O, Ansari K, Doohan FM. Effects of trichothecene mycotoxins on eukaryotic cells: a review. Food Addit Contam. 2005;22:369–78.
42. Chiu TC, Yen JH, Hsieh YN, Wang YS. Reductive transformation of dieldrin under anaerobic sediment culture. Chemosphere. 2005;60:1182–89.
43. Stannard JN, Horecker BL. The in vitro inhibition of cytochrome oxidase by azide and cyanide. J Biol Chem. 1948;1948(172):599–608.
44. McCormick SP. Microbial detoxification of mycotoxins. J Chem Ecol. 2013;39:907–18.
45. Westphal AH, Swaving J, Jacobs L, de Kok A. Purification and characterization of a flavoprotein involved in the degradation of epoxyalkanes by *Xanthobacter* Py2. European J Biochem. 1998;257:160–68.
46. Sergent T, Parys M, Garsou S, Pussemier L, Schneider YJ, Larondelle Y. Deoxynivalenol transport across human intestinal Caco-2 cells and its effects on cellular metabolism at realistic intestinal concentrations. Toxicol Lett. 2006;164:167–76.
47. Völkl A, Vogler B, Schollenberger M, Karlovsky P. Microbial detoxification of mycotoxin deoxynivalenol. J Basic Microbiol. 2004;44:147–56.
48. Zhou G, Zhang H, He Y, He L. Identification of a chitin deacetylase producing bacteria isolated from soil and its fermentation optimization. Afr J Microbiol Res. 2010;4:2597–603.
49. Saier MH, Roseman S. Inducer exclusion and repression of enzyme synthesis in mutants of *Salmonella typhimurium* defective in enzyme I of the phosphoenolpyruvate: sugar phosphotransferase system. J Biol Chem. 1972; 247:972–75.
50. Mesterházy Á. Role of deoxynivalenol in aggressiveness of *Fusarium graminearum* and *F. culmorum* and in resistance to Fusarium head blight. Eur J Plant Pathol. 2002;108:675–84.

Event-specific detection of transgenic potato AV43-6-G7 using real-time and digital PCR methods

Hongwei Gao[1*], Xiaofan Yu[2], Tingting Deng[3], Min Sun[1], Xizhi Xiao[1], Xin Huang[3], Ying Chen[3] and Ronggui Li[2]

Abstract

Background: The isolation of unknown DNA sequences flanked by known sequences is an important task in the event-specific detection of GMOs. None of event-specific detection method was developed based on the junction sequence of an exogenous integrant in the transgenic potato AV43-6-G7.

Results: The flanking sequence between the exogenous fragment and recombinant chromosome of this potato was successfully acquired through exogenous gene 5′-RACE. The event-specific primers and *Taq*man probe were designed to amplify fragments spanning the exogenous DNA and potato genomic DNA. The specific real-time PCR and digital PCR detection methods for AV43-6-G7 potato were established based on primers designed according to the flanking sequences. The detection limit of the qualitative PCR assay was 0.01 % for AV43-6-G7 potato in 100 ng of potato genomic DNA, corresponding to approximately 11.6 copies of the potato haploid genome. The ddPCR assays for Potato AV43-6-G7 achieved a limit of quantification of approximately 58 target copies, with RSD ≤ 25 %. The aLOQ of this system was approximately 1.2 copies.

Conclusions: These results indicated that these event-specific methods would be useful for the identification of potato AV43-6-G7.

Keywords: Transgenic, Potato AV43-6-G7, Flanking sequence, Event Specific, Real-time PCR, Detection

Background

Since the first genetically modified tomato was approved, the number and production of genetically modified (GM) crops have continuously increased [1]. According to the International Service for the Acquisition of Agri-Biotech Applications (ISAAA), 29 types of plants and 390 transgenic events were developed and commercialized worldwide by the end of March 2016 [2]. However, the perceived risks of genetically modified organisms (GMOs) to food safety and the environment have raised concerns [3]. To provide consumers with information on food, many countries have developed legislations to determine the threshold of GMO ingredients. The threshold for mixing GMOs in feed and foodstuffs are 0.9 % in EU, 3 % in Korea, 5 % in Japan, and 1 % in New Zealand and Australia [4]. In China, the mandatory labelling of GMO products is required without setting a clear threshold value according to the Regulations on Administration of Agricultural GMOs Safety [5].

The isolation of unknown DNA sequences flanked by known sequences is an important task in the event-specific detection of GMOs. Both rapid amplification of cDNA ends (RACE) and thermal asymmetric interlaced PCR (TAIL-PCR) are effective methods for this purpose [6, 7]. Fraiturea et al. [8] also used a DNA walking strategy using a genome walking kit on rice products. To comply with GMO labelling regulations and ensure product legality and traceability, event-specific detection methods are essential for GMO identification and quantification. Event-specific detection was primarily performed at the DNA level using polymerase chain reaction (PCR)-based methods because of its high specificity for the flanking sequence of the exogenous integrant. In case of quantitative analysis, the Taqman-based real-time quantitative PCR (qPCR) method is preferred, reflecting the high accuracy (how close a measured value

* Correspondence: ghw75@126.com
[1]Shandong Entry-Exit Inspection and Quarantine Bureau of People's Republic of China, Qingdao, China
Full list of author information is available at the end of the article

is to the actual value) and precision (how close the measured values are to each other) of this method. However, this method has some obvious drawbacks. Notably, PCR inhibitors, such as chloroform, ionic detergents, etc., the experience of the technicians, and the matrix effect of the PCR tubes affect standard curve-based qPCR [9]. Particularly, the significant bias observed when the target is present at low concentrations could seriously limit the use of qPCR [9] Another requirement for qPCR is reference material for GMO at serial concentrations, as standard curves for quantities of endogene and exogene are prepared separately. The lack of certified reference material has been also a hindrance for qPCR. Despite the shortcomings of qPCR, this method is convenient for qualitative detection and cheap in terms of instruments and reagents compared with dPCR.

Digital PCR (dPCR) has rapidly developed since it was first conceived in the 1992 [10]. In dPCR, the reaction mixture is distributed in a large number of partitions containing zero, one or more copies of the target nucleic acid. After end-point amplification, the signal would be positive (1) or negative (0) for partitioned samples [10]. The absolute number of target nucleic acid molecules in the original sample prior to partitioning can be calculated from the ratio of positive to total partitions using Poisson distribution . Compared with the real-time PCR methods, PCR inhibitors do not impact dPCR, and this method does not need reference material for GMO at serial concentrations [11]. Digital PCR also is an effective approach to detect SNP variants in a diluted target sample [12, 13]. The development and application of absolute quantitative detection through the duplex chamber-based digital PCR of genetically modified maize events has been reported [13]. Stevanato demonstrated that dPCR could differentiate all varieties of sugar beet samples carrying specific SNP alleles associated with flowering at low concentrations (1 %) of the target sequence, while qPCR and HRM showed only a moderate frequency (10 %) [12].

Thus, this technique is better suited for international-trade applications. Currently, two designated dPCR technologies are commercially available: micro fluidic chamber digital PCR (cdPCR) and emulsion droplet digital PCR (ddPCR). Both commercial systems enable suitable GMO testing. According to our experience, each sample would cost approximately US $12.1 in ddPCR, while each sample should cost US $150 to US $400 per chip or plate in cdPCR [14]. In addition, ddPCR has higher repeatability than cdPCR according to the results of preliminary experiments using three genes [data not shown].

Potato is one of the first important transgenic plants [15]. The main purposes of gene transfer technology for potato were to improve the quality of potatoes and enhance pest-resistant varieties. Monsanto, a company in US, developed a range of anti-coleopteran pests and transgenic potatoes against leafroll virus prior to 1997 [16]. Recently, the quality of potato starch has been improved in more commercial transgenic potatoes, such as transgenic potato Amflora (EH92-527-1 potato) and transgenic potato AM04-1020 [17, 18]. These transgenic potatoes produce amylose-free starch through the antisense inhibition of the gene encoding granule bound starch synthase I (GBSSI) [19]. These improved functionalities are not only promising for technical applications in paper and textile, but also in the food industry for the production of dairy products, soups, sauces and noodles. The transgenic potato AV43-6-G7 in the present study is a variety of starch potato intended for starch isolation by the starch industry, and the starch might undergo further processing.

In the present study, the exogenous gene fragment and flanking sequence of transgenic potato AV43-6-G7 were obtained using the 5′-RACE method. According to information about the determined sequence, a pair of primers and a Taqman probe were designed based on the exogenous fragment and endogenous chromosomal sequence of potatoes. An event-specific detection method was developed for the transgenic potato AV43-6-G7 using real-time PCR and ddPCR.

Methods

Text materials and DNA preparation

A list of all test materials is provided in Additional file 1 the Supporting Information. All non-powder samples, including cottonseed, rapeseed, corn liquor dregs, and non-transgenic potatoes, were ground into a 60-mesh powder. The genomic DNA was extracted and purified using the Nucleic Acid Extraction System np968 (Tianlong Science and Technology Co., Ltd, Xi'an, China) in accordance with the manufacturer's instructions. This DNA extraction method involves four successive steps: (1) each ground sample (50 mg) was added to 300 μL DNA extraction buffer; (2) the samples were homogenized and incubated with lysis buffer; (3) the samples were centrifuged, and the supernatant collected; and (4) the nucleic acids were extracted using magnetic beads.

The relative sensitivity of the samples was determined according to the procedure below. A series of different percentages 100 %, 10 %, 1 %, 0.1 %, 0.01 %, and 0.001 % (v/v) of the transgenic potato preps were made individually after mixing transgenic potato DNA-containing supernatant with non-transgenic potato DNA-containing supernatant. The 100 % transgenic potato preps were obtained after mixing 1000 μL of transgenic potato DNA-containing supernatant with 0 μL of non-transgenic potato DNA-containing supernatant, and 10 % transgenic potato preps were was obtained after mixing 100 μL of transgenic potato DNA-containing supernatant with 900 μL of non-transgenic potato DNA-containing supernatant, etc. The transgenic potatoes were converted into non-transgenic potatoes at 100 %, 10 %, 1 %, 0.1 %, 0.01 %, and 0.001 % (v/v). The quantity of DNA samples was measured at 260 nm and prepared into solutions of 10 ng/μL for real-time PCR and ddPCR. The DNA purity was evaluated based on A260/A280 ratios using a UV spectrophotometer (BioPhotometer plus, Eppendorf, Germany).

Determination of the 5′-flanking DNA sequence

The 5′-flanking DNA sequence of transgenic potato AV43-6-G7 was determined using the 5′-RACE method. Based on the partial sequence of the lac gene in heterologous DNA in potato AV43-6-G7 [19], the 5′ ends were obtained through the rapid amplification of cDNA ends (RACE) using gene-specific and adapter primers. For 5′-RACE, PCR was performed using the gene-specific outer primer MAV43 R1 and primer AP1. The DNA sample was amplified a 25-μL reaction in a 0.2-mL tube containing 200 nM of primer gene-specific outer primer MAV43 R1 and primer AP1, 2× GC Buffer (TaKaRa DRR02AG, China), 2 mM magnesium chloride, 200 nM each of dATP, dCTP, dGTP, and dTTP, 1 unit of LA Taq (TaKaRa DRR02AG, China) and 100 ng of the extracted DNA template. The PCR program was set at 95 °C for 5 min, followed by 10 cycles at 94 °C for 30 s, 68 °C (less than 0.8 °C per cycle) for 30 s, and 72 °C for 150 s and 30 - cycles at 94 °C for 30 s, 60 °C for 30 s and 72 °C for 150 s, with a final cycle at 72 °C for another 10 min.

Nested PCR was performed using the gene-specific outer primer MAV43 R2 and primer AP2. The DNA sample was amplified in a 25-μL reaction in a 0.2-mL tube containing 200 nM of each primer MAV43 R1 and primer AP1, 2× GC Buffer (TaKaRa DRR02AG, China), 2 mM magnesium chloride, 200 nM each of dATP, dCTP, dGTP and dTTP, 1 unit of LA Taq (TaKaRa DRR02AG, China), and100 ng extracted DNA template. The PCR was conducted at 95 °C for 5 min, followed by 28 cycles at 94 °C for 30 s, 58 °C for 30 s, and 72 °C for 150 s, with a final cycle at 72 °C for another 10 min.

Oligonucleotide primers and probe

The oligonucleotide primers and probe used in this study were designed using Primer Express version 3.0 (Applied Biosystems) and are listed in Table 1. For potato, the beta-fructosidase gene (fru) was selected as an endogenous reference gene [20]. The locations of the designed event-specific primer pairs and probes targeting the DNA sequence are shown in Fig. 1. All primers and probes were synthesized at Sangon Co. Ltd. (Shanghai, China).

Qualitative PCR assays

The potato endogenous beta-fructosidase gene (fru) and event-specific sequence of GM potato AV43-6-G7 were separately amplified using potato AV43-6-G7 DNA and 24 negative non potato AV43-6-G7 DNA samples.

Real-time PCR was performed with a real-time PCR instrument (Master Cycle Realplex4, Eppendorf, Germany). Each DNA sample was amplified in a 25-μL reaction in a 0.2-mL tube containing 200 nM each of primer F, primer R, and probe, 10× Taq Buffer (Transgene, Beijing, China), 2 mM magnesium chloride, 200 nM each of dATP, dCTP, dGTP and dTTP, 1.5 units of Taq DNA polymerase (Transgen, Beijing, China), and 100 ng of extracted DNA template [21]. After initial denaturation at 95 °C for 120 s, the PCR conditions were optimized through 45 cycles of amplification (at 95 °C for 15 s, and 60 °C for 30 s). In each RT-PCR setup, the reaction systems were filled with purified water (non template control, NTC) and negative sample DNA as negative controls. The same baseline was used for the threshold cycle (Ct) to compare the RT-PCR performances of the extracted DNA and negative DNA samples. For sensitivity tests of event-specific primers and probes, the concentration of the DNA template ranged from 0.001 to 100 ng.

Droplet digital PCR reactions and analysis

Droplet digital PCR was performed using a Bio-Rad QX100 droplet system (Bio-Rad, Pleasanton, CA, USA). The reaction mixtures contained 10 μL of 2× ddPCR™ Supermix (Bio-Rad, Pleasanton, CA, USA) for probe, 1 μL of primer and probe, and 5 μL of DNA template, in a total reaction volume of 20 μL. Each concentration was repeated in three parallel reaction cells. The optimized cycling program for ddPCR assays included a denaturation step at 95 °C for 10 min, followed by 45 cycles at 94 °C for30 s, 60 °C for 1 min, and 98°C for 10 min. After PCR cycling, the fluorescence signals were collected using a QX100 droplet reader. The data were acquired and analysed using Quanta Soft software [22].

The absolute sensitivity of the samples was realized after diluting 100 % genomic DNA to a 10 ng/μl solution. Subsequently, the 10 ng/μl solution was diluted 5 times at a 10-fold dilution.

Duplex droplet digital PCR reactions were performed using a FAM-labelled probe for the endogenous gene fru and a VIC-labelled probe for the AV43-6-G7-specific sequence at a concentration of 1×10^{-5} ng/μL [23], with 10 repeat reactions.

The detection data for ddPCR using Bio-Rad platforms was analysed using Bio-Rad analysis software (Bio-Rad). The number of positive partitions for both event-specific and endogenous amplifications for each sample were counted using Bio-Rad platforms. Based on the Poisson distribution, the original copy number of the different samples was calculated using the following equations [24]:

Table 1 Primers and probe in this study

Name	Sequence and labled fluorescence (5′- -3′)	Origin
potato endogenous gene *fru* -F	CTGCCTCCGTCAAGATTTGGTCACT	[12]
potato endogenous gene *fru*-R	CTCTTCCCTTTCTTGATGG	[12]
potato endogenous gene *fru*-P	FAM-ACTTGTAATTCATCAAGCCAT- BHQ1	[12]
MAV43 R1	GCAAGCTTGGCGTAATCATGGTCATA	this study
MAV43-R 2	CGCTCACAATTCCACACAACATACGA	this study
AP1 primer	GTAATACGACTCACTATAGGGC	this study
AP 2 primer	ACTATAGGGCACGCGTGGT	this study
Potato AV43-6-G7 primerF	GGTATCAGGTTCTGGAATAAGACCAA	this study
Potato AV43-6-G7 primerR	TGTCGTGCCAGCTGCATTA	this study
Potato AV43-6-G7 Probe	FAM-CCCGCGCGTTGGCCGAT-BHQ1	this study
Potato AV43-6-G7 Probe used in duplex ddPCR	VIC-CCCGCGCGTTGGCCGAT-BHQ1	this study

$$\text{A (AV43-6-G7 specific sequence)} = -\ln\left[(\text{N}-\text{X})/\text{N}\right] \times \text{N} \tag{1}$$

$$\text{B (endogenous gene fru)} = -\ln\left[(\text{N}-\text{Y})/\text{N}\right] \times \text{N} \tag{2}$$

In the above two equations, A (AV43-6-G7 specific sequence) and B (endogenous gene fru) were the estimated copy numbers of the endogenous and event-specific genes for each panel. N was the total number of partitions; and X and Y were the positive wells for the AV43-6-G7 event-specific genes and the endogenous gene fru. The GMO proportion was calculated using the following equation:

$$\frac{\text{A(AV43-6-G7 specific sequence)}}{\text{B(endogenous gene fru)}} \times 0.62 \tag{3}$$

In equation (3), 0.62 was used as a correction coefficient for the different copy numbers of endogenous gene and screening gene in the genome. Ten samples of pure GM potato AV43-6-G7 were utilized to examine the AV43-6-G7 event-specific sequence and the endogenous gene fru using digital PCR. The average ratio of AV43-6-G7 event-specific sequence and endogenous gene fru was 0.62.

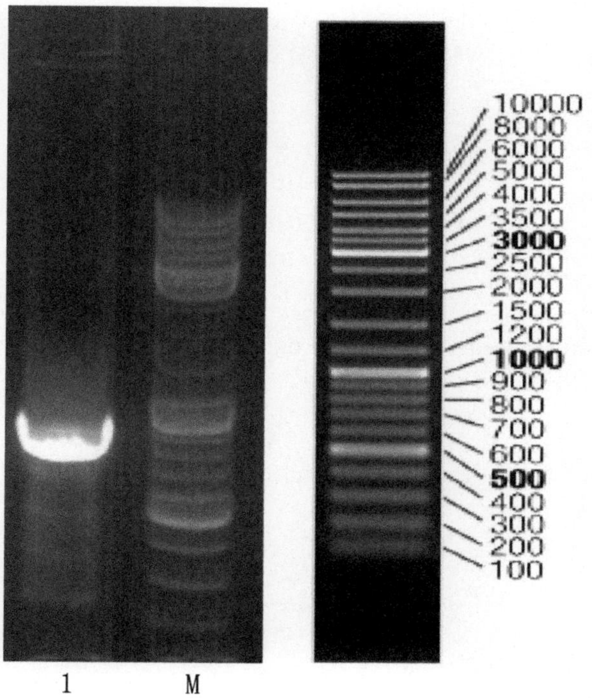

Fig. 1 Electrophoresis of transgenic AV43-6-G7 potato in flanking region using the second PCR and 5′-RACE. 1: products of PCR using primers MAV43-R2 and AP2. M: DNA markers

Results

Characterization of the 5′-flanking region of transgenic AV43-6-G7 potato

According to information of "Overall Application of Amylopectin Potato Event AV43-6-G7 Based on Regulation (EC) on 1829/2003" from the website (http://www.gmo-compass.org/pdf/regulation/potato/AV43-6-G7_application_food_feed_cultivation.pdf) [25], the integrated heterologous DNA in potato AV43-6-G7 is shown in Fig. 1. The DNA fragment (shown in Fig. 1) was derived from the 5′-flanking region using 5′-RACE.

To identify and analyse the junction region of the exogenous integrant and the host genomic DNA of potatoes, 4 specific primers (MAV43 R1, MAV43-R 2, AP1, and AP2) were designed based on the sequence of the Lac operon from M13 mp 19 to amplify the 5′-flanking sequences using 5′-RACE. The RACE-PCR product was sequenced. Figure 2 shows the 5′-junction sequence of an exogenous integrant of the AV43-6-G7 potato. This sequence comprised a 239-bp fragment from M13 mp 19 and an unknown 221-bp DNA sequence.

Because there is little sequence-related information on the potato genome available from the GenBank database, it was impossible to determine whether the unknown DNA

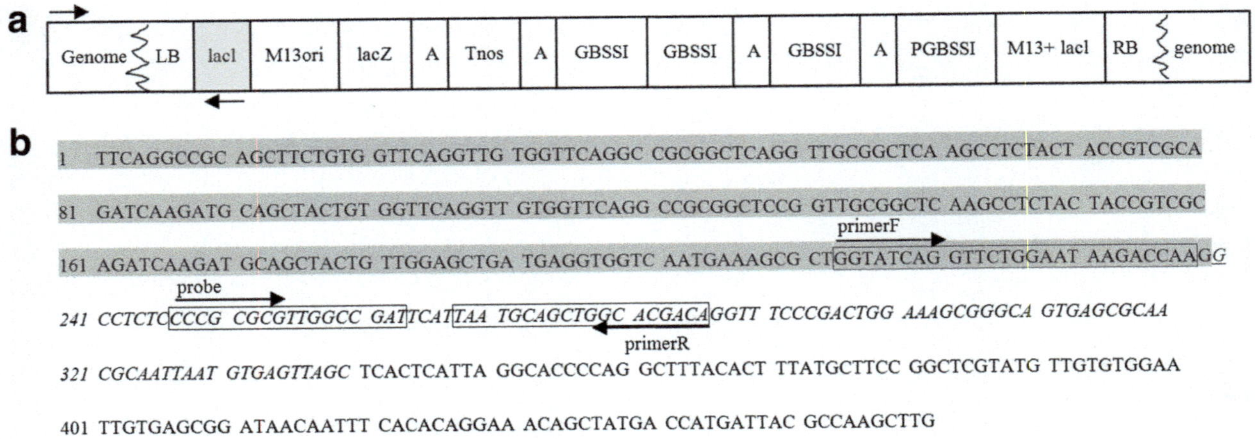

Fig. 2 Frame of transgenic elements in the AV43-6-G7 potato genome. **a** Schematic diagram of the integrated exogenous gene in AV43-6-G7 potato, where 'A' means artificial poly linker sequence; **b** Flanking sequence of the potato genome and left border sequence of exogenous gene. The sequence in the grey background was obtained from the potato genome, while the other sequence was exogenous and obtained from the left border and the *lacI* gene. The sequence of primer F was obtained from the potato genome, and that of primer R and probe was obtained from the AV43-6-G7 potato

sequence (221 bp) was derived from the potato genome. Therefore, 1 pair of specific primers (Table 1) was designed based on the 470-bp DNA sequence, and qualitative PCR was performed using the DNA of AV43-6-G7 potato, GMO PH05-026-0048 potato, wild potato, and 24 other GMO plant lines as templates. As expected, specific DNA amplification was observed in AV43-6-G7 potato, GMO PH05-026-0048 potato, and wild potato, but unobservable in the other 24 GMO plant lines (Fig. 3). Hence, the unknown DNA sequence was derived from the potato genome.

Specific for AV43-6-G7 potato of real-time PCR assay

Based on the revealed 5′-flanking sequence, event-specific primers were designed and used in a qualitative PCR assay of AV43-6-G7 potato. As expected, following the event-specific qualitative PCR assay of AV43-6-G7 potato, a passive real-time PCR amplification curve was obtained for AV43-6-G7 potato, whereas no amplification curve was observed for the other GM lines [26] (Fig. 3).

The amplification curves and standard curve were obtained for GM AV43-6-G7 potato using real-time

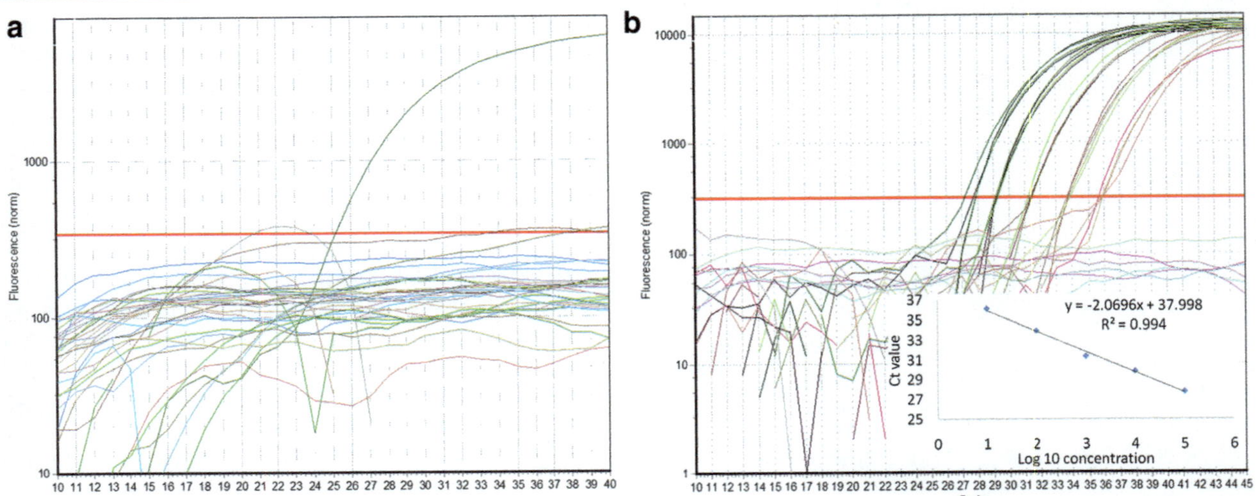

Fig. 3 The amplification curves and standard curve for detecting GM AV43-6-G7 potato using real-time PCR. **a** Specificity Tests of AV43-6-G7 Potato and the Other 24 Plant Samples. **b** Amplification Plots and Standard Curve for the Event-specific Quantitative PCR Assay of AV43-6-G7 Potato. Amplification curves (the amount of the serially diluted genomic DNA of AV43-6-G7 potato was 100, 10, 1, 0.1, 0.01, and 0.001 ng AV43-6-G7 potato in a total of 100 ng DNA, respectively corresponding to 100 %, 10 %, 1 %, 0.1 %, 0.01 %, and 0.001 % of AV43-6-G7 potato haploid genome in a total of 100 ng of DNA per reaction) were generated to quantify AV43-6-G7 potato

PCR. (A) Specificity Tests of AV43-6-G7 Potato and Other 24 Plant Samples. (B) Amplification Plots and Standard Curve for the Event-specific Quantitative PCR Assay of AV43-6-G7 Potato. Amplification curves (the amount of the serially diluted genomic AV43-6-G7 potato DNA was 100, 10, 1, 0.1, 0.01, and 0.001 ng AV43-6-G7 potato DNA in a total 100 ng of DNA, respectively corresponding to 100 %, 10 %, 1 %, 0.1 %, 0.01 %, and 0.001 % of the AV43-6-G7 potato haploid genome in a total of 100 ng DNA per reaction) were generated to quantify AV43-6-G7 potato.

Determination of the LOD of real-time PCR

To test the limit of detection (LOD) of the established method of event-specific real-time PCR, the DNA templates diluted along seven gradients were prepared at concentrations of 100 %, 10 %, 1 %, 0.1 %, 0.01 %, and 0.001 %, corresponding to 100, 10, 1,0.1, 0.01, and 0.001 ng AV43-6-G7 potato haploid genome in a total of 100 ng of DNA per reaction, respectively. A total of 4 parallel reactions were performed among triplicate samples. As expected, AV43-6-G7 potato DNA at 0.01 ng, corresponding to approximately 0.01 % ng of genomic DNA of AV43-6-G7 potato, was detected. The quantitative PCR results showed that the squared regression coefficient (R2) of the standard curve was 0.994 (Fig. 3). The high PCR efficiency and favourable linear relationship between the log DNA quantity and the fluorescence value (Ct) suggested that these primers and probes were suitable for quantitative measurements of AV43-6-G7 potato genomic DNA. Hence, the LOD of the event-specific quantitative PCR assay was 0.01 ng/action genomic DNA.

Determining the LOQ of ddPCR

Given the limitations of ddPCR for the quantification of GMO in food and feed samples, particularly at low target levels and in some complex matrices, generally accepted minimum performance requirements for analytical methods should be satisfied [27]. To avoid biases in comparing real-time PCR with ddPCR, inter-laboratory validation of real-time qPCR assays were alternatively used to validate the ddPCR using minimum adaptation. Therefore, apart from the master mix and settings specific to the QX100 droplet system, the primers and probe nucleotide sequences and concentrations, DNA concentration, and PCR thermo-profile were the same as those used in the real-time PCR assays (Fig. 4).

The absolute limit of quantification (LOQ) is the lowest target copy number in a sample that can be reliably quantified with an acceptable level of precision and accuracy [28]. In the absolute sensitivity analysis, all samples for determining dynamic range were prepared through serial dilution of a single stock AV43-6-G7 potato-specific DNA sample. The genomic DNA of AV43-6-G7 potato was tested along 5 concentration gradients. For each sample, copies of the endogene and flanking sequence were separately measured using ddPCR. For individual endogene

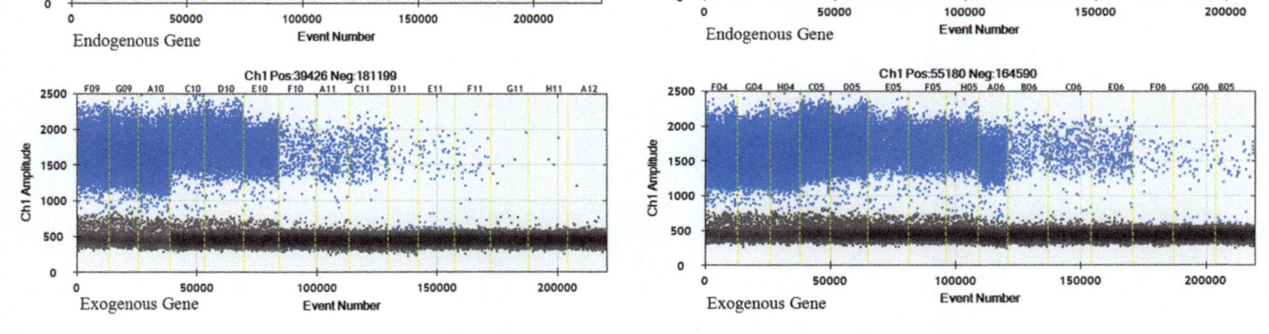

Fig. 4 The digital PCR picture of serial concentration of genomic DNA in AV43-6-G7 potato in absolute sensitivity and relative sensitivity analysis. **a** Analysis of Absolute Sensitivity. DNA was diluted along five concentration gradients, including 10, 1, 0.1, 0.01, and 0.001 ng/μL, from left to right. For each concentration, three parallel samples were tested. **b** Analysis of Relative Sensitivity. Samples with a concentration of 100 %, 50 %, 10 %, 1 %, and 0.1 % of AV43-6-G7 potato haploid genome per reaction concentration are shown from left to right, respectively. Three parallel samples were tested for each concentration

Table 2 Relative sensitivity for the AV43-6-G7 potato

Prepared concentration	AV43-6-G7 potato genome DNA	Total template DNA	Copies of AV43-6-G7 potato genome	Group average concentration	Group RSD
100 %	50 ng/ reaction	50 ng/ reaction	57948.5	100.2 %	0.5 %
50 %	25 ng/ reaction	50 ng/ reaction	28974.5	53.4 %	0.2 %
10 %	5 ng/ reaction	50 ng/ reaction	5795.0	10.6 %	2.8 %
1 %	0.5 ng/ reaction	50 ng/ reaction	579.5	1.1 %	4.0 %
0.1 %	0.05 ng/ reaction	50 ng/ reaction	58.0	0.12 %	8.2 %

targets and for the flanking sequence content, the coefficient of correlation R^2 was 0.9849 and 0.9842, respectively (Additional file 2). The correlation coefficients obtained from ddPCR met the requirements (R^2.0.98) set by the European Union Reference Laboratory for GM Food and Feed for acceptance of a quantitative PCR-based detection method for GMO [28] (Table 2).

In analysing relative sensitivity, the DNA template was diluted at a concentration of 100 %, 50 %, 10 %, 1 %, and 0.1 % of AV43-6-G7 potato haploid genome, respectively. When 0.1 % sample of AV43-6-G7 potato in a total of 50 ng/reaction potato DNA was detected, the experimental values and group RSD (relative standard deviation) were high, thus the accuracy of sample concentrations was higher than 0.1 % AV43-6-G7 potato in a total 50 ng/ reaction potato DNA. The LOQ of each target group in the ddPCR assays was estimated as the lowest copy number within the dynamic with a relative standard deviation (RSD) of the measured copy number ≤25 % [28]. When the nucleic acid concentration is 50 ng per reaction, the 0.1 % concentration of the target nucleic acid can be quantified as 0.12 %, with an RSD of 8.2 %. The results in duplex droplet digital PCR reactions with 0.1 % target nucleic acid were quantified as 0.09 %, with an RSD of 9.7 % [13, 29] (Fig. 5).

This result was consistent with the requirements of the test methods [28]. Because the genome size of the potato is 840 Mb [30], the LOQ for ddPCR was 58 copies in a total of 50 ng DNA/reaction, consistent with a range from 30 to 100 copies in routine qPCR testing [22] (Table 3).

Discussion

This finding demonstrates that the AV43-6-G7 potato/ hmg ddPCR assay can be performed over a wide range of target concentrations to determine the AV43-6-G7 potato content in a given sample and that values of approximately 11.6 copies constitute the range of quantification with ddPCR. The absolute limit of quantification (aLOQ) is the lowest target copy number in a sample that can be reliably quantified with acceptable precision and accuracy [31]. The aLOQ of ddPCR systems for hmg or AV43-6-G7 potato was estimated as the lowest copy number of 1.2 (Table 2), which was within the dynamic range with a

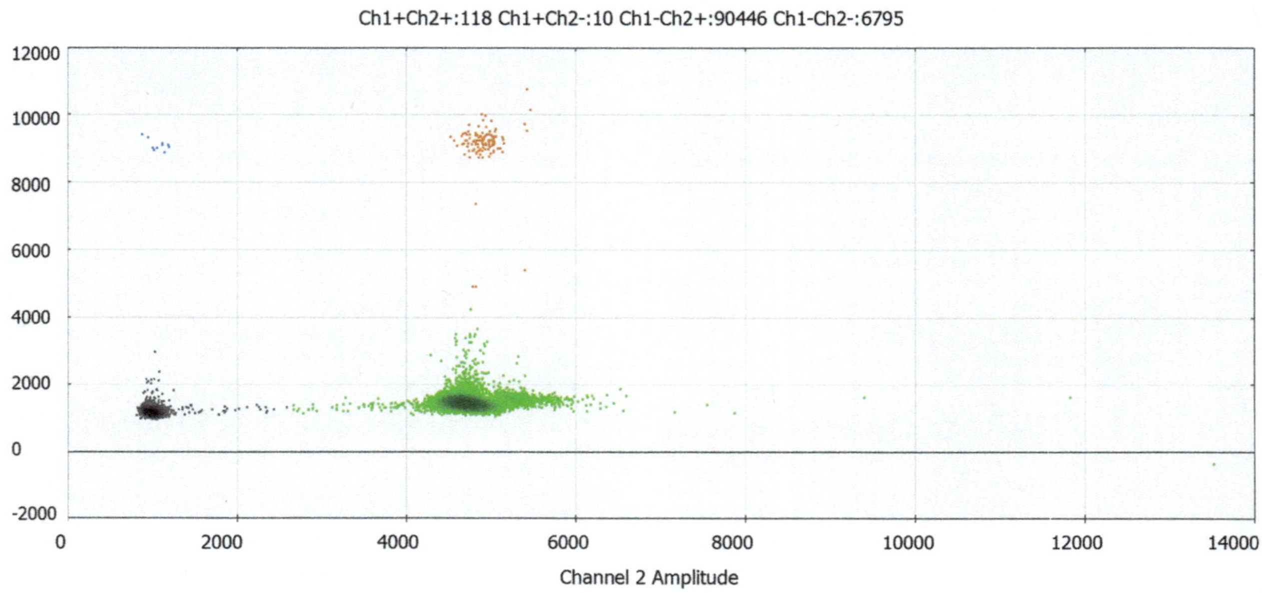

Fig. 5 The amplification hot maps of ddPCR using the Bio-Rad system. For the different colours in C, black indicates negative for both FAM and VIC, the blue and green indicate positive only for FAM or VIC, and orange indicates positive for both FAM and VIC

Table 3 Absolute sensitivity for AV43-6-G7 potato

DNA concentration of target	Copies of AV43-6-G7 potato genome	Group average	Group RSD
10 ng/reaction	11589.5	103.32 %	5.8 %
1 ng/ reaction	1159.0	107.31 %	3.5 %
0.1 ng/ reaction	115.9	93.39 %	8.7 %
0.01 ng/ reaction	11.6	97.29 %	17.6 %
0.001 ng/ reaction	1.2	95.14 %	24.9 %

coefficient of variation (CV) of the measured copy number of 25 % [28]. Based on this criterion, aLOQ was estimated at approximately 1.2 copies for the system. The results showed that ddPCR and real-time PCR have their own advantages. The applicability of ddPCR for GMO detection has previously been investigated and demonstrated as comparable to real-time PCR [32].

Conclusions

In the present study, the 5′-exogene integration sequence was isolated, and event-specific qualitative and quantitative PCR assays were performed for identifying and quantifying Potato AV43-6-G7. The LOD of real-time PCR was 11.6 copies. A total of 58 copies, corresponding to a concentration of 0.1 %, reflected the relative sensitivity. In relative sensitivity experience, the lowest concentration was 0.1 %, which was considered practically sensitive, as this value was lower than the thresholds of 0.9 % in EU. The ddPCR assays of Potato AV43-6-G7 introduced and performed here achieved a limit of quantification of approximately 58 target copies with an RSD ≤ 25 %. The aLOQ of this system was approximately 1.2 copies. These results suggested that these methods are suitable for detecting the GMO content in internationally traded products.

Abbreviations

aLOQ: absolute limit of quantification; cdPCR: chamber digital PCR; CV: coefficient of variation; ddPCR: droplet digital PCR; dPCR: digital PCR; GMO: genetically modified organisms; LOD: limit of detection; LOQ: limit of quantification; NTC: non-template control; qPCR: quantitative PCR; R^2: coefficient of correlation; RACE: rapid-amplification of cDNA ends; RSD: relative standard deviation

Acknowledgements

Not applicable.

Funding

This work was financially supported through grants from the Quality inspection public welfare research Project (201410014) and the National Grand Project of Science and Technology (2016ZX08012-001). The funders had no role in study design, data collection and analysis, decision to publish, or preparation of the manuscript.

Authors' contributions

HG performed the 5'-RACE, designed the primers and probe for real-time PCR, and drafted the manuscript. XY, MS, and XX performed the real-time PCR experiments. TD and YC performed the ddPCR experiments. XH and HL conceived the study, participated in the study design and coordination, and assisted with drafting the manuscript. All authors read and approved the final manuscript.

Competing interests

The authors declare that they have no competing interests.

Author details

[1]Shandong Entry-Exit Inspection and Quarantine Bureau of People's Republic of China, Qingdao, China. [2]Qingdao University Medical College, Qingdao, China. [3]Chinese Academy of Inspection and Quarantine Institute, Beijing, China.

References

1. Redenbaugh K, Hiatt B, Martineau B, Kramer M, Sheehy R, Sanders R, Houck C, Emlay D. Safety Assessment of Genetically Engineered Fruits and VegeTables: ACase Study of the Flavr Savr Tomato. Boca Raton: CRC Press; 1992. p. 288.
2. The International Service for the Acquisition of Agri-Biotech, pplications .ISAAA. http://www.isaaa.org/gmapproval-database/eventslist/default.asp. Accessed 5 Apr 2016.
3. Meyerson LA, Reaser JK. Biosecurity: moving toward a comprehensive approach. BioScience. 2002;52(7):593–600.
4. Broeders S, Keersmaecker S, Roosens N. How to deal with the upcoming challenges in GMO detection in food and feed. Biomed Res Int. 2012;Article ID 402418:11. http://dx.doi.org/10.1155/2012/402418.
5. Zhao J, Sheng JP. Problem and confusion about GMO food labeling. J China Agricultural University. 2015;20(3):1–8. Chinese.
6. Yeku O, Frohman MA. Rapid amplification of cDNA ends (RACE). Methods Mol Biol. 2011;703:107–22. doi:10.1007/978-1-59745-248-98.
7. Liu YG, Chen YL. High-efficiency thermal asymmetric interlaced PCR for amplification of unknown flanking sequences. Biotechniques. 2007;43(5):649–56.
8. Fraiturea MA, Hermanb P, Taverniersc I, Loosec MD, Deforced D, Roosensa NH. An innovative and integrated approach based on DNA walking to identify unauthorised GMOs. Food Chem. 2014;147(15):60–9.
9. Krüger N-J, Buhler C, Iwobi AN, Huber I, Ellerbroek L, Appel B, et al. "Limits of control" – crucial parameters for a reliable quantification of viable campylobacter by real-time PCR. PLoS One. 2014;9(2):e88108. doi:10.1371/journal.pone.0088108.
10. Baker M. Digital PCR, hits its stride. Nat Methods. 2012;9(6):541–4.
11. Burns M, Mburrell A, Afoy C. The applicability of digital PCR for the assessment of detection limits in GMO analysis. Eur Food Res Technol. 2010;231(3):353–62.
12. Stevanato P, Biscarini F. Digital PCR as new approach to SNP genotyping in sugar beet. Sugar Tech. 2015. doi:10.1007/s12355-015-0408-8.
13. Zhu PY, Fu W, Wang CG, Du ZX, Huang KL, Zhu SF, Xu WT. Development and application of absolute quantitative detection by duplex chamber-based digital PCR of genetically modified maize events without pretreatment steps. Anal Chim Acta. 2016;916(15):60–6.
14. Dany M, Dejan S, Mojca M, Kristina G, Jana Z, et al. Quantitative analysis of food and feed samples with droplet digital PCR. PLoS One. 2013;8(5):e62583.
15. Jhansi RS, Usha R. Transgenic plants: Types, benefits, public concerns and future. J Pharm Res. 2013;6(8):879–83.
16. Atoevs E, Fguenthner J, Jjohnson A, Smcintosh C, Kthornton M. An industry perspective of all-native and transgenic potatoes. AgBioForum. 2011;14(1):14–9.
17. Ryffel GU. Making the most of GM potatoes. Nat Biotechnol. 2010;28:318.
18. BASF Plant Science Company GmbH. Application EFSA-GMO-MS-2010-XX Application for authorizations of Amylopectin potato AM04-1020 for food and feed uses, processing and cultivation according to regulation (EC) No 1829/2003.

19. Sonnewald U, Kossmann J. Starches—from current models to genetic engineering. Plant Biotechnol J. 2013;11:223–32.
20. Chaouachi M, El Malki R, Berard A, Romaniuk M, Laval V, Brunel D, Bertheau Y. Development of a real-time PCR method for the differential detection and quantification of four solanaceae in GMO analysis: potato (solanum tuberosum), tomato (solanum lycopersicum), eggplant (solanum melongena), and pepper (capsicum annuum). J Agric Food Chem. 2008;56(6):1818–28.
21. Li P, Jia JW, Jiang LX, Zhu H, Bai L, Wang JB, Tang XM, Pan AH. Event-specific qualitative and quantitative PCR detection of the GMO carnation (Dianthus caryophyllus variety Moonlite based upon the 5′-exogene integration sequence. Genet Mol Res. 2012;11(2):1117–29.
22. David D, Bjørn S, Alexandra BK, Arne HJ, Jana Ž. Multiplex quantification of 12 European union authorized genetically modified maize lines with droplet digital polymerase chain reaction. Anal Chem. 2015;87:8218–26.
23. Huggett JF, Foy CA, Benes V, Emslie K, Garson JA, Haynes R, Hellemans J, Kubista M, Mueller RD, Nolan T, Pfaffl MW, Shipley GL, Vandesompele J, Wittwer CT, Bustin SA. Guidelines for minimum information for publication of quantitative digital PCR experiments. Clin Chim. 2013;59(6):1–12.
24. Zhu PY, Wang CG, Huang KL, Luo YB, Xu WT. A novel pretreatment-free duplex chamber digital PCR detection system for the absolute quantitation of GMO samples. Int J Mol Sci. 2016;17(3):402. doi:10.3390/ijms17030402.
25. GMO compass. Summary Application for amylopectin potato event AV43-6-G7 according to Regulation(EC) on 1829/2003. AVEBE: GM PartII Summary. http://www.gmo-compass.org/pdf/regulation/potato/AV43-6-G7_application_food_feed_cultivation.pdf. Accessed 15 Jun 2015.
26. Li X, Wang XX, Yang J, Liu YM, He YP, Pan LW. A novel quadruplex real-time PCR method for simultaneous detection of Cry2Ae and two genetically modified cotton events (GHB119 and T304-40). BMC Biotechnol. 2014;14:43.
27. Morisset D, Štebih D, Milavec M, Gruden K, Žel J. Quantitative analysis of food and feed samples with droplet digital PCR. PLoS One. 2013;8(5): e62583. doi:10.1371/journal.pone.0062583.
28. EURL-GMFF.Definition of Minimun Performance Requirements for Analytical Methods of GMO Testing European Network of GMO Laboratories (ENGL), http://gmo-crl.jrc.ec.europa.eu/doc/Min_Perf_Requirements_Analytical_methods.pdf. Accessed 22 Jan 2016.
29. Cheng N, Zhu PY, Xu YC, Huang KL, Luo YB, Yang ZS, Xu WT. High-sensitivity assay for Hg (II) and Ag (I) ion detection: a new class of droplet digital PCR logic gates for an intelligent DNA calculator. Biosens Bioelectron. 2016;84(15):1–6.
30. Arumuganathan K, Earle ED. Nuclear DNA content of some important plant species. Plant Mol Biol Reporter. 1991;9:210–20.
31. Moon GS, Shin WS. Establishment of quantitative analysis method for genetically modified maize using a reference plasmid and novel primers. Prev Nutr Food Sci. 2012;17:74–279.
32. Milavec M, Dobnik D, Yang LT, Zhang DB, Gruden K, Žel J. GMO quantification: valuable experience and insights for the future. Anal Bioanal Chem. 2014;406(26):6485–97.

11

Diversity of ABC transporter genes across the plant kingdom and their potential utility in biotechnology

Thomas S. Lane[1,3], Caroline S. Rempe[1,3], Jack Davitt[2], Margaret E. Staton[2], Yanhui Peng[3], Douglas Edward Soltis[4,5,6], Michael Melkonian[7], Michael Deyholos[8], James H. Leebens-Mack[9], Mark Chase[10,11], Carl J. Rothfels[12], Dennis Stevenson[13], Sean W. Graham[14], Jun Yu[15], Tao Liu[16], J. Chris Pires[17], Patrick P. Edger[18], Yong Zhang[19], Yinlong Xie[18,19,20], Ying Zhu[18], Eric Carpenter[21], Gane Ka-Shu Wong[18,21,22] and C. Neal Stewart Jr.[1,3*]

Abstract

Background: The ATP-binding cassette (ABC) transporter gene superfamily is ubiquitous among extant organisms and prominently represented in plants. ABC transporters act to transport compounds across cellular membranes and are involved in a diverse range of biological processes. Thus, the applicability to biotechnology is vast, including cancer resistance in humans, drug resistance among vertebrates, and herbicide and other xenobiotic resistance in plants. In addition, plants appear to harbor the highest diversity of ABC transporter genes compared with any other group of organisms. This study applied transcriptome analysis to survey the kingdom-wide ABC transporter diversity in plants and suggest biotechnology applications of this diversity.

Results: We utilized sequence similarity-based informatics techniques to infer the identity of ABC transporter gene candidates from 1295 phylogenetically-diverse plant transcriptomes. A total of 97,149 putative (approximately 25 % were full-length) ABC transporter gene members were identified; each RNA-Seq library (plant sample) had 88 ± 30 gene members. As expected, simpler organisms, such as algae, had fewer unique members than vascular land plants. Differences were also noted in the richness of certain ABC transporter subfamilies. Land plants had more unique ABCB, ABCC, and ABCG transporter gene members on average ($p < 0.005$), and green algae, red algae, and bryophytes had significantly more ABCF transporter gene members ($p < 0.005$). Ferns had significantly fewer ABCA transporter gene members than all other plant groups ($p < 0.005$).

Conclusions: We present a transcriptomic overview of ABC transporter gene members across all major plant groups. An increase in the number of gene family members present in the ABCB, ABCC, and ABCD transporter subfamilies may indicate an expansion of the ABC transporter superfamily among green land plants, which include all crop species. The striking difference between the number of ABCA subfamily transporter gene members between ferns and other plant taxa is surprising and merits further investigation. Discussed is the potential exploitation of ABC transporters in plant biotechnology, with an emphasis on crops.

Keywords: ABC transporter, Transcriptomics, Computational biology, Taxonomic diversity

* Correspondence: nealstewart@utk.edu
[1]The Graduate School of Genome Science and Technology, University of Tennessee, Knoxville, TN 37996, USA
[3]Department of Plant Sciences, University of Tennessee, Knoxville, TN 37996, USA
Full list of author information is available at the end of the article

Background

The ATP-binding cassette (ABC) transporter family is one of the largest known protein superfamilies in biology, which is represented in all living organisms [1–7]. Shared among members within the ABC transporter protein superfamily is the ability to hydrolyze adenosine triphosphate (ATP), which is used in a wide array of functions, including DNA repair, RNA translocation, and most commonly, active transport of a wide variety of substrates across various types of membranes in cells [3].

ABC transporters have three structural types. Full transporters are composed of two transmembrane domains (TMD) and two nucleotide-binding domains (NBD). Half transporters are composed of one TMD and one NBD, which dimerize in pairs to create virtual full transporters as homodimers or heterodimers. A third type of transporter has no TMDs but two NBDs [5]. The NBD is present in all three structural types and contains many key conserved motifs: Walker A, Q-loop, Walker B, D-loop, switch H-loop, and a signature motif (LSGGQ). The D-loop primarily functions in holding dimers together, the switch H-loop interacts with the transmembrane domain, the Walker A and B motifs form the P-loop, which binds to ATP, and the Q-loop and H-loop contain residues important for interacting with the γ-phosphate of the ATP [3]. Finally, the signature motif (LSGGQ) is exclusively found in ABC proteins, which enables ABC proteins to be distinguished from other ATPases [7].

In plants, a full inventory of ABC transporters was first catalogued for *Arabidopsis thaliana* [8], partially characterized [9, 10]. The current and most widely used classification system for ABC transporter subfamilies in plants is based on protein solubility, presence of TMDs, function, and amino acid sequence [5]. In our current study we used the system described in [5], which is consistent with the Human Genome Organization (HUGO) designation but including another subfamily, subfamily I. Plant ABC transporters fall into eight subfamilies using this nomenclature: A, B, C, D, E, F, G, and I (Table 1). ABCH subfamily members, a ninth subfamily, have not been identified in plants. Soluble proteins lacking TMDs such as the ABCF and ABCE members of the ABC transporter protein superfamily, despite being called such, lack any transport phenomena [5]. Despite the obvious importance of ABC transporters in plants, relative to other organisms there is a dearth of functional studies on individual members in plant as well as –omics studies; typically studies are limited to a few dicot species [8–14].

There are several known functions of ABC transporters of agricultural importance. One of the most important is detoxification. Detoxification mechanisms are of particular interest in herbicide resistance studies, particularly those involving putative glyphosate sequestration into vacuoles [15–17]. The *Arabidopsis* pleiotropic drug resistance 9 (PDR9) is also known to confer auxinic herbicide resistance [18]. Export proteins are also involved in environmental response and plant development such as auxin exporters ABCB19 and ABCB1 [19]. Importers primarily function in the acquisition of substrates important for cellular activity and ultimately the plant's survival. Such is the case of AtPMP2, which imports critical substrates needed in the glyoxylate cycle during germination [20].

The main objective of our study was to survey the diversity of ABC transporters across plants using a new public resource for plant transcriptomes, the One Thousand Plants Transcriptome Project (1KP). 1KP includes raw transcript data as well as assembled reference transcriptomes generated with standardized processing procedures. These data included transcriptomes from two species with extant reference genomes: *Ricinus communis* (castor bean) and *Linum usitatissmum* (flax), which acted as internal controls for matching the 1KP transcriptomes with the published reference genomes. These resources offer a source of transcriptome sequencing data spanning the breadth of diversity of plants [13, 21–24].

Methods

Data acquisition and quality control

All of the data in this project, including raw sequence reads, transcriptome assemblies, and assembly statistics, were gathered from the 1KP collaborative project [21]. The 1KP consortium performed all sample collection, sequencing, quality control, and assembly [13, 21]. A total of 1462 assemblies, assembled with SOAP de novo, were downloaded from the 1KP repository [11, 25]. Because some crucial taxonomic information was missing from accompanying data files, 161 samples were excluded from subsequent analysis. Another six samples were removed because of contamination that was discovered during the quality control step. The remaining 1295 1KP samples were translated into six frames using the Transeq program from EMBOSS (The European Molecular Biology Open Software Suite) [26]. A six-frame translation coupled with searches amongst confirmed ABC transporter protein sequences enabled us to assign each transcript into the correct reading frame for analysis. Since well-annotated genomes exist for only a handful of species analyzed and most species lacked sufficiently large and diverse tissue sequencing results, alternative splice variants were not considered.

HMMER domain e-value selection

The full sets of *Arabidopsis* and rice protein sequences were obtained from Phytozome version 10, including the *Arabidopsis* ABC transporter protein list [5] and rice list

Table 1 A comparison of different nomenclature systems for ABC proteins

HUGO subfamily	Sánchez-Fernández subfamily[a]	ABCISSE family[b,c]	ABCISSE subfamily[b,c]	TC subfamily[d]	Domain organization	Taxa
ABCA	ABC1 homologue (AOH)	Drug and antibiotic resistance (DRA)	ABCA	Cholesterol/phospholipid/ retinal flippase (CPR)	(TMD-NBD)2	eukaryotes (not yeast)
	ABC2 homologue (ATH)				TMD-NBD	eukaryotes
ABCB	Multidrug resistance (MDR)	Drug, peptides and lipid export (DPL)	p-glycoprotein (p-GP)	MDR	(TMD-NBD)2	prokaryotes and eukaryotes
	Transporter associated with antigen processing (TAP)		TAP and multidrug resistance-like protein (MDL)	TAP and mitochondrial peptide exporter (MPE)	TMD-NBD	eukaryotes
	ABC transporter of the mitochondria (ATM)		Heavy metal tolerance (HMT)	HMT	TMD-NBD	eukaryotes
	-		Lipid A-like exporter, putative (LLP)	-	TMD-NBD	prokaryotes and plants
ABCC	Multidrug resistance associated protein (MRP)	Organic anion conjugates and drug export (OAD)	MRP	Conjugate transporter (CT)	(TMD-NBD)2	eukaryotes
ABCD	Peroxisomal membrane protein (PMP)	Fatty acid export (FAE)	-	Peroxisomal fatty acyl-CoA transporter (P-FAT)	TMD-NBD; (TMD-NBD)2	bacteria and eukaryotes
ABCE	RNase L inhibitor (RLI)	RNase L inhibitor (RLI)	-	-	NBD-NBD	archaea and eukaryotes
ABCF	General control non-repressible (GCN)	Antibiotic resistance and translation regulation (ART)	Gene expression regulation (REG)	-	NBD-NBD	bacteria and eukaryotes
ABCG	White-brown complex homologue (WBC)	Eye pigment precursors and drugs (EPD)	WHITE	Eye pigment precursor transporter (EPP)	NBD-TMD	bacteria and eukaryotes
	Pleiotropic drug resistance (PDR)		PDR	PDR	(NBD-TMD)2	plants and fungi
ABCH	-	Drug resistance, bacteriocin and lantibiotic immunity (DRI)	YHIH		NBD-TMD	prokaryotes, slime moulds, echinoderms, insects and fish

References: [a][8], [b][6], [c][4], [d][52], Figure reprinted from [5] with permission

[27] Additional file 1. The hmmscan program from the HMMER package version 3.1 was used to search for the pfam record PF00005 across the full *Arabidopsis* and rice protein data sets [5, 28, 29]. A pfam record, such as the PF00005 domain, is a hidden Markov model based on a multiple sequence alignment of protein domains; in the case of PF00005 this multiple sequence alignment is of the NBD of many different ABC transporter proteins. A cutoff e-value of 3.10E-18 was selected in the two data sets that maximizes true positives (reported actual ABC transporters) and minimizes false negatives (reported as non-ABC transporters).

Comparison of genome and transcriptome data sources

The full set of flax and castor bean proteins were obtained from Phytozome v10 [30, 31]. Three sets of transcriptomic data for flax (shoot sample 1, shoot sample 2, shoot sample 3) and one set of castor bean (mixed tissue sample) were obtained from the 1KP data set [11, 25]. The hmmscan program was used to search for PF00005 pfam domains among all six sets of data using an e-value of 3.10E-18, yielding six sets of putative ABC transporter genes.

ABC transporter classification

Each sample's translated transcriptome was searched using hmmscan [28]. Search results containing the PF00005 pfam domain with an e-value of 3.10E-18 were subjected to further classification using best BLAST searches [32]. The de novo assembled transcriptomes from the 1KP data set were fragmented, transporters containing multiple nucleotide binding domains may only have had one of their binding domains accounted for in the transcriptomes sequencing results. Determination of ABC transport protein family was made strictly using BLAST searches against known ABC transport proteins in Arabidopsis and rice. A specialized database of known plant ABC transporters from *Arabidopsis* and rice was constructed for subfamily classification purposes using legacy BLAST formatdb version 2.2.26 (Additional file 2: Table S1). All known plant ABC transporter subfamilies were represented in our BLAST database.

Each of the 1295 1KP subsamples along with flax and castor bean genome-derived proteomes were used to query the ABC transporter database using legacy BLAST version 2.2.26, in each case the best hit to the ABC transporter database was used to classify the query among one of the eight ABC transporter subfamilies.

Nonparametric comparison of ABC diversity

JMP version 11 [33] was used to perform a one-way analysis of counts by species and performed pairwise nonparametric comparisons using the Wilcoxon method. All subfamily and total ABC transporter gene member classifications were subjected to pairwise analysis between each of the groups: angiosperms, conifers, ferns, lycophytes, mosses, hornworts, liverworts, chlorophytes, charophytes, glaucophytes, rhodophytes, and Chromista.

Results and discussion

ABC protein discovery

The pfam record PF00005 represents the highly conserved ATP-binding domain of ABC transporters. We used this pfam record to identify ABC transporter gene members across an entire transcriptome database as well as full gene sets of *Arabidopsis* and rice (Table 2). ABC transporter protein sequences had much lower expectation values for the PF00005 pfam domain from the hmmscan than did known non-ABC transporter protein sequences (85 % of the top 200 lowest e-value hits were known ABC transporter proteins) (Additional file 3: Table S2). These results were obtained using 233 known ABC transporters and 85,152 known non-ABC transporter protein sequences. While the pfam search was effective in ABC transporter gene member discovery, there was not a clear delineation in e-value scores between known gene family members and other genes. The highest e-value among known ABC transporters was 2.10E-10. However, some non-ABC transporters showed e-values lower than the 2.10E-10 cutoff. Searching among the top five highest e-values we identified the largest gap between e-values of 3.10E-18 and 2.10E-15. This gap represents three orders of magnitude between non-ABC transporters and ABC transporters. Using the 3.10E-18 threshold of significance, we identified 282

Table 2 A summary of HMMER search results

E-value cutoff	True positives	False positives	True negatives	False negatives	Specificity	Sensitivity
3.10E-21	222	48	84,882	11	0.9994	0.9528
3.10E-18	230	52	84,870	3	0.9994	0.9871
2.10E-15	231	52	84,869	2	0.9994	0.9914
2.10E-13	232	55	84,865	1	0.9994	0.9957
2.10E-10	233	64	84,855	0	0.9992	1.0000

Summary of sensitivity and specificity of 85,152 HMMER queries against pfam as a method for selecting an appropriate cutoff e-value for the identification of PF00005 (ABC transporter) domains

sequences as ABC transporters. Of the 282 sequences identified as ABC transporters, 230 had been previously classified as ABC transporters, but 52 were not (putative false positives). Therefore, the cutoff excluded three known ABC transporters (putative false negatives). With the entire set of 85,152 starting sequences, the usage of the pfam PF00005 domain yielded a sensitivity (true positives divided by true positive plus false negatives) of 0.9871 and a specificity of 0.9994 (true negatives divided by true negatives plus false positives). Consequently, an e-value of 3.10E-18 was selected as the optimum, which allowed a total of three false negatives for ABC transporter detection for all further HMMER analysis.

Comparison of genome and transcriptome data sources

The flax and castor bean transcriptomes allowed us to examine how the number of expressed ABC transporters identified in 1KP corresponds to the number of known ABC transporters in their entire genomes, which allowed us to make inferences throughout the plant kingdom, with caveats. ABC transporter genes are known to have variable expression across tissues [11], and thus their identification from RNA-Seq data is unlikely to yield a complete complement of gene family members for an organism.

The 1KP-generated assemblies for flax had 99,855, 95,813, and 101,110 total putative transcripts for each library. The number of transcripts identified by hmmscan as ABC protein transporter gene members was similar between the three flax samples: 82, 91, and 92. These figures represented 42 – 47 % of the number of ABC transporters identified from the reference genome protein sequences. A total of 59 transcripts were putatively identified as ABC transporters from the single available castor bean transcriptome from 1KP, 42 % of the 142 putative genes identified in the genome-sequence derived gene set.

Subfamilies for putative ABC transporters were assigned on the basis of the most similar amino acid sequence from *Arabidopsis* or rice. The distribution of the ABC transporters among the eight subfamilies shared a similar pattern between the 1KP transcriptome samples and the reference genome-derived gene sets (Figs. 1a and b).

In general, the 1KP data per accession contained nearly half the number of unique ABC transporter gene members, by comparison, to their reference genome counterparts as identified by hmmscan searches. In addition, hmmscan found nearly the same number of unique ABC transporters for each of the three 1KP data sets for flax. The relative consistency between corresponding 1KP data samples is encouraging for extension of inferences across data sets in order to estimate the total number of gene family members for all plant taxa.

ABC transporter subfamily classification

The majority of our knowledge about the ABC transporter protein superfamily in plants is from studies limited to a few angiosperm species. In an effort to expand our knowledge of the varying representation of ABC transporter protein subfamilies across the plant kingdom, we classified nearly 97,149 transcripts across 1295 samples into one of the seven ABC transporter HUGO subfamilies (A, B, C, D, E, F, G) and the plant-only subfamily, I [5]. The 1295 1KP samples span a diversity of plant taxa: 816 from angiosperms, 76 from conifers, 71 ferns, 22 lycophytes, 25 liverworts, 41 mosses, 6 hornworts, 123 chlorophytes, 46 charophytes, 5 glaucophytes, and 28 rhodophytes. In general, the results comparing hornworts/glaucophytes and other plants were not significant, likely owing to the small sample sizes in these

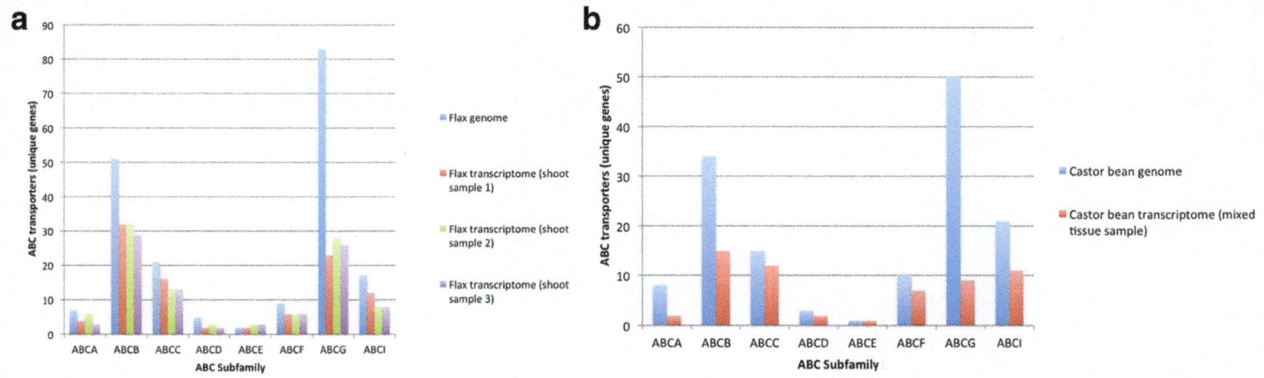

Fig. 1 Flax and castor bean ABC transporter unique gene distribution inferred from RNA-Seq transcriptome data. **a** Plot comparing total ABC transporter unique genes observed in the flax genome and the three 1KP transcriptomes for flax (shoot sample 1, shoot sample 2, shoot sample 3). **b** Plot comparing total ABC transporter unique genes observed in the castor bean genome and the 1KP transcriptomes for castor bean (mixed tissue sample). Searches based on hmmscan searches for the PF00005 pfam domain and subsequent BLAST queries against a custom BLAST database of ABC transporters from *Arabidopsis* and rice

groups. The breadth of samples across various plant groups provides robust statistical support for comparisons amongst groups in most cases. The gymnosperms were mainly represented by conifers, with few other gymnosperms being represented: three Gnetales, four Cycadales, and one Ginkgoales.

The hmmscan analysis with a cutoff of 3.10E-18 found at least 50 unique ABC genes (transcripts) containing PF00005 pfam domains per plant sample (Fig. 2a). In many cases, we recorded outliers, the largest of which, *Blasia sp.* (526 ABC transporters), more than doubled the next largest sample. For ease of interpretation, the

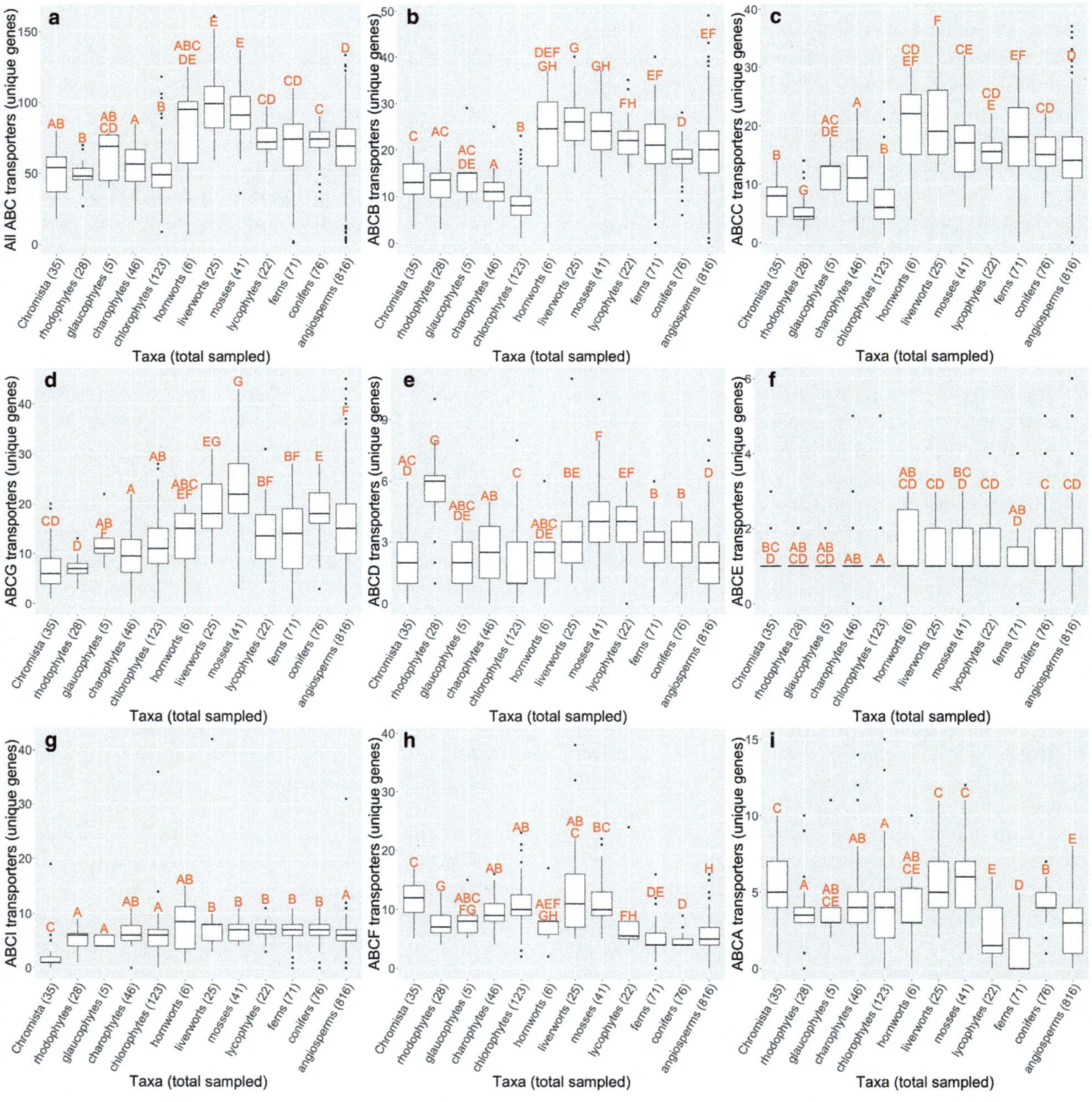

Fig. 2 Number of unique ABC transporter gene members distributed over plant taxa inferred from RNA-Seq transcriptome data. **a** Box and whisker plot of subfamily ABC transporter unique genes among each set of samples in indicated plant groups: Chromista, rhodophytes, glaucophytes, charophytes, chlorophytes, hornworts, liverworts, mosses, lycophytes, ferns, conifers, and anigosperms. **b** ABCB transporter gene members. **c** ABCC transporter gene members. **d** ABCG transporter gene members. **e** ABCD transporter gene members. **f** ABCE transporter gene members. **g** ABCI transporter gene members. **h** ABCF transporter gene members. **i** ABCA transporter gene members. Classifications are based on hmmscan searches for the PF00005 pfam domain. Significant differences between unique gene counts of the twelve groups were determined by a comparison of means using the Wilcoxon method. Filled circles indicate outliers. The total samples per plant group are indicated in parentheses beside each label along the x-axis. Statistical differences are indicated by letter groupings ($p < 0.005$)

Blasia sp. sample was omitted from figures in most cases. Bryophytes, lycophytes, conifers, ferns, and angiosperms had 16 to 43 more ABC transporter transcripts than rhodophytes, chlorophytes, charophytes, glaucophytes, and Chromista. Many observed differences in average number of transporters between plant groups were found to be significant ($p < 0.005$). Although green algae and land plants have been shown to harbor similar numbers of ABC transporter subfamily gene members and total ABC transporter genes [34], the simpler transport systems of green algae, rhodophytes, glaucophytes, and Chromista may contribute to the reduced number of unique ABC transporter genes we observed on our study. Angiosperms, conifers, ferns, chlorophytes, charophytes, glaucophytes, rhodophytes, and Chromista had 22 to 43 fewer unique ABC transporter gene members than bryophytes. The differences in mean number of transporters between these groups is also significant, with exception to hornworts ($p < 0.005$).

Results from BLAST subfamily classifications were used to identify full-length transcripts based on query coverage. Because the ABC transport protein subfamilies split prior to the most recent common ancestor of all plants, it is likely that ABC transport protein subfamily members will closely match an ortholog of the same subfamily. Transcripts with greater than 90 % of the query aligned to the target sequence were classified as full length. All 1295 transcriptome samples contained at least one putative unique ABC transporter gene member with 97,149 ABC transporters identified across all samples. A total of 22,343 ABC transporter sequences were identified as full length. There were, on average, 66 ABC transporter gene members discovered per RNA-Seq library with a standard deviation of 26 gene members (Fig. 2a). Most samples (1052 of 1295) contained at least one of each of the ABC transporter protein subfamily members.

The number of unique ABC transporter subfamily gene members varied between distantly related plants. Among the functionally diverse subfamilies ABCB and ABCC, we saw significantly fewer unique ABC transporter gene members on average in algae (5 to 15 gene members) than in other groups: angiosperms, conifers, ferns, lycophytes, and bryophytes ($p < 0.005$) (Figs. 2b and c). Among ABCG subfamily transporters the differences in the average number of ABC transporter gene members seen between algae was significantly lower (5 to 11 gene members) than the following groups: angiosperms, conifers, and bryophytes ($p < 0.005$) (Fig. 2d).

Several studies have indicated that plant genomes have more ABCB, ABCC and/or ABCG subfamily gene members in comparison to other eukaryotes such as human and yeast [5, 35]. Our results revealed additional variability among plants regarding the size of these subfamilies, specifically in that algae contain relatively fewer unique ABCB, ABCC and ABCG transporter members. The larger number of unique ABCB transporter gene members seen in vascular plant species may have adaptive significance for complex and heterogeneous environments of "higher plants." Land plants' best defense against xenobiotics, such as heavy metals, might come from ABCB transporters' ability to sequester and/or transport foreign chemicals. For example, AtABCB25, an *Arabidopsis* ABCB export protein, allows some tolerance to cadmium and lead when overexpressed [36, 37]. The overexpression data are useful to infer functional significance, but also guide potential utility in biotechnology. Aluminum tolerance is conferred by the expression of ABCB27 in root tips [38]. ABCC transporter proteins have been identified among plants for their role in detoxification and the regulation of stomatal guard cells [39]. The gap in types of ABCC transporter transcript number seen between algae and other plant taxa might be explained by a lack of need by algae for stomata regulation and/or detoxification. While liverworts also lack stomata we do not see significantly fewer ABCC transporter proteins among liverworts. Algae do not require the specialized transport systems of land plants because most cells have direct access to nutrients in their water environment. Such direct access to an aquatic environment may contribute to fewer unique gene members of subfamilies ABCB and ABCC. Only 26 ABC transport proteins were found in *Ceratophyllum demersum,* relative to average 66 ABC transport proteins seen across all 1295 tissues sampled. This finding would appear to support the reduced expression of ABC transport proteins in an aquatic environment (Additional file 4: Table S3). Algae also do not produce a waxy cuticle - so reduced ABCG transcripts may be related to a lack of the ABCG transporter proteins that are necessary for the transport of cuticular lipids [40, 41]. Whether the variation in size of these families is a result of gene expansion is unclear and requires further investigation for evaluation.

Among subfamilies ABCD, ABCE, and ABCI, few significant differences were observed in the average number of gene members across plants with a single exception in the ABCD subfamily (Figs. 2e, f, and g). Significant differences in the average number of ABC transporter gene members detected among ABCD, ABCE, and ABCI were small, ranging from zero to two ($p < 0.005$). The presence of ABCD, ABCE, and ABCI transcripts with very similar gene member averages might suggest that the proteins that result from the translation of these transcripts are essential to life among all plants. Plant ABCD transporters are well known for their import of substrates necessary in the glyoxylate cycle [20]. It seems likely that plants retaining ABCD transporters would have an evolutionary advantage over plants that do not, given the importance of the glyoxylate cycle to survival across all plant taxa. Rhodophytes have nearly double, on average, the number of ABCD subfamily members compared with other plants. We can speculate that since these ABCD transporters are peroxisome-localized and important for fatty acid metabolism, especially long chain fatty acids, and that Rhodophyta produce a high number of long-chain fatty acids [42], that there may be functional significance. Rhodophytes might be good targets to mine ABCD transporters for use to produce novel oils in crops. ABCE proteins are found in archaea, bacteria, and eukaryotes, suggesting that this class of ABC transporter proteins is essential to life and are thus conserved [5]. ABCE1 is a translation termination factor [43]. Once a ribosome reaches a stop codon, ABCE1 helps to detach the ribosome from the mRNA [44], and also plays an important role in ribosome recycling [45]. The maintenance of similar numbers of unique ABCD and ABCE subfamily gene members may be because of the conservation of essential functions provided by those subfamilies. Our understanding of the ABCI subfamily, is expected to be significantly improved as more information regarding the structure and function of the subfamily members becomes available. However, current data suggest the origins of ABCI subfamily transporter genes may be from prokaryotic genomes and so they may result from a movement of genes from mitochondria and plastids to the nuclear genome. As such, the ABCI transporter proteins have been classified as a heterogeneous group composed of multicomponent transporters (reviewed in [5]) so the similarity seen in average number of unique gene members across plant taxa was unexpected.

In the case of ABCF subfamily proteins, we observed significantly higher transcript count averages in Chromista, glaucophytes, rhodophytes, chlorophytes, charophytes, and bryophytes than in all other groups, four to seven more transcripts on average than angiosperms, conifers, ferns, and lycophytes ($p < 0.005$) (Fig. 2h). When directly comparing algae and bryophytes, we found that bryophytes had more unique ABCF subfamily gene members on average ($p < 0.005$). Understanding the increased variation, specifically regarding the ABCF subfamily size among algae and bryophytes is difficult with a lack of functional characterization of ABCF proteins among plants. Studies in yeast and humans have indicated ABCF protein function in the activation of eIF-2α kinase [5]. Activation of eIF-2α kinase is important in the regulation of stress response factors [46]. Indeed one member of the ABCF, ABC50 appears to play an especially-important role in mammalian translation processes in interactions with eiF2 and the ribosome [47, 48]. There is a published patent application in which ABC50 is claimed to be useful to increase protein in plants, among other organisms [49]. We noted an increase in the total number of ABCF protein gene members among algae and bryophytes; we cannot speculate about the potential adaptive significance.

The average number of unique ABCA subfamily transporter gene members was significantly lower in ferns, three to four less gene members on average in most cases ($p < 0.005$) (Fig. 2i). There was a significant difference between the average number of unique ABCA transporter gene members in ferns, conifers, and angiosperms ($p < 0.005$). While little is known about the function of ABCA transporter proteins among plants, they are suspected to be involved in lipid metabolism based on their known function in humans (reviewed in [50]). Upon comparison with *Arabidopsis*, rice was found to be missing an orthologous full length ABCA transporter, which is consistent with previously published results [51]. Our results indicate an occurrence of decreased ABCA transporter genes among ferns relative to other plant groups, possibly from gene loss. While the single gene loss observed in rice and the results observed in fern are likely unrelated, they may point to similar phenomena regarding the importance of ABCA transport proteins in sterol metabolism [52].

Conclusions

ABC transport proteins have diverse functions and are ubiquitous in plants. We catalogued ABC transporters across the plant kingdom from hundreds of transcriptomes. While our current study was constrained by imperfect transcriptome assemblies taxonomic patterns should be of interest to the community to build hypotheses. Specifically, we found significant differences in the number of unique ABC transporter gene members between algae (Chromista, rhodophytes, glaucophytes, chlorophytes, and charophytes), bryophytes (mosses, liverworts, hornworts), lycophytes, ferns, conifers (other gymnosperms were not well represented among the 1KP data), and angiosperms. Algae and bryophytes have fewer unique ABC transporter gene members than lycophytes, ferns, conifers, and angiosperms. The number of unique ABCD, ABCE, and ABCI subfamily protein gene members show few significant differences amongst plants. With regards to plant biotechnology applications, it is noted that these three subfamilies might not provide much novelty. In contrast, the diversity of ABCB and ABCF proteins, known for detoxification and stress resistance, respectively, are attractive genes to screen for improving crops via biotechnology. Especially intriguing are the ABCF members in algae and bryophytes for potential use in crop biotechnology. Given the diversity and function of ABCF members, algae and bryophytes have been largely ignored as sources of potentially useful genes in agriculture.

Additional files

Additional file 1: file S1. Fasta file of ABC transporter proteins from rice and *Arabidopsis*.

Additional file 2: Table S1. Detailed table of HMMER testing results.

Additional file 3: Table S2. Detailed table of classification results and associated metadata.

Additional file 4: Table S3. Filtered BLAST results of transcripts with 90 % or greater query coverage of ABC transporter protein subject.

Acknowledgments
The authors would like to thank Dylan Storey for his suggestions and advice regarding the handling, storage, and analysis of the 1KP data.

Funding
The authors thank the University of Tennessee Genome Science and Technology graduate program for providing funds to TL, whose project was also supported by UT AgResearch and the Racheff endowment. We also thank the financial supporters of the 1KP consortium.

Authors' contributions
TSL designed and performed the data analyses and drafted the manuscript. CSR co-conceived the strategy for separating putative ABC transporter transcripts into respective ABC transporter subfamilies and was involved in the collection of known ABC transporter proteins. JD performed statistical analysis based on data. YP provided insights regarding the form and function of ABC transporters among plant species and provided the initial list of known plant ABC transporters. MES contributed strategies for verifying analyses performed. DES, MM, MD, JHL-M, MC, CJR, DS, SWG, JY, TL, JCP, PPE, YZ, YX, YZ, EC, GK-SW, and CNS are members of the 1KP consortium and were involved in 1KP organization and structure, sample collection and sequencing. CNS was partially responsible for the conception of the study and writing. All authors agree with the contents of the manuscript.

Competing interests
The authors declare that they have no competing interests.

Author details
[1]The Graduate School of Genome Science and Technology, University of Tennessee, Knoxville, TN 37996, USA. [2]Department of Entomology and Plant Pathology, University of Tennessee, Knoxville, TN 37996, USA. [3]Department of Plant Sciences, University of Tennessee, Knoxville, TN 37996, USA. [4]Department of Biology, University of Florida, Gainesville, FL 32611, USA. [5]Florida Museum of Natural History, Gainesville, FL 32611, USA. [6]Genetics Institute, University of Florida, Gainesville, FL 32611, USA. [7]Department of Biology, Cologne Biocenter, University of Cologne, Cologne 50674, Germany. [8]Department of Biology, University of British Columbia, Kelowna, BC V1V 1V7, Canada. [9]Department of Biological Sciences, University of Georgia, Athens, GA 30602, USA. [10]Jodrell Laboratory, Royal Botanic Gardens Kew, Richmond, Surrey TW9 3AE, UK. [11]Department of Biology, Duke University, Durham, NC 27708, USA. [12]Department of Integrative Biology, University of California, University Herbarium, Berkeley, CA 94720-2465, USA. [13]New York Botanical Garden, Bronx, New York, NY 10458, USA. [14]Department of Botany, University of British Columbia, Vancouver, BC, Canada. [15]CAS Key Laboratory of Genome Sciences and Information, Beijing Key Laboratory of Genome and Precision Medicine Technologies, Beijing Institute of Genomics, Chinese Academy of Sciences, Beijing 100101, People's Republic of China. [16]College of Marine Life Sciences, Ocean University of China, Qingdao 266003, People's Republic of China. [17]Bond Life Sciences Center, Division of Biological Sciences, University of Missouri, Columbia, MO, USA. [18]BGI-Shenzhen, Beishan Industrial Zone, Yantian District, Shenzhen 518083, China. [19]Department of Computer Science, HKU-BGI Bioinformatics Algorithms and Core Technology Research Laboratory, University of Hong Kong, Pokfulam, Hong Kong. [20]School of Bioscience and Bioengineering, South China University of Technology, Guangzhou, China. [21]Department of Biological Sciences, University of Alberta, Edmonton, AB T6G 2E9, Canada. [22]Department of Medicine, University of Alberta, Edmonton, AB T6G 2E1, Canada.

References
1. Higgins CF. ABC transporters: from microorganisms to man. Annu Rev Cell Biol. 1992;8:67–113.
2. Jones PM, George AM. The ABC transporter structure and mechanism: perspectives on recent research. Cell Mol Life Sci. 2004;61:682–99.
3. Davidson AL, Dassa E, Orelle C, Chen J. Structure, function, and evolution of bacterial ATP-binding cassette systems. Microbiol Mol Biol Rev. 2008;72:317–64.
4. Bouige P, Laurent D, Piloyan L, Dassa E. Phylogenetic and functional classification of ATP-binding cassette (ABC) systems. Curr Protein Pept Sci. 2002;3:541–59.
5. Verrier PJ, Bird D, Burla B, Dassa E, Forestier C, Geisler M, Klein M, Kolukisaoglu U, Lee Y, Martinoia E et al. Plant ABC proteins - a unified nomenclature and updated inventory. Trends Plant Sci. 2008;13:151–9.
6. Dassa E, Bouige P. The ABC of ABCs: a phylogenetic and functional classification of ABC systems in living organisms. Res Microbiol. 2001;152:211–29.
7. Davies TGE, Coleman JOD. The *Arabidopsis thaliana* ATP-binding cassette proteins: an emerging superfamily. Plant Cell Env. 2000;23:431–43.
8. Sanchez-Fernandez R, Davies TG, Coleman JO, Rea PA. The *Arabidopsis thaliana* ABC protein superfamily, a complete inventory. J Biol Chem. 2001; 276:30231–44.
9. Schneider E, Hunke S. ATP-binding-cassette (ABC) transport systems: functional and structural aspects of the ATP-hydrolyzing subunits/domains. FEMS Microbiol Rev. 1998;22:1–20.
10. van den Brule S, Smart CC. The plant PDR family of ABC transporters. Planta. 2002;216:95–106.
11. Nguyen VN, Moon S, Jung KH. Genome-wide expression analysis of rice ABC transporter family across spatio-temporal samples and in response to abiotic stresses. J Plant Physiol. 2014;171:1276–88.
12. Cakir B, Kilickaya O. Whole-genome survey of the putative ATP-binding cassette transporter family genes in *Vitis vinifera*. PLoS One. 2013;8(11):e78860.
13. Matasci N, Hung L-H, Yan Z, Carpenter EJ, Wickett NJ, Mirarab S, Nguyen N, Warnow T, Ayyampalayam S, Barker M et al. Data access for the 1,000 Plants (1KP) project. Gigascience. 2014;3:17.
14. Adolfo G, Ruocco M, Di Donato A, Frusciante L, Lorito M, Scala F, Ercolano MR. Genetic variability and evolutionary diversification of membrane ABC transporters in plants. BMC Plant Biol. 2015;15:51.
15. Peng Y, Abercrombie LL, Yuan JS, Riggins CW, Sammons RD, Tranel PJ, Stewart Jr CN. Characterization of the horseweed (*Conyza canadensis*) transcriptome using GS-FLX 454 pyrosequencing and its application for expression analysis of candidate non-target herbicide resistance genes. Pest Manag Sci. 2010;66:1053–62.
16. Ge X, d'Avignon DA, Ackerman JJ, Sammons RD. Rapid vacuolar sequestration: the horseweed glyphosate resistance mechanism. Pest Manag Sci. 2010;66:345–8.
17. Hanson BD, Shrestha A, Shaner DL. Distribution of glyphosate-resistant horseweed (*Conyza canadensis*) and relationship to cropping systems in the Central Valley of California. Weed Sci. 2009;57:48–53.
18. Ito H, Gray WM. A gain-of-function mutation in the Arabidopsis pleiotropic drug resistance transporter PDR9 confers resistance to auxinic herbicides. Plant Physiol. 2006;142:63–74.
19. Yang H, Murphy AS. Functional expression and characterization of Arabidopsis ABCB, AUX 1 and PIN auxin transporters in *Schizosaccharomyces pombe*. Plant J. 2009;59:179–91.
20. Theodoulou FL, Holdsworth M, Baker A. Peroxisomal ABC transporters. FEBS Lett. 2006;580:1139–55.
21. The One Thousand Transcriptomes Project [https://sites.google.com/a/ualberta.ca/onekp/]. Accessed 28 May 2016.
22. Chan AP, Crabtree J, Zhao Q, Lorenzi H, Orvis J, Puiu D, Melake-Berhan A, Jones KM, Redman J, Chen G. Draft genome sequence of the oilseed species *Ricinus communis*. Nat Biotechnol. 2010;28:951–6.

23. Allan G, Williams A, Rabinowicz PD, Chan AP, Ravel J, Keim P. Worldwide genotyping of castor bean germplasm (*Ricinus communis* L.) using AFLPs and SSRs. Genet Res Crop Evol. 2007;55:365–78.
24. Wang Z, Hobson N, Galindo L, Zhu S, Shi D, McDill J, Yang L, Hawkins S, Neutelings G, Datla R et al. The genome of flax (*Linum usitatissimum*) assembled *de novo* from short shotgun sequence reads. Plant J. 2012;72:461–73.
25. Luo R, Liu B, Xie Y, Li Z, Huang W, Yuan J, He G, Chen Y, Pan Q, Liu Y et al. SOAPdenovo2: an empirically improved memory-efficient short-read *de novo* assembler. Gigascience. 2012;1(1):18.
26. Rice P, Longden I, Bleasby A. EMBOSS: The European molecular biology open software suite. Trends Genet. 2000;16:276–7.
27. Crouzet J, Trombik T, Fraysse AS, Boutry M. Organization and function of the plant pleiotropic drug resistance ABC transporter family. FEBS Lett. 2006;580:1123–30.
28. Finn RD, Clements J, Eddy SR. HMMER web server: interactive sequence similarity searching. Nucleic Acids Res. 2011;39(Web Server issue):W29–37.
29. Punta M, Coggill PC, Eberhardt RY, Mistry J, Tate J, Boursnell C, Pang N, Forslund K, Ceric G, Clements J. The Pfam protein families database. Nucleic Acids Res. 2012;40(Database issue):D290–301.
30. Phytozome [http://phytozome.jgi.doe.gov/pz/portal.html]. Accessed 28 May 2016.
31. Goodstein DM, Shu S, Howson R, Neupane R, Hayes RD, Fazo J, Mitros T, Dirks W, Hellsten U, Putnam N, Rokhsar DS. Phytozome: a comparative platform for green plant genomics. Nucleic Acids Res. 2012;40(D1):D1178–86.
32. Altschul SF, Gish W, Miller W, Myers EW, Lipman DJ. Basic local alignment search tool. J Mol Biol. 1990;215:403–10.
33. JMP®, Version 11. SAS Institute Inc., Cary, NC, 1989–2007.
34. Merchant SS, Prochnik SE, Vallon O, Harris EH, Karpowicz SJ, Witman GB, Terry A, Salamov A, Fritz-Laylin LK, Marechal-Drouard L. The *Chlamydomonas* genome reveals the evolution of key animal and plant functions. Science. 2007;318:245–50.
35. Geisler M, Murphy AS. The ABC of auxin transport: the role of p-glycoproteins in plant development. FEBS Lett. 2006;580:1094–102.
36. Kim DY, Bovet L, Kushnir S, Noh EW, Martinoia E, Lee Y. AtATM3 is involved in heavy metal resistance in *Arabidopsis*. Plant Physiol. 2006;140:922–32.
37. Chen S, Sanchez-Fernandez R, Lyver ER, Dancis A, Rea PA. Functional characterization of AtATM1, AtATM2, and AtATM3, a subfamily of Arabidopsis half-molecule ATP-binding cassette transporters implicated in iron homeostasis. J Biol Chem. 2007;282:21561–71.
38. Larsen PB, Cancel J, Rounds M, Ochoa V. Arabidopsis *ALS1* encodes a root tip and stele localized half type ABC transporter required for root growth in an aluminum toxic environment. Planta. 2007;225:1447–58.
39. Klein M, Burla B, Martinoia E. The multidrug resistance-associated protein (MRP/ABCC) subfamily of ATP-binding cassette transporters in plants. FEBS Lett. 2006;580:1112–22.
40. Pighin JA, Zheng H, Balakshin LJ, Goodman IP, Western TL, Jetter R, Kunst L, Samuels AL. Plant cuticular lipid export requires an ABC transporter. Science. 2004;306:702–4.
41. Panikashvili D, Savaldi-Goldstein S, Mandel T, Yifhar T, Franke RB, Hofer R, Schreiber L, Chory J, Aharoni A. The Arabidopsis *DESPERADO/AtWBC11* transporter is required for cutin and wax secretion. Plant Physiol. 2007;145:1345–60.
42. Lang I, Hodak L, Friedl T, Feussner I. Fatty acid profiles and their distribution patterns in microalgae: a comprehensive analysis of more than 2000 strains from the SAG culture collection. BMC Biol. 2011;11:124.
43. Khoshnevis S, Gross T, Rotte C, Baierlein C, Ficner R, Krebber H. The iron-sulphur protein RNase L inhibitor functions in translation termination. EMBO Rep. 2010;11:214–9.
44. Sarmiento C, Nigul L, Kazantseva J, Buschmann M, Truve E. AtRLI2 is an endogenous suppressor of RNA silencing. Plant Mol Biol. 2006;61:153–63.
45. Barthelme D, Dinkelaker S, Albers S-V, Londei P, Ermler U, Tampé R. Ribosome recycling depends on a mechanistic link between the FeS cluster domain and a conformational switch of the twin-ATPase ABCE1. Proc Natl Acad Sci U S A. 2011;108:3228–33.
46. Kimball SR. Eukaryotic initiation factor eIF2. Intl J Biochem Cell Biol. 1999;31:25–9.
47. Tyzack JK, Wang X, Belsham GJ, Proud CG. ABC50 interacts with eukaryotic initiation factor 2 and associates with the ribosome in an ATP-dependent manner. J Biol Chem. 2000;275:34131–9.
48. Paytubi S, Wang X, Lam YW, Izquierdo L, Hunter MJ, Jan E, Hundal HS, Proud CG. ABC50 promotes translation initiation in mammalian cells. J Biol Chem. 2009;284:24061–73.
49. Berger SA. Increasing protein production by increasing ABC50 expression or activity. 2012. US Patent Publication US 20120094329 A1.
50. Kaminski WE, Piehler A, Wenzel JJ. ABC A-subfamily transporters: structure, function and disease. Biochim Biophys Acta. 2006;1762:510–24.
51. Garcia O, Bouige P, Forestier C, Dassa E. Inventory and comparative analysis of rice and Arabidopsis ATP-binding cassette (ABC) systems. J Mol Biol. 2004;343:249–65.
52. Saier MH. A functional-phylogenetic classification system for transmembrane solute transporters. Microbiol Mol Biol Rev. 2000;64:354.

12

Biochemical characterization of the xylan hydrolysis profile of the extracellular endo-xylanase from *Geobacillus thermodenitrificans* T12

Martinus J.A. Daas[1†], Patricia Murciano Martínez[2†], Antonius H.P. van de Weijer[1], John van der Oost[1], Willem M. de Vos[1], Mirjam A. Kabel[2] and Richard van Kranenburg[1,3*]

Abstract

Background: Endo-xylanases are essential in degrading hemicellulose of various lignocellulosic substrates. Hemicellulose degradation by *Geobacillus* spp. is facilitated by the hemicellulose utilization (HUS) locus that is present in most strains belonging to this genus. As part of the HUS locus, the *xynA* gene encoding an extracellular endo-xylanase is one of the few secreted enzymes and considered to be the key enzyme to initiate hemicellulose degradation. Several *Geobacillus* endo-xylanases have been characterized for their optimum temperature, optimum pH and generation of degradation products. However, these analyses provide limited details on the mode of action of the enzymes towards various substrates resulting in a lack of understanding about their hydrolytic potential.

Results: A HUS-locus associated gene (*GtxynA1*) from the thermophile *Geobacillus thermodenitrificans* T12 encodes an extracellular endo-xylanase that belongs to the family 10 glycoside hydrolases (GH10). The *GtxynA1* gene was cloned and expressed in *Escherichia coli*. The resulting endo-xylanase (termed *Gt*XynA1) was purified to homogeneity and showed activity between 40 °C and 80 °C, with an optimum activity at 60 °C, while being active between pH 3.0 to 9.0 with an optimum at pH 6.0. Its thermal stability was high and *Gt*XynA1 showed 85% residual activity after 1 h of incubation at 60 °C. Highest activity was towards wheat arabinoxylan (WAX), beechwood xylan (BeWX) and birchwood xylan (BiWX). *Gt*XynA1 is able to degrade WAX and BeWX producing mainly xylobiose and xylotriose. To determine its mode of action, we compared the hydrolysis products generated by *Gt*XynA1 with those from the well-characterized GH10 endo-xylanase produced from *Aspergillus awamori (Aa*XynA). The main difference in the mode of action between *Gt*XynA1 and *Aa*XynA on WAX is that *Gt*XynA1 is less hindered by arabinosyl substituents and can therefore release shorter oligosaccharides.

Conclusions: The *G. thermodenitrificans* T12 endo-xylanase, *Gt*XynA1, shows temperature tolerance up to 80 °C and high activity to a variety of xylans. The mode of action of *Gt*XynA1 reveals that arabinose substituents do not hamper substrate degradation by *Gt*XynA1. The extensive hydrolysis of branched xylans makes this enzyme particularly suited for the conversion of a broad range of lignocellulosic substrates.

Keywords: *Geobacillus*, Thermophile, Endo-xylanase, Xylan, Biotechnology

* Correspondence: r.vankranenburg@wur.nl
†Equal contributors
1Laboratory of Microbiology, Wageningen University, Stippeneng 4, Wageningen 6708 WE, The Netherlands
3Corbion, Arkelsedijk 46, Gorinchem 4206 AC, The Netherlands
Full list of author information is available at the end of the article

Background

Many *Geobacillus* species contain a hemicellulose utilization (HUS) locus in their genome that encodes for multiple hydrolytic enzymes and sugar transporters responsible for the degradation of hemicellulose [1, 2]. The most well characterized HUS locus belongs to *G. stearothermophilus* T-6 and its proposed degradation mechanism is based on a limited number of secreted hydrolytic enzymes and further processing of the generated degradation products intracellularly [3]. One of these secreted hydrolytic enzymes is an endo-xylanase that cleaves the xylan backbone of hemicellulose into shorter oligosaccharides [3, 4].

The xylan backbone is composed of xylosyl residues linked by ß-(1 → 4)-glycosidic bonds, and is typically decorated with a varying degree of substitutions containing (4-*O*-methyl)-α-D-glucuronic acids, α-L-arabinofuranosyl residues and acetyl residues [5]. The degree of substitution and polymerization is dependent on the origin of the substrate. Substrates from grasses and annual plants are rich in arabinoxylans, whereas hardwoods mostly consist of (4-*O*-methyl)-α-D-glucuronoxylan. This variety of substitutions requires multiple enzymes acting in synergy to fully degrade the xylans. Endo-(1 → 4)-ß-D-xylanases (E.C. 3.2.1.8) hydrolyse the ß-(1 → 4)-xylosidic bonds to release xylose and (substituted) xylo-oligosaccharides of various lengths that can be degraded further by other hemicellulases.

Endo-xylanases belonging to glycoside hydrolase family 10 and 11 are the best characterized and described [6]. Endo-xylanases belonging to family 10 (GH10) can tolerate the presence of substituents, such as arabinosyl residues, better than those belonging to glycoside hydrolase family 11 (GH11) [7]. An example of the mode of action of two endo-xylanases, belonging to families GH10 and GH11, from *Aspergillus awamori* (*Aa*) towards an arabinoxylan model substrate has been described [8]. The *Aa*GH10 was found to be able to cleave the xylan backbone next to a xylosyl residue substituted with one or two arabinosyl residues. In contrast, xylanases from family GH11 are hindered by arabinosyl substituents and cleave the xylan backbone only next to an unsubstituted xylosyl residue, producing longer oligosaccharides in comparison to those obtained from GH10 [8].

Endo-xylanases from *Geobacillus* species belong to the GH10 family and have been described in multiple studies [9–16]. These studies mainly focused on the characterization of activity ranges at higher temperatures and thermo-stability but lack detailed analyses of the mode of action of the enzyme on its substrate. Although some studies have described xylo-oligosaccharides formation upon hydrolysis of xylan by the endo-xylanase, the position of hydrolysis and the possible hindrance of substituents present on the xylan backbone have not been characterized [9, 12, 14, 16].

In this paper, we describe the cloning and overproduction in *E.coli* of *Gt*XynA1, a GH10 family endo-xylanase isolated from *Geobacillus thermodenitrificans*. We also determined the enzyme's mode of action by identifying the generated hydrolysis products and propose a model for the degradation of substituted xylans.

Methods

Media, strains and cultivation methods

Wheat arabinoxylan (WAX) was of medium viscosity and supplied by Megazyme (Wicklow, Ireland). The chemical composition of this type of WAX has previously been described [17]. Beechwood xylan (BeWX) and all chemicals were purchased from Sigma-Aldrich (St Louis, MO, USA) unless otherwise specified. *Aa*XynA was kindly provided by the laboratory of Food Chemistry (Wageningen University) as described elsewhere [8].

Luria-Bertani (LB) medium contained per L: 10 g NaCl, 10 g tryptone (Oxoid), 5 g yeast extract (Roth). LB2 contains per liter: 10 g tryptone (Oxoid), 5 g yeast extract (Roth), 10 g sodium chloride and salts mix consisting of 1 g NH_4Cl; 3 g NaCl; 1.50 g Na_2SO_4; 0.08 g $NaHCO_3$; 1 g KCl; 1.8 g $MgCl_2 \times 6H_2O$; 0.30 g $CaCl_2 \times 2H_2O$. pH was set to 6.6 at room temperature and the medium was autoclaved for 20 min at 121 °C, after which 1 mL K_2HPO_4 (250 g/L) was added.

For plasmid propagation and protein expression *E. coli* strains DH5α and BL21(DE3) were used respectively. *E. coli* strains were grown in LB medium at 37 °C, pH 7.0 with an agitation speed of 150 RPM.

Geobacillus thermodenitrificans T12 was isolated from compost [18]. Strain T12 was grown in LB2 medium at 65 °C, 150 RPM and pH 7.0.

Cloning of *Gt*XynA1

Genomic DNA of strain T12 was obtained using the MasterPure™ Gram Positive DNA Purification Kit (Epicentre) according to manufacturer's protocol. The XynA coding sequence was cloned without the signal peptide, which was determined by SignalP4.1 to consist of the first 28 amino acids of the protein. Primers used for amplification of the *xynA* gene (GenBank accession number KX962565) were XynA-FW (5'-GCGCTCATG AAAACTGAACAATCATACGCTAAAAAG-3') and Xy nA-RV (5'-GCGCCTCGAGCTTATGATCGATAATAGC CCAATACG-3') that contain the *Bsp*HI and *Xho*I restriction sites, respectively. PCR amplification was performed on 50 ng of genomic DNA using PhusionHF DNA polymerase (Thermo Scientific) in the following conditions; 1 cycle 98 °C 5 min., 30 cycles of 98 °C 30 s, 50 °C 30 s and 72 °C 45 s followed by a final elongation step at 72 °C for 5 min. For enzyme characterization studies *Gt*XynA1 was cloned into pET24d(+) using the restriction sites of *Bsp*HI and *Xho*I. The reverse primer was designed in such a way that the stop codon was removed to include the His-tag from the pET24d vector. Plasmids were purified from *E. coli* DH5α using the GeneJET™ Plasmid Miniprep Kit (Thermo Scientific) and subsequently transformed into *E. coli* BL21(DE3).

Overproduction and purification of *Gt*XynA1
Protein production was induced by adding 0.1 mM IPTG to a 250 mL culture at the start of inoculation. Culture was provided with 50 μg/mL kanamycin as antibiotics and incubated O/N at 37 °C at an agitation speed of 150 RPM. Cells were then centrifuged (4000×*g*, 15 min, 4 °C) and cell pellet was resuspended in 25 mL phosphate buffer at a pH of 7.4. Cell suspension was disrupted using French Press and cell free extract (CFE) was obtained by centrifugation (30,000×*g*, 20 min, 4 °C). Filtered CFE was applied to a HisPrep FF 16/10 Ni-NTA Sepharose column (GE Healthcare) equilibrated with 50 mM sodium phosphate buffer (pH 7.4) containing 300 mM NaCl. Bound *Gt*XynA1 was eluted using a linear gradient of 0–500 mM imidazole in the same buffer. Purified *Gt*XynA1 was then desalted by applying the protein solution to a HiPrep 26/10 desalting column (GE Healthcare) equilibrated with 50 mM sodium phosphate buffer (pH 7.4). Recombinant xylanase was eluted at a rate of 5 mL/min and stored at–20 °C for further research. Protein purity was assayed by SDS-PAGE analysis using a 10% (*w/v*) polyacrylamide gel according to the method of [19]. A PageRuler™ protein ladder was used and bands were visualized using PageBlue™ Protein Staining Solution (Thermo Scientific).

Enzyme activity assays
Wheat arabinoxylan (WAX), Beechwood xylan (BeWX) and Birchwood xylan (BiWX) were incubated with the purified *Gt*XynA1 in 10 mM NaOAc buffer, pH 6.0 (1 mL, 10 mg substrate dry matter) at 60 °C for 24 h. The enzymes were dosed at 0.1% (w/w) protein based on substrate dry matter. Next, 2 μl of 4 M HCl was added to stop the enzyme activity and the sample was centrifuged (10,000×*g*, 10 min, 10 °C) prior to analysis.

The pH and temperature optima were tested by incubating WAX with *Gt*XynA1 in a range of pH 3 to 9. For the pH range of 3 to 8 we used 50 mM sodium citrate buffer and for the pH range of 8 to 10 we used 200 mM Tris–HCl buffer. Buffers were adjusted for correct pH at the temperature of incubation. Optimum temperature was determined in a range from 30 to 100 °C. Incubation time, enzyme dosage and stopping the reaction were as described above.

All samples collected were submitted to the PAHBAH-assay, HPSEC, HPAEC and RP-UHPLC-UV-MS.

Analysis of hydrolysis products
The supernatants obtained were analysed for the amount of reducing ends present by the PAHBAH reducing assay in duplicate [20].

HPSEC
High performance size exclusion chromatography (HPSEC) of WAX and BiWX before and after incubation with *Gt*XynA1 was performed on an Ultimate 3000 HPLC system (Thermo Scientific, Sunnyvale, CA, USA) equipped with a set of three TSK-gel columns (6.0 mm × 15.0 cm per column) in series (SuperAW4000, SuperAW3000, SuperAW25000, Tosoh Bioscience, Stuttgart, Germany) in combination with a PWX-guard column (Tosoh Bioscience). HPSEC was controlled by the Chromeleon software (Thermo Scientific). Elution took place at 40 °C with 0.2 M sodium nitrate at a flow rate of 0.6 mL/min. The eluate was monitored using a refractive index (RI) detector (Shoko Scientific Co., Yokohama, Japan). Calibration was made by using pullulan series (Polymer Laboratories, Union, NY, USA) with a molecular weight in the range of 0.18–788 kDa.

HPAEC
Oligosaccharides and monosaccharides released after enzymatic incubation of WAX and BeWX with *Gt*XynA1 were analysed by HPAEC as described elsewhere [21].

RP-UHPLC-UV-MS
Xylo-oligosaccharides produced by the enzymes were 2-AA labelled and analysed by RP-UHPLC-UV-MS as previously described by Murciano Martinez et al. [21].

Results
In this study, we cloned and overexpressed the *Geobacillus thermodenitrificans* T12 *xynA1* gene in *E.coli* and determined the mode of action of the *Gt*XynA1 xylanase enzyme on wheat arabinoxylan (WAX), beechwood xylan (BeWX) and birchwood xylan (BiWX). The *xynA* gene is 1224 bp long and encodes a protein of 407 amino acid residues, which belongs to GH family 10. The amino acid sequence contains a signal peptide of 28 amino acids and two catalytic residues were predicted at positions H264 and D295. Phylogenetic analysis revealed high amino acid identity to several *Geobacillus* endo-xylanases (Table 1).

Cloning, expression and characterization of *Gt*XynA1
The gene coding for *Gt*XynA1 (1224 bp) was PCR-amplified, cloned into the pET24d vector and successfully expressed in *E.coli* BL21(DE3). The endo-xylanase was purified to apparent homogeneity, as it appeared as a 50 kDa single band on SDS-PAGE after visualization with PageBlue stain (Additional file 1: Figure S1;

Table 1 Sequence identity of several characterized endo-xylanases to the *G. thermodenitrificans* T12 endo-xylanase

Origin	DNA identity (%)	AA identity (%)	Reference
G. thermodenitrificans T12	-	-	This study
G. thermodenitrificans JK1	99	100	[11]
G. thermodenitrificans TSAA1	99	100	[16]
G. sp. TC-W7	99	99	[13]
G. sp. 71	91	91	[10]
G. stearothermophilus T-6	84	84	[4]

Additional file 2: Figure S2, lane 6). The purified endoxylanse *Gt*XynA1 showed highest activity at pH 6.0 in a broad activity range from pH 4.0 to 9.0 (Fig. 1a). *Gt*XynA1 showed activity between 40 °C to 80 °C with an optimum at 60 °C (Fig. 1b). The thermostability of *Gt*XynA1 was tested in the range of 30 °C to 80 °C. After 1 h at 50 and 60 °C the enzyme retained 90 and 45% residual activity, respectively, whereas the residual activity after 1 h at 70 °C was only 18% (Fig. 2).

Mode of action

Activity of *Gt*XynA1 towards wheat arabinoxylan (WAX), birchwood xylan (BiWX), and beechwood xylan (BeWX) was determined as outlined in materials and methods. Degradation profiles of BiWX and BeWX were similar (data not shown), therefore only the BeWX degradation patterns will be further explained. WAX contains β-(1→4) linked xylosyl residues substituted with α-(1→2) linked arabinosyl residues or with both α-(1→2) and α-(1→3) linked arabinosyl residues. BiWX and BeWX contain β-(1→4) linked xylosyl residues substituted with α-(1→2) linked 4-*O*-methylglucuronic acid (UA_{me}). Both WAX and BeWX were incubated with *Gt*XynA1 at its optimum pH and temperature for 24 h (Fig. 3). Both WAX and BeWX were degraded to oligomers and monomers, as deduced from the disappearance of soluble polymers (WAX around 55 kDa, BeWX around 12 kDa; Fig. 3) and the increase of lower molecular weight products of around 1 kDa after 24 h incubation. The difference in molecular weight of the polymeric structures analysed at 0 h incubation is because WAX is completely soluble and BeWX is only partly soluble [22]. *Gt*XynA1 incubated with WAX and BeWX released a series of linear and substituted xylo-oligosaccharides that were identified based on the known HPAEC elution of WAX digested with the well-characterized GH10 endo-xylanase from *Aspergillus awamori* (*Aa*XynA) [8]. The oligomeric profile released by *Gt*XynA1 is comparable to the profile produced by *Aa*XynA, supporting that *Gt*XynA1 is a GH10 endo-xylanase (Fig. 4). Only one substituted oligosaccharide was produced from BeWX, eluting at 16 min (Fig. 4), which was identified as a xylotriose substituted by 4-*O*-methylglucuronic acid located at the non-reducing end (Fig. 4b') after 2-AA labelling and analysis by UPLC-UV-MS. The UV chromatogram at 340 nm of 2-AA labelled degradation products from BeWX and the full MS spectra overlapped. The two predominant masses seen in the full MS spectra were 724 (m/z) and 666 (m/z) (data not shown), and they overlapped with UV chromatogram peaks eluting at 13 and 16.8 min, respectively. Fig. 4 also shows that *Gt*XynA1 is able to degrade WAX and BeWX to smaller-sized linear xylo-oligosaccharides, mainly xylose and xylobiose. This observation is substantiated on the integrated peak area of linear xylo-oligosaccharides (Table 2). On the contrary, *Aa*XynA is able to degrade WAX and BeWX producing mainly xylobiose and xylotriose, but also xylotetraose, xylopentaose and xyloheaxaose (Table 2).

When comparing the diversity of substituted xylo-oligosaccharides produced (Fig. 4), it is clear that *Gt*XynA1 produces shorter oligomers from both WAX and BeWX compared to *Aa*XynA. Substituted xylo-oligosaccharides are identified in a range of DP from 2 to 6 by both HPAEC (Fig. 4a) and MS (not shown), suggesting that *Aa*XynA is less tolerant towards 4-*O*-methylglucuronic acid substituted xylan.

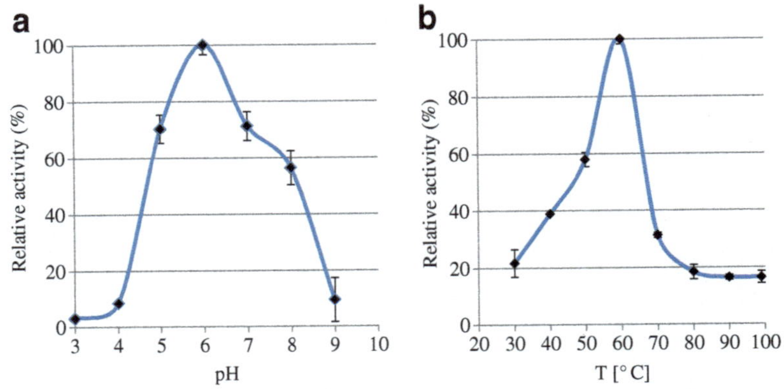

Fig. 1 pH (**a**) and temperature (**b**) profiles of the incubated *Gt*XynA1 with wheat arabinoxylan (WAX)

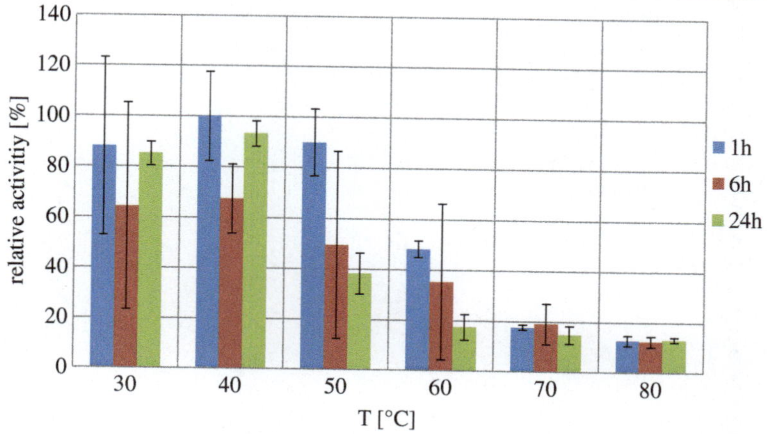

Fig. 2 Temperature stability of *Gt*XynA1. The temperature stability was determined by incubating *Gt*XynA1 at temperatures between 30 and 80 °C for 1 h, 6 h, and 24 h after which the residual activity was measured against wheat arabinoxylan at 60 °C and a pH of 6.0

Degradation products formed during the degradation of WAX by *Aa*XynA that where eluting at 15, 17 and 25 min are absent in the profile of *Gt*XynA1 incubated on WAX (Fig. 4c, d). These xylo-oligosaccharides are characteristic because of the presence of two arabinosyl residues located at the *O*-2 and *O*-3 of the same xylosyl residue. Hence, *Gt*XynA1 cleaves next to an adjacent non-substituted residue, from the non-reducing end when a single substituted residue is present. In case the xylosyl residue is doubly substituted by arabinosyl residues, *Gt*XynA1 needs two adjacent unsubstituted residues at the non-reducing end. In case the cleavage takes place from the reducing end, the enzyme can cleave directly next to the substituted (single or doubly substituted) residue in most cases (Fig. 4d, minutes 18, 22, 25.5). However, when the substituted residue (single or doubly substituted), is followed by a doubly substituted xylosyl residue, the enzyme cleaves next to an unsubstituted residue (Fig. 4d, minutes 23, 27). On the contrary, the presence of more than one doubly or single substituted xylosyl residues obstructs the action of *Aa*XynA. Based on the information provided by the HPAEC chromatograms and RP-UV-UHPLC-MS profile on the linear and substituted xylo-oligosaccharides produced by the enzymes, an schematic mode of action of *Gt*XynA1 is proposed in Fig. 4b* and d*.

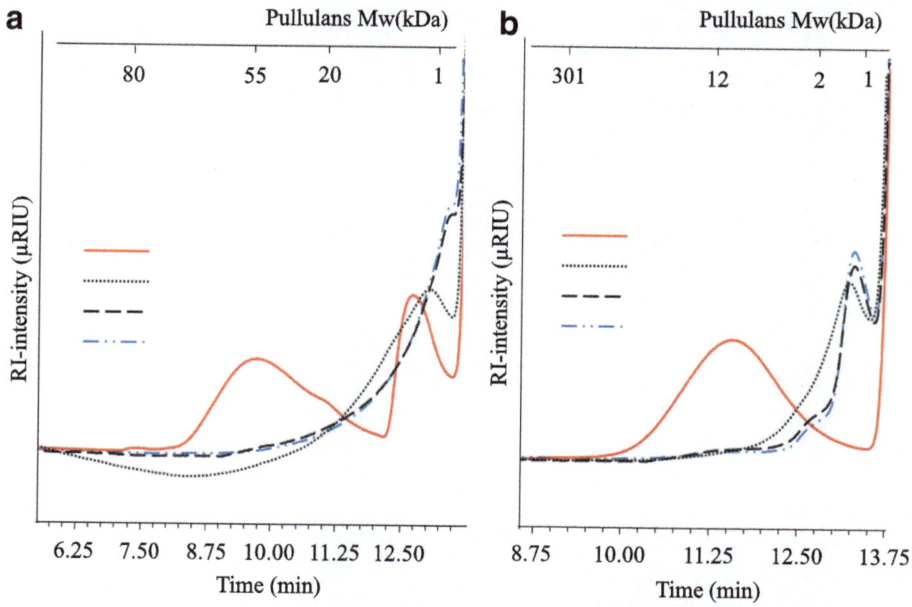

Fig. 3 Molecular weight distributions of wheat arabinoxylan (WAX) (**a**) and beechwood xylan (BeWX) (**b**) incubated with *Gt*XynA1 for 0, 1, 6 and 24 h

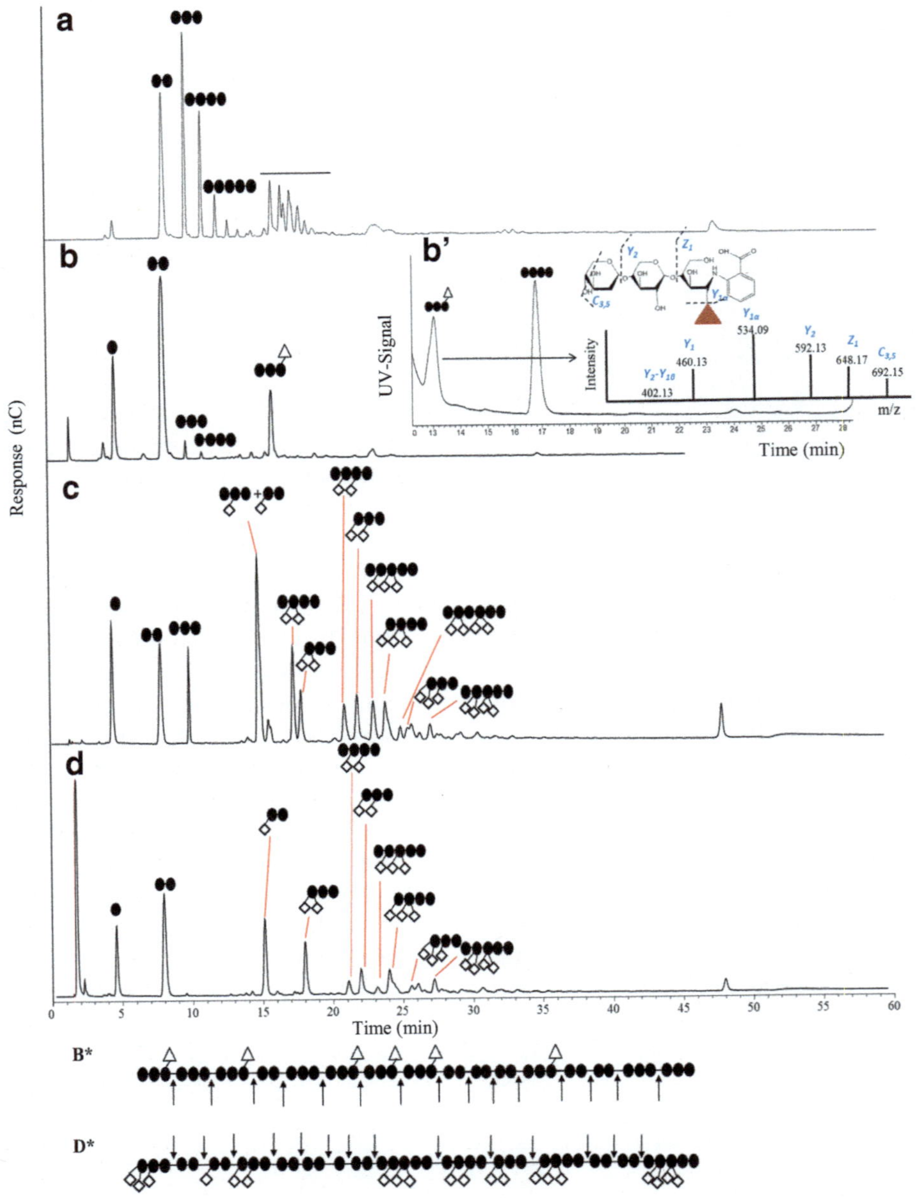

Fig. 4 HPAEC elution pattern of with *Aa*XynA and *Gt*XynA1 incubated with beech wood xylan (BeWX) (**a**, **b** respectively) and wheat arabinoxylan (WAX) (c, d respectively) for 24 h, and respective cleavage pattern of GtXynA1 on BeWX (**b***) and WAX (**d***). • = xylosyl residue, ◊ = arabinosyl residue, ▲ = 4-*O*-methylglucuronic acid. (**b**) shows the identification of the main substituted xylo-oligosaccharide from B. The structure is identify based on its UV signal and the MS^2 fragmentation pattern

Discussion

The ability of *Geobacillus* species to degrade xylan relies on the activity of the hemicellulose utilization (HUS) locus. The HUS-locus contains a variety of xylan degrading enzymes, sugar transporters and metabolic pathways for the complete utilization of (substituted) xylans. Xylan degradation initiates by the action of an extracellular endo-xylanase (XynA1) that cleaves the xylan backbone, thereby producing (substituted) xylo-oligosaccharides. The linear oligosaccharides generated by the action of XynA1 can then enter the cell through an ABC-transporter that is encoded by the *xynEFG* operon whereas the substituted oligosaccharides enter the cell *via* other substrate specific ABC transporters [1, 3, 23]. In this research, we cloned and expressed *Gt*XynA1, the endo-xylanase of *G. thermodenitrificans* T12 and analysed its mode of action.

The purified endo-xylanase *Gt*XynA1 showed highest activity at pH 6.0 in a broad activity range from pH 4.0 to 9.0 (Fig. 1a), which is in line with previous reports on endo-xylanases from geobacilli [11, 13, 16]. *Gt*XynA1

Table 2 HPAEC integrated peak area of linear xylo-oligosaccharides produced from WAX and BeWX after 24 h incubation with *Gt*XynA1 and *Aa*XynA

Enzyme tested	*Gt*XynA1		*Aa*XynA	
Substrate	WAX	BeWX	WAX	BeWX
Quantification	peak area			
Xylose (•)	87.8	165.6	4.4	17
Xylobiose (••)	192.7	457.9	17.5	157.1
Xylotriose (•••)	*	15.5	10.6	131.7
Xylotetraose (••••)	*	6.9	4.3	81.1
Xylopentaose (•••••)	*	*	1.2	25.3
Xylohexaose (••••••)	*	*	*	10.2

• = xylosyl residue; * = not detected

showed activity between 40 °C to 80 °C with an optimum at 60 °C (Fig. 1b) where previous reports on *Geobacillus* endo-xylanases demonstrated optimum temperatures from 60 °C up to 75 °C [11, 13, 14, 16]. The thermostability of *Gt*XynA1 was tested in the range of 30 °C to 80 °C. After 1 h at 50 and 60 °C the enzyme retained 90 and 45% residual activity, respectively, whereas the residual activity after 1 h at 70 °C was only 18% (Fig. 2). The endo-xylanase characterized from *Geobacillus stearothermophilus* T-6 remains 90% active (1 h time point) after incubation at 70 °C [4]. This higher temperature stability of the T-6 endo-xylanase may be caused by its production from the native strain instead of a recombinant expression system in *E.coli*. A lower thermostability of an endo-xylanase produced in *E. coli* compared to the same protein produced in *Geobacillus* has been reported for *Geobacillus* sp. TC-W7 [13], which produces an endo-xylanase highly similar to *Gt*XynA1 (Table 1). Alternatively, the difference in thermostability between *Gt*XynA1 and the T-6 endo-xylanase may be (partly) caused by differences in their amino acid composition. The *Gt*XynA1 enzyme was more thermostable in comparison with other bacterial endo-xylanases such as XynC from *Bacillus subtilis* 168, that shows around 20% residual activity at 50 °C and no residual activity at 60 °C [24]. The thermostability of *Gt*XynA1 has great advantage for the degradation of xylan at higher temperatures (60 °C) over enzymes produced by most mesophilic organisms.

In previous studies, *G. stearothermophilus* T-6 was grown on 4-*O*-methyl –D-glucurono-D-xylan and the main end products found were xylotriose substituted with 4-O-methyl–D-glucuronic acid together with xylobiose and xylose [25]. Although we found the same end products in the present study, we also detected xylotriose and xylotetraose as minor end products (Fig. 4). Only one substituted oligosaccharide was produced, identified as a xylotriose substituted by 4-*O*-methylglucuronic acid located at the non-reducing end (Fig. 4b').

When comparing the mode of action on WAX, the main difference between *Gt*XynA1 and *Aa*XynA is that the former enzyme is less hindered by arabinosyl substituents. This is deduced from the difference in degradation products formed between the incubations of *Gt*XynA1 and *Aa*XynA on WAX. Similar end products were found in previous studies using TLC analysis but, no quantification of the linear end products and no identification of substituted end products was performed [14, 26]. Formed xylo-oligosaccharides are taken up by the cell *via* specific sugar transporters [3]. These transporters are encoded by the operon *xynEFG*, located in the HUS locus, and have a preference for short xylo-oligosaccharides within a DP of 2 to 6. More specifically,

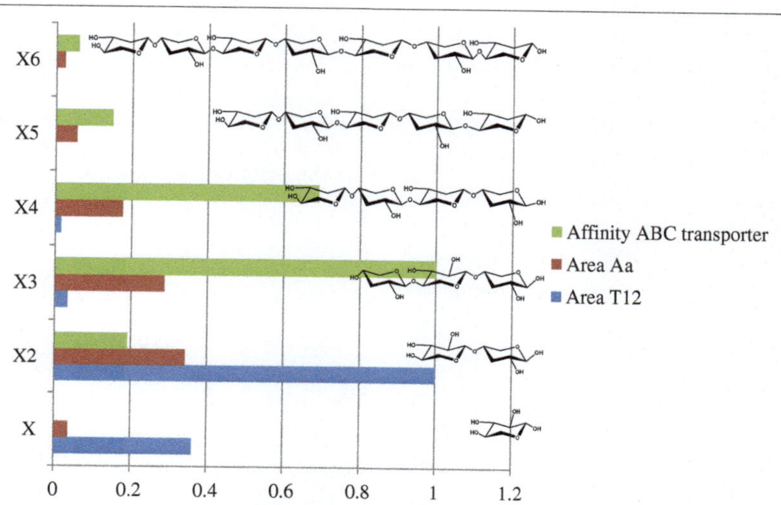

Fig. 5 Peak area ratio of linear end products after 24 h incubation on beech wood xylan in comparison to oligosaccharide transporter (*xynEFG*) affinity. Green: ratio of affinity of the XynEFG oligosaccharide transporter of *G. stearothermophilus* towards different oligosaccharides. Red: ratio of the end product area after a 24 h incubation of *Aa*XynA on beechwood xylan. Blue: ratio of the end product area after a 24 h incubation of *Gt*XynA1 on beechwood xylan

highest affinity is towards trisaccharides and for other xylo-oligosaccharides the affinity is as follows: X2 < X3 > X4 > X5 > X6 [1]. The relative quantity of different xylo-oligosaccharides formed by the action of *Gt*XynA1 on BeWX shows that mostly xylobiose and xylose is formed (Table 2, Fig. 5). The discrepancy between end products formed by *Gt*XynA1 and the affinity of the sugar transporter XynEFG is most likely caused by the 24 h incubation as *Gt*XynA1 is also active towards xylotriose. Previous studies demonstrated the formation of xylotriose within 30 min of incubation and prolonged incubation resulted in a decrease of xylotriose and an increase in xylobiose and xylose [14, 26].

The mode of action of *Gt*XynA1 creates linear and branched xylo-oligosaccharides of which most are in the DP range of 2 to 6. These small branched oligosaccharides are believed to enter the cell *via* specific ABC-transporters [3]. For most of these transporters their affinity towards linear oligosaccharides was determined [1, 23, 25]. However, for the arabino-oligosaccharide transporter AbnEFJ it was shown that the binding constants of branched oligosaccharides where 2 orders of magnitude higher than those obtained for linear oligosaccharides [23]. It is likely that also other ABC transporters that are involved in the uptake of oligosaccharides have higher affinity for branched oligosaccharides in comparison to linear oligosaccharides. The possibility of transporting branched oligosaccharides enables the *Geobacillus* species to use natural substrates in a rapid and efficient way.

Conclusions

*Gt*XynA1 is a GH10 endo-xylanase isolated from *G. thermodenitrificans* T12 and is active towards WAX and BeWX, showing extensive degradation of the soluble polymeric structures to linear xylo-oligosaccharides in a DP range from 1 to 4 and to relatively short substituted xylo-oligosaccharides. These degradation products are taken up by the cells *via* specific ABC transporters. We have demonstrated that *Gt*XynA1 is able to cleave the xylan backbone of WAX next to an arabinose substituent on the reducing end and the need for one to two free xylose units on the non-reducing end. In contrast, the hydrolysis of BeWX by *Gt*XynA1 occurs next to a 4-*O*-methylglucuronic acid located at the non-reducing end. The thermophilic origin and its extensive degradation of xylans from different origins demonstrates the potential of *Gt*XynA1 for biomass conversion processes at elevated temperatures.

Additional files

Additional file 1: Figure S1. FPLC purification of *Gt*XynA1. Protein fraction not bound to the nickel column eluted with the first 65 mL eluent. Bound *Gt*XynA1 protein was removed from the nickel column using a imidazole gradient (yellow line) which increased from 0 mM to 500 mM over a time span of 20 min (A). Fractions 22–25 were pooled and used for desalting the purified *Gt*XynA1 (B). Fractions 8–10 of the desalting column were pooled and then used for further experiments.

Additional file 2: Figure S2. 10% SDS-PAGE of purified endo-xylanase from *G. thermodenitrificans* T12 followed by PageBlue staining. Lane 1: Protein marker; Lane 2: Pellet fraction; Lane 3: cell-free extract; Lane 4: non-binding protein fraction from FPLC; Lane 5: Purified recombinant *Gt*XynA1; Lane 6: Purified and desalted recombinant *Gt*XynA1; Lane 7: Protein marker.

Abbreviations

GH: Glycosyl hydrolase; SP: Signal peptide; HPSEC: High Pressure Size Exclusion Chromatography; HPAEC: High-Performance Anion-Exchange Chromatography; MALDI-TOF MS: Matrix-Assisted Laser Desorption Ionization – Time Of Flight Mass Spectrometry; BeWX: BeechWood glucuronoXylan; BiWX: BirchWood glucuronoXylan; CFE: Cell free extract; DP: Degree of Polymerization; UHPLC-UV-MS: Ultra High Performance Liquid Chromatography Ultra Violet Mass Spectrometry; 2-AA: Anthranilic Acid.

Acknowledgements

We thank Leandra Vermeulen for her excellent technical assistance.

Funding

This work was financially supported by BE-Basic (www.be-basic.org) and Corbion.

Authors' contributions

MD designed and performed the cloning and overexpression in *E.coli*, protein purifications, participated in the design of the enzyme assays and drafted the manuscript. PM designed and carried out the enzyme assays, analysed the data and helped drafting the manuscript. AW assisted in the cloning and overexpression of the enzyme, enzyme assays, prepared corresponding data and revised the manuscript. WdV, JvdO, MK and RvK coordinated the study and revised the manuscript. All authors read and approved on the final manuscript.

Competing interests

The authors declare to have no conflict of interest. RvK is employed by Corbion (Gorinchem, The Netherlands).

Author details

[1]Laboratory of Microbiology, Wageningen University, Stippeneng 4, Wageningen 6708 WE, The Netherlands. [2]Laboratory of Food Chemistry, Wageningen University, Bornse Weilanden 9, Wageningen 6708 WG, The Netherlands. [3]Corbion, Arkelsedijk 46, Gorinchem 4206 AC, The Netherlands.

References

1. Shulami S, Zaide G, Zolotnitsky G, Langut Y, Feld G, Sonenshein AL, Shoham Y. A two-component system regulates the expression of an ABC transporter for xylo-oligosaccharides in *Geobacillus stearothermophilus*. Appl Environ Microbiol. 2007;73(3):874–84.
2. De Maayer P, Brumm PJ, Mead DA, Cowan DA. Comparative analysis of the *Geobacillus* hemicellulose utilization locus reveals a highly variable target for improved hemicellulolysis. BMC Genomics. 2014;15(1):836.
3. Shulami S, Shenker O, Langut Y, Lavid N, Gat O, Zaide G, Zehavi A, Sonenshein AL, Shoham Y. Multiple regulatory mechanisms control the expression of the *Geobacillus stearothermophilus* gene for extracellular xylanase. J Biol Chem. 2014;289(37):25957–75.

4. Khasin A, Alchanati I, Shoham Y. Purification and characterization of a thermostable xylanase from *Bacillus stearothermophilus* T-6. Appl Environ Microbiol. 1993;59(6):1725–30.
5. Ebringerová A, Hromádková Z, Heinze T. Hemicellulose. Adv Polym Sci vol. 2005;186:1–67.
6. Shallom D, Shoham Y. Microbial hemicellulases. Curr Opin Microbiol. 2003;6(3):219–28.
7. Beaugrand J, Chambat G, Wong VWK, Goubet F, Rémond C, Paës G, Benamrouche S, Debeire P, O'Donohue M, Chabbert B. Impact and efficiency of GH10 and GH11 thermostable endoxylanases on wheat bran and alkali-extractable arabinoxylans. Carbohydr Res. 2004;339(15):2529–40.
8. Kormelink FJM, Gruppen H, Viëtor RJ, Voragen AGJ. Mode of action of the xylan-degrading enzymes from *Aspergillus awamori* on alkali-extractable cereal arabinoxylans. Carbohydr Res. 1993;249(2):355–67.
9. Bhalla A, Bischoff KM, Uppugundla N, Balan V, Sani RK. Novel thermostable endo-xylanase cloned and expressed from bacterium *Geobacillus* sp. WSUCF1. Bioresour Technol. 2014;165(C):314–8.
10. Canakci S, Cevher Z, Inan K, Tokgoz M, Bahar F, Kacagan M, Sal FA, Belduz AO. Cloning, purification and characterization of an alkali-stable endoxylanase from thermophilic *Geobacillus* sp. 71. World J Microbiol Biotechnol. 2012;28(5):1981–8.
11. Gerasimova J, Kuisiene N. Characterization of the novel xylanase from the thermophilic *Geobacillus thermodenitrificans* JK1. Microbiol (Russian Federation). 2012;81(4):418–24.
12. Irfan M, Guler HI, Belduz AO, Shah AA, Canakci S. Cloning, purification and characterization of a cellulase-free xylanase from *Geobacillus thermodenitrificans* AK53. Appl Biochem Microbiol. 2016;52(3):277–86.
13. Liu B, Zhang N, Zhao C, Lin B, Xie L, Huang Y. Characterization of a recombinant thermostable xylanase from hot spring thermophilic *Geobacillus* sp. TC-W7. J Microbiol Biotechnol. 2012;22(10):1388–94.
14. Marcolongo L, La Cara F, Morana A, Di Salle A, del Monaco G, Paixão SM, Alves L, Ionata E. Properties of an alkali-thermo stable xylanase from *Geobacillus thermodenitrificans* A333 and applicability in xylooligosaccharides generation. World J Microbiol Biotechnol. 2015;31(4):633–48.
15. Mitra S, Mukhopadhyay BC, Mandal AR, Arukha AP, Chakrabarty K, Das GK, Chakrabartty PK, Biswas SR. Cloning, overexpression, and characterization of a novel alkali-thermostable xylanase from *Geobacillus* sp. WBI J Basic Microbiol. 2015;55(4):527–37.
16. Verma D, Anand A, Satyanarayana T. Thermostable and alkalistable endoxylanase of the extremely thermophilic bacterium *Geobacillus thermodenitrificans* TSAA1: Cloning, expression, characteristics and its applicability in generating xylooligosaccharides and fermentable sugars. Appl Biochem Biotechnol. 2013;170(1):119–30.
17. Van Gool MP, Vancsó I, Schols HA, Toth K, Szakacs G, Gruppen H. Screening for distinct xylan degrading enzymes in complex shake flask fermentation supernatants. Bioresour Technol. 2011;102(10):6039–47.
18. Daas MJA, van de Weijer AHP, de Vos WM, van der Oost J, van Kranenburg R. Isolation of a genetically accessible thermophilic xylan degrading bacterium from compost. Biotechnol Biofuels. 2016;9(1):1–13.
19. Laemmli UK. Cleavage of structural proteins during the assembly of the head of bacteriophage T4. Nature. 1970;227(5259):680–5.
20. Lever M. A new reaction for colorimetric determination of carbohydrates. Anal Biochem. 1972;47(1):273–9.
21. Murciano Martínez P, Bakker R, Harmsen P, Gruppen H, Kabel M. Importance of acid or alkali concentration on the removal of xylan and lignin for enzymatic cellulose hydrolysis. Ind Crops Prod. 2015;64:88–96.
22. van Gool MP, van Muiswinkel GCJ, Hinz SWA, Schols HA, Sinitsyn AP, Gruppen H. Two GH10 endo-xylanases from *Myceliophthora thermophila* C1 with and without cellulose binding module act differently towards soluble and insoluble xylans. Bioresour Technol. 2012;119:123–32.
23. Shulami S, Raz-Pasteur A, Tabachnikov O, Gilead-Gropper S, Shner I, Shoham Y. The L-arabinan utilization system of *Geobacillus stearothermophilus*. J Bacteriol. 2011;193(11):2838–50.
24. St. John FJ, Rice JD, Preston JF. Characterization of XynC from *Bacillus subtilis* subsp. *subtilis* strain 168 and analysis of its role in depolymerization of glucuronoxylan. J Bacteriol. 2006;188(24):8617–26.
25. Shulami S, Gat O, Sonenshein AL, Shoham Y. The glucuronic acid utilization gene cluster from *Bacillus stearothermophilus* T-6. J Bacteriol. 1999;181(12):3695–704.
26. Anand A, Kumar V, Satyanarayana T. Characteristics of thermostable endoxylanase and β-xylosidase of the extremely thermophilic bacterium *Geobacillus thermodenitrificans* TSAA1 and its applicability in generating xylooligosaccharides and xylose from agro-residues. Extremophiles. 2013;17(3):357–66.

13

Sequential and counter-selectable cassettes for fission yeast

Hanna Amelina[†], Vera Moiseeva[†], Laura Catharine Collopy, Siân Rosanna Pearson, Christine Anne Armstrong and Kazunori Tomita[*]

Abstract

Background: Fission yeast is one of the most commonly used model organisms for studying genetics. For selection of desirable genotypes, antibiotic resistance cassettes are widely integrated into the genome near genes of interest. In yeasts, this is achieved by PCR amplification of the cassette flanked by short homology sequences, which can be incorporated by homology directed repair. However, the currently available cassettes all share the same *tef* promoter and terminator sequences. It can therefore be challenging to perform multiple genetic modifications by PCR-based targeting, as existing resistance cassettes in strains can be favored for recombination due to shared homology between the cassettes.

Results: Here we have generated new selection cassettes that do not recombine with those traditionally used. We achieved this by swapping the *tef* promoter and terminator sequences in the established antibiotic resistance MX6 cassette series for alternative promoters and/or terminators. The newly created selection cassettes did not recombine with the *tef*-containing MX6 cassettes already present in the genome, allowing for sequential gene targeting using the PCR-based method. In addition, we have generated a series of plasmids to facilitate the C-terminal tagging of genes with desired epitopes. We also utilized the anti-selection gene *HSV-TK*, which results in cell death in strains grown on the drug 5-Fluoro-2′-deoxyuridine (FdU, Floxuridin or FUDR). By fusing an antibiotic resistance gene to *HSV-TK*, we were able to select on the relevant antibiotic as well as counter-select on FdU media to confirm the desired genomic modification had been made. We noted that the efficiency of the counter selection by FdU was enhanced by treatment with hydroxyurea. However, a number of DNA replication checkpoint and homologous recombination mutants, including *rad3Δ*, *cds1Δ*, *rad54Δ* and *rad55Δ*, exhibited sensitivity to FdU even though those strains did not carry the *HSV-TK* gene. To remove counter-selectable markers, we introduced the Cre-loxP irreversible recombination method. Finally, utilizing the negative selectable markers, we showed efficient induction of point mutations in an endogenous gene by a two-step transformation method.

Conclusions: The plasmid constructs and techniques described here are invaluable tools for sequential gene targeting and will simplify construction of fission yeast strains required for study.

Keywords: Schizosaccharomyces pombe, DNA replication, Point mutation, Gene disruption and insertion, Thymidine kinase, FUdR, HA, Flag, PK tagging, Zeocin

* Correspondence: k.tomita@ucl.ac.uk
[†]Equal contributors
Chromosome Maintenance Group, UCL Cancer Institute, University College London, Paul O'Gorman Building, Huntley Street, London WC1E 6DD, UK

Background

Fission yeast *Schizosaccharomyces pombe* is a popular model organism for fundamental biological research [1]. A haploid life cycle along with efficient homologous recombination system and simple reproductive processes make it an excellent system to study genetics. In yeast, genes can be deleted and replaced with selection markers readily, simply by inducing homology directed repair with a selection cassette amplified with primers that contain short, flanking sequences homologous to a target gene [2, 3]. A strain of induced genotype can be isolated simply by plating transformants onto corresponding selection plates. Generated strains can be further crossed to isolate multiple genetic alternations that can be selected for using combination of appropriate markers.

In fission yeast, commonly used drug selection cassettes are *kanMX6*, *natMX6*, *hygMX6* (*hphMX6*) and *bleMX6* [2, 4, 5]. These cassettes are composed of a promoter and terminator from the fungus *Ashbya gossypii* and genes conferring selectable resistance to G418 (geneticin or neomycin), nourseothricin (ClonNat or Nat), hygromycin B and bleomycin respectively. Auxotrophic gene cassettes, such as *ura4*$^+$ from *S. pombe* and *LEU2* from *Saccharomyces cerevisiae* have also been extensively used in fission yeast [6, 7]. However, when using auxotrophic cassettes, strains must have mutations in the corresponding endogenous gene, which could affect cell metabolism, stress response and cell growth. Thus, use of antibiotic resistance markers has become standard for yeast organisms, as the working strain does not require any particular auxotrophic genetic background. However, as these antibiotic cassettes all share the same promoter and terminator sequences, it is difficult to perform sequential gene targeting due to the shared homology [5]. This issue is usually solved by genetic cross between two mutant strains. However, this strategy is not feasible for mutants dysfunctional in mating, meiosis or germination. In some cases the mating is successful, but it is difficult to obtain desired genotype in the offspring if genetic distance is too close or if desired genotypes are not associated with selectable marker or not readily isolated by the marker. Importantly, genetic crossing also mixes epigenetic status of genome. For example, telomere length will change if strains harboring distinct mean length of telomeres are crossed. These risks are avoidable when sequential gene targeting is performed. Therefore, *ura4*$^+$, *arg3*$^+$ and *his3*$^+$ and *LEU2* cassettes are still used, as these auxotrophic markers contain their own distinct promoters and terminators [6–9].

The recently developed genome editing technology, CRISPR (Clustered regularly interspaced short palindromic repeats) and Cas9 (CRISPR-associated protein-9 endonuclease), has become available to use in fission yeast [10]. In this system, the Cas9 nuclease is targeted to specific regions of genomic DNA by its CRISPR RNA component, a so-called guide RNA that binds to the target DNA, resulting in Cas9-mediated DNA double-strand breaks (DSBs) at the target genomic locus. This system does not require selection cassettes, as cells confer constitutive DNA DSBs until the target DNA sequence is lost or altered. Because a selection cassette is not required, this method is ideal for the sequential targeting of multiple genes/loci; in particular for gene disruption and for introducing point mutations or insertions using a designed DNA repair template. However, although the CRISPR/Cas9 strategy is widely used for genetic engineering in animals and humans, the use of this method to mutate genes in fission yeast currently holds a number of limitations. It is often necessary to design a number of guide RNAs, which need to be cloned into a fission yeast CRISPR/Cas9 plasmid [10]. Target sequences must be unique, 20 bases long, and are limited to loci upstream of a protospacer adjacent motif (–NGG for Cas9). This means that the designed DNA repair template will recombine only at the DSB site, and the substitution of the designed point mutation needs to be close to the DSB site [11]. It is also noted that adenine and thymine rich sequences cannot be targeted. Hence, whereas gene disruption, single point mutations and insertions are relatively straightforward for the CRISPR/Cas9 method, replacement of a large genomic DNA region with the designed DNA template, such as wide range multiple mutagenesis of the whole gene sequence [12], is not feasible. Thus, traditional homologous recombination-based genetic engineering remains the most robust method to perform genetic manipulations in fission yeast.

Here, we report optimization of the classical PCR-based targeting method by modification of well-known *MX6* markers, which enables sequential knock in/out genes of interest. We also generated the C-terminal tagging vectors for PCR-based gene targeting. In addition, we generated counter selectable drug resistance markers by fusing the herpes simplex virus thymidine kinase (HSV-TK) with antibiotic resistance genes. This can replace the traditionally used counter-selectable *ura4*$^+$ cassette. Furthermore, introduction of irreversible mutant *loxP* (locus of X(cross)-over P1) sites allows recycling of selection cassettes. Finally, using the generated constructs, we report a reliable two-step mutagenesis method, suitable for mutagenesis of essential genes. Altogether, we have demonstrated innovative techniques that address current difficulties in the construction of required fission yeast strains.

Results and discussion

Cassettes for PCR-based sequential targeting

Although PCR-based gene targeting by drug selection markers is the conventional approach, it is not efficient if a transforming strain already contains other drug selection markers. All *MX6* cassettes share 300 bp of the *tef* promoter and 200 bp of the *tef* terminator, which are significantly larger than the 80–100 bp sequences introduced in primers to target genes of interest. Therefore, there is a high chance of recombination between the *MX6* cassettes [5]. In order to overcome this problem, *CMVneo* and *CMVzeo* cassettes (Life Technology) were cloned to replace the *MX6* cassette of the plasmid pFA6a-kanMX6 [2]. Resulting plasmids, pFA6a-neoCV and pFA6a-zeoCV, contain neomycin/kanamycin and Zeocin^{TM} resistant genes, respectively, flanked by the cytomegalovirus promoter (*Pcmv*) and the simian virus 40 terminator (*Tsv40*) (Fig. 1a). These cassettes share only 20 bases of the primer annealing sites with the MX6 cassettes, which reduces the risk of unwanted recombination between selection markers. In addition, the *tef* promoter in the *natMX6* cassette, which provides resistance to nourseothricin (Nat), was replaced with the *CMV* promoter (*Pcmv*), resulting in the plasmid named pFA6a-natCX. The *tef* terminator of the *hygMX6* cassette, which provides resistance to Hygromycin B (Hyg), was replaced with either the *Tsv40* terminator or the *S. cerevisiae LEU2* terminator (*Tscleu2*), resulting in the plasmids named pFA6a-hygMV and pFA6a-hygML, respectively (Fig. 1a). The *zeoCV* cassette acts as an equivalent to *bleMX6* [4], and provides resistance to Zeocin^{TM}, bleomycin and phleomycin (Fig. 1b). We confirmed that replacement of the promoter or terminator did not impair the function of the drug resistant cassettes. Thus, the newly generated selection cassettes can be used as an alternative to existing MX6 cassettes.

Using the generated cassettes, we tested if they could be used for sequential gene deletion. Two genes, *crb2*$^{+}$ and *rdh54*$^{+}$, were sequentially targeted in strains bearing both a *kanMX6* and *natMX6* cassette. Crb2 is the fission yeast ortholog of human 53BP1 [13], and Rdh54 is the equivalent of Rad54, which is required for meiotic recombination [14]. To delete *crb2*$^{+}$, the *hygMV* cassette was amplified by PCR using the *crb2 100 bp* primer set (Table 1, crb2 top and bot) and pFA6a-hygMV plasmid as a template. A strain bearing both *kanMX6* and *natMX6* [*h*^{90} *ura4-D18 taz1-YFP:kanMX6 sid4-mCherry:natMX6 hht1-Cerulean:ura4*$^{+}$] was transformed with the *crb2::hygMV* PCR fragment, and the transformant colonies were selected for resistance to hygromycin B (Hyg). The presence of *kanMX6* and *natMX6* cassettes was determined by resistance to G418

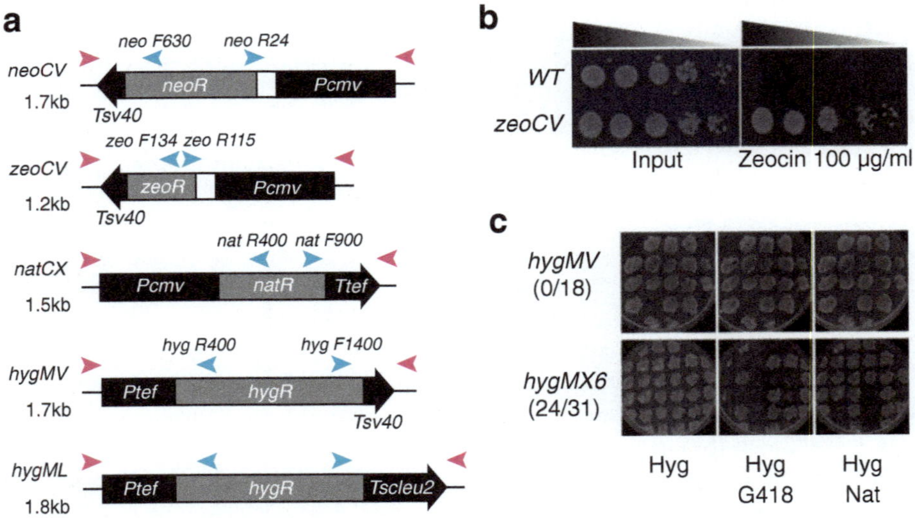

Fig. 1 Selection markers for sequential gene targeting. **a** Schematic representation of sequential targeting cassettes pFA6a-neoCV, pFA6a-zeoCV, pFA6a-natCX, pFA6a-hygMV and pFA6a-hygML. The region used as a PCR template is shown. The 100 base *Top (Tag)* primer anneals to the left end (*pink arrowhead on the left*) and *Bot* primer anneals to the right end to amplify indicated cassettes (*pink arrowheads on the right*). The 100 base primers used in this study are listed in Table 1. Cyan arrows represent diagnostic primers used for screening of correct targeting (Table 2). Sequences of primers are shown in Table 2. Backbone vector region (pFA6a) encodes ampicillin resistant cassette and *ColE1* bacteria replication origin, and the cassettes were inserted between *PacI* and *PmeI* sites. *Black box* and *black arrow* in the cassette represent indicated promoter and terminator, respectively. *White box* in *neoCV* and *zeoCV* indicates the *em7* promoter for *E. coli*. The size of the cassette is shown on the left. Transcription direction is toward left for *neoCV* and *zeoCV* cassettes and toward right for other cassettes. **b** Wild-type and *rdh54Δ* cells in which the *zeoCV* cassette replaced *rdh54* were cultured in YES rich media, normalised and serially diluted. Five microliter of diluted fractions were spotted on YES (input) or YES containing 100 μg/ml of Zeocin^{TM} and incubated at 32 °C for 3 days. Only cells containing the *zeoCV* cassette grew on YES with Zeocin. Deletion of the *rdh54* gene required for meiosis does not impair cell growth and mitotic DNA damage repair. **c** Cells carrying *kanMX6* or *natMX6* cassettes were transformed with *hygMV* (Top) or *hygMX6* (*Bottom*) and selected for on YES plates containing 100 μg/ml hygromycin (Hyg). Eighteen and 31 colonies that were randomly picked for *hygMV* and *hygMX6 transformations respectively,* were re-streaked on YES plates containing 100 μg/ml Hyg (*Left panel*). Cells were then replica-plated to YES plates containing 100 μg/ml Hyg plus either 100 μg/ml G418 (*Middle panel*) or 100 μg/ml ClonNat (Nat) (*Right panel*). In case of *hygMV* transformation, all transformed cells retained resistance to G418 and ClonNat, whereas in case of *hygMX6, only seven out of 31 did*

Table 1 One hundred base primers for PCR-based targeting fragments used in this study

Name	Sequence
crb2 Top	CCCTGGTTAAATTTGTAGTTCTGACAATTGTGAGGTATTTTAGATGTTTTCAATATTTTGTTTGAAAGTTTAACAATATTCGGATCCCCGGGTTAATTAA
crb2 Bot	CTAAAATTAATAAAAAGCTAAATTAATGAGAGTGAAACTCAGGGGGAGTTAGTAAAAATAACTATATCAAAAAACCAAAAGAATTCGAGCTCGTTTAAAC
rdh54 Top	ACTTGAACACCACACGTTCGGTCTCATAGTATTGTTGAGTAAATAAACACAGTTACCAAAGAGAATTGAAACCTTACTTTCGGATCCCCGGGTTAATTAA
rdh54 Bot	TGCTCATAGAGATTGCCCAGTATCGAAGGCTTGCTCAAAAGCGTTGTTTCTAATGGGCAAGGAAAAACCGCTTACGCCTCGAATTCGAGCTCGTTTAAAC
tpz1 Tag	CTCTGAGGCCTGTGAAATGTGTCGGCTTGGGCTACCTCATGGATCATTCTTTGAGCTATTGCGAGATTGGAAAAAAATAGAGGAGTTTCGAAACAAAAGCCGGATCCCCGGGTTAATTAA
tpz1 Bot	TTGGTCCGTTGTAAGCCATTTCACTGTATGTCTGTAACAGTTAACTTCCGTACTTAGTAAAATGTTAGTAAAAAAGGAAGATATGTGATACAGCAATTGAGAATTCGAGCTCGTTTAAAC
tpz1 Top	ATCAACAGACTTCAGTCAGCACTGTTACTTATTAAAAAAAAGTTGATTTTTATATAAAAGTTAGCTGCGTTAAACAGTGCACGGATCCCCGGGTTAATTAA

Under line encodes annealing site for the template. Top primer contains CGGATCCCCGGGTTAATTAA at the 3′ end and 80–100 base upstream sequence of the target site at the 5′ end. Bot primer contains GAATTCGAGCTCGTTTAAAC at the 3′ end and 80–100 base downstream complement sequence of the target site at the 5′ end

Table 2 Primers for diagnostic PCR

Name	Sequence
neo R24	GTGCAATCCATCTTGTTCAATC
neo F630	TTTCTGGATTCATCGACTGTGG
zeo R115	CGAAGTCGTCCTCCACGAAG
zeo F134	ACGTGACCCTGTTCATCAGC
nat F900	GGGGTTCACCCTCTGCGGCC
nat R400	GGGACACTGGTGCGGTACCG
hyg F1400	CCGTCTGGACCGATGGCTGT
hyg R400	CTTCTCGACAGACGTCGCGG
kan F800	GGATTCAGTCGTCACTCATGGTG
kan R276	ATGCATCATCAGGAGTACGG
TK F1040	TACCGACGATCTGCGACCTG
TK R90	ACTTCCGTGGCTTCTTGCTG
adh1 R540U	AGGAATACGGATACGATGGAG
Ptef R81	ACATGGGGATGTATGGGCTA
Ttef F1	CAGTACTGACAATAAAAAGATTCTTG
crb2 R280D	CAACCATTCGAAACCTGCTAC
rdh54 R265D	GAAAGCCAGAAACAGACAAGC
tpz1 F842U	GGTAGTAGGCTAAATGTGAGTTG
tpz1 R400D	CCCTTAGAAGATAAGCTCAACC

and ClonNat (Nat), respectively. The *hygMV* cassette was therefore successfully integrated into the genome without recombining with the MX6 cassettes present, conferring resistance to multiple antibiotics (Fig. 1c: hygMV panel). Deletion of *crb2* by *hygMV* was confirmed by diagnostic PCR using a forward primer within the hyg^R gene and a reverse primer downstream *of crb2* (for sequences see Table 2, *hyg F1400* and *crb2 R280D).* In contrast, transformation with the *crb2::hygMX6* PCR fragment resulted in replacement of *kanMX* or *natMX6* markers, assessed by gain of sensitivity to G418 and Nat, respectively (Fig. 1c: hygMX6 panel). Next, the $rdh54^+$ gene was deleted in a similar fashion, using a *rdh54::zeoCV* fragment generated by PCR using the *rdh54 100 bp* primer set and pFA6a-zeoCV plasmid as a template. Like *hygMV, zeoCV* did not recombine with any of existent marker cassettes (data not shown). The *rdh54* deletion in the resulting strain [h^{90} *ura4-D18 crb2::hygMV rdh54::zeoCV taz1-YFP:kanMX6 sid4-mCherry:natMX6 hht1-Cerulean:ura4*$^+$] was confirmed by diagnostic PCR using primers *rdh54 R265D* and *zeo R115* (Table 2). We also confirmed that *neoCV, hygML* and *natCX* did not recombine with the MX6 cassettes (data not shown). Therefore, we conclude that replacement of the *tef* promoter and/or terminator sequences prevents recombination of newly introduced drug selection cassette with the MX6 cassettes already existing in a strain.

HSV-TK chimera selection markers

Until recently, the *ura4*$^+$ auxotrophic marker was the only marker in the fission yeast system to be both selectable and counter-selectable. Cells carrying *ura4*$^+$ gene can be eliminated by growing on media supplemented with 5-Fluoroorotic acid (5-FOA), as Ura4 converts 5-FOA to toxic fluorodeoxyuridine [7]. Alternatively, the thymidine kinase from herpes simplex virus (HSV-TK, or TK) is available for use as a negative selection marker [15]. TK phosphorylates 5-Fluoro-2'-deoxyuridine (FdU, Floxuridin or FUDR) to give 5-Fluoro-2'-deoxyuridine-5-monophosphate (FdUMP). FdUMP blocks activity of thymidylate synthase, resulting in inhibition of 2'-deoxythymidine-5-triphosphate (dTTP) synthesis, leading to loss of the dTTP pool and cell death. Hence, strains expressing a *TK* cassette cannot grow in the presence of FdU [15, 16]. Unlike the *ura4*$^+$ cassette, the *TK* cassette can be used with any genotype background. Although fission yeast carrying the *TK* cassette are sensitive to FdU, a positive selectable marker is also needed in order to isolate cells carrying *TK* [15]. This issue was solved by a fusion of the *TK* gene with a positive selectable marker in mammalian systems [17, 18].

We decided to create fused selection markers for fission yeast, which express fusion antibiotic resistance genes with the *TK* gene that serve for both positive and negative selection by chemical treatments. We first inserted the *TK* gene upstream of the gene coding sequences in *kanMX6* and *natMX6*, giving *TKanMX6* and *TKnatMX6*. To assess function of the fusion genes, the TK-fusion cassettes were inserted at the *leu1*$^{+}$ locus. As expected, cells carrying *TKanMX6* and *TKnatMX6* were resistant to G418 and Nat, respectively. However, these cells grew in the presence of 100 μg/ml FdU, suggesting that the fused *TK* gene is not functional (see the *TKnatMX6* in Fig. 2b). We speculated that the level of TK, expressed by the *tef* promoter, was not sufficient to convert FdU for a toxic amount of FdUMP, as previously published the *TK* cassette for fission yeast uses the stronger *adh1* promoter [15]. We therefore replaced the *tef* promoter with a stronger *CMV* promoter (giving *TKnatCX*, Fig. 2a), which indeed enhanced cells sensitivity to FdU (Fig. 2b). Furthermore, an even greater sensitivity was obtained with replacement of the *S. pombe adh1*

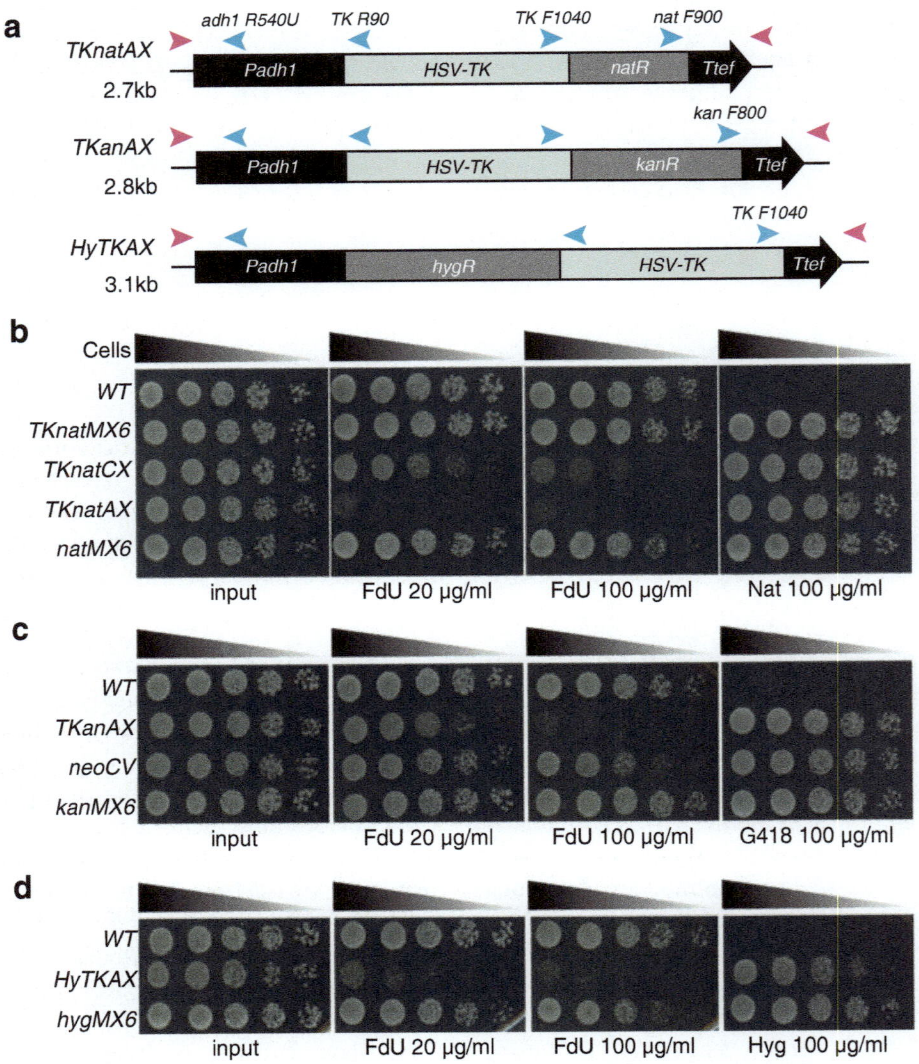

Fig. 2 TK-fusion counter-selectable cassettes. **a** Schematic representation of counter-selectable cassettes *HyTKAX*, *TKanAX* and *TKnatAX*. Region used as a PCR template is shown. The backbone vector is pFA6a and the cassettes were inserted between PacI PmeI sites. *Top* primers anneal to the left end (*pink arrowhead on the left*) and *Bot* primer anneals to the right end to amplify indicated cassettes (*pink arrowheads on the right*). *Cyan arrows* represent diagnostic primers used for screening of correct targeting, and sequence of primers is shown in Table 2. The *TK* fusion genes are transcribed by the *S. pombe adh1* promoter and terminated at the *tef* terminator. The size of the cassette is shown on the left. Transcription direction is toward right. **b**, **c** and **d** FdU sensitivity of cells carrying the *HSV-TK* fusion cassettes. Indicated cells were cultured in YES rich media, normalised and sequentially diluted five times. Five microliter of diluted fractions were spotted on YES containing the indicated drugs and incubated at 32 °C for 3 days. **b** Cells carrying the *TKnatAX* cassette are hyper sensitive to 20 μl/ml FdU. **c** Cells carrying the *TKanAX* cassette are sensitive to 100 μl/ml FdU. **d** Cells carrying the *HyTKAX* cassette are sensitive to 20–100 μl/ml FdU

promoter, giving *TKnatAX* (Fig. 2a and b). The success of this finding was reiterated with the *TK-kan*^{R} fusion gene, giving *TKanAX* (Fig. 2a). Cells carrying *TKanAX* were fully resistant to G418 and displayed an intermediate sensitivity to FdU (Fig. 2c). Finally, *HSV-TK* was inserted after the *hyg*^{R} gene coding sequence, similar to the *HyTK or HygR-TK* fusion genes used in mammalian systems [17], giving *HyTKAX* (Fig. 2a). Cells carrying *HyTKAX* were resistant to hygromycin B and sensitive to FdU (Fig. 2d). In conclusion, fusions of the yeast antibiotic markers with the *TK* gene are functional, allowing both positive and negative selections of cells carrying the fusion cassettes.

In yeast, low sensitivity to FdU is partially caused by poor uptake of thymidine. This problem can be overcome by ectopic expression of human equilibrative nucleoside transporter (ENT), which is highly efficient at the uptake of thymidine analogs [19]. However, this strategy requires cells to carry the ENT expression cassette. Toxicity of FdU in the presence of TK is caused by impaired deoxythymidine-5-monophosphate (dTMP) synthesis and reduced dTTP levels [16]. Hence, we anticipated that reduction of the deoxynucleotide (dNTP) pool by hydroxyurea (HU) might further reduce the dTTP level, leading to increased sensitivity to FdU. To test this possibility, cells carrying *TKnatAX*, *TKanAX* and *HyTKAX* were spotted on YES media containing increasing concentrations of HU (1, 2.5 and 5 mM) along with 100 μg/ml FdU (Fig. 3). The growth of cells carrying *TKanAX* was significantly suppressed by FdU in the presence of HU, as represented by microcolonies. Taken together, we recommend the following concentrations for FdU anti-selections: 20 μg/ml FdU for *TKnatAX*, 100 μg/ml FdU for *HyTKAX* and 100 μg/ml FdU plus 5 mM HU for *TKanAX*.

Mutants defective in DNA replication checkpoint and homologous recombination are sensitive to FdU

Although the *TK* cassette is a useful tool, we have come across a number of fission yeast mutants that exhibited sensitivity to FdU while not containing the *TK* cassette. Specifically, our data suggest that mutants defective in DNA replication and DNA damage response pathways are sensitive to FdU. It has previously been reported that *rad3Δ* cells, which lack the DNA damage checkpoint protein Rad3 (a fission yeast ATR homolog [20]), are sensitive to the deoxythymidine variant, 5'-Ethyl-2'-deoxyuridine in the presence of TK [21]. Surprisingly, we identified that *rad3Δ* cells that did not express TK were also sensitive to 20 μg/ml FdU (Fig. 4a). In response to DNA damage and DNA replication problems, Rad3 activates Chk1 and Cds1, respectively [20]. Interestingly, *cds1Δ* cells were sensitive to 100 μg/ml FdU, whereas *chk1Δ* cells were only mildly sensitive to FdU (Fig. 4b). Moreover, homologous recombination repair mutant *rad54Δ* cells also exhibited sensitivity to 100 μg/ml FdU (Fig. 4b). Whilst Rad54 is essential for homologous recombination, the Rad55-Rad57 complex acts in parallel to Swi5-Sfr1 and upstream of Rad54 [22]. As such, *rad55Δ* cells exhibited only mild sensitivity. Although it remains unclear how FdU can be converted to FdUMP under DNA replication stress, our data suggest that mutants defective in DNA replication and DNA damage response pathways are sensitive to FdU. Therefore, we recommend to test working strains for the sensitivity to FdU prior to performing counter-selection screening.

The Cre/loxP system can be utilized for the recycling of counter-selectable markers

Another way to address the limited availability of selection markers is to recycle the cassette using the Cre/loxP approach [23–25]. In this system, Cre recombinase recombines two short *loxP* sequence sites to generate crossover products. Hence, a *loxP*-flanked (floxed) gene of interest or selection marker can be deleted *via* expression of Cre recombinase. To generate conditional pop-out markers for *TKanAX*, *TKnatAX* and *HyTKAX*, mutant variants of *loxP* sequences - *lox71* and *lox66* - were introduced at the start and end of the cassettes. Resulting cassettes were named

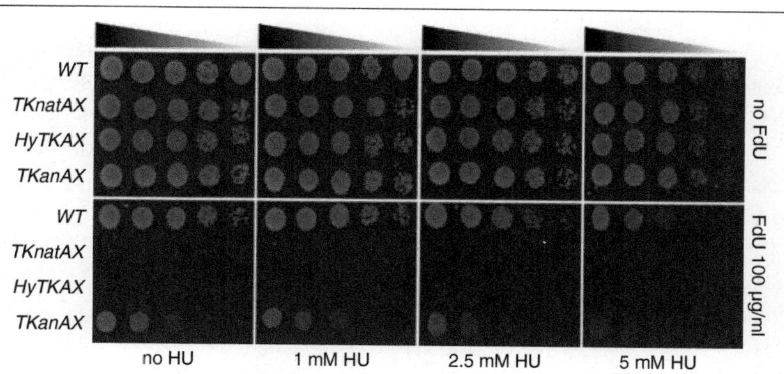

Fig. 3 Concomitant treatment with HU enhances sensitivity to FdU. Indicated cells were cultured in YES, normalised and serially diluted five times with YES. Five microliter of diluted fractions were spotted on rich media containing the indicated drugs and incubated at 32 °C for 3 days. Cells carrying *TKanAX* grew poorly in YES containing 5 mM HU and 100 μg/ml FdU

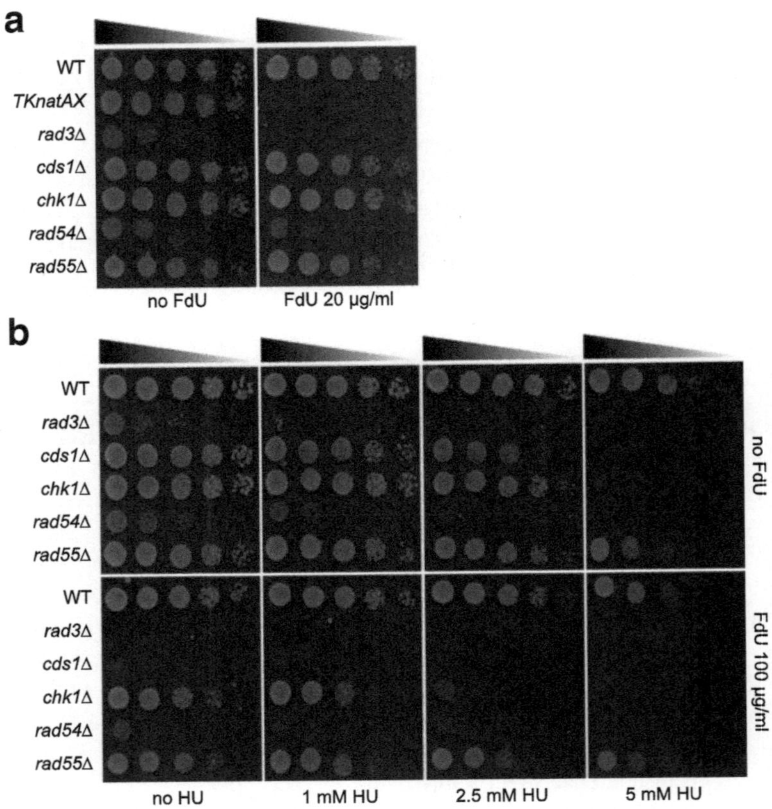

Fig. 4 *rad3Δ*, *cds1Δ* and *rad54Δ* are sensitive to FdU. Indicated cells were cultured in YES, normalised and 5 times sequentially diluted with YES. Five microliter of diluted fractions were spotted on rich media containing the indicated drugs and incubated at 32 °C for 3 days. **a** *rad3Δ* cells are hyper sensitive to 20 µg/ml FdU. **b** *cds1Δ* and *rad54Δ* cells are sensitive to 100 µg/ml FdU. *chk1Δ* and *rad55Δ* cells are mildly but significantly sensitive to FdU, compared to wild type cells

FTKanAX, *FTKnatAX* and *FHyTKAX*, respectively (Fig. 5a, Table 3). Although the *lox71* and *lox66* sequences are substrates for Cre recombinase, the crossover results in the generation of wild type *loxP* and double mutated *loxP* sequences; the latter is no longer recognized by the recombinase [26]. Therefore, the scar of a *lox71-lox66* recombined site will not crossover with a newly introduced *loxP*-containing marker in the genome.

We also generated a series of the Cre expression vectors, encoding a fusion protein, comprised of the three tandem hemagglutinin (HA) genes, a nuclear localization signal (nls) sequence and Cre recombinase (3xHA-nls-Cre), under the control of a *CMV* promoter (Fig. 5b). To maximize versatility, the Cre expression vector carries a series of the selection markers; *kanMX6*, *TKkanAX*, *hygMX6*, *HyTKAX*, *natMX6*, *TKnatAX* or the *aur*^{R} cassette, which encodes gene resistance to aureobasidin A [27], as well as auxotrophic markers, including *ura4*$^+$, *arg3*$^+$, *leu1*$^+$ and *ade6*$^+$ (Table 4). Altogether, a choice of three 'floxed' markers and a series of the Cre expression vectors would maximize the utility for genome engineering.

Using the Cre expression vector, the efficiency of the *FTKnatAX* cassette removal was assessed. Cells carrying the *FTKnatAX* cassette were transformed with the Cre expression vector, and were directly plated on FdU plates. Total 162 colonies were formed on the FdU plate after Cre expression, and all were lost the resistance to Nat. Ten clones were randomly selected and their loss of the *FTKnatAX* cassette was confirmed by PCR. Thus, FdU counter selection along with Cre expression efficiently eliminates cells retaining the *FTKnatAX* Cassette.

Tandem epitopes C-terminal tagging plasmids, pNX3 series

In addition to creating a series of plasmids useful for the sequential modification of genes, we have also created similar plasmids for C-terminal tagging of gene products. Tandem repeats of epitope tags enhance the detection efficiency of target proteins, which is important for their visualization, especially if they have low-abundance. However, larger tags can potentially interfere with protein function. Therefore, the size of the tag needs to be optimized for each protein of interest. To address this, three tandem genes coding for either HA, PK (V5) and FLAG epitopes, flanked by *Nhe*I and *Xba*I restriction sites were synthesized, and cloned between *Pac*I and *Asc*I sites of the pFA6a-kanMX6 C-terminal tagging plasmid [2] (Fig. 6a

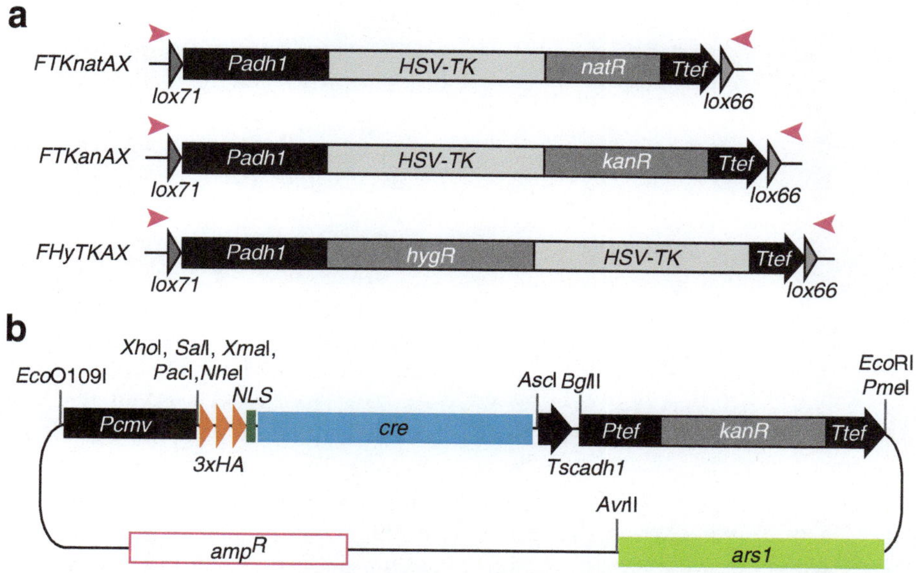

Fig. 5 Cre expression vector and loxP cassette. **a** Schematic representation of *loxP*-flanked (floxed) TK-fusion cassettes *FHyTKAX*, *FTKanAX* and *FTKnatAX*. Region used as a PCR template is shown. Forward direction of *lox71* and *lox66* sequences are inserted into the *TK* cassettes shown in Fig. 2a. **b** Schematic representation of pNXRVa-HACre. Fusion protein of three tandem HA epitope tag, NLS (nuclear localization signal) and Cre recombinase is expressed by the constitutively active *CMV* promoter. *kanMX6* is inserted between *Bgl*II and *Pme*I sites. The genomic DNA fragment containing the early replication origin *ARS1* is inserted downstream of the *kanMX6* cassette. A number of unique restriction enzyme sites are indicated

and b). PCR-based amplification of the C-terminal tagging plasmid for gene targeting results in addition of a peptide - Arg-Ile-Pro-Gly-Leu-Ile-Asn-Ala-Ser - which acts as a linker between the target protein and the epitopes (Fig. 6A). The resulting plasmids were named pNX3 (plasmid *Nhe*I-*Xba*I version three).

The pNX3 plasmid series was extended by replacing *kanMX6* with other drug resistance cassettes [*hygMX6*, *natMX6*, *neoCV*, *hygML*, *natCX* and *zeoCV*] or $ura4^+$ cassette. The cleavage site of *Nhe*I is compatible with that of *Xba*I, which allows ligation of epitope tags. For example, 3xPK can be sequentially subcloned to generate 6xPK, 9xPK and 12xPK-tagging plasmids. A list of the pNX3 plasmid series created is shown in Table 5. All plasmids were verified by DNA sequencing, and some of the constructs have been used elsewhere [12, 28].

Table 3 The targeting cassette vectors

Name	Promoter/gene/terminator	Selection/counter
pFA6a-kanMX6	*Ptef/kanR/Ttef*	G418
pFA6a-neoCV	*Pcmv/neo^R/Tsv40*	G418
pFA6a-zeoCV	*Pcmv/zeo^R/Tsv40*	Zeocin
pFA6a-hygMX6	*Ptef/hph/Ttef*	G418
pFA6a-hygMV	*Ptef/hph/Tsv40*	Hygromycin B
pFA6a-hygML	*Ptef/hph/Tscleu2*	Hygromycin B
pFA6a-natMX6	*Ptef/nat^R/Ttef*	ClonNat
pFA6a-natCX	*Pcmv/nat^R/Ttef*	ClonNat
pFA6a-TKnatCX	*Pcmv/TK-nat^R/Ttef*	ClonNat/FdU
pFA6a-TKnatAX	*Padh/TK-nat^R/Ttef*	ClonNat/FdU
pFA6a-TKanAX	*Padh/TK-kan^R/Ttef*	G418/FdU + HU
pFA6a-HyTKAX	*Padh/hph-TK/Ttef*	Hygromycin B/FdU
pFA6a-FTKnatAX	*lox71-Padh/TK-nat^R/Ttef-lox66*	ClonNat/FdU
pFA6a-FTKanAX	*lox71-Padh/TK-kan^R/Ttef-lox66*	G418/FdU + HU
pFA6a-FHyTKAX	*lox71-Padh/hph-TK/Ttef-lox66*	Hygromycin B/FdU

Using PK-tagging plasmids from this series and the *tpz1 Tag* and *Bot* 100 base primer set, we successfully inserted the gene encoding three, six, nine and 12 tandem-repeats of the PK epitope at the end of the coding gene of $tpz1^+$ to endogenously express Tpz1 with the PK tagged at the C-terminus (Fig. 6c). Tpz1 is a component of telomere shelterin complex and is a homolog of human TPP1. Tpz1 both positively and negatively regulates telomere length by interacting with different factors [28, 29]. One interacting partner binds to the C-terminal ends of Tpz1, which is required for formation of shelterin to suppress telomere elongation [29–31]. Deletion of *tpz1* leads to telomere deprotection and cell death [31]. As expected, the efficiency of detection of Tpz1 was increased with a number of PK repeats, and the observed difference in band sizes was due to a different number of 3xPK repeats in each sample (molecular weight of the 3xPK is 4.2 kDa). The slower migrating band in each lane, that presumably corresponds to SUMOylated Tpz1 [32], also became readily detectable (Fig. 6c). To determine whether the C-terminal fusion of PK-tags impaired the activity of Tpz1,

Table 4 The Cre-expression vectors

Name	Marker cassette	Selection
pNXRVa-HACre	*kanMX6*	G418
pNXRVat-HACre	*TKanAX*	G418
pNXRVb-HACre	*hygMX6*	Hygromycin B
pNXRVbt-HACre	*HyTKAX*	Hygromycin B
pNXRVc-HACre	*natMX6*	ClonNat
pNXRVct-HACre	*TKnatAX*	ClonNat
pNXRVd-HACre	*ura4*$^{+}$	-uracil
pNXRVg-HACre	*arg3*$^{+}$	-arginine
pNXRVh-HACre	*leu1*$^{+}$	-leucin
pNXRVj-HACre	*aur1*^{R}	Aureobasidin A
pNXRVk-HACre	*ade6*$^{+}$	-adenine

we measured telomere length of cells bearing Tpz1 fused to a different number of repeats of PK epitopes (Fig. 6d). Whereas three and six tandem PK tags did not affect telomere length homeostasis, nine tandem PK tags led to slightly elongated telomeres. Thus, we conclude that, although larger tags make proteins and their modifications more visible, they may increase the risk of interfering with function.

Two-step transformation for gene editing at the endogenous locus

A traditional two-step method can utilize the *ura4*$^{+}$ cassette to introduce point mutations [7]. First, the gene of interest is replaced by the *ura4*$^{+}$ cassette in a strain where the endogenous *ura4*$^{+}$ gene region has been deleted (*ura4-D18*). Next, the inserted *ura4*$^{+}$ cassette is replaced with the mutagenized gene allele. The successfully transformed cells, which lost *ura4*$^{+}$, are initially selected using 5-FOA. We previously utilized this method with success to mutagenize the *rap1* gene [12]. This strategy requires the *ura4-D18* mutation in the strain and further epitope tagging may be necessary to determine the stability of the mutant product downstream. To mutagenize essential genes, a diploid strain should be used in which only one allele of the target gene is replaced with the *ura4*$^{+}$ cassette. However, because the wild type allele has better homology, the targeting construct favors the wild type allele over the deleted allele. Thus, although treatment with 5-FOA can reduce occurrence of false positives, the presence of the wild type allele makes difficult to target the *ura4*$^{+}$ replaced locus in diploid cells.

To address this issue, we propose a new method, which utilizes anti-selection markers. In this method, the gene of interest is replaced with a negative selection cassette from the TK-fusion series. The second step uses a plasmid from the pNX3 series to introduce the desired modification to the gene. Briefly, the gene of interest together with its endogenous promoter is subcloned into a pNX3 plasmid and mutagenized as desired. The TK-fusion cassettes contain the *tef* terminator but not the *tef* promoter. The shared homology between the *tef* terminator and the promoter region of the gene in the pNX3 plasmid and the *TKnat/kan* cassettes permits insertion of the modified gene at the deleted allele *via* homology directed repair.

As an example, we utilized this method to introduce point mutations in the *tpz1*$^{+}$ gene. As *tpz1*$^{+}$ is essential gene, we first deleted one *tpz1* allele in a diploid strain using the PCR-based *TKnatCX* recombination method (Fig. 7a). The cloned *tpz1*$^{+}$ in the pNX3a-3HA plasmid (pNX3a-tpz1-HA3) was mutagenized at the position c.322–323 to give the mutation K75A. The resulting construct, pNX3a-tpz1-K75A-HA3, was digested with the restriction enzymes, and the *tpz1* heterozygous diploid strain was transformed with this targeting construct (Fig. 7b). The resulting transformants were selected by G418 resistance. Most of the positive colonies lost their Nat resistance as successful insertion of the mutant gene would result in removal of the *TKnatCX* cassette. In our experience, success rate of the gene replacement was 96 % (25 cells lost the Nat resistance out of 26 cells resistant to G418). Diagnostic PCR using primer sets shown in Fig. 7c confirmed that all of the transfomants that lost Nat resistance had the correct replacement of *tpz1(K75A)-3xHA:kanMX6* at the *TKnatCM* site. Expression of Tpz1(K75A)-3xHA was confirmed by Western blot. Finally, the resulting heterozygous diploid strain (h^{-}/h^{+} *ade6-M210/M216 tpz1(K75A)-3xHA:-kanMX6/+*) was starved to induce sporulation, and the offspring was selected on the G418 to isolate a *tpz1(K75A)* haploid strain. Expression level, function of Tpz1(K75A), and the phenotype of this mutant strain were reported previously [28]. Therefore we showed that this method can be successfully and reliably used to target and modify genes to give stable products.

Conclusions

We have demonstrated that the replacement of promoter and/or terminator sequences in drug resistance cassettes can eliminate the need to replace existing markers in working strains. This enables sequential genetic manipulations to be performed without making changes to the genetic and epigenetic background of strains. In addition, we have shown that the TK-fusion markers can be used to replace genes and subsequently swapped in order to counter-select for undesirable genotypes. While PCR-based gene targeting remains a powerful and cost-effective method, the modified plasmids and optimized techniques presented in this study would enable researchers to perform their desired genetic manipulations more efficiently and reliably.

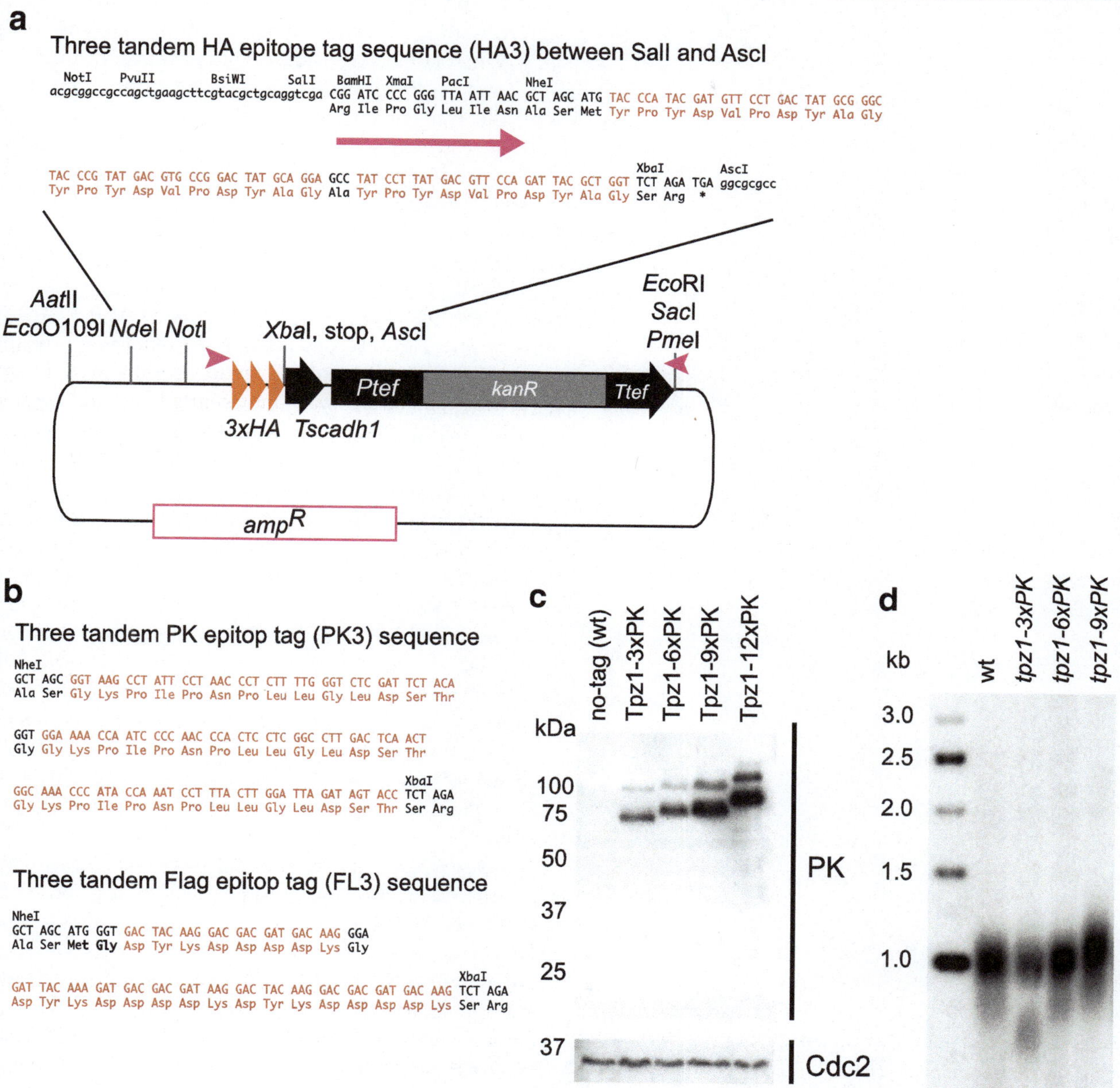

Fig. 6 The COOH-terminus tagging plasmid. **a** Schematic representation of pNX3a-HA3 and sequence of three tandem HA and multi cloning sites. The HA encoding gene is inserted between *PacI AscI* sites. The 100 base *Tag* primer anneals to the left end (*pink arrowhead on the left* in plasmid image and long arrow in HA3 sequence) and *Bot* primer anneals to the right end to amplify indicated cassettes (*pink arrowheads on the right* in plasmid image). A number of unique restriction enzyme sites are indicated. **b** Sequence of three tandem PK (*top*) and FLAG (*bottom*) fragments. pNX3a-PK3 and pNX3a-FL3 plasmids were generated by replacing HA3 sequence between *NheI* and *XbaI* sites in pNX3a-HA3 (**a**) with indicated PK and FLAG sequences, respectively. **c** Detection efficiency of PK-tagged Tpz1. Western blot shows detection of PK epitope fused Tpz1. No obvious non-specific bands were detected. Proteins were extracted from cells and subjected to SDS-PAGE. Anti-V5 antibody (AbD Serotec) was used to detect PK fused Tpz1 protein. Anti-Cdc2 antibody (anti-PSTAIRE) (Santa Cruz) was used as a control for loading. **d** Telomere length homeostasis is slightly impaired with the nine tandem PK tagging of Tpz1. Genomic DNA was harvested from cells cultured over 2 weeks after generation of strains, and digested with *EcoRI* and separated in 1 % agarose gel. Telomere containing fragments were detected with the synthetic telomeric DNA probe

Methods

Yeast media

All media and supplements were purchased from FORMEDIUM™. Fission yeast were grown at 32 °C.

Plasmid constructions

All plasmids generated are listed in Tables 3, 4 and 5 and are available from *Addgene* (www.addgene.com). Oligos used for cloning are listed in Table 6. To create pFA6a-

Table 5 The C-terminal tagging vectors, pNX3 series

Name	Epitope tag	Marker cassette	Selection
pNX3a-PK3	3xPK	kanMX6	G418
pNX3a-PK6	6xPK	kanMX6	G418
pNX3a-PK9	9xPK	kanMX6	G418
pNX3a-PK12	12xPK	kanMX6	G418
pNX3a-FL3	3xFlag	kanMX6	G418
pNX3a-HA3	3xHA	kanMX6	G418
pNX3a-Myc13	13xMyc	kanMX6	G418
pNX3b-PK3	3xPK	hygMX6	Hygromycin B
pNX3b-PK6	6xPK	hygMX6	Hygromycin B
pNX3b-PK9	9xPK	hygMX6	Hygromycin B
pNX3b-PK12	12xPK	hygMX6	Hygromycin B
pNX3b-FL3	3xFlag	hygMX6	Hygromycin B
pNX3b-HA3	3xHA	hygMX6	Hygromycin B
pNX3b-Myc13	13xMyc	hygMX6	Hygromycin B
pNX3c-PK3	3xPK	natMX6	CloneNat
pNX3c-PK6	6xPK	natMX6	CloneNat
pNX3c-PK9	9xPK	natMX6	CloneNat
pNX3c-PK12	12xPK	natMX6	CloneNat
pNX3c-FL3	3xFlag	natMX6	CloneNat
pNX3c-HA3	3xHA	natMX6	CloneNat
pNX3c-Myc13	13xMyc	natMX6	CloneNat
pNX3d-PK3	3xPK	$ura4^+$	-uracil
pNX3d-PK6	6xPK	$ura4^+$	-uracil
pNX3d-PK9	9xPK	$ura4^+$	-uracil
pNX3d-PK12	12xPK	$ura4^+$	-uracil
pNX3d-FL3	3xFlag	$ura4^+$	-uracil
pNX3d-HA3	3xHA	$ura4^+$	-uracil
pNX3d-Myc13	13xMyc	$ura4^+$	-uracil
pNX3a2-PK3	3xPK	neoCV (rev)	G418
pNX3a2-FL3	3xFlag	neoCV (rev)	G418
pNX3a2-HA3	3xHA	neoCV (rev)	G418
pNX3a2-Myc13	13xMyc	neoCV (rev)	G418
pNX3b2-PK3	3xPK	hygML	Hygromycin B
pNX3b2-FL3	3xFlag	hygML	Hygromycin B
pNX3b2-HA3	3xHA	hygML	Hygromycin B
pNX3b2-Myc13	13xMyc	hygML	Hygromycin B
pNX3c2-PK3	3xPK	natCX	CloneNat
pNX3c2-HA3	3xHA	natCX	CloneNat
pNX3c2-Myc13	13xMyc	natCX	CloneNat
pNX3s2-PK3	3xPK	zeoCV (rev)	Zeocin
pNX3s2-FL3	3xFlag	zeoCV (rev)	Zeocin
pNX3s2-HA3	3xHA	zeoCV (rev)	Zeocin
pNX3s2-Myc13	13xMyc	zeoCV (rev)	Zeocin

(rev): neoCV and zeoCV cassettes are inserted in reverse direction

neoCV and pFA6a-*zeoCV* plasmids, the *CMVneo* and *CMVzeo* cassettes were amplified from pCMVneo and pCMVzeo plasmids, respectively, using primers *Pcmv top-PmeI* and *Tsv40 bqt-BglII*. The *kanMX6* cassette of pFA6a-kanMX6 was digested with *Bgl*II and *Pme*I and replaced with the *CMVneo* and *CMVzeo* cassettes. To create pFA6a-natCX, the *CMV* promoter was amplified from pCMVpuro plasmid using primers *Pcmv top-BglII* and *Tsv40 bot-PmeI*, and the *tef* promoter of the *natMX6* cassette in pFA6a-natMX6 was digested with *Bgl*II and *Nco*I (blunt ended) and replaced by *Bgl*II and *Kpn*I digested (blunt ended) *Pcmv* PCR fragment. To create pFA6a-hygMV, the terminator of the *sv40* was subcloned from pSV2-hyg plasmid. pSV2-hyg was digested with *PfM*I and *Hpa*I and self-ligated to remove the gene encoding large tumor antigen, and the *hyg* resistant gene with *Tsv40* was cloned using primers, *Hyg^R top-HindIII* and *Tsv40 bot-PmeI*. The *tef* terminator along with a part of the *hyg* resistant gene of pFA6a-hygMX6 was replaced with the *hyg^R–Tsv40* PCR product using *Sac*II and *Pme*I sites, and resulting plasmid was named pFA6a-hygMV. To replace the terminator of the *hygMX6* cassette with the *S. cerevisiae LEU2* gene, the terminator of *LEU2* was cloned from the *S. cerevisiae* genomic DNA using primers, *Tscleu2 top-ClaI-PstI*, and *Tscleu2 bot-PmeI*, and inserted between *Pst*I and *Cla*I sites of pFA6a. The *tef* promoter and the *hyg* resistant gene were cloned using primers, *Amp rev* and *Hyg bot-NheI-ClaI*, and inserted at *Bgl*II-*Cla*I sites of pFA6a. The resulting plasmid pFA6a-Ptef-hyg was digested with *Cla*I and the *LEU2* terminator was inserted to generate pFA6a-hygML.

To create pFA6a-TKanMX6 and pFA6a-TKnatMX6, the *TK* gene was cloned using primers, *HSVtk top-SpeI-NcoI-HindIII* and *HSVtk bot-XmaI-NcoI*, digested and inserted at *Nco*I site of pFA6a-kanMX6 and pFA6a-natMX6. To create pFA6a-HyTKMX6, the *TK* PCR product was digested with *Nco*I and *Xma*I (blunt ended) and replaced *nat^R* gene of pFA6a-natMX6 using *Nco*I and *Sph*I (blunt ended) sites. The resulting plasmid was named pFA6a-HSVTKMX6. pFA6a-hygML was digested with *Nhe*I and *Pme*I, and *Tscleu2* was replaced with the *Spe*I and *Pme*I digested *HSV-TK-Ttef* fragment from pFA6a-HSVTKMX6. To create pFA6a-TKnatCX, the *CMV* promoter was cloned from pCMVzeo using primers *Pcmv top-BglII* and *zeo R115* and digested with *Bgl*II and *Msc*I. TKnatMX6 was digested with *Bgl*II and *Spe*I (blunt ended), and the *tef* promoter was replaced with the *CMV* promoter.

The *adh1* promoter was amplified from *S. pombe* genomic DNA using primers, *Padh1 top-PmeI* and *bot-HindIII-NheI*, and cloned into pJET vector (ThermoFisher). pJET-Padh1 was digested with *Pme*I and *Cla*I (within pJET), and the *natCM* cassette of pFA6a-natCM was replaced with *Padh1* using *Nru*I and *Cla*I digest to generate pFA6a-Padh1. To create pFA6a-TKanAX and pFA6a-TKnatAX, pFA6a-TKanMX6 and pFA6a-TKnatMX6 were

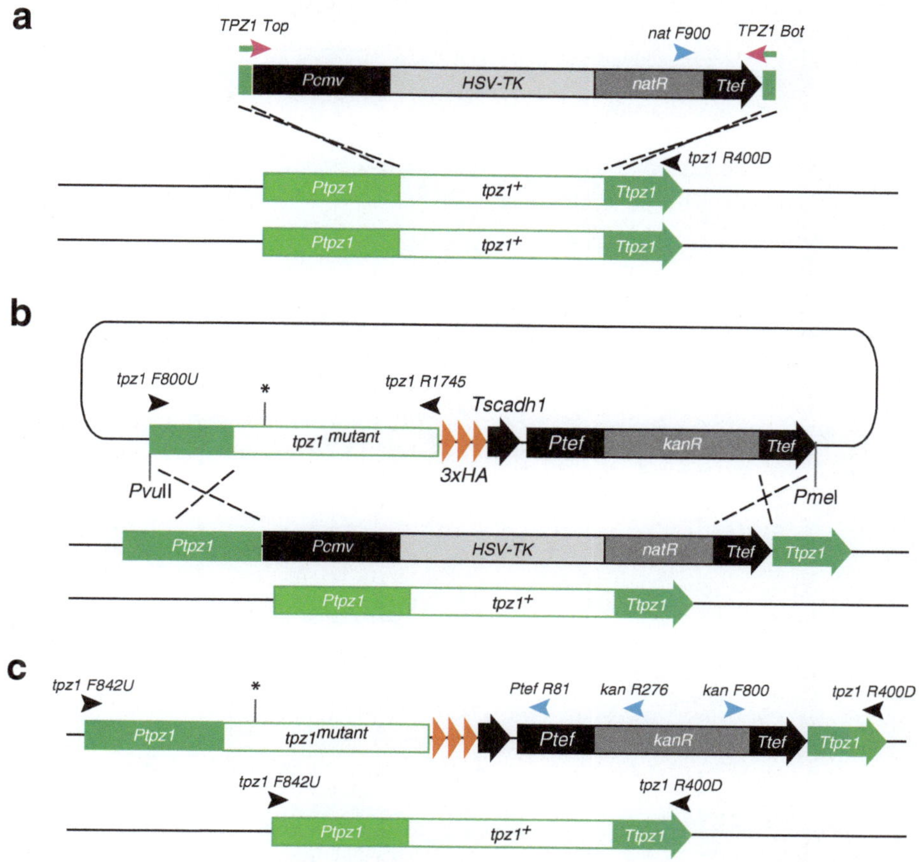

Fig. 7 Two-step mutagenesis. Schematic diagram of the two-step gene replacements and induction of point mutation of the *tpz1* gene. List of diagnostic primers is shown in Table 2. **a** PCR based *tpz1* gene deletion by the *TKnatCX* targeting fragment generated using the *tpz1 Top and Bot* 100 base primer set (Table 1). 80–100 base of homologous sequences in the 100 base primer set targets 5′ and 3′ UTR regions of *tpz1*$^{+}$ of one of endogenous *tpz1*$^{+}$ alleles. Replacement of *tpz1*$^{+}$ by *TKnatCX* was confirmed by amplification of DNA fragment using diagnostic PCR primers, *nat F900* and *tpz1 R400D*. **b** Replacement of the *TKnat* cassette by the *tpz1* mutant gene. The DNA fragment containing the promoter region and the gene of *tpz1*$^{+}$ was amplified using primers *tpz1 F800U* and *R1745* and cloned between *Eco*O109I and *Nhe*I sites of pNX3a-HA3. The *tpz1* gene was mutated to generate *K75A* mutation using the sight-directed mutagenesis method. *Pvu*II and *Pme*I digested *tpz1(K75A)-3xHA:kanMX6* fragment from pTpz1a4-K75A-HA3 can only recombine with *TKnatCX* deleted *tpz1* allele, as right arm homology *Ttef* sequence is only present in the *TKnatCX* cassette. **c** Integration of *tpz1-K75A* mutation and 3xHA tagging. Correct replacement was confirmed by amplification of DNA fragments using diagnostic PCR primer sets, *kan F800* and *tpz1 R400D*, and *tpz1 F842U* and *Ptef R81* (or *kan R276*)

digested with *Spe*I, and the *TKan* and *TKnat* genes were inserted at *Spe*I site of pFA6a-Padh1 in forward direction. To create pFA6a-HyTKAX, the *HyTK-Ttef* fragment was amplified from pFA6a-HyTKMX6 using primers, *Hyg top-HindIII* and *100 bp-bot*, and cloned into pFA6a between *Hin*dIII and *Pme*I sites, and *Padh1* from pFA6a-Padh1 was inserted at *Hin*dIII site in forward direction.

The Cre expression plasmid was created as follows. The *CMV* promoter was isolated from pAUR224 (TAKARA/Clontech) using *Tsp*45I (blunt ended) and *Hin*dIII, and was inserted at *Nde*I (blunt ended) and *Hin*dIII site of pFA6a-13xmyc-hygMX6. The three tandem HA epitope encoding DNA fragments (*3xHA*) were synthesized by annealing of oligos, *3xHA top* and *bot*, followed by DNA polymerase reaction. The 3xHA DNA fragment was generated by digest with *Pac*I and *Asc*I and replaced 13xmyc gene. The Cre recombinase encoding gene was cloned using primer *nls-Cre top-XbaI* that includes *sv40* nuclear localization signal (NLS) at 5′ linker and *Cre bot-XbaI*, and was digested and inserted at XbaI site downstream of 3xHA. The resulting plasmid carried the cassette expressing 3xHA-nls-Cre from the *CMV* promoter and was named pNXVa-HA3nlsCre. The DNA fragment containing the *S. pombe* replication origin, *ars1*, was isolated by *Avr*II digest of the *S. pombe* plasmid pKAN1. pFA6a-kanMX6 was digested with *Eco*RI and *Sap*I, and the *ars1* containing *Avr*II fragment was inserted following blunt ending of the cleaved sites. The resulting plasmid that carried *kanMX6* and *ARS1* was named pNXRa. The Cre expression cassette was isolated by *Eco*O109I and *Bgl*II digest and inserted into pNXRa to generate pNXRVa-HA3nlsCre.

Table 6 Oligos used for cloning and plasmid construction

Name	Sequence
Pcmv top-PmeI	atcatgaccggtttaaacCGTTACATAACTTACGGTAA
Pcmv top-BglII	atcaccggtgctagcagatctcgcgaCGTTACATAACTTACGGTAA
Tsv40 bot-BglII	atccctaggctgcagatctGCAGTGAAAAAAATGCTTTA
Tsv40 bot-PmeI	agctgtacagatatcgagctcgtttaaacGCAGTGAAAAAAATGCTTTA
Hyg top-HindIII	atcctgaagcttacatgtccATGGGTAAAAAGCCTGAACTCACCGCGACG
Hyg bot-NheI-ClaI	atcgattagctagcTACTTCTACACAGCCATCGGTCC
Tscleu2 top-ClaI-PstI	atcgattagctgcagtAATCCTTGCTTAAAAAGATTCTCT
Tscleu2 bot-PmeI	agctgtacagatatcgagctcgtttaAACTCCATCAAATGGTCAGG
HSVtk top-SpeI-NcoI-HindIII	atcaagcttccatggaaactagtGCGTTCGACCAGGCTGCGCGTT
HSVtk bot-XmaI-NcoI	atcgattaacccatggccCGGGCAAACGTGCGCG
Padh1 top-PmeI	tcaggcccttgaattcgtttaaacCCTACAACAACTAAGAAAATGGCTATCATGCGGAAG
Padh1 bot-HindIII-NheI	atcgctagcAAgcTTCTCTTGCTTAAAGAAAAGAAAAGCGAAGGCACCTGTCCACCACCC
100 bp-bot	GAATTCGAGCTCGTTTAAAC
Tscadh1 For-AscI	tgaggCGCGCCACTTCTAAATAAGCG
Tscadh1 R110-BglII	aagatctCCTAGCGGATCTGCCGGTAG
Amp rev	CGACACGGAAATGTTGAATACTC
ori-AflIII	CCTTTTGCTGGCCTTTTGCTCACAT
lox71-Padh1 top-BglII	acggttaattaagatctaccgttcgtatagcatacattatacgaagttatactagtcatatgCCTACAACAACTAAGAAAATGGCTATCA
lox66-Ttef bot-PmeI	atcgaattcgagctcgtttaaactaccgttcgtataatgtatgctatacgaagttataggcctaGGATGGCGGCGTTAGTATCG
nls-Cre top-XbaI	atttaattaactctagaTGTACTCCACCAAAGAAGAAGAGAAAGGTTGCCcctagcATGGCCAATTTACTGACCGTACACCA
Cre bot-XbaI	aggctagcggcgcgccTCAtctagaATCGCCATCTTCCAGCAGGC
ura4 F530U-BglII	ggcatagatctgagctcgatatcAAGCTTAGCTACAAATCCCACTGGCTATATGT
ura4 R190D-PmeI	ttacctgagctcgtttaaacAAGCTTGTGATATTGACGAAACTTTTTGACATCTAAT
leu1 F130U-BglII	aagcttagatctCCATACGATATCCCAATCTGTAG
leu1 R200D-PmeI	atcgagctcgtttaaacGGATGTCGTAAATCAATTCCATGC
arg3 F130U-BglII	aagatcTCGGCTATATGCAATCTCAC
arg3 R130D-PmeI	aagtttaaacAGCCTGTGTCCTGCGCATA
ade6 F300U-BglII	aagatcTAAGGTATAACGACAACAAACGTTGC
ade6 R150D-PmeI	aagtttaaaCTGCTTCACAGCACATTATTCAGGATTCTT
aur1 F333U-BglII	aagtcagcggcgcgccaagatctGTCAGGAAAGCTTTTTTGCCTCT
aur1 R280D-PmeI	agaattcgagctcgtttaaaCAAATACCTATACATCACACTGGAA
3xHA top	atccccgggttaattaacgctagcatgTACCCATACGATGTTCCTGACTATGCGGGCTACCCGTATGACGTGCCGGACTATGCAGGAGCCTATCCTTATGACGTTCCAGA
3xHA bot	aggctagcggcgcgccTCAtctagaACCAGCGTAATCTGGAACGTCATAAGGATAGGCTCCTGCATAGTCCGGCACGTCATACGGGTAGCCCGCATAGTCAGGAACATCG
3xPK-H6 top	ggatccatatgTTAATTAACGCTAGCGGTAAGCCTATTCCTAACCCTCTTTTGGGTCTCGATTCTACAGGTGGAAAACCAATCCCCAACCCACTCCTCGGCCTTGACTCAACTGGCAAAC
3xPK-H6 bot	agatctggcgcgcctaggatggtgatggtgatgatgtctagaGGTACTATCTAATCCAAGTAAAGGATTTGGTATGGGTTTGCCAGTTGAGTCAAGGCCGAGGAGTGGGTTGGGGATTGG
3xFlag top	taacgctagcatgGGTGACTACAAGGACGACGATGACAAGGGAGATTACAAAGATGACGACGATAAGGACTACAAGGACGACGATGACAAGtctagaTGAgg
3xFlag bot	cgcgccTCAtctagaCTTGTCATCGTCGTCCTTGTAGTCCTTATCGTCGTCATCTTTGTAATCTCCCTTGTCATCGTCGTCCTTGTAGTCACCcatgctagcgttaat
tpz1 F800U-PvuII-EcoO	atcaggccctcagCTGTTCAGCACGGTACCAAG
tpz1 R1745-NheI	gttgcagctagcGCTTTTGTTTCGAAACTCCTCTAT

Uppercase at 3′ end of oligo anneals to target and lowercase at 5′ end is a linker containing restriction sites and additional sequence

Auxotrophic markers *ura4*⁺, *leu1*⁺, *arg3*⁺ and *ade6*⁺ were cloned from *S. pombe* genomic DNA using corresponding primer sets listed in Table 6, and cloned into pJET vector. To generate Cre-expression vectors with various selection cassettes, auxotrophic markers *leu1*⁺, *arg3*⁺, *ura4*⁺and *ade6*⁺ and drug selection markers *TkanAX*, *hygMX6*, *HyTkAX*, *natMX6*, *TKnatAX* and *aur1*^{R} from the plasmid pAUR224 were amplified with primers containing with *Bgl*II and *Pme*I sites and cloned into pJET vector (ThermoFisher). The cloned cassettes were digested with *Bgl*II and *Pme*I and replaced with *kanMX6* of pNXRVa-HA3nls-Cre. For the systematic reason, the drug resistant cassettes for G418, hygromycin B, clonNat, and aureobasidin A and auxotrophic markers for uracil, arginine, leucine, and adenine were assigned as 'a', 'b', 'c', 'j', 'd', 'g', 'h' and 'k', respectively. Through the cloning process, we found point mutation (G201T) in the Zeocin resistant gene that substitutes Tryptophan at the position 67 to Cysteine (W67C), but the cassette retained resistance to Zeocin (Fig. 1b).

The series of the C-terminal tagging plasmids, pNX3 was constructed as follows. The terminator of the *S. cerevisiae adh1* was cloned from the genomic DNA using primers, *Tscadh1 For-AscI* and *Tscadh1 R110-BglII*, and inserted into pNXRa after digestion with *Asc*I and *Bgl*II. *ARS1* sequence was removed by digest of pNXRa with *Eco*RI and *Avr*II and ligation following blunt ending of the cleaved sites. Unlike pFA6a C-tag plasmids, this resulting plasmid, named pNX3a, lacks *I-SceI* after *Tadh1* and *ClaI*, *EcoRV*, *SpeI*, *SfiI*, *NotI*, *SacII*, *HpaI* and *SapI* after kanMX6. To generate pNX3a-HA3, the *3xHA* fragment was produced by digest as mentioned above and inserted into pNX3a following *PacI* and *AscI* digest. Likewise, the three tandem *PK* fragment (3xPK) was generated by annealing of oligos, *3xPK-H6 top* and *bot*, followed by DNA polymerase reaction. To generate pNX3a-PK3, the 3xPK fragment was obtained by digest and replaced 3xHA of pNX3a-HA3 after *PacI-XbaI* digest. Three tandem *FLAG* fragment (3xFL) was generated by annealing of oligos, *3xFlag top* and *bot*, followed by 5' end phosphorylation. To generate pNX3a-FL3, the 3xFL fragments were inserted into pNX3a after *PacI* and *AscI* digest. The *kanMX6* cassette was replaced by other selection markers using *BglII* and *PmeI* digest. All generated plasmids for the C-terminus tagging are listed in Table 5.

PCR amplification for gene targeting fragments

For amplification of the deletion fragments using pFA6a series, the primer set (*Top* and *Bot*) was used. *Top* primer is composed of 80–100 base forward sequence upstream of the start codon of the target gene at the five prime end and the vector annealing sequence CGGATCCCCGGGTTAATTAA at the three prime end. *Bot* primer is composed of the complement sequence of downstream 80–100 base sequence after the stop codon of the target gene at the five prime end and vector annealing sequence GAATTCGAGCTCGTTTAAAC at the three prime end.

For C-terminus tagging using pNX3 series, the C-tag specific Tag primer and the deletion Bot primer were used. The annealing sequence of Tag primer is the same as Top primer for deletion but targeting sequence is 80–100 base upstream from the end of the gene excluding the stop codon. These oligos were synthesized by Integrated DNA Technologies Inc. (IDT) using standard desalting purification service. Details of primer design have been described previously [2].

Three hundred microliter of PCR mixtures contained 30 μl of 10xbuffer I, 6 μl of 10 mM dNTPs and 1.5 μl of long template polymerase mixture (Roche) with 0.75 μl of pFA6a or pNX3 template vector (approximately 50–500 ng/μl, obtained using standard miniprep kit), and 1.5 μl each of 100 μM primers. Premix was aliquoted into 50 μl and subjected to PCR [94 °C 30 sec, 50 °C 2 min, 68 °C 5 min for 5 cycles, followed by 94 °C 30 sec, 55 °C 1 min, 68 °C 5 min for 30 cycles, finishing with 68 °C 10 min]. PCR products were combined, ethanol precipitated, and reconstituted in 10 μl of TE (10 mM Tris–HCl 1 mM EDTA pH8.0).

PCR-based gene targeting

Fission yeast transformation was performed as described [2] with minor changes. Logarithmically growing cells ~10^7 cells/ml (O.D_{600} = 0.5–0.8) were harvested, and washed once with LiOAc solution (100 mM Lithium Acetate pH7.5 in TE). Cell pellets were suspended in equal volume of LiOAc solution to make competent cells. 50 μl of the competent cells were mixed with the gene targeting PCR product, 130 μl of 40 % PEG solution (40 % of PEG4000 in LiOAc solution) was added to the cells and mixed gently by tapping. Cells were incubated at 32 °C for 1–4 h. After adding 21.5 μl of DMSO, and the cells were heat-shocked at 42 °C for 5 min, precipitated briefly and washed once with YES media. The transformed cells were directly plated on EMM media for auxotrophic selection or YES/EMM media for Aureobasidin B (TAKARA/Clontech) selection. For other drug selection, transformed cells were recovered in YES media for 4 h in shaking incubator or overnight on the plate, and transferred to YES plate containing G418 (FORMEDIUM), Zeocin (Invivogen), Hygromycine B (FORMEDIUM and Roche) or ClonNat/Nourseothricin (WERNER BioAgents). YES can be replaced with PMG media.

Cre-loxP counter selection

To remove *loxP*-flanked *TK* fusion cassette, cells were transformed with any Cre expression vector (see transformation procedure in previous section). After heat- shock process, cells were directly plated on FdU containing YES plate. FdU-resistant colonies were re-streaked on a fresh FdU plate to isolate a single colonies. Loss of *TK* fusion cassette and the Cre expression vector were confirmed by sensitivity to the drugs. Alternatively, pNXRVj-HACre was used for Cre expression, the transformants carrying Cre vector were selected by resistance to aureobasidin B. The resistant colonies were re-streaked on YES plates, and colonies that lost the drug resistance were isolated.

Southern blot

Telomere Southern blotting was performed as described previously [28]. Genomic DNA was prepared 2 weeks after generation of the strains unless indicated. Equal amounts of *EcoRI* digested DNA fragments were separated on a 1 % agarose gel and subjected to Southern blotting with a telomere probe.

Funding

This work is supported mainly by the European Research Council (281722-HRMCB) and Cancer Research UK (C36439/A12097), and partly by the Cancer Research UK - UCL Centre. These funding bodies had no role in design, collection, and analysis of the study and preparation of the manuscript.

Authors' contributions

KT, HA and VM were responsible for study design, experimental work, data analysis. KT and SRP constructed plasmids and verified by sequencing. HA and VM performed stress and drug resistance assay. VM, HA, LCC, SRP, CAA and KT performed strain construction using generated plasmids and confirmed their functions. LCC performed Tpz1 PK tagging and strain characterization. SRP and VM performed Cre/loxP experiment. LCC, CAA and SRP conducted the two-step point mutation of the *tpz1* gene. KT drafted and all authors were responsible for manuscript preparation. All authors read and approved the final manuscript.

Competing interests

The authors declare that they have no competing interests.

References

1. Hoffman CS, Wood V, Fantes PA. An ancient yeast for young geneticists: a primer on the schizosaccharomyces pombe model system. Genet. 2015;201(2):403–23.
2. Bahler J, Wu JQ, Longtine MS, Shah NG, McKenzie 3rd A, Steever AB, Wach A, Philippsen P, Pringle JR. Heterologous modules for efficient and versatile PCR-based gene targeting in Schizosaccharomyces pombe. Yeast. 1998;14(10):943–51.
3. Siam R, Dolan WP, Forsburg SL. Choosing and using Schizosaccharomyces pombe plasmids. Methods. 2004;33(3):189–98.
4. Hentges P, Van Driessche B, Tafforeau L, Vandenhaute J, Carr AM. Three novel antibiotic marker cassettes for gene disruption and marker switching in Schizosaccharomyces pombe. Yeast. 2005;22(13):1013–9.
5. Sato M, Dhut S, Toda T. New drug-resistant cassettes for gene disruption and epitope tagging in Schizosaccharomyces pombe. Yeast. 2005;22(7):583–91.
6. Beach D, Nurse P. High-frequency transformation of the fission yeast Schizosaccharomyces pombe. Nature. 1981;290(5802):140–2.
7. Grimm C, Kohli J, Murray J, Maundrell K. Genetic engineering of Schizosaccharomyces pombe: a system for gene disruption and replacement using the ura4 gene as a selectable marker. Mol Gen Genet MGG. 1988;215(1):81–6.
8. Burke JD, Gould KL. Molecular cloning and characterization of the Schizosaccharomyces pombe his3 gene for use as a selectable marker. Mol Gen Genet MGG. 1994;242(2):169–76.
9. Waddell S, Jenkins JR. arg3+, a new selection marker system for Schizosaccharomyces pombe: application of ura4+ as a removable integration marker. Nucleic Acids Res. 1995;23(10):1836–7.
10. Jacobs JZ, Ciccaglione KM, Tournier V, Zaratiegui M. Implementation of the CRISPR-Cas9 system in fission yeast. Nat Commun. 2014;5:5344.
11. Paquet D, Kwart D, Chen A, Sproul A, Jacob S, Teo S, Olsen KM, Gregg A, Noggle S, Tessier-Lavigne M. Efficient introduction of specific homozygous and heterozygous mutations using CRISPR/Cas9. Nature. 2016;533(7601):125–9.
12. Amelina H, Subramaniam S, Moiseeva V, Armstrong CA, Pearson SR, Tomita K. Telomere protein Rap1 is a charge resistant scaffolding protein in chromosomal bouquet formation. BMC Biol. 2015;13:37.
13. Saka Y, Esashi F, Matsusaka T, Mochida S, Yanagida M. Damage and replication checkpoint control in fission yeast is ensured by interactions of Crb2, a protein with BRCT motif, with Cut5 and Chk1. Genes Dev. 1997;11(24):3387–400.
14. Catlett MG, Forsburg SL. Schizosaccharomyces pombe Rdh54 (TID1) acts with Rhp54 (RAD54) to repair meiotic double-strand breaks. Mol Biol Cell. 2003;14(11):4707–20.
15. Kiely J, Haase SB, Russell P, Leatherwood J. Functions of fission yeast orp2 in DNA replication and checkpoint control. Genet. 2000;154(2):599–607.
16. Longley DB, Harkin DP, Johnston PG. 5-fluorouracil: mechanisms of action and clinical strategies. Nat Rev Cancer. 2003;3(5):330–8.
17. Lupton SD, Brunton LL, Kalberg VA, Overell RW. Dominant positive and negative selection using a hygromycin phosphotransferase-thymidine kinase fusion gene. Mol Biol Cell. 1991;11(6):3374–8.
18. Schwartz F, Maeda N, Smithies O, Hickey R, Edelmann W, Skoultchi A, Kucherlapati R. A dominant positive and negative selectable gene for use in mammalian cells. Proc Natl Acad Sci U S A. 1991;88(23):10416–20.
19. Hodson JA, Bailis JM, Forsburg SL. Efficient labeling of fission yeast Schizosaccharomyces pombe with thymidine and BUdR. Nucleic Acids Res. 2003;31(21):e134.
20. Martinho RG, Lindsay HD, Flaggs G, DeMaggio AJ, Hoekstra MF, Carr AM, Bentley NJ. Analysis of Rad3 and Chk1 protein kinases defines different checkpoint responses. EMBO J. 1998;17(24):7239–49.
21. Hua H, Kearsey SE. Monitoring DNA replication in fission yeast by incorporation of 5-ethynyl-2'-deoxyuridine. Nucleic Acids Res. 2011;39(9):e60.
22. Akamatsu Y, Dziadkowiec D, Ikeguchi M, Shinagawa H, Iwasaki H. Two different Swi5-containing protein complexes are involved in mating-type switching and recombination repair in fission yeast. Proc Natl Acad Sci U S A. 2003;100(26):15770–5.
23. McNabb DS, Pak SM, Guarente L. Cassette for the generation of sequential gene disruptions in the yeast Schizosaccharomyces pombe. BioTechniques. 1997;22(6):1134–9.
24. Iwaki T, Takegawa K. A set of loxP marker cassettes for Cre-mediated multiple gene disruption in Schizosaccharomyces pombe. Biosci Biotechnol Biochem. 2004;68(3):545–50.
25. Toh-e A. Construction of a marker gene cassette which is repeatedly usable for gene disruption in yeast. Curr Genet. 1995;27(4):293–7.
26. Zhang Z, Lutz B. Cre recombinase-mediated inversion using lox66 and lox71: method to introduce conditional point mutations into the CREB-binding protein. Nucleic Acids Res. 2002;30(17):e90.
27. Hashida-Okado T, Yasumoto R, Endo M, Takesako K, Kato I. Isolation and characterization of the aureobasidin A-resistant gene, aur1R, on Schizosaccharomyces pombe: roles of Aur1p + in cell morphogenesis. Curr Genet. 1998;33(1):38–45.
28. Armstrong CA, Pearson SR, Amelina H, Moiseeva V, Tomita K. Telomerase activation after recruitment in fission yeast. Current Biol. 2014;24(17):2006–11.
29. Jun HI, Liu J, Jeong H, Kim JK, Qiao F. Tpz1 controls a telomerase-nonextendible telomeric state and coordinates switching to an extendible state via Ccq1. Genes Dev. 2013;27(17):1917–31.
30. Harland JL, Chang YT, Moser BA, Nakamura TM. Tpz1-Ccq1 and Tpz1-Poz1 interactions within fission yeast shelterin modulate Ccq1 Thr93 phosphorylation and telomerase recruitment. PLoS Genet. 2014;10(10):e1004708.
31. Miyoshi T, Kanoh J, Saito M, Ishikawa F. Fission yeast Pot1-Tpp1 protects telomeres and regulates telomere length. Science. 2008;320(5881):1341–4.
32. Miyagawa K, Low RS, Santosa V, Tsuji H, Moser BA, Fujisawa S, Harland JL, Raguimova ON, Go A, Ueno M, et al. SUMOylation regulates telomere length by targeting the shelterin subunit Tpz1(Tpp1) to modulate shelterin-Stn1 interaction in fission yeast. Proc Natl Acad Sci U S A. 2014;111(16):5950–5.

14

Characterization of an L-arabinose isomerase from *Bacillus coagulans* NL01 and its application for D-tagatose production

Wending Mei[1], Lu Wang[1], Ying Zang[2], Zhaojuan Zheng[1] and Jia Ouyang[1,3*]

Abstract

Background: L-arabinose isomerase (AI) is a crucial catalyst for the biotransformation of D-galactose to D-tagatose. In previous reports, AIs from thermophilic bacterial strains had been wildly researched, but the browning reaction and by-products formed at high temperatures restricted their applications. By contrast, AIs from mesophilic *Bacillus* strains have some different features including lower optimal temperatures and lower requirements of metallic cofactors. These characters will be beneficial to the development of a more energy-efficient and safer production process. However, the relevant data about the kinetics and reaction properties of *Bacillus* AIs in D-tagatose production are still insufficient. Thus, in order to support further applications of these AIs, a comprehensive characterization of a *Bacillus* AI is needed.

Results: The coding gene (1422 bp) of *Bacillus coagulans* NL01 AI (BCAI) was cloned and overexpressed in the *Escherichia coli* BL21 (DE3) strain. The enzymatic property test showed that the optimal temperature and pH of BCAI were 60 °C and 7.5 respectively. The raw purified BCAI originally showed high activity in absence of outsourcing metallic ions and its thermostability did not change in a low concentration (0.5 mM) of Mn^{2+} at temperatures from 70 °C to 90 °C. Besides these, the catalytic efficiencies (k_{cat}/K_m) for L-arabinose and D-galactose were 8.7 mM^{-1} min^{-1} and 1.0 mM^{-1} min^{-1} respectively. Under optimal conditions, the recombinant *E. coli* cell containing BCAI could convert 150 g L^{-1} and 250 g L^{-1} D-galactose to D-tagatose with attractive conversion rates of 32 % (32 h) and 27 % (48 h).

Conclusions: In this study, a novel AI from *B. coagulans* NL01was cloned, purified and characterized. Compared with other reported AIs, this AI could retain high proportions of activity at a broader range of temperatures and was less dependent on metallic cofactors such as Mn^{2+}. Its substrate specificity was understood deeply by carrying out molecular modelling and docking studies. When the recombinant *E. coli* expressing the AI was used as a biocatalyst, D-tagatose could be produced efficiently in a simple one-pot biotransformation system.

Keywords: L-arabinose isomerase, *Bacillus coagulans*, D-tagatose, Biotransformation

* Correspondence: hgouyj@njfu.edu.cn
[1]College of Chemical Engineering, Nanjing Forestry University, Nanjing 210037, People's Republic of China
[3]Key Laboratory of Forest Genetics & Biotechnology of the Ministry of Education, Nanjing, People's Republic of China
Full list of author information is available at the end of the article

Background

D-tagatose is a natural rare ketohexose that possesses 92 % of the sweetness, but only 38 % of the calories of sucrose [1]. Since it attained GRAS (Generally Recognized As Safe) status under U.S. Food and Drug Administration (FDA) regulations, D-tagatose has become a promising functional sweetener on the food market. Until now, it has been used in the productions of confectionery, soft drinks and health foods for improving the flavors and reducing the calories. It also shows positive attributes in treatment of type II diabetes and hyperglycemia. Currently, one mature method for D-tagatose production is the direct isomerization of D-galactose into D-tagatose with metal hydroxides as the chemical catalysts under basic conditions [2]. This process was applied into commercial food grade D-tagatose production by Arla Food Company between 2002 and 2006 [3]. Nevertheless, it has been gradually dismissed because of the drastic reaction conditions and high cost of the subsequent purification steps. Another method is the enzymatic process that mainly depends on L-arabinose isomerase (AI, EC 5.3.1.4). AIs can isomerize D-galactose to D-tagatose in one step at a milder environment. The method has some significant advantages over the chemical process, such as a lower alkali dosage and less unexpected by-products [3].

In microorganisms, AIs are encoded by the *araA* genes which is an important component of the L-arabinose operon and responsible for the conversion of L-arabinose to L-ribulose [4]. AIs can also recognize D-galactose and catalyze it to D-tagatose due to the similar molecular conformations of D-galactose and L-arabinose (Fig. 1). But many studies showed that the catalytic efficiency (k_{cat}/K_m) of AIs for L-arabinose was generally 10 times (or higher) than that for D-galactose and some AIs even did not show any catalytic efficiency for D-galactose [5]. Since the AI from *Lactobacillus gayonii* was presumed to be a catalyst for D-galactose isomerization in 1993 [6], about 40 types of microbial AIs have been identified and characterized so far. A large proportion of them achieved the maximal enzyme activities under moderate high temperatures (50 °C to 80 °C) and neutral-alkaline conditions (pH 7.0 to 8.0). In addition, most studies on AIs were dependent on metallic ions. Mn^{2+} and Co^{2+} always served as the activating factors of AIs and can significantly improve the enzyme activities and thermal stabilities [7]. Metallic ions in reaction mixtures are difficult to be removed in industry. Thus, their presence will reduce the purity and safety of target products.

Several reports indicated that high reaction temperature is a favorable condition for D-tagatose formation [8, 9]. Numerous AIs from thermophile strains have been reported since the year 2000 including *Thermotoga maritima*, *Thermotoga neapolitan*, *Anoxybacillus flavithermus*, *Bacillus stearothermophilus* US100 [8–11]. Their optimal temperatures were within 80 to 95 °C. High operation temperature causes browning reaction and formation of by-products in catalysis, which is a big obstacle for isolation of the target product [5]. Therefore, many researchers changed their interests to AIs from mesophilic *Bacillus* strains such as *Bacillus licheniformis*, *Bacillus subtilis*, *Bacillus halodurans* and *Bacillus coagulans* [12–15]. These enzymes could adapt to moderate temperatures from 50 °C to 60 °C. They possessed inherently high k_{cat}/K_m values for the natural substrate L-arabinose. Therefore, their applications mainly focused on the synthesis of L-ribulose. Only *B. halodurans* AI was reported to show a k_{cat}/K_m of 0.4 mM^{-1} min^{-1} toward

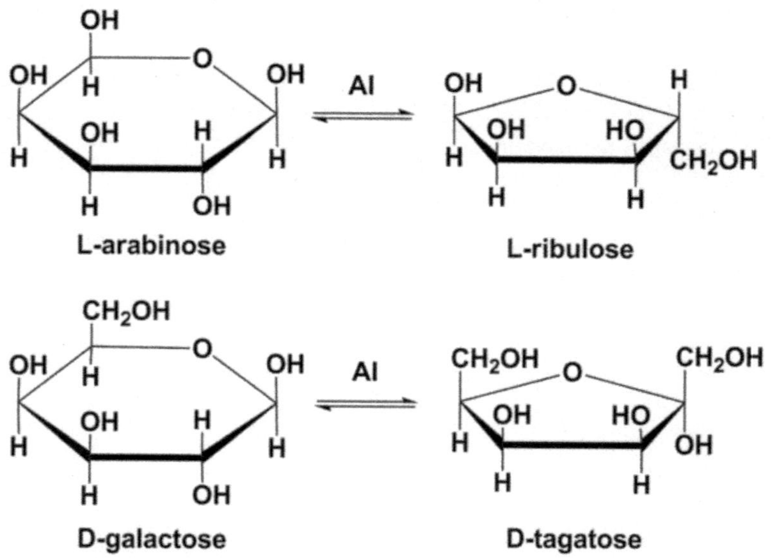

Fig. 1 Isomerization of L-arabinose to L-ribulose and isomerization of D-galactose to D-tagatose catalyzed by AI

D-galactose. Relevant data about the properties of *Bacillus* AIs in D-tagatose production are insufficient and needed to be complemented.

In this study, an *araA* gene from the *B. coagulans* NL01 was cloned and expressed in *Escherichia coli* BL21 (DE3). The biochemical properties of purified *B. coagulans* AI (BCAI) were comprehensively studied. Whole cells of *E. coli* expressing BCAI were used to produce D-tagatose under high D-galactose concentrations in order to test its actual bioconversion capacity.

Results and discussion

Over-expression and purification of BCAI

The *araA* gene contained an open reading frame (ORF) of 1422 base pairs encoding a protein of 473 amino acids. Protein sequence alignments showed that, among the AIs with high activities toward D-galactose, BCAI was mostly similar to *Lactobacillus sakei* AI and *Pediococcus pentosaceus* AI with identities of 68.8 and 67.5 % respectively. By contrast, the identities to other AIs without obvious D-galactose activities from *Bacillus* strains including *B. halodurans* (57.1 %) and *B. substilis* (55.6 %) were lower (Fig. 2). Expression was induced upon addition of IPTG in Luria Bertani (LB) medium. The activity of BCAI crude extract was 2.3 U mg^{-1} for L-arabinose and 0.3 U mg^{-1} for D-galactose. Then, the crude extract was subjected to heat treatment (60 °C) and purified by HisTrap HP 5 mL column. SDS-PAGE showed distinct bands with a molecular mass around 55 kDa (expected size: 53.5 kDa, Additional file 1: Figure S1). The electrophoretically pure AI showed a specific activity of 8.0 U mg^{-1} toward L-arabinose (Table 1) and 1.1 U mg^{-1} toward D-galactose. BCAI was highly similar to the AI from *B. coagulans* 2-6 with an identity of 96.0 %, but no reports indicated that *B. coagulans* 2-6 AI possessed the same D-galactose activity [14].

Effects of temperature and pH on activity of BCAI

The effects of temperature were determined at 40 to 90 °C and pH 7.5 (Fig. 3a). BCAI displayed its maximal activity at 60 °C and retained above 85 % of the activity at 50 to 70 °C. Even at 80 to 90 °C, it still preserved above 60 % of the maximal activity. Compared with the AIs from *B. stearothermophilus* IAM11001, *Lactobacillus fermentum* CGMCC2921 and *T. mathranii*, BCAI was less sensitive to temperature change and could adapt to a broader range of temperatures [16, 17].

To investigate the effect of pH, enzyme assays were carried out at a series of pH from 2.2 to 9.0. The relative activity of BCAI reached the maximal value at pH 7.5 (Fig. 3b) and decreased by less than 10 % at pH 8.0 to 9.0. By contrast, the activity was weaker at acidic conditions. It decreased severely when the pH dropped to 5.0 as most of moderate alkaline AIs did previously, possibly because some side chain groups close to its substrate binding sites were difficult to ionize under this condition [18].

It has been experimentally proved that the optimum pH (pH_{opt}) of AIs is affected by some crucial residues with polar groups, for example the E268 residue of *B. halodurans* AI (BHAI, pH_{opt} =8.0) and the equivalent D269 of *L. fermentum* AI (LFAI, pH_{opt} =6.5) [13, 17]. Modifications of the two residues to lysine (K) resulted that the pH_{opt} of BHAI and LFAI decreased to 7.0 and 5.0, respectively [19, 20]. Protein sequence alignment showed D268 in BCAI was the counterpart of E268 of BHAI and D269 of LFAI (Fig. 2a). It could be presumed that if the D268 residue was changed to lysine, the pH_{opt} of BCAI would probably decrease to a lower value.

Effects of metallic ions on activity and thermostability of BCAI

After BCAI was treated by ethylenediamine tetraacetic acid (EDTA), a dramatic loss (60 %) of activity was observed in an enzyme assay at 60 °C and pH 7.5. The EDTA-treated enzyme was then incubated in Tris-HCl (pH 7.5) solutions with different types of divalent metallic ions (Mg^{2+}, Ca^{2+}, Mn^{2+}, Fe^{2+}, Co^{2+}, Ni^{2+}, Cu^{2+}). Enzyme assays showed that, except Cu^{2+}, all other divalent ions could serve as activators (Fig. 4a). 0.5 mM Mn^{2+} and 0.5 mM Co^{2+} respectively boosted the activity by 270 and 190 % and the combination of them finally resulted in a 370 % increase. Previous researches had demonstrated that Mn^{2+} or Co^{2+} could assist AIs to transfer to correct substrate-binding conformations at elevated temperatures [21]. They significantly boosted the activities of AIs from other strains such as *T. maritima, L. fermentum* CGMCC2921 and *G. thermodenitrificans* [9, 17, 22], but some of them required higher Mn^{2+} or Co^{2+} concentrations (≥2 mM). In this study, when the Mn^{2+} concentration was increased from 0.5 mM to 4 mM, the BCAI activity changed very slightly (Fig. 4b). 0.5 mM Mn^{2+} was completely sufficient for BCAI to maintain a high activity. Further increase of Mn^{2+} by many times could not boost BCAI activity effectively but also would influence the quality of products. For purified BCAI without EDTA treatment, which will be called raw purified BCAI below, the activity was 92.6 and 72.5 %, i.e. very close to the activity of EDTA-treated enzyme measured in 0.5 mM Co^{2+} and Mn^{2+} solutions (Fig. 4a), possibly because it originally bonded metal ions or other types of ligands in the *E. coli* cells.

As EDTA treatment would not be used in practical applications of AIs, the thermal stability test was carried out using the raw purified BCAI. Figure 5a showed it was perfectly stable at 60 °C. After an incubation of 2 h at this temperature, it preserved 90 % of the initial

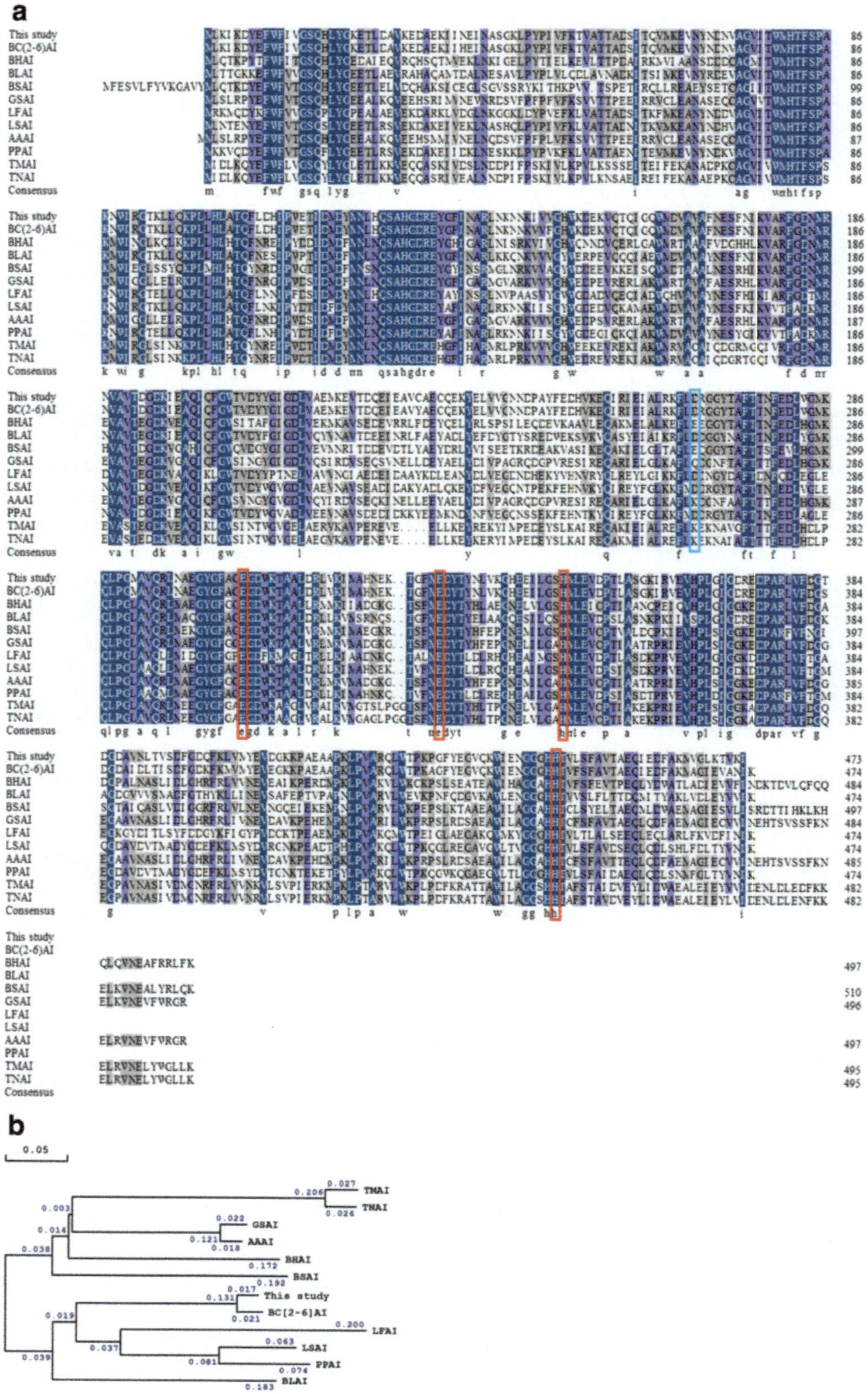

Fig. 2 (See legend on next page.)

(See figure on previous page.)
Fig. 2 Multiple sequence alignment (**a**) and phylogenetic tree (**b**) of different bacterial AIs using DNAMAN multiple alignment tool. The amino acid sequence of BCAI (This study) was aligned with other AI sequences from different bacterial sources including *B. coagulans* 2-6 (BC [2–6]AI, AEH54185.1), *B. halodurans* (BHAI, WP_010898034.1), *B. licheniformis* (BLAI, WP_011198012.1), *B. subtilis* (BSAI, WP_003237722), *G. stearothermophilus* (GSAI, ABY84698.1), *L. fermentum* CGMCC2921 (LFAI, 4LQL_A), *L. sakei* 23 K (LSAI, YP_396468.1), *A. acidocaldarius* (AAAI, AAY68209.1), *P. pentosaceus* PC-5 (PPAI, AEM17146.1), *T. maritima* (TMAI, AKE26215.1), *T. neapolitana* (TNAI, AAK18729.1). Identical and similar amino acid residues were respectively typed *dark blue* and *purple*. The putative catalytic residues which could also bind metal ions were marked with *red boxes*. Residues that can influence the optimal pH of AIs were marked with *blue box*. The phylogenetic tree was generated using Neighbor-Joining method on the basis of the alignment above. The numbers in diagram b showed the lengths of branches that represented evolutionary relationships

activity. When incubated at 70 °C or higher temperatures, the enzymatic activity declined very quickly by more than 80 % during the first 30 min. Some reports showed that adding Mn^{2+} could enhance the thermostability of EDTA-treated AIs [8, 9]. But in this study, a low concentration of Mn^{2+} did not make the same enhancement effect on the raw purified BCAI. As shown in Fig. 5b, after incubation with 0.5 mM Mn^{2+} at temperatures between 60 and 90 °C for 2 h, no positive changes appeared on the residual activities of the enzyme.

From the two perspectives of enzyme activity and thermostability, it seemed that external Mn^{2+} was not essential for the raw purified BCAI. Since a low amount of metallic ions can increase the purity and safety of products, BCAI will show its unique value in food-grade D-tagatose production.

Kinetic parameters of BCAI and molecular docking studies

Kinetic constants of BCAI were measured by following the rules of Lineweaver-Burck plots (Table 2). The catalytic efficiencies (k_{cat}/K_m) of BCAI were 8.7 mM^{-1} min^{-1} and 1.0 mM^{-1} min^{-1} for L-arabinose and D-galactose respectively. According to the ratio (8.7: 1) between the k_{cat}/Km for L-arabinose and D-galactose, L-arabinose was obviously a more favorable substrate than D-galactose for BCAI. On the other hand, the k_{cat}/Km for D-galactose of BCAI was noticeable because many other AIs from *Bacillus* strains had not been reported to show any catalytic efficiency for D-galactose [5, 13, 14].

Molecular modelling and docking techniques were used for gaining a deeper understanding of the substrate specificity. In previous studies, the monomers of *B. licheniformis* AI (BLAI) and *P. pentosaceus* PC-5 AI (PPAI) had been comparatively modeled by using the crystal structure of *E. coli* AI as a template [23, 24]. However, the crystallographic analysis on *L. fermentum* CGMCC2921 AI (LFAI) and *E. coli* AI (ECAI) indicated that they were hexamers [25, 26]. Native-PAGE showed that the total molecular mass of BCAI was around 300 kDa, which meant that BCAI was also a hexamer (expected size: 320 kDa, Additional file 1: Figure S2). It could be assumed that substrate catalyses of these enzymes might be affected by subunit interactions. Therefore, monomer models were not accurate enough when protein subunit interactions were considered. In this study, half of the BCAI structure (a trimer, Additional file 2) was constructed on the base of LFAI (PDB ID: 4LQL, identity: 59.5 %) and ECAI (PDB ID: 4F2D, identity: 44.0 %) crystal structures. The present choice represents a compromise for achieving better conformations of the active sites and reducing computation size as well (Fig. 6a). The superimposition of the obtained BCAI structure with 4LQL and 4F2D ensured the conservation of the putative catalytic amino acids (E306, E331, H348 and H447) (Fig. 6b). The structure energy was sufficiently minimized through a 1000-step Conjugate Gradient Descent until the RMS Gradient reached 0.1.

Although molecular docking studies had been implemented on BLAI before, the result was not very representative because BLAI did not possess any D-galactose activity in experiments. Most AIs such as BCAI were able to catalyze L-arabinose and D-galactose simultaneously. As shown in Fig. 6c, two hydrogen bonds (2.15 Å, 2.34 Å) existed between the C2 hydroxyl group of L-arabinose and the oxygens of E306. One hydrogen bond (1.96 Å) was found between the C1 hydroxyl group and the oxygen of E331. According to the presumed catalytic mechanism of *E. coli* AI and *B. licheniformis* AI [26, 27], L-arabinose was firstly transformed to an enediol intermediate and then L-ribulose was formed. During the first step, protons were transferred through C1 and C2 of the substrate with the assistances of E306 and E331. The docking result here confirmed that E306 and E331 of BCAI played important roles of targeting the C2 and C1 hydroxyl groups of L-arabinose. The hydrogen bond

Table 1 Purification of BCAI expressed in recombinant *E. coli*

Purification step	Protein(mg)	Total activity(U)	Specific activity(U mg^{-1})	Yield(%)	Purification fold
Crude extract	39.4	89.9	2.3	100	1
Heat treatment	15.7	78.4	5.0	87.2	2.2
HisTrap HP 5 mL column	2.3	18.3	8.0	20.3	3.5

Then enzyme activity was measured using L-arabinose as a substrate

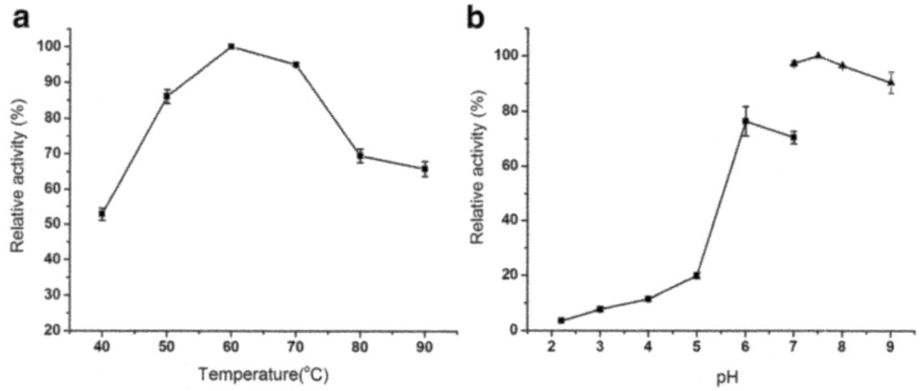

Fig. 3 Effects of temperature (**a**) and pH (**b**) on BCAI activity. Effect of temperature was determined at pH 7.5 and effect of pH was measured at 60 °C. L-arabinose was used as the substrate. The activities at optimal temperature, pH were defined as 100 %

interactions between L-arabinose and the residues were sufficiently strong. By contrast, the interactions for D-galactose were weaker. Among the two hydrogen bonds (2.35 Å, 1.94 Å) found in Fig. 6d, only the bond between the C3 hydroxyl group of D-galactose and E306 could promote the reaction according to the putative mechanism exposed above. E331 residue did not orientate the C1 hydroxyl group of D-galactose correctly, which would cause difficulties on proton transfer and slowed formation of enediol intermediate. This could be an explanation for why the formation of D-tagatose was always slower than that of L-ribulose in most AI catalyses.

Meanwhile, the C-DOCKER energies for the docking poses of L-arabinose and D-galactose were -9.39 kcal/mol and -7.07 kcal/mol respectively. A lower value indicates a more favorable binding, thus further confirming that D-galactose is poorer fit in the active site pocket of BCAI than L-arabinose.

Conversion of D-galactose to D-tagatose by using whole cells of recombinant *E. coli*

Since BCAI could isomerize D-galactose to D-tagatose, the feasibility of D-tagatose production was further studied. It was complicated to use purified enzyme as biocatalyst in industry. Instead, whole cells of recombinant *E. coli* was constructed and selected as a suitable biocatalyst for D-tagatose production.

To increase the efficiency of D-tagatose production, the biocatalytic conditions were optimized. The effect of cell concentration on D-tagatose production was firstly investigated. As shown in Fig. 7a, the highest conversion rate was obtained at 4.8 g DCW L^{-1}. Then, the effect of reaction temperature was evaluated within a temperature range from 40 °C to 80 °C. Figure 7b showed that the conversion rate reached a maximum (40.8 %) at 60 °C. It was consistent with the optimum temperature of purified BCAI. Although *E. coli* cells suffer from viability loss and cellular structure

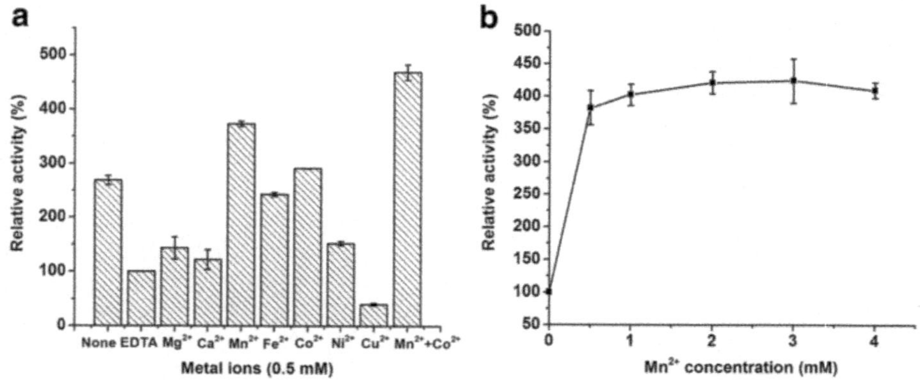

Fig. 4 Effects of various divalent metal ions at 0.5 mM (**a**) and Mn^{2+} concentration (**b**) on BCAI activity. Enzyme assay was carried out at 60 °C and pH 7.5. None: purified BCAI without EDTA treatment; EDTA: purified BCAI treated by EDTA; The EDTA treated enzyme was used when determining of the effects of metal ions and Mn^{2+} concentration. Mn^{2+} + Co^{2+}: the mixture of 0.5 mM Mn^{2+} and 0.5 mM Co^{2+}

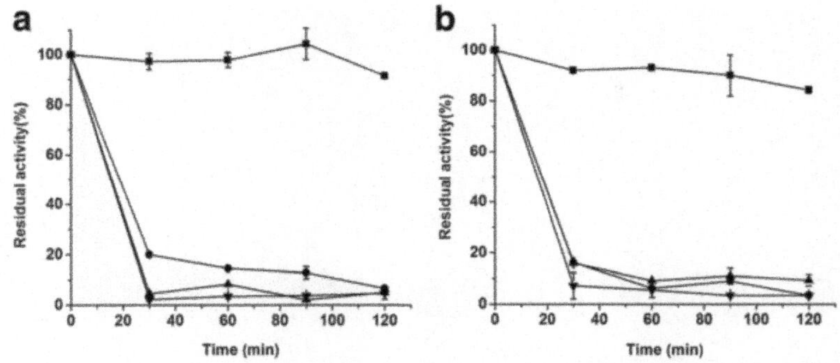

Fig. 5 Thermal profiles of BCAI without addition of metallic ions (**a**) and in the presence of 0.5 mM Mn^{2+} (**b**). The residual activities were determined at 60 °C (■), 70 °C (●), 80 °C (▲) and 90 °C (▼). The activity of a standard reaction at 60 °C and 0 min was defined as 100 %

transition when the temperature is higher than 55 °C [21], these serious damages to *E. coli* cells did not suppresses the transportation of D-galactose and D-tagatose at 60 °C. The effect of substrate concentration was tested in the range from 50 g L^{-1} to 250 g L^{-1} D-galactose (4.8 g L^{-1} cells and 60 °C). With the increase of D-galactose concentration, the amount of D-tagatose kept rising without obvious substrate inhibition, whereas the conversion rate decreased from 31.4 to 17.0 % (Fig. 7c). In order to achieve a higher yield of D-tagatose, 250 g L^{-1} galactose was firstly selected for the following experiment. Meanwhile, 150 g L^{-1} D-galactose was contained, because a relatively higher conversion rate was also expected.

Based on the above experiments, the time course of tagatose production at 150 g L^{-1} and 250 g L^{-1} galactose were performed under the optimal conditions. After 32 h biotransformation, the concentrations of D-tagatose were 48.1 g L^{-1} and 55.5 g L^{-1} respectively (Fig. 7d). The conversion rates were 32.1 and 22.2 % respectively. During the next 16 h, the conversion rate for 250 g L^{-1} D-galactose rose up a little to 27.2 %. The achieved conversion rates were attractive for industrial D-tagatose production. Although immobilized AIs from *G. stearothermophilus*, *T. mathranii* and *T. neapolitana* [28–30] had been used to produce D-tagatose before, the process in this study was easier to operate and due to the enzyme purification and immobilization steps were eliminated. It was a one-pot bioconversion process and introductions of high cell density cultivation and continuous reactors could hopefully improve its feasibility in the future.

Conclusions

In this study, an AI from *B. coagulans* NL01 was comprehensively studied. It showed a broad adaptability to moderate high temperatures. Its original dependency on added metallic ions such as Mn^{2+} was considerably low. Besides, molecular modelling of BCAI trimer combined with docking studies was used to understand its substrate specificity more deeply. Finally, a simple bioconversion system was established using whole cells of recombinant *E. coli* harboring BCAI as the biocatalyst. Attractive D-galactose conversion rates and D-tagatose productions were obtained.

Table 2 Kinetic parameters of AIs

Source	L-arabinose			D-galactose			
	V_{max} (U mg^{-1})	K_m (mM)	k_{cat}/K_m (min^{-1} mM^{-1})	V_{max} (U mg^{-1})	K_m (mM)	k_{cat}/K_m (min^{-1} mM^{-1})	
B. coagulans NL01	43.7	269.8	8.7	6.8	355.1	1.0	This study
B. halodurans	33.1	36	51	1.3	167	0.4	[13]
B. licheniformis	NR	369	34	NR	NR	NR	[15]
B. substilis	NR	120	121	ND	ND	ND	[12]
G. thermodenitrificans	86	142	48	NR	NR	0.5	[22]
P. pentosaceus PC-5	ND	ND	ND	7.8	66	2.9	[24]
A. flavithermus	NR	78.5	0.7	NR	25.2	5.2	[11]
T. neapolitana	119	116	58.1	14.3	250	3.2	[36]

Kinetic parameters of BCAI were determined by using 12.5 to 700 mM substrate (L-arabinose or D-galactose) at standard enzyme assay conditions (60 °C, pH 7.5 and 20 min)

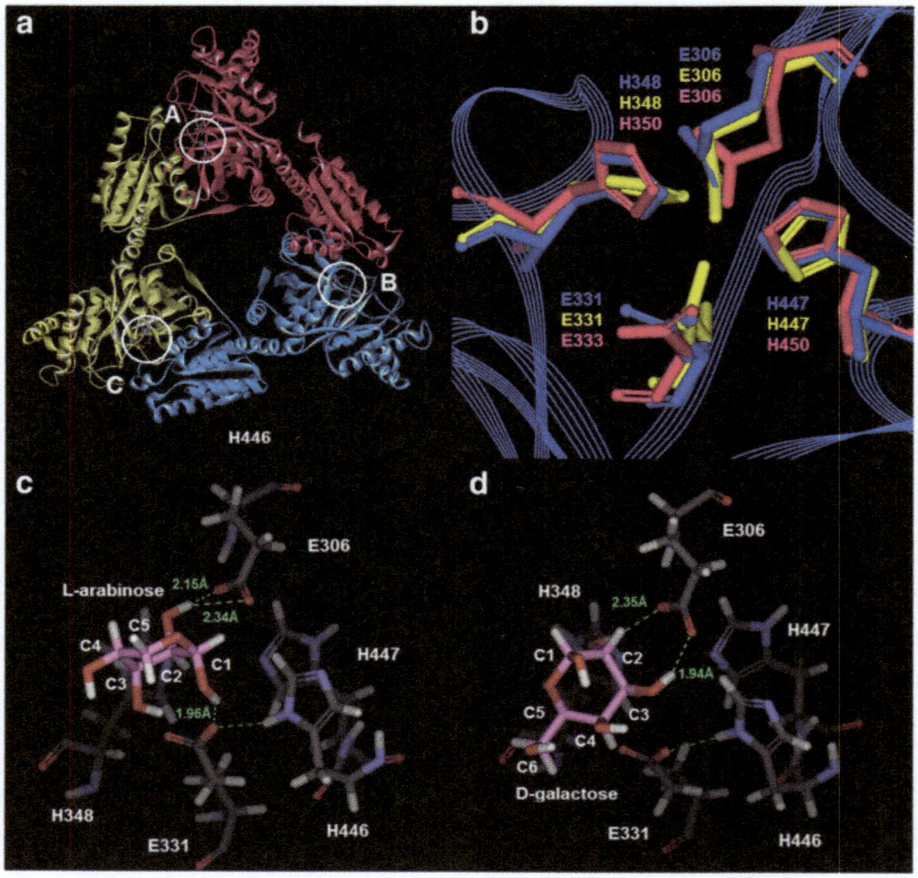

Fig. 6 Model of *B. coagulans* NL01 AI trimer (**a**). The locations of the three substrate active sites (A,B,C) are close to the interface of two identical protein subunits. The model was constructed with the MODELLER module of Discovery Studio 4.0 package. Superimposition of the putative catalytic residues of BCAI (blue), *L. fermentum* CGMCC2921 AI (yellow, PDB ID: 4LQL) and *E. coli* AI (pink, PDB ID: 4F2D) (**b**). The complexes of BCAI active sites with L-arabinose (**c**) and D-galactose (**d**). The intermolecular H bonds are represented by green dotted lines. L-arabinose, D-galactose molecules and amino acid residues are displayed in stick form and colored according to elemental types

Methods

Strains, plasmids and reagents

B. coagulans NL01 was stored in our lab and preserved at -80 °C [31]. *E. coli* BL21 (DE3) was used as the expression host. Plasmid pEASY-Blunt (TransGen Biotech, China) and pETDuet-1 (Novagen) were used for gene cloning and gene expression respectively. *FastPfu* DNA polymerase was purchased from TransGen Biotech (China). HisTrap HP 5 mL column was from GE Health Life Science (USA). D-galactose, D-tagatose and L-arabinose were acquired from TCI (Japan) and L-ribulose was acquired from Carbosynth (United Kingdom).

Construction of recombinant *E. coli*

The *B. coagulans* L01 strain was cultured in LB medium for 12 h. Its genomic DNA was extracted by using TIANamp Bacteria DNA Kit (TIANGEN, Beijing) and then was used as the template DNA of PCR amplication. The primers used for cloning the *araA* gene were 5′-CGCGGATCCGATGTTGAAAATAAAAGA-3′ (forward primer) and 5′-CCGGAATTCTGTTAAAGAAGTGCATC-3′ (reverse primer). The underlined sequences represent restriction sites *Bam*H I and *Eco*R I respectively. The PCR product was ligated with pEASY-Blunt cloning vector. The resulting recombinant plasmid was sequenced by BGI Tech. (Shanghai, China). Then, both the recombinant cloning plasmid and the expression vector pETDuet-1 were digested with *Bam*H I and *Eco*R I, and the *araA* gene was cloned into the multiple cloning sites of pETDuet-1 to generate the recombinant expression plasmid, pETDuet-*araA*. Finally, the plasmid was transformed into *E. coli* BL21 (DE3) for expression.

Overexpression of the *araA* gene and enzyme purification

The *E. coli* BL21 (DE3) harboring pETDuet-*araA* gene was grown in LB medium with shaking at 37 °C until OD_{600} reaches 0.6-0.8. IPTG was added into the medium with a final concentration of 0.5 mM for the

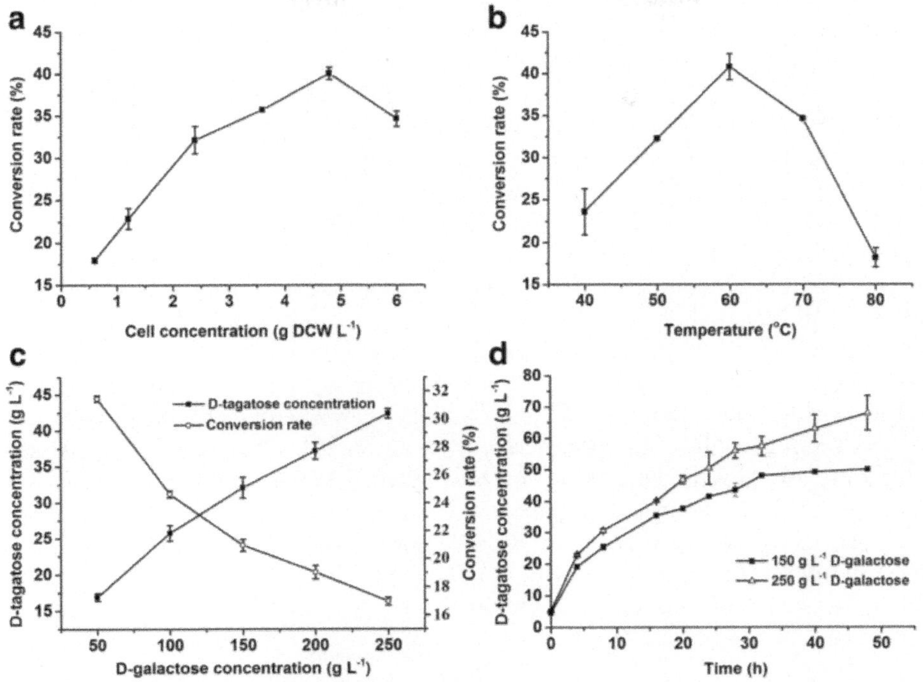

Fig. 7 Effect of cell concentration on the conversion rate of D-tagatose (**a**). Reaction conditions: 18 g L^{-1} D-galactose, 50 mM Tris-HCl, 60 °C, 15 h. Effect of temperature on the conversion rate of D-tagatose (**b**). Reaction conditions: 18 g L^{-1} D-galactose, 50 mM Tris-HCl, 4.8 g L^{-1} *E. coli* cells, 15 h. Effect of D-galactose concentration on the conversion rate of D-tagatose (**c**). Symbols: (■) the D-tagatose concentration; (○) the conversion rate of D-tagatose. Reaction conditions: 4.8 g L^{-1} *E. coli* cells, 50 mM Tris-HCl, 60 °C, 15 h. Time course of D-tagatose conversion (**d**). Symbols: (■) 150 g L^{-1} initial D-galactose; (△) 250 g L^{-1} initial D-galactose. Reaction conditions: 150 g L^{-1} and 250 g L^{-1} D-galactose, 4.8 g L^{-1} *E. coli* cells, 50 mM Tris-HCl, 60 °C, 48 h

recombinant protein expression. After incubation at 25 °C and 200 rpm for 8 h, cells were harvested by centrifugation and resuspended in phosphate buffer solution (PBS, 50 mM, pH 7.4). Cell disruption was carried out by sonication and the obtained solution was centrifugated at 10,000 × g for 15 min at 4 °C to remove insoluble cell debris. The supernatant was used as crude cell extract. For the following purification, the crude extract was heated at 60 °C in order to remove host proteins. Then it was filtered by a 0.22-μm filtering membrane and loaded on a HisTrap HP 5 mL column and equilibrated with binding buffer (20 mM sodium phosphate, 500 mM sodium chloride, and 20 mM imidazole, pH 7.4). The target protein (BCAI) was eluted with 60 % binding buffer and 40 % elution buffer (20 mM sodium phosphate, 500 mM sodium chloride, 500 mM imidazole, pH 7.4). Purity of the protein was assessed by 11.25 % SDS-PAGE. Estimation of molecular mass of multimeric protein was carried out by using 4–16 % Native-PAGE. Gels were visualized by Coomassie Blue R250 staining. The expected band size of BCAI monomer and hexamer was predicted by Compute pI/Mw tool (http://web.expasy.org/compute_pi/). Purified BCAI was stored at 4 °C for biochemical property studies.

Enzyme assay and protein determination

The activity of BCAI was measured by determining the amount of formed keto sugar (L-ribulose or D-tagatose). Under standard conditions, 1 mL reaction mixture contained 100 mM L-arabinose or D-galactose, 50 mM Tris-HCl buffer (pH 7.5), 1 mM $MnCl_2$ and 100 μL of enzyme solution at a suitable concentration. The reaction mixture was incubated at 60 °C for 20 min. Samples were cooled on ice for stopping the reaction. The amount of L-ribulose or D-tagatose was determined by cysteine-carbazole-sulfuric-acid method and the absorbance at 560 nm [32]. One unit of AI acitivity was defined as the amount of enzyme producing 1 μmol keto sugar per min under the conditions above. The protein concentration was determined by the Bradford (Sigma) method using bovine serum albumin for calibration.

Effects of temperature, pH and metallic ion on purified BCAI

The effect of temperature on activity of purified BCAI was determined by testing the activities at temperatures from 40 °C to 90 °C at pH 7.5. The effect of pH was determined by testing the activities at pH 2 to 9 and the optimal temperature obtained above.

Two buffer systems, disodium hydrogen phosphate-citric acid (50 mM, pH 2.2 to 7.0) and Tris-HCl (50 mM, pH 7.0 to 9.0) were used to get desired pH ranges.

To investigate the effect of metal ions on BCAI activity, purified BCAI was dialyzed against 50 mM Tris-HCl buffer (pH 7.5) containing 10 mM EDTA at 25 °C for 3 h. Then, the buffer was changed to 50 mM Tris-HCl (pH 7.5) for another dialysis of 36 h. Metallic ions were added into reaction mixture containing EDTA-treated BCAI at a final concentration of 0.5 mM (or 0.5 to 4 mM when studying effect of Mn^{2+} concentration). Enzyme assay was carried out at the standard condition without adding other metallic ions.

To investigate the thermostability, the raw purified BCAI was divided into two groups. One is incubated without Mn^{2+} and the other is incubated in presence of 0.5 mM Mn^{2+}. The incubations are at 60 to 90 °C and pH 7.5 for 120 min. The enzyme activity was measured under the standard condition without addition of Mn^{2+}.

Determination of kinetic parameters

Kinetic parameters were determined using a 50 mM Tris-HCl buffer (pH 7.5), and 12.5 to 700 mM substrate (L-arabinose or D-galactose) and incubation for 20 min at 60 °C. The reaction was stopped by cooling on ice and the amount of L-ribulose or D-tagatose was determined by cysteine-carbazole-sulfuric-acid method.

Molecular modelling and docking studies

The template structures for comparative modelling were searched from RCSB PDB database (http://www.rcsb.org/). The structure of BCAI trimer was constructed with the MODELLER program and validated by the Profiles-3D tool in Discovery Studio 4.0 package (DS 4.0, BIOVIA, San Diego, CA). Then the protein structure typed with CHARMm [33] force field and the substrate structures typed with MMFF94 force field [34] were subjected to energy minimizations using Conjugate Gradient Descent algorithm. Then, the substrate molecules were docked into the binding pocket of BCAI by using CDOCKER module [35]. The docking poses with the lowest interaction energy were selected for the analysis of orientation and binding interaction.

Optimization of D-tagatose transformation conditions using recombinant *E. coli* cells containing BCAI

10 mL reaction mixtures were prepared in a 50 mL centrifuge tube containing 50 mM Tris-HCl buffer (pH 7.5). The optimal cell concentration was determined by adding 0.6 to 6 g L^{-1} recombinant *E. coli* cells to the reaction mixtures and carrying out an incubation at 60 °C for 15 h. The optimal reaction temperature was determined by incubating the reaction mixtures at 40 °C to 80 °C with the optimal concentration of *E. coli* cells. Then, a gradient of D-galactose concentrations (50 to 250 g L^{-1}) were set for investigating their effect on the conversion rate. The time courses were tested with the selected D-galactose concentrations at the optimal conditions. Samples were taken periodically and analyzed by a high-performance liquid chromatography (HPLC) system (Agilent 1200 series, USA).

Analytical methods

The amount of D-galactose and D-tagatose was measured by a HPLC equipped with a Waters Sugar-pak1 column (6.5 × 300 mm) and a refractive index detector (SHIMADZU). Deionized water was used as mobile phase at a flow rate of 0.4 mL min^{-1} and a column temperature of 80 °C [24].

Abbreviations

AI, L-arabinose isomerase; BCAI, *Bacillus coagulans* NL01 AI; DCW, dry cell weight; EDTA, ethylenediaminetetraacetic acid; HPLC, high-performance liquid chromatography; IPTG, isopropyl β-D-thiogalactopyranoside; PCR, polymerase chain reaction

Acknowledgements

We thank for Dr. Bingfang He from Nanjing Tech University for offering Discovery Studio Package 4.0 software.

Funding

The research project was financially supported by the National Natural Science Foundation of China (51561145015, 31300487), the Natural Science Foundation of Jiangsu Province of China (BK20130970) and the Priority Academic Program Development of Jiangsu Higher Education Institutions (PAPD).

Authors' contributions

WM participated in the design of the experiments, data analysis and writing the manuscript. LW carried out the experiments of whole cell catalysis; ZY participate in the experiments of molecular biology. ZZ participated in data analysis and modifying the manuscript; JO coordinated the whole project and completed the manuscript. All authors read and proved the final manuscript.

Competing interests

The authors declare that they have no competing interests.

Author details

[1]College of Chemical Engineering, Nanjing Forestry University, Nanjing 210037, People's Republic of China. [2]College of Forestry, Nanjing Forestry University, Nanjing 210037, People's Republic of China. [3]Key Laboratory of Forest Genetics & Biotechnology of the Ministry of Education, Nanjing, People's Republic of China.

References

1. Donner T, Wilber JF, Ostrowski D. D-tagatose, a novel hexose: acute effects on carbohydrate tolerance in subjects with and without type 2 diabetes. Diabetes Obes Metab. 1999;1(5):285–91.
2. Kim P. Current studies on biological tagatose production using L-arabinose isomerase: a review and future perspective. Appl Microbiol Biot. 2004;65:243–9.
3. Salonen N, Salonen K, Leisola M, Nyyssola A. D-Tagatose production in the presence of borate by resting *Lactococcus lactis* cells harboring *Bifidobacterium longum* L-arabinose isomerase. Bioproc Biosyst Eng. 2013; 36(4):489–97.
4. Heath E, Horecker B, Smyrniotis P, Takagi Y. Pentose fermentation by *Lactobacillus plantarum* II. L-arabinose isomerase. J Biol Chem. 1958;231(2):1031–7.
5. Xu Z, Li S, Feng X, Liang J, Xu H. L-Arabinose isomerase and its use for biotechnological production of rare sugars. Appl Microbiol Biot. 2014; 98(21):8869–78.
6. Cheetham P, Wootton A. Bioconversion of D-galactose into D-tagatose. Enzyme Microb Tech. 1993;15(2):105–8.
7. Boudebbouze S, Maguin E, Rhimi M. Bacterial L-arabinose isomerases: industrial application for D-tagatose production. Recent Pat DNA Gene Seq. 2011;5(3):194–201.
8. Rhimi M, Bejar S. Cloning, purification and biochemical characterization of metallic-ions independent and thermoactive L-arabinose isomerase from the *Bacillus stearothermophilus* US100 strain. Biochim Biophys Acta. 2006; 1760(2):191–9.
9. Lee DW, Jang HJ, Choe EA, Kim BC, Lee SJ, Kim SB, Hong YH, Pyun YR. Characterization of a thermostable L-arabinose (D-galactose) isomerase from the hyperthermophilic eubacterium *Thermotoga maritima*. Appl Environ Microb. 2004;70(3):1397–404.
10. Kim BC, Lee YH, Lee HS, Lee DW, Choe EA, Pyun YR. Cloning, expression and characterization of L-arabinose isomerase from *Thermotoga neapolitana*: bioconversion of D-galactose to D-tagatose using the enzyme. Fems Microbiol Lett. 2002;212(1):121–6.
11. Li Y, Zhu Y, Liu A, Sun Y. Identification and characterization of a novel l-arabinose isomerase from *Anoxybacillus flavithermus* useful in D-tagatose production. Extremophiles. 2011;15(3):441–50.
12. Kim JH, Prabhu P, Jeya M, Tiwari MK, Moon HJ, Singh RK, Lee JK. Characterization of an L-arabinose isomerase from *Bacillus subtilis*. Appl Microbiol Biotechnol. 2010;85(6):1839–47.
13. Lee DW, Choe EA, Kim SB, Eom SH, Hong YH, Lee SJ, Lee HS, Lee DY, Pyun YR. Distinct metal dependence for catalytic and structural functions in the L-arabinose isomerases from the mesophilic *Bacillus halodurans* and the thermophilic *Geobacillus stearothermophilus*. Arch Biochem Biophys. 2005; 434(2):333–43.
14. Zhou X, Wu JC. Heterologous expression and characterization of *Bacillus coagulans* l-arabinose isomerase. World J Microb Biot. 2012;28(5):2205–12.
15. Prabhu P, Tiwari MK, Jeya M, Gunasekaran P, Kim IW, Lee JK. Cloning and characterization of a novel L-arabinose isomerase from *Bacillus licheniformis*. Appl Microbiol Biot. 2008;81(2):283–90.
16. Cheng L, Mu W, Jiang B. Thermostable L-arabinose isomerase from *Bacillus stearothermophilus* IAM 11001 for D-tagatose production: gene cloning, purification and characterisation. J Sci Food Agric. 2010;90(8):1327–33.
17. Xu Z, Qing Y, Li S, Feng X, Xu H, Ouyang P. A novel L-arabinose isomerase from *Lactobacillus fermentum* CGMCC2921 for D-tagatose production: Gene cloning, purification and characterization. J Mol Catal B-Enzym. 2011;70(1–2):1–7.
18. Lee SJ, Lee SJ, Lee YJ, Kim SB, Kim SK, Lee DW. Homologous alkalophilic and acidophilic L-arabinose isomerases reveal region-specific contributions to the pH dependence of activity and stability. Appl Environ Microbiol. 2012;78(24):8813–6.
19. Lee SJ, Lee DW, Choe EA, Hong YH, Kim SB, Kim BC, Pyun YR. Characterization of a thermoacidophilic L-arabinose isomerase from *Alicyclobacillus acidocaldarius*: role of Lys-269 in pH optimum. Appl Environ Microbiol. 2005;71(12):7888–96.
20. Xu Z, Li S, Feng X, Zhan Y, Xu H. Function of aspartic acid residues in optimum pH control of L-arabinose isomerase from *Lactobacillus fermentum*. Appl Microbiol Biot. 2013;98(9):3987–96.
21. Lee J, Kaletunc G. Evaluation of the Heat Inactivation of *Escherichia coli* and *Lactobacillus plantarum* by differential scanning calorimetry. Appl Environ Microb. 2002;68(11):5379–86.
22. Kim HJ, Oh DK. Purification and characterization of an L-arabinose isomerase from an isolated strain of *Geobacillus thermodenitrificans* producing D-tagatose. J Biotechnol. 2005;120(2):162–73.
23. Prabhu P, Jeya M, Lee JK. In silico studies on the substrate specificity of an L-arabinose isomerase from *Bacillus licheniformis*. Bioorg Med Chem Lett. 2010;20(15):4436–9.
24. Men Y, Zhu Y, Zhang L, Kang Z, Izumori K, Sun Y, Ma Y. Enzymatic conversion of D-galactose to D-tagatose: cloning, overexpression and characterization of L-arabinose isomerase from *Pediococcus pentosaceus* PC-5. Microbiol Res. 2014; 169(2-3):171–8.
25. Xu Z, Li S, Liang J, Feng X, Xu H. Protein purification, crystallization and preliminary X-ray diffraction analysis of L-arabinose isomerase from *Lactobacillus fermentum* CGMCC2921. Acta Crystallogr F. 2015;71(Pt 1):28–33.
26. Manjasetty BA, Chance MR. Crystal structure of *Escherichia coli* L-arabinose isomerase (ECAI), the putative target of biological tagatose production. J Mol Biol. 2006;360(2):297–309.
27. Prabhu P, Jeya M, Lee JK. Probing the molecular determinant for the catalytic efficiency of L-arabinose isomerase from *Bacillus licheniformis*. Appl Environ Microbiol. 2010;76(5):1653–60.
28. Liang M, Chen M, Liu X, Zhai Y, Liu XW, Zhang H, Xiao M, Wang P. Bioconversion of D-galactose to D-tagatose: continuous packed bed reaction with an immobilized thermostable L-arabinose isomerase and efficient purification by selective microbial degradation. Appl Microbiol Biot. 2011;93(4):1469–74.
29. Kim HJ, Ryu SA, Kim P, Oh DK. A feasible enzymatic process for D-tagatose production by an immobilized thermostable L-arabinose isomerase in a packed-bed bioreactor. Biotechnol Prog. 2003;19(2):400–4.
30. Lim BC, Kim HJ, Oh DK. Tagatose production with pH control in a stirred tank reactor containing immobilized L-arabinose rom *Thermotoga neapolitana*. Appl Biochem Biotech. 2008;149(3):245–53.
31. Ouyang J, Cai C, Chen H, Jiang T, Zheng Z. Efficient non-sterilized fermentation of biomass-derived xylose to lactic acid by a thermotolerant *Bacillus coagulans* NL01. Appl Biochem Biotech. 2012;168(8):2387–97.
32. Dische Z, Borenfreund E. A new spectrophotometric method for the detection and determination of keto sugars and trioses. J Blol Chem. 1951; 192(2):583–7.
33. Brooks BR, Brooks 3rd CL, Mackerell Jr AD, Nilsson L, Petrella RJ, Roux B, Won Y, Archontis G, Bartels C, Boresch S, et al. CHARMM: the biomolecular simulation program. J Comput Chem. 2009;30(10):1545–614.
34. Halgren TA. Merck molecular force field. I. Basis, form, scope, parameterization, and performance of MMFF94. J Comput Chem. 1996;17(5-6):490–519.
35. Wu G, Robertson DH, Brooks CL, Vieth M. Detailed analysis of grid-based molecular docking: A case study of CDOCKER-a CHARMm-based MD docking algorithm. J Comput Chem. 2003;24(13):1549–62.
36. Hong YH, Lee DW, Lee SJ, Choe EA, Kim SB, Lee YH, Cheigh CI, Pyun YR. Production of D-tagatose at high temperatures using immobilized *Escherichia coli* cells expressing L-arabinose isomerase from *Thermotoga neapolitana*. Biotechnol Lett. 2007;29(4):569–74.

15

Penicillin production in industrial strain *Penicillium chrysogenum* P2niaD18 is not dependent on the copy number of biosynthesis genes

Sandra Ziemons, Katerina Koutsantas, Kordula Becker, Tim Dahlmann and Ulrich Kück*

Abstract

Background: Multi-copy gene integration into microbial genomes is a conventional tool for obtaining improved gene expression. For *Penicillium chrysogenum*, the fungal producer of the beta-lactam antibiotic penicillin, many production strains carry multiple copies of the penicillin biosynthesis gene cluster. This discovery led to the generally accepted view that high penicillin titers are the result of multiple copies of penicillin genes. Here we investigated strain P2niaD18, a production line that carries only two copies of the penicillin gene cluster.

Results: We performed pulsed-field gel electrophoresis (PFGE), quantitative qRT-PCR, and penicillin bioassays to investigate production, deletion and overexpression strains generated in the *P. chrysogenum* P2niaD18 background, in order to determine the copy number of the penicillin biosynthesis gene cluster, and study the expression of one penicillin biosynthesis gene, and the penicillin titer. Analysis of production and recombinant strain showed that the enhanced penicillin titer did not depend on the copy number of the penicillin gene cluster. Our assumption was strengthened by results with a penicillin null strain lacking *pcbC* encoding isopenicillin N synthase. Reintroduction of one or two copies of the cluster into the *pcbC* deletion strain restored transcriptional high expression of the *pcbC* gene, but recombinant strains showed no significantly different penicillin titer compared to parental strains.

Conclusions: Here we present a molecular genetic analysis of production and recombinant strains in the P2niaD18 background carrying different copy numbers of the penicillin biosynthesis gene cluster. Our analysis shows that the enhanced penicillin titer does not strictly depend on the copy number of the cluster. Based on these overall findings, we hypothesize that instead, complex regulatory mechanisms are prominently implicated in increased penicillin biosynthesis in production strains.

Keywords: *Penicillium chrysogenum*, Beta-lactam antibiotics, Production strain, Gene cluster amplification

Background

Fungi can produce diverse secondary metabolites with antibacterial activity against numerous microorganisms. Among these metabolites, penicillin represents the starting point of the discovery of highly effective antibiotics, a milestone in therapeutic medicine [1, 2]. To date, only the filamentous ascomycete *Penicillium chrysogenum* is used industrially to obtain economically relevant penicillin titers [3].

In the first reaction of penicillin biosynthesis, the three precursor amino acids L-α-aminoadipic acid, L-cysteine, and L-valine are condensed to the tripeptide δ-(L-α-aminoadipyl)-L-cysteinyl-D-valine (ACV). This step is catalyzed by ACV synthetase, a single multifunctional enzyme with non-ribosomal peptide synthetase activity that is coded by the *pcbAB* gene (synonym, *acvA*). The second step is characterized by the oxidative ring closure of the linear ACV tripeptide, leading to the formation of a bicyclic ring comprising the β-lactam and thiazolidine ring. This reaction is catalyzed by the isopenicillin N synthase, encoded by the *pcbC* gene (synonym, *ipnA*). The resulting compound,

* Correspondence: ulrich.kueck@rub.de
Lehrstuhl für Allgemeine und Molekulare Botanik, Ruhr-Universität Bochum, ND7/131, Universitätsstraße 150, 44780 Bochum, Germany

isopenicillin N, is the first bioactive intermediate of the penicillin biosynthesis pathway. In the third reaction of penicillin biosynthesis, the hydrophilic L-α-aminoadipate side chain of isopenicillin N is exchanged for a hydrophobic phenylacetyl or phenoxyacetyl group, resulting in the formation of penicillin G and penicillin V, respectively. This final step is catalyzed by the acyl-coenzyme A: isopenicillin N acyltransferase, and the corresponding gene is *penDE* (synonym, *aatA*) (for an overview see [2, 4, 5]).

All three penicillin biosynthesis genes occur in a single cluster that is structurally conserved in pro- and eukaryotic microbial producers. This shared characteristic supports the hypothesis that fungi have acquired these genes from bacteria through horizontal gene transfer [6, 7].

The progenitor of all industrially used *P. chrysogenum* strains is strain NRRL 1951 (= CBS 307.48), which was isolated in 1943 from a moldy cantaloupe in Peoria, IL. Since then, this strain and its descendants have been subjected to strong mutagenic treatments during strain improvement programs. This pressure has not only resulted in sharply increased antibiotic production but also in increased copy number of the penicillin biosynthesis cluster in some high-production strains, several of which harbor as many as 50 copies of the cluster [3, 8–10]. For example, in the high producer AS-P-78, a 106.5-kb DNA region comprising the pen cluster is amplified in tandem repeats of five or six copies linked by conserved hexanucleotide sequences, whereas wild-type strains contain a single copy of this region [11]. Fierro et al. [11] proposed that the amplification occurred by mutation-induced site-specific recombination at the conserved hexanucleotide sequences. The amplified region is not identical in the different high-producing strains tested, although the mechanism of amplification is probably similar.

Another descendant of strain NRRL 1951, obtained by X-ray and UV mutagenesis, is the former industrial strain *P. chrysogenum* P2 (ATCC 48271) [12], which shows a 85-fold increased penicillin titer compared to its ancestor. This strain was used for conventional mutagenesis to construct P2niaD18, a nitrate reductase-deficient derivative [13]. Recently, whole genome sequencing of this strain revealed that chromosome I carries a tandem repeat duplication of the penicillin biosynthesis cluster comprising genes *pcbAB*, *pcbC*, and *penDE* [14].

Here, we performed pulsed-field gel electrophoresis (PFGE) to further determine the size of the duplicated region. The PFGE revealed that a genomic region of about 110 kb, which harbors the pen cluster, is duplicated in the high-producer strain compared to the wild-type strain. Most strikingly, the loss of one of these copies did not result in decreased penicillin production, thus indicating that the copy number is not responsible for high production in P2niaD18. Although the penicillin biosynthesis pathway is well-studied and the enzymes involved are characterized in detail [2, 4], little is known about the complex regulatory mechanisms behind this process. Our results indicate that instead, regulation of penicillin biosynthesis may have a far more important effect on the amount of penicillin that the fungus produces and therefore represents an important starting point for targeted strain improvement programs.

Methods

Strains and culture conditions

All *P. chrysogenum* strains used in this study are listed in Table 1. Strain P2niaD18 [13], whose genome was recently determined by high-throughput sequencing [14], served as the fungal recipient for all experiments. Like all commonly used industrial strains, P2niaD18 is a derivate of the former industrial strain Q176, which have underdone multiple rounds of conventional mutagenesis [15]. Based on this strain, a marker-free deletion strain of gene *Pcku70* (ΔPcku70) was generated [16]. ΔPcku70 served as a recipient for the construction of knockout mutants based on the FLP/*FRT* recombination system. All *P. chrysogenum* strains were grown in liquid complex medium or minimal medium at 27 °C and 120 rpm or grown on solid medium as already described [17]. To inoculate shake flasks and solid medium, we used spores collected from 7-day-old cultures grown on medium M322. Transformation of individual *P. chrysogenum* strains was performed as described previously [13, 18], and selection of transformants was done by growth on solid medium supplemented with 200 μg ml^{-1} nourseothricin, 40 μg ml^{-1} phleomycin, or 700 μg ml^{-1} pyrithiamine.

Recombinant plasmids were generated using either standard laboratory techniques [19] or the In-Fusion® HD Cloning Kit (Clontech) according to the manufacturer's instructions, with *Escherichia coli* strain XL1-Blue MRF' as host for general plasmid construction and maintenance [20].

Construction of plasmids

All plasmids used in this study are listed in Table 2. For generation of a *pcbC* deletion plasmid, the 5' and 3' regions of *PcvelB* in plasmid pKOvelB (a derivative of pD-Phleo) were replaced by 1-kb 5' and 3' flanking regions of *pcbC*, via *Sph*I and *Mlu*I (5' flank) and *Nhe*I and *Not*I restriction sites (3' flank), respectively, resulting in plasmid pKOpcbCFRTble. In an alternative approach, the 5' flank of *PcvosA* in plasmid pKO-vosA [21] was replaced by *pcbC*-specific flanks, using the *Sfu*I and *Nde*I restriction sites. The 3' *PcvosA* flank was replaced by the 3' *pcbC* flank using the In-Fusion® HD Cloning Kit (Clontech) according to the manufacturer's instructions, resulting in plasmid pKOpcbCFRTnat. For complementation by homologous integration, the *pcbC* open reading frame (ORF)

Table 1 *P. chrysogenum* strains used in this study

Strains	Relevant genotypes	Source
NRRL 1951 (= CBS 307.48)	Wild type, isolated from moldy cantaloupe; parent of most high yielding penicillin producing strains	[50]
P2niaD18	Penicillin producer; *niaD*⁻	[13]
ΔpcbC T7-1	*ΔPcku70::FRT; ΔpcbC::FRT::PtrpC::phleo::FRT; niaD*⁻	This study
ΔPcku70 EK2	*ΔPcku70::PtrpC::nat1*	[31]
ΔPcku70.1 T1	*ΔPcku70::FRT::PtrpC::ble::FRT*	[16]
ΔPcku70.1 T17	*ΔPcku70::FRT::PtrpC::ble::FRT*	[16]
ΔPcku70.2 T1-1	*PtrpC::Pcflp; PptrA::ptrA; ΔPcku70::FRT*	[16]
ΔPcku70.2 T17-1	*PtrpC::Pcflp; PptrA::ptrA; ΔPcku70::FRT*	[16]
ΔPcku70.2 T1-2	*ΔPcku70::FRT; niaD*⁻	[16]
ΔPcku70.2 T17-2	*ΔPcku70::FRT; niaD*⁻	[16]
ΔpcbC::pcbC	*ΔPcku70::FRT; ΔpcbC:: pcbC::FRT:: PtrpC::nat1::FRT; niaD*⁻	This study
ΔpcbC::pcbC-gfp	*ΔPcku70::FRT; ΔpcbC::FRT::PtrpC::phleo::FRT; Pgpd::pcbC::egfp::TtrpC; PtrpC::nat1; niaD*⁻	This study
ΔpcbC::pPCPV1	*ΔPcku70::FRT; ΔpcbC::FRT::PtrpC::phleo::FRT; PpenDE::penDE::TpenDE; PpcbC::pcbC::TpcbC::TpcbC; PpcbAB::pcbAB::TpcbAB; niaD*⁺	This study

was introduced behind the 5' *pcbC* flank of plasmid pKOpcbCFRTnat by using the *Nde*I restriction site. To achieve complementation by ectopic integration of a *pcbC-egfp* fusion construct, the *pcbC* ORF was integrated using the *Nco*I and *Not*I sites of p1783-1nat. Finally, for complementation of the *pcbC* null mutant with the complete penicillin biosynthesis cluster, plasmid pPCPV1 was used. This plasmid has a size of 39.8 kb and also carries the bacterial ampicillin resistance gene and the *A. nidulans niiA* and *niaD* genes, a 24.5 kb fragment with the penicillin biosynthesis gene cluster (*pcbAB, pcbC, penDE*) from *P. chrysogenum* (Kamerewerd and Kück, unpublished). The nucleotide sequence of the penicillin gene cluster is identical in NRRL1951 (Dahlmann, unpublished data), Wisconsin 1255-54 [22], and P2niaD18 [23], and the same result was reported for another high titer strain BW1901 [8].

Construction of knockout mutants and complementation strains

To generate a *pcbC* null mutant, strain ΔPcku70.2 was used as a recipient. After restriction of plasmid pKOpcbCFRTble with *Pvu*II, the knockout cassette harboring the 5' and 3' flanking regions of pcbC, two FRT sites, and a phleomycin resistance cassette was introduced into the genome of ΔPcku70.2 by homologous recombination. The recombination event was verified by PCR and Southern analysis. Sequences of oligonucleotides used in these studies are given in Table 3.

For complementation by homologous recombination, the *Pvu*II fragment of ppcbCFRTnat, harboring the complementation cassette with the 5' and 3' flanking regions of *pcbC*, the *pcbC* ORF, and a nourseothricin resistance cassette, was introduced into the *pcbC* null mutant. For ectopic complementation of a *pcbC-egfp* fusion construct, ΔpcbC was transformed with plasmid pGFP-pcbC

Table 2 Plasmids used in this study

Plasmid	Characteristics	Source
pD-Phleo	*trpC(p)::ble*	[46]
pKOvelB	1 kb 5' flank region and 1 kb 3' flank region of *PcvelB* with *FRT* sites in pD-Phleo	This study
pKOpcbCFRTble	Replacement of *PcvelB*-specific flanks by *pcbC*-specific flanks in pKOvelB	This study
pKOvosA	*5'PcvosA* flank, *5'FRT*, *PtrpC*, *nat1* resistance gene, *3'FRT* sequence, *3'PcvosA* flank	[16]
pPTRII_PcFLP	*trpC* promoter, *Pcflp* gene, *ptrA* resistance gene of *A. oryzae*, *AMA1* sequences of *A. nidulans*	[16]
pKOpcbCFRTnat	Replacement of *PcvosA*-specific flanks by *pcbC*-specific flanks in pKOvosA	This study
ppcbCFRTnat	Introduction of *pcbC* ORF behind 5' *pcbC* flank in pKOpcbCFRTnat	This study
pGFP-pcbC	*Pgpd::pcbC::egfp::Ttrpc, PtrpC::nat1*	This study
pPCPV1	Plasmid harbouring the complete penicillin biosynthesis cluster and the *niaD* gene	Kamerewerd and Kück, unpublished

Table 3 Oligonucleotides used in this study

Name	Sequence (5'→3')	Specificity
3'pcbC_s	TTCGTCGAGAACGGTGAAGC	downstream region of *pcbC* 3'flank used for homologous recombination
Phleo_s	TCCTGCGCCTGATACAGAAC	3'region phleomycin resistance cassette
PtrpC_a	TGTTGACCTCCACTAGCTCC	5'region P*trpC*
5'pcbC_a	TAGTGGCCGAGAAGCCTATC	upstream region of *pcbC* 5' flank used for homologous recombination
pcbC_s	TATACCATGGATGGCTTCCACCCCCAAGG	*pcbC*
pcbC_a	TACTCCATGGTAGTCTGGCCGTTCTTGTTG	*pcbC*
SSU1	ATCCAAGGAAGGCAGCAGGC	SSUr RNA
SSU2	TGGAGCTGGAATTACCGCG	SSUr RNA
PcpcbC-RT_s	CCCTCCCGTTCTTCGTCAATC	*pcbC* (qRT-PCR)
PcpcbC-RT_a	CTGCAGATAGTAGCCGTACGA	*pcbC* (qRT-PCR)

resulting in the construction of ΔpcbC::pcbC-gfp. Strains of both complementation variants were verified by PCR and Southern analysis.

For complementation of ΔpcbC with the complete penicillin biosynthesis cluster, plasmid pPCPV1 was transformed into the deletion mutant. The *niaD* gene on pPCPV1 served as a selection marker, thus complementing the nitrate reductase deficiency of the deletion mutant.

Preparation and analysis of nucleic acids

Isolation of fungal genomic DNA was carried out as described previously [24, 25], and DNA was isolated from hyphal cells grown at 27 °C and at 120 rpm for 72 h in liquid media. Southern blotting was performed with a GeneScreen hybridization transfer membrane (PerkinElmer, USA), hybridized with [α-^{32}P]dCTP-labeled probes using standard methods [19].

Pulsed-field gel electrophoresis (PFGE)

Protoplasts were treated as described previously to isolate intact chromosomes [26]. The CHEF Mapper system (Bio-Rad, Richmond, CA) was used to separate large DNA fragments [27], which were obtained by hydrolysis with the rare-cutter endonucleases *Pac*I, *Pme*I, and *Swa*I, as specified in the results section. Pulse times were done for 18 h at 6 V/cm with initial switching intervals of 10 s and final switching intervals of 20 s.

qRT-PCR for quantification of the *pcbC* transcript

RNA extraction for quantitative reverse transcriptase PCR (qRT-PCR) was performed using the RNeasy® Plus Universal Midi Kit (QIAGEN, Hilden, Germany) according to the instructions provided by the manufacturer. qRT-PCR analysis was performed as described previously [13, 28]. Amplification of the SSUrRNA (small subunit ribosomal RNA) was used as a reference for normalization. Sequences of oligonucleotides used for qRT-PCR are given in Table 3.

Penicillin bioassay

For a penicillin bioassay, 100 ml of liquid complex medium were inoculated with 1×10^7 spores. Cultures were incubated for 96 h at 27 °C and 120 rpm. After being harvested, supernatants were used to perform the penicillin bioassay, with *Staphylococcus aureus* as the indicator organism. The obtained mycelia were used to measure the dry weight. All experiments were performed in triplicate from at least two independent isolates.

Results

High-producer P2niaD18 has a duplicated penicillin biosynthesis cluster

Fungi have relatively small genomes on the order of about 30–40 Mb, and can be separated on a single gel by PFGE. This analysis can be extended by using rare cutting restriction enzymes, thus allowing determination of the chromosomal structure of a region of interest [29]. This approach is in particular reasonable for *P. chrysogenum*, which has only four chromosomes, because the two larger chromosomes with a size above 10 Mb are difficult to separate electrophoretically [30]. Here, we performed PFGE (Fig. 1A-C) to compare the copy number of the penicillin biosynthesis cluster in the high producer P2niaD18 with the wild-type strain NRRL 1951, the progenitor of all industrially used penicillin producers. P2niaD18 is a derivatives of P2, a former producer strain of Nippon Kayaku Co. Laboratories, and among the original strains of the Panlabs series [12]. P2 and Wisconsin 54-1255 are two independently derived derivatives from the very early penicillin production strain Q176, which was later used for further strain improvement programs [10, 12].

PFGE was conducted with chromosomal DNA of both strains cut with the three different rare cutting enzymes *Swa*I, *Pac*I, and *Pme*I. After PFGE, Southern analysis with four different probes (*adh-like* (Pc21g21650), *pcbC* (Pc21g21380), *rco3* (Pc21g21590), and *exo84* (Pc21g21980))

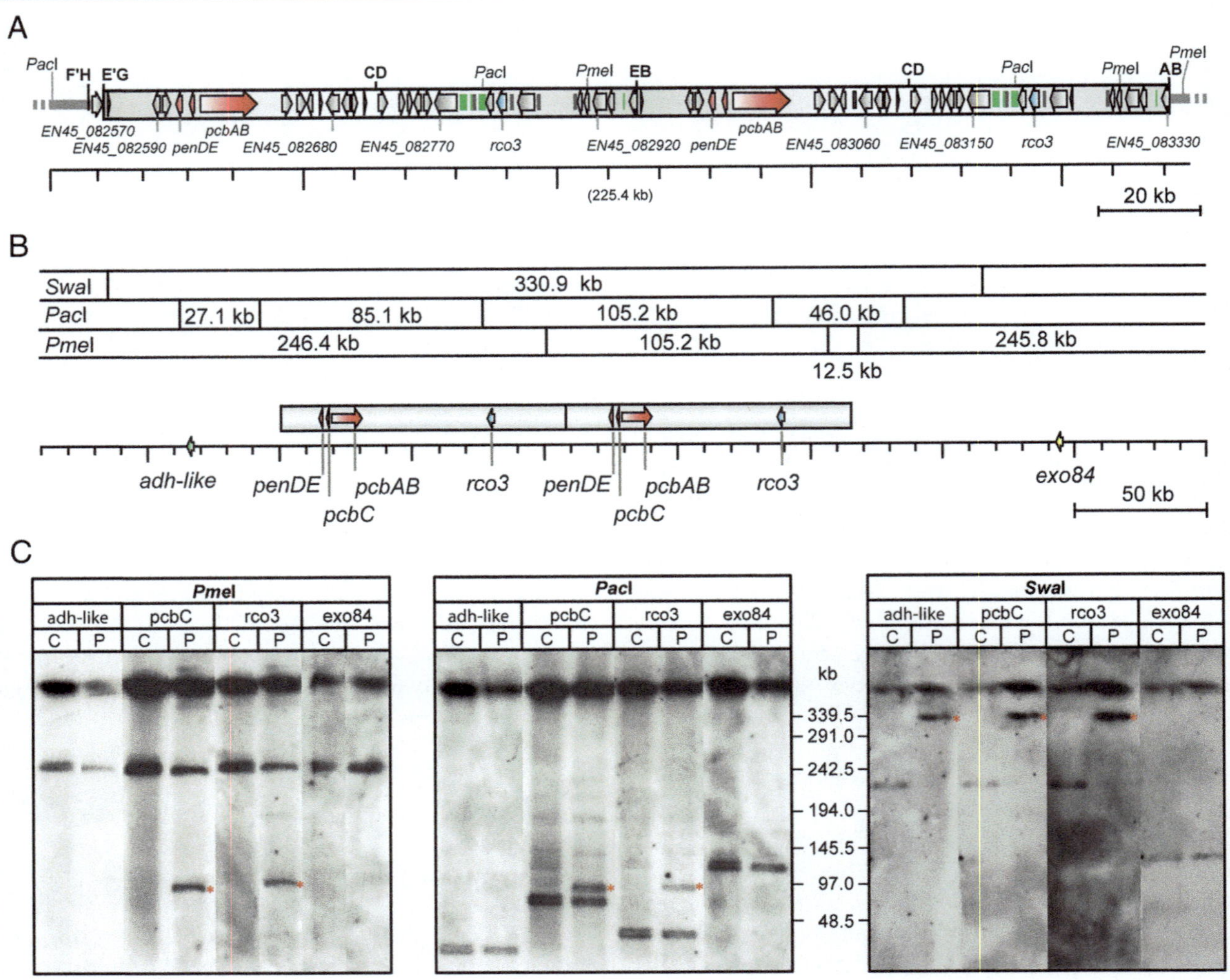

Fig. 1 Determination of the copy-number of different *P. chrysogenum* strains using PFGE. (A) Genetic map of the duplicated 110 kb pen cluster region, based on the recently published genome sequence [23]. The penicillin biosynthesis genes are marked in red and recombination sites are indicated in bold face (**A-H**) and are identical to those reported for other Penicillium strains [9]. (B) Simplified map of the penicillin genes and adjacent regions showing the probes (*adh-like, pcbC, rco3* and *exo84*) for Southern hybridization experiments. In addition restriction sites for *PmeI*, *PacI* and *SwaI* are given. (C) Pulsed field gel electrophoresis (PFGE) of *PmeI*, *PacI* or *SwaI* restricted genomic DNA. For these experiments, the original cantaloupe strain NRRL 1951 [C] and the high-producer P2niaD18 [P] were used. The duplicated area is given as a *grey bar* in AB, and fragments that indicate the proposed duplications are marked with a *red asterisk* in C

revealed several differences (Fig. 1B, C). For example, after restriction with both *Pme*I and *Pac*I, an additional signal with a size of about 97 kb was present in the genomic DNA of P2niaD18 after hybridization with the probes *pcbC* and *rco3*. After restriction with *Swa*I, a shift in the fragment size from about 230 kb to 340 kb occurred in P2niaD18 for all probes except *exo84* (Fig. 1C). These data confirmed that a genomic region of about 110 kb, which harbors the pen cluster, is duplicated in the high-producer strain compared to the wild-type strain. The recent genome sequence of P2niaD18 revealed that the two copies of the pen cluster are identical at the nucleotide level, and the corresponding gene map is given in Fig. 1A.

Generation of a *pcbC* null mutant and different complementation strains

In the next set of experiments, we constructed strains lacking the *pcbC* gene (Fig. 2a-c). ΔPcku70 EK2 served as parental strain for a series of derivatives, which are displayed in Fig. 2c and are described further in the following section. Previously, strain P2niaD18 was used to generate a ΔPcku70 strain for optimized homologous recombination, which was designated ΔPcku70 EK2 [31]. Both strains have a penicillin V titer of 3 g/L, when grown for 96 h in liquid shaking cultures [28]. ΔPcku70 EK2 was transformed with a flipper cassette to generate a marker-free ΔPcku70 strain [16] in two steps. First, the nourseothricin resistance cassette of ΔPcku70 EK2 was

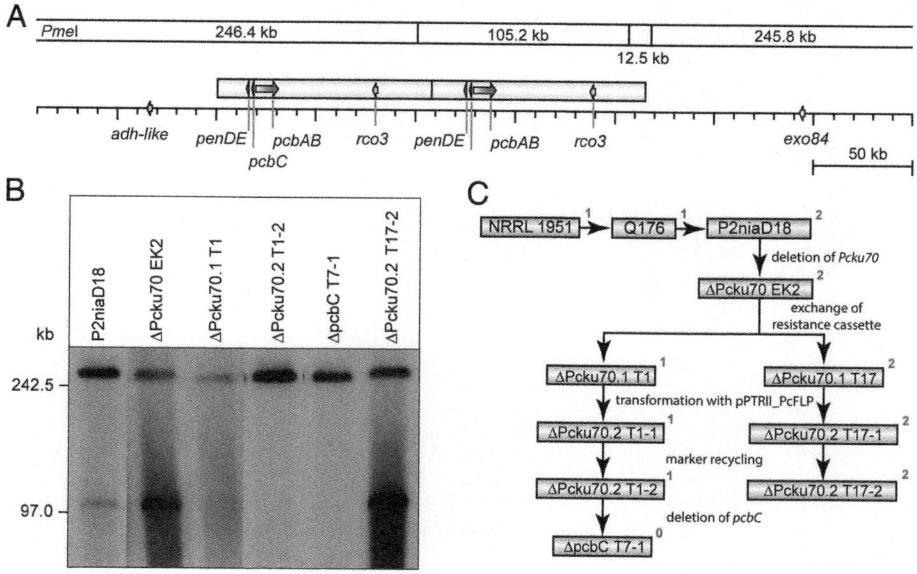

Fig. 2 PFGE of P2niaD18 and its derivatives indicate loss of one cluster copy. **a** *Pme*I restriction map of the duplicated penicillin cluster. **b** PFGE was performed using restriction enzyme *Pme*I and a genomic fragment comprising the *pcbC* gene as a probe. For P2niaD18, this results in two signals with the size of 246.2 kb and 97 kb (**a**, **b**), thus indicating the duplication as proposed in **a**. During the generation of a marker-free ΔPcku70 strain, the extra copy of the penicillin cluster was lost in T1 and its derivatives. In contrast, T17 and its derivatives still carry both copies (**c**). In ΔpcbC, all copies of the *pcbC* gene were deleted, however a signal still occurs since the probe comprises the gene, the adjacent promotor region, and part of the flanking *penDE* gene. **c** Genealogy of penicillin production strains used in this study. The copy number of the penicillin gene cluster is given in *red*

replaced by a phleomycin resistance cassette flanked by *FRT* sites, resulting in two independent isolates, named ΔPcku70.1 T1 and ΔPcku70.1 T17 (Fig. 2c). Subsequently, both isolates were transformed with the free replicating plasmid pPTRII_PcFLP, which carries the *Pcflp* gene coding for the FLP recombinase, and isolates were named ΔPcku70.2 T1-1 and ΔPcku70.2 T17-1. Induction of the flipper-recombinase gene resulted in an excision of the phleomycin resistance cassette and the following loss of the free replicating plasmid results in the marker-free strains ΔPcku70.2 T1-2 and ΔPcku70.2 T17-2 [16]. PFGE revealed, after restriction with *Pme*I and subsequent hybridization with probe *pcbC* (comprising not only the *pcbC* ORF but also the promoter region and part of the *penDE* gene), that the primary transformant ΔPcku70.1 T1, as well ΔPcku70.2 T1-2 had lost one copy of the pen cluster (Fig. 2a, b). In contrast, ΔPcku70.2 T17-2 still carries both copies. In the next step, we used ΔPcku70.2 T1-2 to construct a marker-free *pcbC* null mutant. This strain, designated ΔpcbC T7-1, was obtained by homologous recombination using the flipper cassette (Additional file 1: Figure S1A).

The deletion of ΔpcbC was further confirmed by PCR (Additional file 1: Figure S1B) and Southern hybridization using the 3' flanking region of *pcbC* as a probe (Additional file 1: Figure S1C). As an example, three independently derived transformants (T7, T9, and T10) are shown.

To generate penicillin-producing strains from the above-described null mutant, the *pcbC* gene under control of its native promoter was used for homologous recombination (Additional file 2: Figure S2A). The approach generated ΔpcbC::pcbC, of which five individual transformants (T1-T5) were tested. The site-specific integration of the *pcbC* genes was verified by PCR (Additional file 2: Figure S2B) and Southern hybridization (Additional file 2: Figure S2C). Additionally, we generated a penicillin-producing strain (ΔpcbC::pcbC-gfp) by ectopic integration of a *pcbC-egfp* fusion construct under control of the constitutive *gpd* promoter. Southern hybridization revealed at least three ectopic integrations of this construct (Additional file 2: Figure S2C, right lane).

Our next step was to determine whether the copy number in these strains affected penicillin biosynthesis, so we conducted a halo assay using the indicator bacterium *Staphylococcus aureus*. The size of the halo was measured in relation to the dry weight of the mycelium. Most strikingly, no differences were detectable between the reference strain P2niaD18 and the strain ΔPcku70.2 T17-2, which both harbored two copies of the cluster, and two strains T1 (ΔPcku70.1) and T1-1 (ΔPcku70.2), which lack one copy (Fig. 3). As expected, disruption of the *pcbC* gene in ΔPcku70.2 T1-2 yielded a penicillin non-producer (ΔpcbC T7-1), verifying that this strain is indeed single copy with respect to the penicillin biosynthesis gene cluster. Complementation of the null

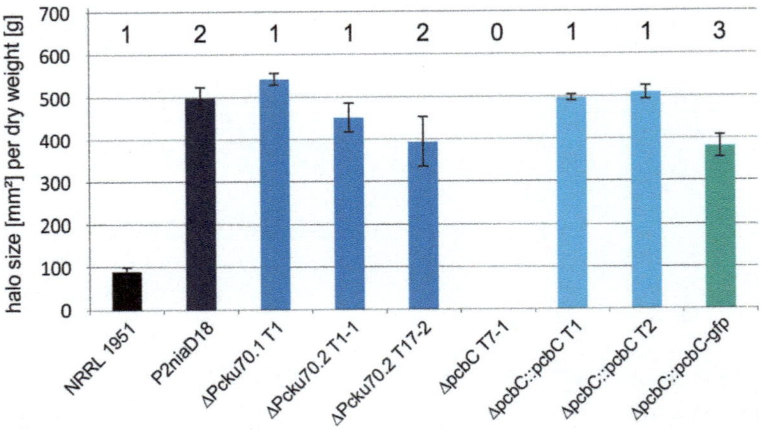

Fig. 3 Quantification of penicillin production of strains with different copy numbers of the *pcbC* gene. The strains were grown for 96 h in shaking cultures. The diameter of each halo was measured to calculate the area and is given in relation to the dry weight of the respective culture. Standard deviations were determined from representative isolates, which were measured in triplicate. The copy number of each strain is displayed above the corresponding column. For all strains with a copy number of 1 or 2, the number indicates both the copy number of the *pcbC* gene and the complete cluster. For ΔpcbC T7-1, one copy of both *pcbAB* and *penDE* is still present, whereas *pcbC* is missing. To generate ΔpcbC::pcbC-gfp, the ΔpcbC recipient was complemented with three ectopic integrations of a *pcbC-egfp* fusion construct. For comparison, wild type strain NRRL 1951, which carries one copy of the penicillin cluster, is given

mutant with one copy of the *pcbC* gene under its native promoter resulted in penicillin biosynthesis comparable to the reference strains. Additionally, an ectopic integration of at least three copies of a *pcbC-egfp* fusion construct under control of the constitutive *gpd* promoter (ΔpcbC::pcbC-gfp) complemented the penicillin defect of the null mutant to an extent similar to that of the native complementation constructs (Fig. 3). For comparison, the titer of the wild type strain NRRL 1951 is given, which has a titer of about 20% in our plate assays compared with the other penicillin producing strains.

We then complemented the ΔpcbC mutant with plasmid pPCPV1, which carries the complete penicillin biosynthesis cluster of 26 kb, together with the *niaD* gene, thus allowing complementation of the nitrate reductase deficiency of P2niaD18 and its descendants. After restriction of the chromosomal DNA with *Pme*I, followed by PFGE, the subsequent Southern hybridization with a probe comprising *pcbC* revealed that this construct integrated in all cases in the genomic area that harbors the native cluster. However, some transformants (T1, T4) showed a single integration of the plasmid whereas T2 and T3 carried two copies of the plasmid (Fig. 4), thus resulting in complementation strains with either one or two intact copies of the cluster. A penicillin bioassay again revealed no significant differences between the complementation strains, independent from

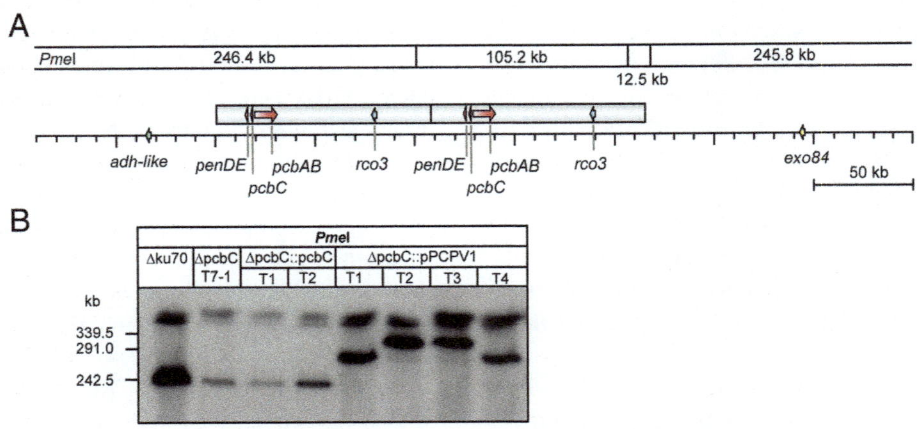

Fig. 4 Generation of *pcbC* complementation strains by complementation with the complete penicillin biosynthesis cluster. The *pcbC* null mutant was complemented with plasmid pPCPV1 harboring the complete biosynthesis cluster (**a**). PFGE proved that integration of the plasmid occurred in the genomic region of the penicillin cluster and further showed single (T1, T4) or double (T2, T3) integration of the plasmid (**b**). The gene content of individual strains is indicated above each lane, for comparison strain **Δ**ku70 (**Δ**Pcku70 EK2) is given

the copy number (Fig. 5). All four complementation strains even slightly exceeded the penicillin production observed in the reference strain ΔPcku70.2 T1-2.

Quantification of the *pcbC* transcript in deletion and complementation strains, carrying different copies of the penicillin cluster

Finally, we tested whether the transcriptional expression level reflects the penicillin titer observed in the recipient and recombinant strains. All strains were grown for 3 days in shaking flasks with rich Complete Culture Medium (CCM). This time point represents the early phase of β-lactam production and is characterized by an expressional switch from genes related to vegetative growth to those involved in secondary metabolite formation [17]. As shown in Fig. 6, we quantified the *pcbC* transcript from three biological replicates in relation to the SSUrRNA. Both pcbC null mutants (ΔpcbC T7.1 + T10) gave negative results, and the single copy strain ΔPcku70.1 T1 showed an almost zero expression level. The results of the quantitative PCR analysis correspond roughly to the copy number of the *pcbC* gene. The four independently derived transformants ΔpcbC::pPCPV1 T1-T4 have similar transcript levels although distinct by the copy number of their penicillin gene cluster. This analysis of relative log2-fold expression ratios of the *pcbC* transcript support our previous results that copy number and thus transcript level have only a minor effect on penicillin titers.

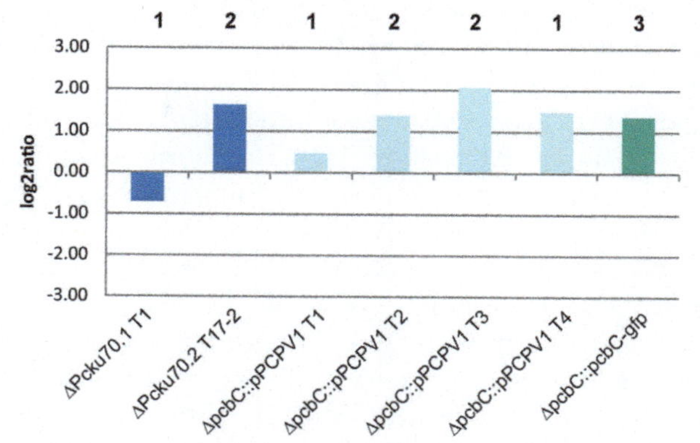

Fig. 6 qRT-PCR analysis to quantify the *pcbC* transcript. RNAs, isolated from strains as indicated, were used for quantification. The procedure is described in the Material and Methods section, with oligonucleotides as listed in Table 3. SSUrRNA served as reference for normalization. The copy number of each strain is displayed above the corresponding column

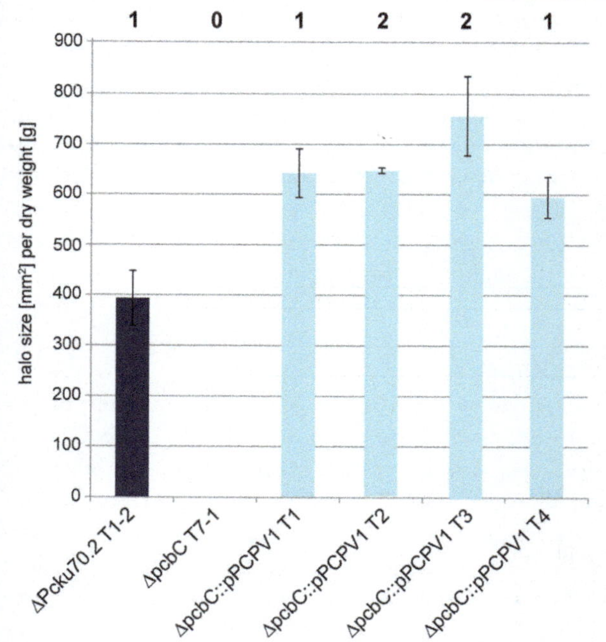

Fig. 5 Quantification of penicillin production of complementation strains with different copy numbers of the penicillin biosynthesis cluster. Standard deviations were determined from representative isolates, which were measured in triplicate. The copy number of each strain is displayed above the corresponding column

Discussion

The three penicillin biosynthesis genes *pcbAB*, *pcbC*, and *penDE* are clustered in a single 18-kb region in wild-type strains of the filamentous fungi *P. chrysogenum* and *Aspergillus nidulans*. Previous chromosome separation and DNA hybridization analysis showed that production strains from *P. chrysogenum* have up to 14 copies of a 56.8-kb region carrying further protein coding genes that are not characterized in detail [15, 32]. This high copy number was suggested to be relevant for the high penicillin titer observed [11]. Our analysis, however, indicates that regulatory genes unrelated to the penicillin biosynthesis gene cluster are responsible for increased penicillin production, at least in the P2 line of production strains.

Theilgaard et al. [33] showed that penicillin production in the low-producing, single gene copy strain Wisconsin 54-1255 could be increased by integration of additional copies of the three penicillin biosynthesis genes. However, for increased titer, all three genes had to be integrated; other combinations with only one or two of the genes did not result in higher penicillin production. The authors proposed that amplification of all three biosynthesis genes is responsible for the high penicillin titer of production strains. More recently, it was found that the amount of penicillin V increases with penicillin biosynthetic gene cluster number but with saturation at high copy numbers. This study was done in industrial strains with a Wisconsin 54-1255 background [34]. Remarkably, in that study, the protein level of the acyltransferase, the gene product of *penDE*, was saturated already at low cluster copy numbers, suggesting that the acyltransferase reaction presents a bottleneck in the biosynthesis process.

Several high-production strains have been described that comprise multiple copies of the penicillin cluster [3]. For example, strain AS-P-78 carries between five and nine copies [11, 15]. Another strain, BW 1890, harbors between 8 and 16 copies of the cluster [32]. The Panlabs strain P2, which is also a derivative of Q176, was thought to carry between 5 and 14 copies [11, 15, 22]. However, genome analysis and results from our work revealed that at least strain P2niaD18, a nitrate reductase–deficient derivative that emerged from P2 by conventional mutagenesis, harbors only two copies of the penicillin biosynthesis cluster [14]. Still, this strain produces high amounts of penicillin, indicating that copy number is not the sole factor in increased production rates. Our data strongly support this conclusion, revealing that even the loss of one of the two copies did not result in significantly decreased penicillin biosynthesis. In addition, complementation strains with one or two copies of the cluster yielded no significant differences in titer.

The amplified regions might even be responsible for the genetic instability of strains with multiple copies of the biosynthetic gene clusters. Harris et al. [35] described, for example, that after protoplasting, gene clusters are easily lost in industrial strains derived from Wisconsin54-1255. Similar observations were made for the yeast *Yarrowia lipolytica*. To test gene amplification in the rDNA of *Y. lipolytica*, several plasmids were transformed into the yeast cells [36]. Among other elements, these plasmids harbored the reporter gene *XPR2* encoding alkaline extracellular protease (AEP). Plasmid copy number was stable for strains containing fewer than 10 copies per cell. However, for higher copy numbers, multiple integrations were stable only when AEP synthesis was not induced, while in inducing medium, the stability of the multiple integrations was dramatically affected. After AEP induction, a reduced growth rate was observed, suggesting that the increased secretory pathway cargo load influenced cell growth.

These data together with our observation that one of two penicillin biosynthesis clusters was randomly lost support the hypothesis that the cluster copies are easily lost and that high-copy strains are unstable. P2niaD18 is a strain with only two copies but nevertheless capable of high penicillin production, indicating that other factors have an important influence on penicillin biosynthesis.

To date, several regulators of penicillin biosynthesis are already known in filamentous fungi ([2], reviewed in [5]). So far, a pathway specific regulator of the penicillin biosynthesis cluster has not yet been described. In addition to positively acting global regulators like the pH-dependent transcriptional activator PACC and the CCAAT binding complex AnCF [37–40], proteins of the velvet family have become of special interest as repressors in recent years [41]. These regulatory proteins play a key role in coordinating secondary metabolism and differentiation processes such as sexual and asexual sporulation in various filamentous fungi (reviewed in [42]). In *Aspergillus nidulans*, VeA forms a heterotrimeric complex with VelB, another protein of the velvet family, and the global regulator LaeA under dark conditions to control sexual development and secondary metabolism [43, 44]. In addition, VelB interacts with VosA, a third velvet-like protein, to form a subcomplex that is essential for asexual and sexual spore formation as well as trehalose biogenesis [44, 45]. Meanwhile, homologs of velvet components have been identified in numerous other filamentous fungi (for an overview see [42]. In *P. chrysogenum*, the velvet proteins control hyphal morphogenesis, conidiophore development, and penicillin biosynthesis. Most importantly, distinct velvet proteins either activate or repress biosynthesis of penicillin [21, 46]. Interestingly, in another industrial fungus, the ascomycete *Acremonium chrysogenum*, a velvet homologue has a regulatory role on beta-lactam antibiotic production [47]. In this fungus, at least seven genes for the biosynthesis of the beta-lactam antibiotic cephalosporin C are located on two different clusters on different chromosomes [23]. Thus, simple amplification of a single cluster will not increase cephalosporin C biosynthesis. Molecular analysis by different investigators has already shown that global regulators are responsible for high titer of cephalosporin C biosynthesis (for review see [48]). Recently, we found a rather unexpected regulation of gene expression. The mating type locus encoded MAT1-1-1 transcription factor is known for its role in sexual identity. However, recent investigations showed a transcriptional control of wide range of genes with biotechnological relevance including those regulating penicillin production. Compared with control strains, mutants lacking the mating type locus showed a significant reduction in penicillin biosynthesis throughout a time course [28, 49].

Conclusions

This report reveals that a high copy number of the three structural genes and an increased *pcbC* transcript level are not strict prerequisites for increased penicillin production in the production strain *Penicillium chrysogenum* P2niaD18. Most strikingly, a loss of one of the two identical copies of the cluster did not significantly influence the amount of penicillin produced. These data imply that copy number is not the limiting factor for increased penicillin biosynthesis in the strains investigated and we anticipate that instead, wide domain regulatory factors *in trans* are involved in this process and are thus important targets for future strain improvement.

Abbreviations

ACV: δ-(L-α-aminoadipyl)-L-cysteinyl-D-valine; AEP: Alkaline extracellular protease; ATCC: American Type Culture Collection; CBS: Centraalbureau voor Schimmelcultures; CCM: Complete culture medium; FRT: FLP recognition target; NRRL: ARS culture collection; ORF: Open reading frame; PCR: Polymerase chain reaction; PFGE: Pulsed-field gel electrophoresis; qRT-PCR: quantitative reverse transcription PCR; SSUrRNA: Small subunit ribosomal RNA; UV: Ultraviolet

Acknowledgments

We are grateful to Ingeborg Godehardt, Stefanie Mertens, and Susanne Schlewinski for excellent technical assistance and Gabriele Frenßen-Schenkel for the artwork. We further thank Drs. Hubert Kürnsteiner, Ivo Zadra, Ernst Friedlin, and Thomas Specht (Sandoz GmbH, Kundl, Austria) for useful discussions.

Funding

This work was funded by the Christian Doppler Society (Vienna, Austria) and by Sandoz GmbH (Kundl, Austria).

Authors' contributions

SZ and UK participated in the design of the experiments, data analysis and writing the manuscript, KK constructed all plasmids, transformed bacterial and fungal strains, and carried out penicillin bioassays, KB performed the qRT-PCR experiments, and participated in data analysis and modifying the manuscript; TD carried out penicillin bioassays, participated in data analysis, and revising the manuscript; UK coordinated the whole project and completed the manuscript. All authors read and proved the final manuscript.

Competing interests

The authors declare that they have no competing interests.

References

1. Peñalva MA, Rowlands RT, Turner G. The optimization of penicillin biosynthesis in fungi. Trends Biotechnol. 1998;16(11):483–9.
2. Brakhage AA, Spröte P, Al-Abdallah Q, Gehrke A, Plattner H, Tüncher A. Regulation of penicillin biosynthesis in filamentous fungi. Adv Biochem Engin/Biotechnol. 2004;88:45–90.
3. van den Berg MA. Functional characterisation of penicillin production strains. Fungal Biol Rev. 2010;24:73–8.
4. Evers ME, Trip H, van den Berg MA, Bovenberg RAL, Driessen AJM. Compartmentalization and transport in beta-lactam antibiotics biosynthesis. Adv Biochem Engin/Biotechnol. 2004;88:111–35.
5. Brakhage AA. Molecular regulation of beta-lactam biosynthesis in filamentous fungi. Microbiol Mol Biol Rev. 1998;62(3):547–85.
6. Carr LG, Skatrud PL, Scheetz 2nd ME, Queener SW, Ingolia TD. Cloning and expression of the isopenicillin N synthetase gene from *Penicillium chrysogenum*. Gene. 1986;48(2–3):257–66.
7. Ingolia TD, Queener SW. Beta-lactam biosynthetic genes. Med Res Rev. 1989;9(2):245–64.
8. Newbert RW, Barton B, Greaves P, Harper J, Turner G. Analysis of a commercially improved *Penicillium chrysogenum* strain series: involvement of recombinogenic regions in amplification and deletion of the penicillin biosynthesis gene cluster. J Ind Microbiol Biotechnol. 1997;19(1):18–27.
9. Wang FQ, Zhong J, Zhao Y, Xiao J, Liu J, Dai M, et al. Genome sequencing of high-penicillin producing industrial strain of *Penicillium chrysogenum*. BMC Genomics. 2014;15 Suppl 1:S11. doi:10.1186/1471-2164-15-S1-S11.
10. Nielsen J. Physiological engeneering aspects of *Penicillium chrysogenum*. Singapore: World Scientific; 1997.
11. Fierro F, Barredo JL, Díez B, Gutierrez S, Fernández FJ, Martín JF. The penicillin gene cluster is amplified in tandem repeats linked by conserved hexanucleotide sequences. Proc Natl Acad Sci U S A. 1995;92(13):6200–4.
12. Lein J. The Panlabs *Penicillium* strain improvement program. In: Vanek Z, Hostálek Z, editors. Overproduction of microbial metabolites. Stoneham: Butterworths; 1986. p. 105–40.
13. Hoff B, Pöggeler S, Kück U. Eighty years after its discovery, Fleming's *Penicillium* strain discloses the secret of its sex. Eukaryot Cell. 2008;7(3):465–70.
14. Specht T, Dahlmann TA, Zadra I, Kürnsteiner H, Kück U. Complete sequencing and chromosome-scale genome assembly of the industrial progenitor strain P2niaD18 from the penicillin producer *Penicillium chrysogenum*. Genome Announc. 2014;2(4). doi:10.1128/genomeA.00577-14.
15. Barredo JL, Díez B, Alvarez E, Martín JF. Large amplifcation of a 35-kb DNA fragment carrying two penicillin biosynthetic genes in high penicillin producing strains of Penicillium chrysogenum. Curr Genet. 1989;16:453–9.
16. Kopke K, Hoff B, Kück U. Application of the *Saccharomyces cerevisiae* FLP/FRT recombination system in filamentous fungi for marker recycling and construction of knockout strains devoid of heterologous genes. Appl Environ Microbiol. 2010;76(14):4664–74.
17. Wolfers S, Kamerewerd J, Nowrousian M, Sigl C, Zadra I, Kürnsteiner H, et al. Microarray hybridization analysis of light-dependent gene expression in *Penicillium chrysogenum* identifies bZIP transcription factor PcAtfA. J Basic Microbiol. 2015;55(4):480–9.
18. Janus D, Hoff B, Kück U. Evidence for Dicer-dependent RNA interference in the industrial penicillin producer *Penicillium chrysogenum*. Microbiology. 2009;155(12):3946–56.
19. Sambrook J, Russell DW. Molecular cloning: a laboratory manual. 3rd ed. Cold Spring Harbor: Cold Spring Harbor Laboratory Press; 2001.
20. Jerpseth B, Greener A, Short JM, Viola J, Kretz PL. XL1-Blue MRF' *E. coli* cells: *McrA*$^-$, *McrCB*$^-$, *mcrF*$^-$, *Mmr*$^-$, *HsdR*$^-$ derivative of XL1-Blue cells. Strateg Mol Biol. 1992;5:81–3.
21. Kopke K, Hoff B, Bloemendal S, Katschorowski A, Kamerewerd J, Kück U. Members of the *Penicillium chrysogenum* velvet complex play functionally opposing roles in the regulation of penicillin biosynthesis and conidiation. Eukaryot Cell. 2013;12(2):299–310.
22. van den Berg MA, Albang R, Albermann K, Badger JH, Daran JM, Driessen AJM, et al. Genome sequencing and analysis of the filamentous fungus *Penicillium chrysogenum*. Nat Biotechnol. 2008;26(10):1161–8.
23. Terfehr D, Dahlmann TA, Specht T, Zadra I, Kürnsteiner H, Kück U. Genome sequence and annotation of *Acremonium chrysogenum*, producer of the b-lactam antibiotic cephalosporin C. Genome Announc. 2014;2(5). doi:10.1128/genomeA.00948-14.
24. Jekosch K, Kück U. Loss of glucose repression in an *Acremonium chrysogenum* beta-lactam producer strain and its restoration by multiple copies of the *cre1* gene. Appl Microbiol Biotechnol. 2000;54(4):556–63.
25. Schmitt EK, Bunse A, Janus D, Hoff B, Friedlin E, Kürnsteiner H, et al. Winged helix transcription factor CPCR1 is involved in regulation of β-lactam biosynthesis in the fungus *Acremonium chrysogenum*. Eukaryot Cell. 2004;3(1):121–34.
26. Walz M, Kück U. Polymorphic karyotypes in related *Acremonium* strains. Curr Genet. 1991;19(2):73–6.
27. Walz M, Kück U. Transformation of *Sordaria macrospora* to hygromycin B resistance: characterization of transformants by electrophoretic karyotyping and tetrad analysis. Curr Genet. 1995;29(1):88–95.
28. Böhm J, Hoff B, O'Gorman CM, Wolfers S, Klix V, Binger D, et al. Sexual reproduction and mating-type-mediated strain development in the penicillin-producing fungus *Penicillium chrysogenum*. Proc Natl Acad Sci U S A. 2013;110(4):1476–81.
29. Kempken F, Kück U. Transposons in filamentous fungi-facts and perspectives. Bioessays. 1998;20(8):652–9.
30. Dahlmann TA, Böhm J, Becker K, Kück U. Sexual recombination as a tool for engineering industrial *Penicillium chrysogenum* strains. Curr Genet. 2015; 61(4):679–83.
31. Hoff B, Kamerewerd J, Sigl C, Zadra I, Kück U. Homologous recombination in the antibiotic producer *Penicillium chrysogenum*: strain DPcku70 shows up-regulation of genes from the HOG pathway. Appl Microbiol Biotechnol. 2010;85(4):1081–94.
32. Smith DJ, Bull JH, Edwards J, Turner G. Amplifcation of the isopenicillin N synthetase gene in a strain of *Penicillium chrysogenum* producing high levels of penicillin. Mol Gen Genet. 1989;216:492–7.
33. Theilgaard H, van den Berg MA, Mulder CA, Bovenberg RAL, Nielsen J. Quantitative analysis of *Penicillium chrysogenum* Wis54-1255 transformants overexpressing the penicillin biosynthetic genes. Biotechnol Bioeng. 2001;72:379–88.

34. Nijland JG, Ebbendorf B, Woszczynska M, Boer R, Bovenberg RA, Driessen AJ. Nonlinear biosynthetic gene cluster dose effect on penicillin production by *Penicillium chrysogenum*. Appl Environ Microbiol. 2010;76(21):7109–15.
35. Harris DM, van der Krogt ZA, Klaassen P, Raamsdonk LM, Hage S, van den Berg MA, et al. Exploring and dissecting genome-wide gene expression responses of *Penicillium chrysogenum* to phenylacetic acid consumption and penicillinG production. BMC Genomics. 2009;10:75. doi:10.1186/1471-2164-10-75.
36. Le Dall MT, Nicaud JM, Gaillardin C. Multiple-copy integration in the yeast *Yarrowia lipolytica*. Curr Genet. 1994;26(1):38–44.
37. Tilburn J, Sarkar S, Widdick DA, Espeso EA, Orejas M, Mungroo J, et al. The *Aspergillus* PacC zinc finger transcription factor mediates regulation of both acid- and alkaline-expressed genes by ambient pH. EMBO J. 1995;14(4):779–90.
38. Then Bergh K, Brakhage AA. Regulation of the *Aspergillus nidulans* penicillin biosynthesis gene *acvA* (*pcbAB*) by amino acids: implication for involvement of transcription factor PACC. Appl Environ Microbiol. 1998;64(3):843–9.
39. Then Bergh K, Litzka O, Brakhage AA. Identification of a major *cis*-acting DNA element controlling the bidirectionally transcribed penicillin biosynthesis genes *acvA* (*pcbAB*) and *ipnA* (*pcbC*) of *Aspergillus nidulans*. J Bacteriol. 1996;178(13):3908–16.
40. van Heeswijck R, Hynes MJ. The *amdR* product and a CCAAT-binding factor bind to adjacent, possibly overlapping DNA sequences in the promoter region of the *Aspergillus nidulans amdS* gene. Nucleic Acids Res. 1991;19(10):2655–60.
41. Spröte P, Brakhage AA. The light-dependent regulator velvet A of *Aspergillus nidulans* acts as a repressor of the penicillin biosynthesis. Arch Microbiol. 2007;188(1):69–79. doi:10.1007/s00203-007-0224-y.
42. Bayram Ö, Braus GH. Coordination of secondary metabolism and development in fungi: the velvet family of regulatory proteins. FEMS Microbiol Rev. 2012;36(1):1–24.
43. Bayram O, Krappmann S, Ni M, Bok JW, Helmstaedt K, Valerius O, et al. VelB/VeA/LaeA complex coordinates light signal with fungal development and secondary metabolism. Science. 2008;320(5882):1504–6. doi:10.1126/science.1155888.
44. Sarikaya Bayram Ö, Bayram Ö, Valerius O, Park HS, Irniger S, Gerke J, et al. LaeA control of velvet family regulatory proteins for light-dependent development and fungal cell-type specificity. PLoS Genet. 2010;6(12), e1001226. doi:10.1371/journal.pgen.1001226.
45. Ni M, Yu JH. A novel regulator couples sporogenesis and trehalose biogenesis in *Aspergillus nidulans*. PLoS One. 2007;2(10), e970. doi:10.1371/journal.pone.0000970.
46. Hoff B, Kamerewerd J, Sigl C, Mitterbauer R, Zadra I, Kürnsteiner H, et al. Two components of a velvet-like complex control hyphal morphogenesis, conidiophore development, and penicillin biosynthesis in *Penicillium chrysogenum*. Eukaryot Cell. 2010;9(8):1236–50.
47. Dreyer J, Eichhorn H, Friedlin E, Kürnsteiner H, Kück U. A homologue of the *Aspergillus velvet* gene regulates both cephalosporin C biosynthesis and hyphal fragmentation in *Acremonium chrysogenum*. Appl Environ Microbiol. 2007;73(10):3412–22.
48. Bloemendal S, Kück U. Biosynthesis and molecular genetics of fungal secondary metabolites. Heidelberg, New York, Tokyo: Springer Verlag; 2014.
49. Becker K, Beer C, Freitag M, Kück U. Genome-wide identification of target genes of a mating-type alpha-domain transcription factor reveals functions beyond sexual development. Mol Microbiol. 2015;96(5):1002–22.
50. Raper KB, Alexander DF, Coghill RD. Penicillin: II. Natural variation and penicillin production in *Penicillium notatum* and allied species. J Bacteriol. 1944;48(6):639–59.

16

Improvement of betulinic acid biosynthesis in yeast employing multiple strategies

Chen Zhou[1,2†], Jing Li[1,2†], Changfu Li[1] and Yansheng Zhang[1*]

Abstract

Background: Betulinic acid (BA) is a lupane-type triterpene which has been considered as a promising agent to cure melanoma with no side effects. Considering that BA is naturally produced in small quantities in plants, we previously reported the success in engineering its production in yeast. In the present study, we attempted to improve BA biosynthesis in yeast by the use of different strategies.

Results: We first isolated a gene encoding a lupeol C-28 oxidase (LO) from *Betula platyphylla* (designated as *BPLO*). BPLO showed a higher activity in BA biosynthesis compared to the previously reported LOs. In addition, two yeast platforms were compared for engineering the production of BA, which demonstrated that the WAT11 strain was better to host BA pathway than the CEN.PK strain. Based on the WAT11-chassiss, the Gal80p mutant was further constructed. The mutant produced 0.16 mg/L/OD_{600} of BA, which was 2.2 fold of that produced by the wild type strain (0.07 mg/L/OD_{600}).

Conclusions: This study reported our efforts to improve BA production in yeast employing multiple strategies, which included the identification of a novel LO enzyme with a higher activity in BA biosynthesis, the evaluation of two yeast strains for hosting the BA pathway, and the up-regulation of the expression of the BA pathway genes by managing yeast GAL gene regulon circuit.

Keywords: Betulinic acid, Betulin, Biosynthesis, *Saccharomyces cerevisiae*

Background

Betulinic acid (BA) belongs to lupane-type triterpenes and its derivatives show a wide range of health benefits, such as anti-cancer [1], anti-HIV [2], anti-virus [3], and anti-malarial activities [4]. In particular, BA shows a selectively inhibitory activity against melanoma cells with no side effects on regular ones [5]. Despite these promising activities, commercial application of this compound is limited largely due to its short supply from nature. BA accumulates at low quantities in many plants including *Syzygium jambos* [6], *Ziziphus jujube* [7], *Diospyros kaki* Thunb [8], and *Betula platyphylla* [9]. Of these BA-producing plant species, *Betula platyphylla* is the major source of BA for the drug market. The production of BA has heavily relied on the phytochemical extraction from the *Betula platyphylla* barks. In China, around 230,000 t of the *Betula platyphylla* barks per year are consumed for the extraction of BA (this statistics data was retrieved from an annual report from the national statistics department of China). Apparently, the overexploitation on the *Betula platyphylla* barks would be problematic for continually supplying BA in a long term aspect.

The biosynthetic pathway of BA has been elucidated before [10, 11]. Cyclization of 2,3-oxidosqualene by lupeol synthase to produce lupeol is the initial committed step toward BA biosynthesis, and the lupeol synthase gene from *Arabidopsis thaliana* (*AtLup1*) has been isolated [12]. Lupeol is then successively oxidized at its C28 position to yield BA by a cytochrome P450 enzyme, lupeol C-28 oxidase (LO). We previously reported that an amyrin oxidase from *Catharanthus roseus* (*CrAO*) showed the LO activity [11]. By the combinatory expression of *CrAO* and *AtLup1*, we succeeded in engineering the BA production in *S. cerevisiae* via the yeast endogenous 2,3-oxidosqualene [10, 11, 13]. In addition to *CrAO* from *Catharanthus roseus*, the genes encoding other LO

* Correspondence: zhangys@wbgcas.cn
[†]Equal contributors
[1]CAS Key Laboratory of Plant Germplasm Enhancement and Specialty Agriculture, Wuhan Botanical Garden, Chinese Academy of Sciences, Wuhan 430074, China
Full list of author information is available at the end of the article

enzymes had been isolated from several plant species, including *CYP716A12* from *Medicago truncatula* and *CYP716A15* from *Vitis vinifera* [14]. Despite that *Betula platyphylla* bark is the major source for the production of BA, the LO gene of this plant remains not isolated. Therefore, here we reported on the isolation and characterization of the LO gene from *Betula platyphylla* bark (designated as *BPLO*). Relative to the previously identified LOs described above, BPLO showed a higher BA-producing activity in yeast.

After the success of engineering BA biosynthesis in yeast, the next challenge is to boost its production. When foreign pathways are introduced into a microbial host, its existing metabolic network would tightly control carbon fluxes to foreign compounds. The yeast strain that we previously used for engineering the BA production is a laboratory *S. cerevisiae* strain WAT11 [10]. The yeast WAT11 strain has a genomically integrated *Arabidopsis thaliana* NADPH-CYP reductase as a redox partner for the CYPs [15], and is usually used to characterize plant cytochrome P450 enzymes [16, 17]. The *S. cerevisiae* CEN.PK is another yeast strain which has often been employed for synthesizing a wide range of products in industry [18, 19]. The *S. cerevisiae* CEN.PK strain contains a significantly higher content of ergosterol than other *S. cerevisiae* strains [20]. Genomic sequencing revealed that the CEN.PK strain contains a number of SNPs in the coding regions of ergosterol and fatty acid pathway enzymes [21]. The authors suggested that those SNPs might be responsible for the high-level ergosterol accumulation of this strain. Both ergosterol and BA are cyclized from 2,3-oxidosqualene, a common precursor for the biosynthesis of sterols and triterpenes. A higher accumulation of ergosterol in the CEN.PK strain prompted us to test whether it is better than other yeast strains to engineer the BA production. Therefore, in this study, the BA pathway was integrated into the genomes of both the WAT11 and CEN.PK strains and their BA-producing abilities were compared.

In our previous attempts to engineer the BA production in yeast, the BA pathway was expressed in either a constitutive or a galactose-inducible manner, and galactose-inducible promoters (*GAL1* or *GAL10* promoter) appeared to be stronger to direct the expression of the BA pathway genes [10, 13]. Therefore, in this study, all the BA pathway genes were cloned under galactose-inducible promoters. The induction of GAL genes is controlled by a protein complex of Gal3p, Gal4p and Gal80p [22]. Gal4p is a transcriptional activator that binds to the upstream activation sequences of GAL genes to activate the transcriptions of the genes. When glucose is used as the sole carbon source, the activation domain of Gal4p is masked by the repressor Gal80p, thereby preventing the transcriptions of GAL genes. In the presence of galactose, the signal transducer Gal3p binds to Gal80p to expose the activating domain of Gal4p, leading to the activation of GAL genes. It has been reported that a yeast mutant lacking Gal80p produced higher levels of carotenoids when the carotenoid pathway genes were controlled by galactose-inducible promoters [23]. Here, we also reported that the Gal80p mutant increased the BA production, which was up to 2.2-folds of that produced by the wild yeast strain.

Results

Isolation and functional analysis of BPLO

A 1082 bp-cDNA fragment was amplified using the degenerate primers 1/2 from the bark of *B. platyphylla* where BA primarily accumulates [9]. The resulting cDNA fragment showed sequence identities to the genes coding for the previously reported LO enzymes, and therefore its full-length cDNA (designated as *BPLO*) was further cloned by RACE-PCR technique using primers 3–6. BPLO was also officially named as CYP716A180 by the standard P450 nomenclature committee [24]. During our biochemical characterization of BPLO, the *BPLO* gene was also isolated by another group from the Northeast Forestry University of China, and was named as *ATH1* (GenBank accession no. KJ452328.1). The predicted protein sequence of BPLO showed 79–81 % identity at the amino acid level with the previously published LOs, including CrAO from *C. roseus*, CYP716A12 from *M. truncatula* and CYP716A15 from *V. vinifera*.

To examine the activity of BPLO, the *BPLO* gene was co-expressed in the WAT11 yeast strain with *AtLup1*. The yeast cells transformed with the empty vector or expressing *AtLup1* alone served as the control. The WAT11 strain contains an *Arabidopsis thaliana* cytochrome P450: NADPH reductase ATR1 [15] that is required for the activity of plant P450 enzymes. Compared to the empty vector control, the expression of *AtLup1* alone led to the production of lupeol and an unknown peak 1. Due to the absence of the authentic standard, we were not able to determine the identity of the peak1, but it was assumed to be 3α,20-dihydroxylupane (lupanediol), as it was previously found as the second major product of the AtLup1, in addition to lupeol [25]. When *AtLup1* and *BPLO* were co-expressed, lupeol was successively oxidized at its C-28 position by BPLO to yield betulin (BN) and betulinic acid (BA) which were not produced by the control yeast cells (Fig. 1). The identities of BN and BA products were confirmed by comparisons with their authentic standards (Additional file 1: Figure S1). These data clearly suggested that BPLO functions as a LO enzyme. Interestingly, BPLO also seemed to be able to convert the peak 1 to the products of peaks 2 and 3 with the peak 2 being negligible (Fig. 1). The peak 2 and peak 3 could be the oxidized alcohol and acid products of lupanediol, respectively, which

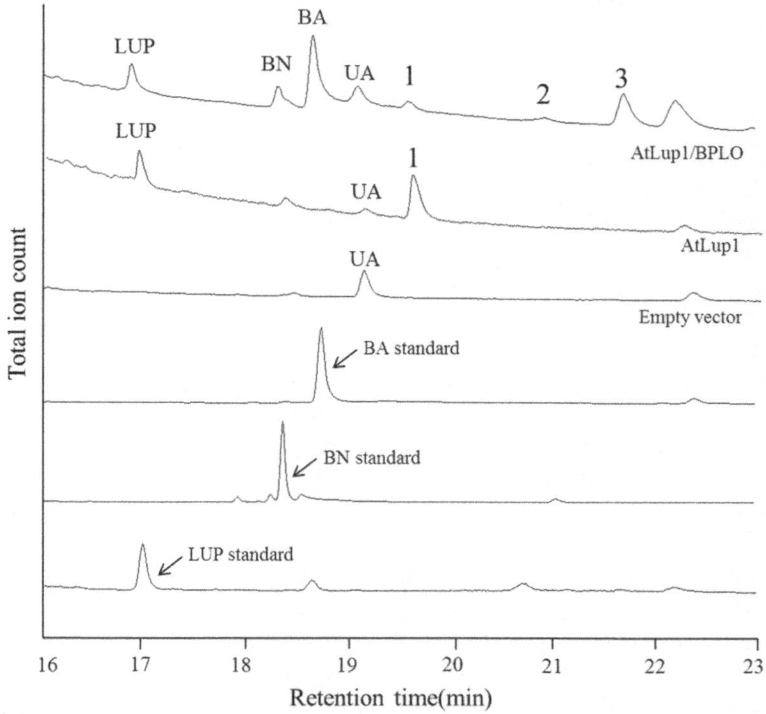

Fig. 1 Functional analysis of *BPLO* by its co-expression with *AtLup1* in yeast cells. GC–MS analysis was shown for the products from the yeast cells expressing AtLup1/BPLO yielding lupeol (LUP), betulin (BN), betulinic acid (BA), and the unknown products (peaks 2–3), and the cells expressing *AtLup1* alone yielding lupeol and the unknown peak 1. The identities of LUP, BN and BA were confirmed with their corresponding chemical standards by comparing their mass fragmented products (Additional file 1: Figure S1). The identities of the peaks 1–3 were not determined due to the absence of their chemical standards, and their MS spectrums were shown in Additional file 1: Figure S2

was presumably catalyzed by BPLO at the C-28 position of lupanediol. The mass spectrums of the trimethylsilyl-derivatives of the peaks 1–3 were shown in Additional file 1: Figure S2.

To examine the relevance of the *BPLO* transcript to BA biosynthesis in *Betula platyphylla*, we have investigated the accumulation pattern of the *BPLO* transcripts and its enzymatic products (BN and BA) across different tissues of *Betula platyphylla*. The *BPLO* transcript was mostly observed in the roots and barks while absent in the leaves (Fig. 2a). On the other hand, GC-MS analysis showed the accumulation of the BPLO products (BN and BA) in the roots and barks, whereas none of them was detected in the leaves (Fig. 2b). Therefore, the *BPLO* gene expression pattern is overall consistent with the BA accumulation in vivo. However, a higher level of the

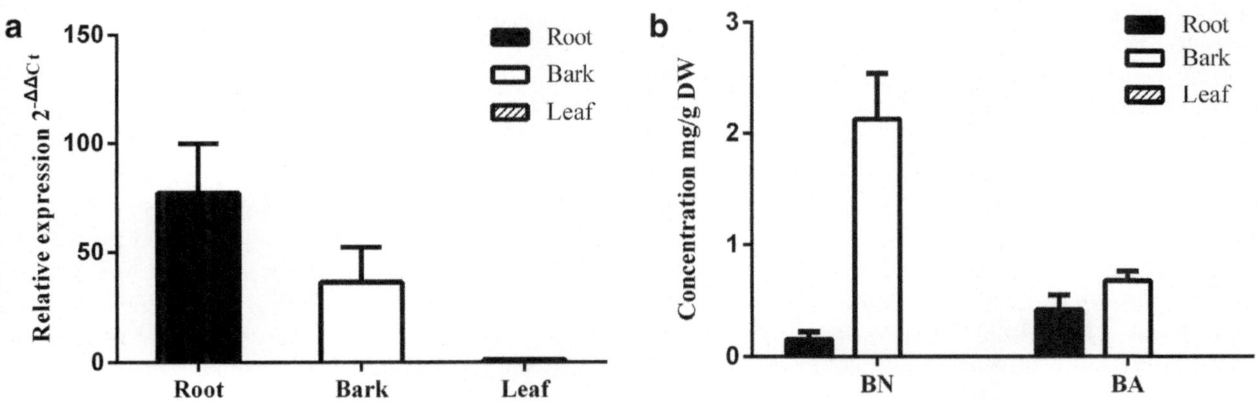

Fig. 2 The accumulation pattern of the BPLO transcripts and its enzymatic products in *B. platyphylla* tissues. **a** the BPLO transcripts measured by qRT-PCRs; **b** the BPLO enzymatic products (BN and BA) measured by GC-MS analysis. BN, betulin; BA, betulinic acid. Error bars represent the standard errors (SE) of the means calculated from three biological replicates

BPLO transcript was detected in the roots relative to the barks whereas the barks contain higher levels of the BPLO products than the roots. This inconsistence may be caused by the transport of BPLO products from the roots to the barks or less precursor lupeol in the roots supplied for BPLO. We also examined the level of lupeol in these tissues, but it was not detectable probably due to its complete conversion to BN and BA.

BPLO showed a higher BA-producing activity than the previously reported LOs

To evaluate the BA-producing activity between BPLO, CrAO and CYP716A15, these LOs were initially expressed in the WAT11 strain alone, and each resultant yeast culture was then fed with 50 μM lupeol as the LO substrate. After galactose induction, very low levels of BA were produced in all the three yeast cultures (Additional file 1: Figure S3A), indicating that the substrate lupeol was indeed taken up by the yeast strain but at a very low efficiency. Interestingly, even under this condition, BPLO seemed to produce higher levels of BA than CrAO and CYP716A15 (Additional file 1: Figure S3B). To further compare their activities, the three LO genes were then co-expressed with the lupeol synthesizing gene *AtLup1* in the WAT11 strain. After induction with 2 % galactose for 60 h, the yeast cultures were all acidified to pH 2.0 and extracted with ethyl acetate for GC-MS analysis. Once again, BPLO produced the most BA while the lowest amount of BN in comparisons with the CrAO and CYP716A15 (Fig. 3). Thus, these results strongly suggested that BPLO displayed a higher BA-producing activity than the previously characterized CrAO and CYP716A15. On the other hand, like the results shown in Fig. 1, the unknown products (peaks 1–3) were also observed in this experiment. Interestingly, compared to CrAO and CYP716A15, BPLO resulted in relatively higher levels of the peak 3 while lower amounts of the peak 2 (Fig. 3a), which pattern was very similar to that of these LOs' activities on the BA pathway. This data may also reflect the truth that BPLO catalyzes a higher C28-oxidizing activity toward lupeol or lupeol-like compounds than CrAO and CYP716A15.

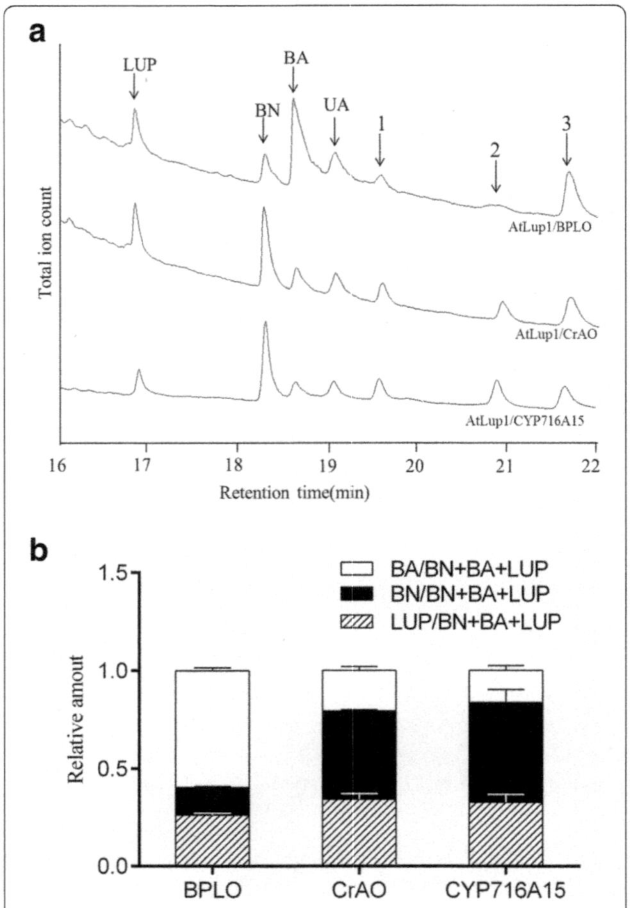

Fig. 3 Comparison of the BA-producing activities between BPLO and the previously reported LOs (CrAO and CYP716A15) in yeast. The LO genes were individually co-expressed with *AtLup1* in the WAT11 yeast strain. **a** Total ion chromatograms were shown for the products by expressing BPLO, CrAO and CYP716A15; **b** the relative amounts of the products lupeol (LUP), betulin (BN), and betulinic acid (BA) produced by the transgenic yeast cells expressing the individual LO. "LUP/BN + BA + LUP" represents the ratio of the yield of the total products (LUP, BN and BA) divided by the LUP yield. Similarly, "BN/BN + BA + LUP" and "BA/BN + BA + LUP" represent the ratios of the total product yield divided by the yield of BN and BA, respectively. Error bars represent the standard errors of the means calculated from two biological replicates

The WAT11 strain was better than the CEN.PK strain to host BA pathway

The yeast strain CEN.PK was demonstrated to be a triterpene hyper producer [26], which prompted us to expect that the strain CEN. PK would be a better host to engineer the biosynthesis of BA. Thus, the strain CEN.PK was then investigated for its BA-producing ability compared to the WAT11 yeast strain. The WAT11 strain contains the gemonically integrated ATR1 [15] that is required for the activity of BPLO. The exactly same ATR1 expression cassette was amplified from the WAT11 strain and integrated into the chromosome of the CEN.PK strain to generate the strain CEN.PK-ATR1 via an integrative expression vector pRS406. For a fair comparison, the empty vector pRS406 was also integrated into the WAT11 strain to give the strain WAT11-406. The BA pathway was then integrated into both the WAT11-406 and CEN.PK-ATR1 strains, generating the strains of WAT11-406-LB and CEN.PK-ATR1-LB, respectively. The integrated yeast strains were cultured for assessing their BA-producing abilities. There were no significant differences in their growth curves of both strains (Fig. 4a). After the galactose-induction for 9 days, the yeast cells and medium were partitioned by

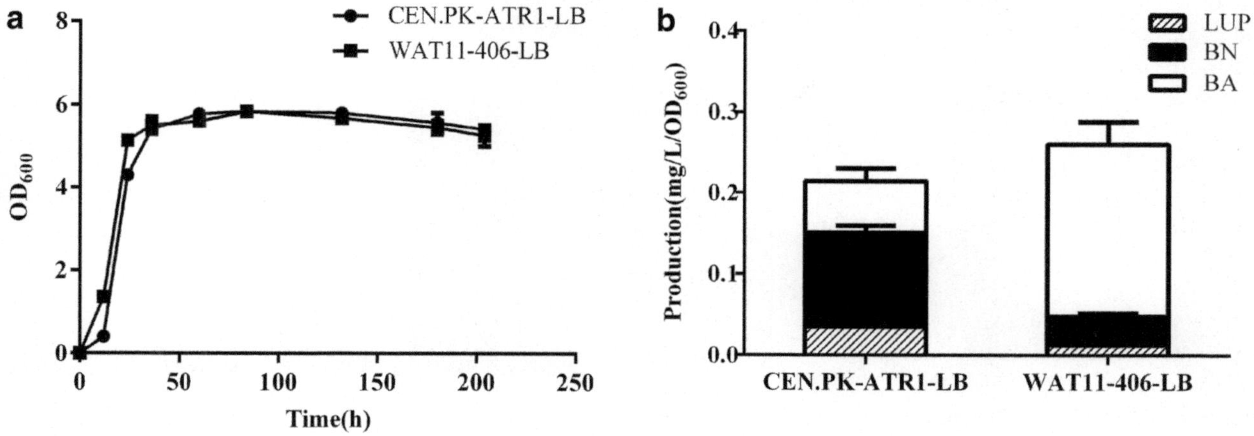

Fig. 4 Comparison of CEN.PK and WAT11 strains for their BA-producing abilities. **a** the growth properties of the both strains measured by checking the optical density at the wave length of 600; **b** the BA-producing abilities of the both strains measured by calculating the yields of lupeol (LUP), betulin (BN) and betulinic acid (BA). Error bars represent the standard errors of the means calculated from three biological replicates

centrifugation and separately extracted with ethyl acetate for GC-MS analysis. The BPLO products (BN and BA) were majorly detected in the medium whereas very low amounts of them accumulated within the cells of both strains (Additional file 1: Figure S4), suggesting that both strains were able to highly transport the products out of cells. For the products in the medium, WAT11 strain accumulated about 3-fold higher BA levels and relatively lower amount of BN compared to the CEN.PK strain (Fig. 4b), suggesting that the WAT11 strain had a more efficient conversion of BN to BA than the CEN.PK strain. On the other hand, compared to the WAT11 strain, a significantly higher level of lupeol accumulated in the CEN.PK strain (Fig. 4b). The total yields of lupeol, BN and BA would reflect the carbon fluxes through the BA pathway. The data of the Fig. 4 indicated that WAT11 strain provided greater metabolic fluxes through the BA pathway than the CEN.PK strain. Thus, further engineering of the BA production of this study was performed based on the WAT11 stain.

The inactivation of Gal80p improved the BA yield based on the WAT11 strain

In this study, the BA pathway genes were all cloned under the control of galactose-inducible promoters and their transcriptions are galactose-inducible. The induction of GAL genes is controlled by the protein complex of Gal3p, Gal4p and Gal80p [22]. The expression level of GAL genes could be increased in the Gal80 mutant when using non-inducing carbon source [27]. It has also been reported that the loss of Gal80p function in yeast improved carotenoid production when the carotenoid biosynthetic pathway genes were cloned under galactose inducible promoters [23]. Based on the BA-producing strain of this study, we also investigated the effect of the Gal80 mutation on the BA production. Using homologous and loxp sites-based specific recombinations, the Gal80p inactivation was made based on the WAT11-LB strain, which was confirmed by genomic DNA PCRs (Additional file 1: Figure S5). The mutant strain was then independently cultured using 2 % glucose or 2 % galactose as the carbon sources. As a control, the corresponding wild type strain WAT11-LB was cultured in the same ways. On 2 % glucose, the growth property of the mutant resembled that of the wild strain (Fig. 5a) and the BA production was hardly detected whether or not the Gal80p was inactivated (Fig. 5b). On 2 % galactose, the BA production from the mutant was 1.2 fold higher than that of the wild strain, suggesting that the Gal80p disruption improved the expressions of the BA pathway genes, which were cloned under the control of galactose inducible promoters. This assumption was proved by the gene expression analysis of the *BPLO* gene from both the mutant and the wild strain by real-time PCRs (Additional file 1: Figure S6).

Discussion

The success of engineering the BA production in yeast cells has previously been reported by our group [10, 11, 13]. In the present study, we attempted to increase the BA yields in yeast using multiple strategies, which include the identification of a higher BA-producing enzyme, the assessment of two different yeast strains, and the increase of the BA pathway gene transcriptions. The bark of white birch is the major source for the BA production, while the BA-producing enzymes in this species have never been reported before this study. Using a homology-based strategy, we have successfully isolated the gene encoding a novel LO enzyme (BPLO) from white birch bark. Compared with the previously reported

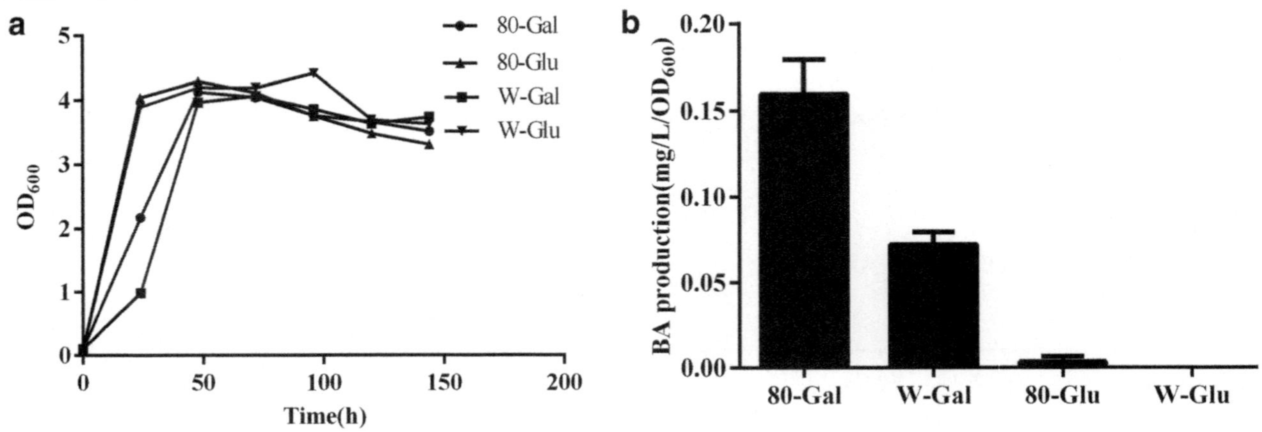

Fig. 5 Yeast growth properties **a** and betulinic acid production **b** were compared between the wild type strain (WAT11-LB) and the Gal80p mutant (WAT11-LB-ΔGal80). 80-Gal, the mutant strain cultured under 2 % galactose; W-Gal, the wild type strain cultured under 2 % galactose; 80-Glu, the mutant strain cultured under 2 % glucose; W-Glu, the wild type strain cultured under 2 % glucose. Error bars represent the standard errors of the means calculated from three biological replicates

LOs, BPLO produced the most BA while the lowest BN (Fig. 3), suggesting that BPLO exhibits a higher activity of converting BN to BA than the other LOs.

Previously, the *S. cerevisiae* CEN.PK strain was revealed to contain higher levels of ergosterol than other *S. cerevisiae* strains [20]. We also observed that the concentration of ergosterol in CEN.PK strain is higher than that in the WAT11 strain (data not shown). Ergosterol is derived from 2,3-oxidosqualene which is also a precursor for the biosynthesis of BA, thus, we initially anticipated that the CEN.PK strain might be a better strain for the BA-production. However, the comparison of the yield of the total products (LUP, BN and BA) between both strains (Fig. 4b) indicated that the WAT11 strain had greater metabolic fluxes through the BA-pathway than the CEN.PK strain. Moreover, the ratios between the individual products suggested that the WAT11 strain was more efficient than the strain CEN.PK in the conversion of BN to BA (Fig. 4b). Given the relatively comparable cell densities of both strains (Fig. 4a), it appeared that the better strain for engineering the BA-production is the WAT11 rather than the CEN.PK. Both strains differ in many respects, the mechanism by which the WAT11 strain performed better for the BA-production is not clear.

In the current study, the BA pathway genes were cloned under the control of galactose inducible promoters. It is well accepted that a protein complex of Gal3p, Gal4p and Gal80p involve in managing the GAL gene transcriptions, and of which the Gal80p acts as an inhibitor in transcribing the GAL genes [22]. On the basis of the WAT11-LB strain which bears the BA-pathway, the disruption of the Gal80p gene caused a higher transcript level of the *BPLO* gene on 2 % galactose (Additional file 1: Figure S6), which ultimately increased the BA levels (Fig. 5). When the Gal80p was inactivated, its inhibiting effects on galactose-dependent gene inductions might be alleviated, resulting in relatively higher transcriptions of the BA pathway genes in the mutant than those in the wild strain. However, when 2 % glucose was used as the sole carbon source, the BA production was inhibited, and this inhibition seemed not to be unlocked upon the Gal80p disruption (Fig. 5b). This data is reasonable because that the GAL gene regulation via the Gal80p depends on the presence of galactose while not glucose [23].

Conclusions

The lupeol C-28 oxidase from *B. platyphylla* (BPLO) exhibited higher activities in BA biosynthesis, comparing to the previously reported LOs. The yeast strain WAT11 strain was better to host BA biosynthesis than the CEN.PK strain. Based on the BA-producing WAT11 strain, the Gal80p mutant improved the BA biosynthesis up to 2.2 folds relative to the wild type strain.

Methods

BPLO cDNA isolation and its gene expression analysis

All the primers used in this study were shown in Additional file 1: Table S1. For the homology-based cloning of *B. platyphylla* lupeol C-28 oxidase cDNAs, degenerate primers 1/2 were designed based on the highly conserved domains of the LOs including CYP716A15 from *V. vinifera* (GenBank accession no. AB619802.1), CYP716A12 from *M. truncatula* (GenBank accession no. FN995113.1) and CrAO from *C. roseus* (GenBank accession no.JN565975). RT-PCRs were performed with the cDNA template that was prepared from the bark of 4 month-old *B. platyphylla* seedling. A 1082-bp amplified product was obtained and cloned into pMD18-T vector for sequencing.

Based on DNA sequence analysis, the amplified product was found to show about 80 % amino acid identity to the previously identified LOs [11, 14], indicating that the gene product may code for the *B. platyphylla* LO (designated as *BPLO*). The full length cDNA of *BPLO* was then recovered by RACE-PCRs using primers 3–6. To examine the *BPLO* gene expression pattern in *B. platyphylla*, cDNA was synthesized from equal amounts of total RNAs that were extracted from its different tissues. The *B. platyphylla* actin gene (GenBank accession no. EU588981.1) was chosen as an internal standard to normalize the variation of the cDNA preparations. For each sample, qRT-PCR was performed with four technical replicates on three biological replicates using FastStart Universal SYBR Green Master (Roche, Mannheim, Germany). The *BPLO* transcript was amplified with primers 7/8 while the primers 9/10 were used for amplifying the *B. platyphylla* actin gene. The thermal cycling conditions were set as follows: 95 °C for 10 min, followed by 40 cycles of 95 °C for 15 s, and 60 °C for 60 s.

Functional analysis of BPLO in comparisons with the previously published LOs

The Open Reading Frame (ORF) of *BPLO* was amplified with primers 11/12 from the birch barks by RT-PCRs. Similarly, the ORFs of other LO cDNAs, including *CYP716A15* from *V. vinifera* and *CrAO* from *C. roseus*, were amplified from their host plant leaves using primers 13–16. The ORFs of these LO cDNAs were then cloned into a yeast expression vector pESC-TRP by standard enzyme digestions and ligations. For in vivo functional analysis, these plasmids were co-transformed into *S. cerevisiae* WAT11 strain [15] with the plasmid pESC-HIS-AtLup1, which gives rise to the supply of lupeol [12]. These transgenic yeast strains were initially grown up at 30 °C in an appropriate SD drop out medium with 2 % (w/v) glucose. Then yeast cultures were collected by centrifugation, washed three times in sterile water, and re-suspended to an OD_{600} of 0.8 in 30 mL of SD dropout medium containing 2 % (w/v) galactose. After 60 h of the induction, the growth medium were collected by centrifugation, all acidified to pH 2.0 using 2 M HCl, and extracted with ethyl acetate. The ethyl acetate fractions were dried by a rotary evaporator and derivatized using BSTFA at 80 °C for 30 min prior to GC-MS analysis. For the extraction from the cell pellets, yeast cells were washed with 0.1 mM Tris-HCl buffer (pH 9.0), re-suspended in 0.5 ml of lysis buffer consisting of 50 mM potassium phosphate (pH 7.0), and broken by glass beads. The cell-free extracts were then acidified to pH 2.0 and extracted with ethyl acetate for GC-MS analysis as described above. The contents of lupeol, BN and BA were measured using standard curves.

Assessments of the WAT11 and CEN.PK yeast strains for their BA-producing abilities

For this experiment, BPLO was chosen to reconstitute the BA pathway as it exhibited a higher BA-producing activity in comparisons with the other LOs, which was concluded from the functional analysis described above. To compare the *S. cerevisiae* WAT11 strain and CEN.PK strain for their BA-producing abilities, the BA pathway genes (*BPLO* and *AtLup1*) were expressed in both strains in an integrative manner. To construct the integrative expression vectors, the BA expression cassette "Gal1 promoter-AtLup1-CYC1 terminator-Gal10 promoter-BPLO-ADH1 terminator" was amplified from the plasmid pESC-TRP-AtLup1-BPLO with primers 17/18 and cloned into a yeast integration expression vector YIplac204 under *Pst*I and *Xba*I sites, yielding the construct YIplac204-AtLup1-BPLO. The expression cassette of the plant cytochrome P450 reductase, "Gal10-CYC1 promoter-ATR1-CPR1 terminator", was amplified from the WAT11 genomic DNA [15] using primers 19/20 and inserted into a yeast integration expression vector pRS406 to give the construct pRS406-ATR1. For the integrative expression in the CEN.PK strain, the construct pRS406-ATR1 was linearized by *Stu*I digestion and the linear DNA molecule was then transformed into the CEN.PK strain to give the strain CEN.PK-ATR1. Based on the strain CEN.PK-ATR1, the linearized construct YIplac204-AtLup1-BPLO by *EcoRV* digestion was transformed to generate the strain CEN.PK-ATR1-LB. For the integration expression in the WAT11 strain, the empty vector pRS406 was first linearized with *Stu*I and transformed into the WAT11 strain to give the strain WAT11-406. The construct YIplac204-AtLup1-BPLO was then linearized with *EcoRV* and transformed into the WAT11-406 strain to give the strain WAT11-406-LB. Those integrated yeast strains were grown in SD-Ura-Trp-double dropout liquid medium. Relevant metabolites including lupeol, BN and BA were extracted and analyzed by GC-MS as described above, and their concentrations were quantified using ursolic acid as an internal standard.

Creation of GAL80p mutant based on the strain WAT11-LB strain

The selection marker "LoxP-kanMX-loxP" was amplified from the plasmid pUG6 [28] using primers 21/22. A 393-bp of 5'-Gal80p target and 393-bp of 3'-Gal80p target were amplified from the WAT11 strain genomic DNA using primers 23–26. The two flanking targets were then linked to the selection marker "LoxP-kanMX-loxP" by overlapping PCRs with primers 23/24 to form the Gal80p disruption cassette, in which the selection marker was positioned between the two targets. The correctness of the Gal80p disruption cassette was confirmed by cloning it into the vector pMD18-T for sequencing. After digestions with *BamH*I and *EcoR*I from the T-vector, the linear Gal80p disruption cassette was gel-purified and transformed into the strain WAT11-LB. After selections with 200 mg/L G418, stable transformants containing the disruption cassette were obtained. The selection marker was then removed by a recombinase which was expressed by transforming the recombinase expression vector pYPT2 [29] into the transformants. Finally, the repressor Gal80p was successfully disrupted, yielding the strain WAT11-LB-△Gal80.

Metabolite extraction from *B. platyphylla* tissues

To examine the BA accumulation pattern in *B. platyphylla* tissues, 0.1 g of dried plant materials (roots, barks and leaves) were extracted with 5 mL of 95 % ethanol for 30 min. The total extract was centrifuged and the pellet was re-extracted twice with 95 % ethanol. The combined supernatants were evaporated and re-dissolved in 2 mL of methanol for GC-MS analysis.

GC-MS analysis

GC-MS analysis was performed in an Agilent 7890A GC machine (Agilent Technologies, USA) as described previously [11]. Ursolic acid (UA) was used as an internal standard to quantify the BA production. Compounds were identified by comparing their retention times and mass spectrums with those of corresponding authentic standards.

Additional file

Additional file 1: Table S1. Primers used in this study. **Figure S1.** The mass fragmented patterns of the products lupeol (LUP), betulin (BN) and betulinic acid (BA) as well as their corresponding chemical standards. **Figure S2.** The mass spectrums of the unknown peaks 1–3 shown in the Fig. 1 of the main text. **Figure S3.** GC-MS analysis of the products extracted from the lupeol C28-oxidase (LO) alone-expressed yeast cultures fed with lupeol. **Figure S4.** The production of betulinic acid (BA) was mostly found in the yeast culture mediums of both CEN.PK-ATR1-LB and WAT11-406-LB strains with relatively much less BA being detected inside the cells. **Figure S5.** Confirmation of the Gal80p gene disruption by diagnostic PCRs. **Figure S6.** Comparison of the *BPLO* transcripts between the wild type strain WAT11-LB and the mutant WAT11-LB-△Gal80 under 2 % galactose as the carbon source.

Funding

This work was supported by the National Science and Technology Program of China during the 25th-Year Plan period (Project No.2012AA02A704) and the Grant from the Hubei Renyue Pharmaceutical Co., LTD (Project No. 492014–026053).

Authors' contributions

YZ designed this work and revised this manuscript. CZ performed the experiments, analyzed the data and drafted the manuscript. JL carried out the genes isolation work. CL coordinated the study. All authors read and approved the final manuscript.

Competing interests

The authors declare that they have no competing interests.

Author details

[1]CAS Key Laboratory of Plant Germplasm Enhancement and Specialty Agriculture, Wuhan Botanical Garden, Chinese Academy of Sciences, Wuhan 430074, China. [2]University of Chinese Academy of Sciences, Beijing 100049, China.

References

1. Pisha E, Chai H, Lee IS, Chagwedera TE, Farnsworth NR, Cordell GA, Beecher CWW, Fong HHS, Kinghorn AD, Brown DM, et al. Discovery of betulinic acid as a selective inhibitor of human-melanoma that functions by induction of apoptosis. Nat Med. 1995;1:1046–51.
2. Fujioka T, Kashiwada Y, Kilkuskie RE, Cosentino LM, Ballas LM, Jiang JB, Janzen WP, Chen IS, Lee KH. Anti-AIDS agents,11. Betulinic acid and platanic acid as anti-HIV principles from *Syzigium claviflorum*, and the anti-HIV activity of structurally related triterpenoids. J Nat Prod. 1994;57:243–7.
3. Ryu SY, Lee C-K, Lee CO, Kim HS, Zee OP. Antiviral triterpenes from *Prunella vulgaris*. Arch Pharmacal Res. 1992;15:242–5.
4. Steele JCP, Warhurst DC, Kirby GC, Simmonds MSJ. In vitro and in vivo evaluation of betulinic acid as an antimalarial. Phytother Res. 1999;13:115–9.
5. Zuco V, Supino R, Righetti SC, Cleris L, Marchesi E, Gambacorti-Passerini C, Formelli F. Selective cytotoxicity of betulinic acid on tumor cell lines, but not on normal cells. Cancer Lett. 2002;175:17–25.
6. Chang CW, Wu TS, Hsieh YS, Kuo SC, Chao PDL. Terpenoids of *Syzygium formosanum*. J Nat Prod. 1999;62:327–8.
7. Schühly W, Heilmann J, Calis I, Sticher O. New triterpenoids with antibacterial activity from *Zizyphus joazeiro*. Planta Med. 1999;65:740–3.
8. Recio MC, Giner RM, Manez S, Gueho J, Julien H, Hostettmann K, Rios J. Investigations on the steroidal anti-inflammatory activity of triterpenoids from Diospyros leucomelas. Planta Med. 1995;61:9–12.
9. Pînzaru SC, Leopold N, Kiefer W. Vibrational spectroscopy of betulinic acid HIV inhibitor and of its birch bark natural source. Talanta. 2002;57:625–31.
10. Li J, Zhang Y. Increase of betulinic acid production in *Saccharomyces cerevisiae* by balancing fatty acids and betulinic acid forming pathways. Appl Microbiol Biotechnol. 2014;98:3081–9.
11. Huang L, Li J, Ye H, Li C, Wang H, Liu B, Zhang Y. Molecular characterization of the pentacyclic triterpenoid biosynthetic pathway in *Catharanthus roseus*. Planta. 2012;236:1571–81.
12. Herrera JBR, Bartel B, Wilson WK, et al. Cloning and characterization of the *Arabidopsis thaliana* lupeol synthase gene. Phytochemistry. 1998;49:1905–11.
13. Li J, Zhang Y. Modulating betulinic acid production in *Saccharomyces cerevisiae* by managing the intracellular supplies of the co-factor NADPH and oxygen. J Biosci Bioeng. 2015;119:77–81.
14. Fukushima EO, Seki H, Ohyama K, Ono E, Umemoto N, Mizutani M, Saito K, Muranaka T. CYP716A subfamily members are multifunctional oxidases in triterpenoid biosynthesis. Plant Cell Physiol. 2011;52:2050–61.
15. Urban P, Mignotte C, Kazmaier M, Delorme F, Pompon D. Cloning, yeast expression, and characterization of the coupling of two distantly related *Arabidopsis thaliana* NADPH-cytochrome P450 reductases with P450 CYP73A5. J Biol Chem. 1997;272:19176–86.
16. Teoh KH, Polichuk DR, Reed DW, Nowak G, Covello PS. *Artemisia annua* L. (*Asteraceae*) trichome-specific cDNAs reveal CYP71AV1, a cytochrome P450 with a key role in the biosynthesis of the antimalarial sesquiterpene lactone artemisinin. FEBS Lett. 2006;580:1411–6.

17. Carelli M, Biazzi E, Panara F, Tava A, Scaramelli L, Porceddu A, Graham N, Odoardi M, Piano E, Arcioni S, et al. *Medicago truncatula* CYP716A12 is a multifunctional oxidase involved in the biosynthesis of hemolytic saponins. Plant Cell. 2011;23:3070–81.
18. van Dijken JP, Bauer J, Brambilla L, Duboc P, Francois JM, Gancedo C, Giuseppin ML, Heijnen JJ, Hoare M, Lange HC, et al. An interlaboratory comparison of physiological and genetic properties of four *Saccharomyces cerevisiae* strains. Enzyme Microb Technol. 2000;26:706–14.
19. Nijkamp JF, van den Broek M, Datema E, de Kok S, Bosman L, Luttik MA, Daran-Lapujade P, Vongsangnak W, Nielsen J, Heijne WH, et al. De novo sequencing, assembly and analysis of the genome of the laboratory strain *Saccharomyces cerevisiae* CEN.PK113-7D, a model for modern industrial biotechnology. Microb Cell Fact. 2012;11:36.
20. Daum G, Tuller G, Nemec T, Hrastnik C, Balliano G, Cattel L, Milla P, Rocco F, Conzelmann A, Vionnet C, et al. Systematic analysis of yeast strains with possible defects in lipid metabolism. Yeast. 1999;15:601–14.
21. Otero JM, Vongsangnak W, Asadollahi MA, Olivares-Hernandes R, Maury J, Farinelli L, Barlocher L, Osteras M, Schalk M, Clark A, Nielsen J. Whole genome sequencing of *Saccharomyces cerevisiae*: from genotype to phenotype for improved metabolic engineering applications. BMC Genomics. 2010;11:723.
22. Bhat PJ, Oh D, Hopper JE. Analysis of the GAL3 signal transduction pathway activating GAL4 protein-dependent transcription in *Saccharomyces cerevisiae*. Genetics. 1990;125:281–91.
23. Xie W, Liu M, Lv X, et al. Construction of a controllable β-carotene biosynthetic pathway by decentralized assembly strategy in *Saccharomyces cerevisiae*. Biotechnol Bioeng. 2014;111:125–33.
24. Nelson DR. The cytochrome P450 homepage. Human Genomics. 2009;4:59–65.
25. Segura MJR, Meyer MM, Matsuda SPT. *Arabidopsis thaliana* LUP1 converts oxidosqualene to multiple triterpene alcohols and a triterpene diol. Org Lett. 2000;2:2257–9.
26. Madsen KM, Udatha GD, Semba S, et al. Linking genotype and phenotype of *Saccharomyces cerevisiae* strains reveals metabolic engineering targets and leads to triterpene hyper-producers. PLoS One. 2011;6:e14763.
27. Shiba Y, Ono C, Fukui F, et al. Effect of ethanol on the production of carboxypeptidase Y using the GAL10 promoter in a *Saccharomyces cerevisiae* gal80 mutant. J Biosci Bioeng. 2000;89:426–30.
28. Guldener U, Heck S, Fiedler T, Beinhauer J, Hegemann JH. A new efficient gene disruption cassette for repeated use in budding yeast. Nucleic Acids Res. 1996;24:2519–24.
29. Sun Z, Meng H, Li J, Wang J, Li Q, Wang Y, Zhang Y. Identification of novel knockout targets for improving terpenoids biosynthesis in *Saccharomyces cerevisiae*. PLoS One. 2014;9:e112615.

17

Distinctive molecular and biochemical characteristics of a glycoside hydrolase family 20 β-N-acetylglucosaminidase and salt tolerance

Junpei Zhou[1,2,3,4†], Zhifeng Song[2†], Rui Zhang[1,2,3,4], Rui Liu[2], Qian Wu[1,2,3,4], Junjun Li[1,2,3,4], Xianghua Tang[1,2,3,4], Bo Xu[1,2,3,4], Junmei Ding[1,2,3,4], Nanyu Han[1,2,3,4] and Zunxi Huang[1,2,3,4*]

Abstract

Background: Enzymatic degradation of chitin has attracted substantial attention because chitin is an abundant renewable natural resource, second only to lignocellulose, and because of the promising applications of N-acetylglucosamine in the bioethanol, food and pharmaceutical industries. However, the low activity and poor tolerance to salts and N-acetylglucosamine of most reported β-N-acetylglucosaminidases limit their applications. Mining for novel enzymes from new microorganisms is one way to address this problem.

Results: A glycoside hydrolase family 20 (GH 20) β-N-acetylglucosaminidase (GlcNAcase) was identified from *Microbacterium* sp. HJ5 harboured in the saline soil of an abandoned salt mine and was expressed in *Escherichia coli*. The purified recombinant enzyme showed specific activities of 1773.1 ± 1.1 and 481.4 ± 2.3 μmol min^{-1} mg^{-1} towards *p*-nitrophenyl β-N-acetylglucosaminide and N,N'-diacetyl chitobiose, respectively, a V_{max} of 3097 ± 124 μmol min^{-1} mg^{-1} towards *p*-nitrophenyl β-N-acetylglucosaminide and a K_i of 14.59 mM for N-acetylglucosamine inhibition. Most metal ions and chemical reagents at final concentrations of 1.0 and 10.0 mM or 0.5 and 1.0% (v/v) had little or no effect (retaining 84.5 – 131.5% activity) on the enzyme activity. The enzyme can retain more than 53.6% activity and good stability in 3.0–20.0% (w/v) NaCl. Compared with most GlcNAcases, the activity of the enzyme is considerably higher and the tolerance to salts and N-acetylglucosamine is much better. Furthermore, the enzyme had higher proportions of aspartic acid, glutamic acid, alanine, glycine, random coils and negatively charged surfaces but lower proportions of cysteine, lysine, α-helices and positively charged surfaces than its homologs. These molecular characteristics were hypothesised as potential factors in the adaptation for salt tolerance and high activity of the GH 20 GlcNAcase.

Conclusions: Biochemical characterization revealed that the GlcNAcase had novel salt–GlcNAc tolerance and high activity. These characteristics suggest that the enzyme has versatile potential in biotechnological applications, such as bioconversion of chitin waste and the processing of marine materials and saline foods. Molecular characterization provided an understanding of the molecular–function relationships for the salt tolerance and high activity of the GH 20 GlcNAcase.

Keywords: β-N-Acetylglucosaminidase, Activity, N-Acetylglucosaminide tolerance, Salt tolerance, *Microbacterium*

* Correspondence: huangzunxi@163.com
†Equal contributors
[1]Engineering Research Center of Sustainable Development and Utilization of Biomass Energy, Ministry of Education, Yunnan Normal University, Kunming 650500, People's Republic of China
[2]College of Life Sciences, Yunnan Normal University, No. 768 Juxian Street, Chenggong, Kunming, Yunnan 650500, People's Republic of China
Full list of author information is available at the end of the article

Background

β-N-acetylglucosaminidases (GlcNAcases, EC 3.2.1.52) are glycoside hydrolases (GHs) that cut the O-glycosidic bonds formed by N-acetylglucosamine (GlcNAc) residues from the non-reducing terminals of oligosaccharides, such as chitooligosaccharides and muropeptides [1, 2]. Chitooligosaccharides can be produced from the degradation of chitin by hydrochloric acid [3] or chitinases (EC 3.2.1.14) [2]. Chitin is second only to lignocellulose in natural abundance, but more than 80,000 tons of chitin from marine sources per year is unutilized chitinous waste [2]. GlcNAcases hydrolyse chitooligosaccharides to produce GlcNAc, which can be further used for the production of bioethanol [4] and single cell protein [5], the treatment of ulcerative colitis and other gastrointestinal inflammation disorders [6] and pharmaceutical therapy for osteoarthritis [7] and tumours [8]. Muropeptides can be derived from the peptidoglycan of the cell wall [1] and flagella [9]. The hydrolysis of muropeptides by GlcNAcases indicates that the enzymes function in cell wall recycling [1] and flagellum assembly of bacteria [9]. Furthermore, GlcNAcases also exhibit many additional important biological functions and wide range of industrial applications, such as catabolism of ganglioside storage in Tay-Sachs disease [10], induction of chitinolytic enzymes [11] and synthesis of biologically important oligosaccharides [12, 13].

According to a homology comparison of amino acid sequences, GlcNAcases have been classified into GH 3, 20, 73 and 84 (http://www.cazy.org/). With the rapid development of genome sequencing technology, many GlcNAcases have been identified in the genomes of animal tissues, insects, plants, bacteria and fungi in recent years. GH 20 GlcNAcases are mainly found in microorganisms, especially bacteria (http://www.cazy.org/GH20.html). However, the molecular and biochemical characteristics of the majority of these GH 20 GlcNAcases have not been reported (http://www.cazy.org/GH20.html).

Yunnan Province is known as "the Kingdom of Nonferrous Metals", "the Kingdom of Plants" and "the Kingdom of Animals" in China. In these natural habitats containing metal ions in high concentrations or special products from plants or animals, novel microorganisms and their enzymes may have to evolve, indicating biotechnological potential. Heijing is called the "town of salt" in Yunnan and has produced salts since ancient China. We previously sampled saline soil from an old mine in the town and reported novel enzymes from bacteria isolated from the soil [14–17]. One example is the GH 32 invertase InvHJ14, showing novel low-temperature and alkaline activity and sucrose tolerance [15]; another is the GH 27 α-galactosidase AgaAHJ8, showing novel salt-protease tolerance and transglycosylation activity, from a potential novel species of *Pontibacter* [16].

In this study, a novel GH 20 GlcNAcase, called HJ5Nag, was mined from the genome of *Microbacterium* sp. HJ5—a new isolate from the saline soil. The purified recombinant HJ5Nag (rHJ5Nag), which was expressed in *E. coli*, showed better tolerance to salts and GlcNAc and higher activity than most GlcNAcases reported in the literature, including the GH 3 GlcNAcase we previously reported [18]. Potential molecular adaptations for salt tolerance and the high activity of the enzyme were presumed.

Methods

Strains, vectors and reagents

Strain HJ5 was isolated from the saline soil of an old salt mine located in Heijing Town. The details of the strain isolation and identification were described in our previous study [14]. The pure culture was deposited in the Strains Collection of the Yunnan Institute of Microbiology under registration no. YMF 4.00007. *E. coli* BL21 (DE3) was purchased from TransGen (Beijing, China) and was used for gene expression.

The following vectors and reagents were used: plasmid isolation and genomic DNA kits (Tiangen, Beijing, China); dNTPs and DNA polymerases (TaKaRa, Otsu, Japan); isopropyl-β-D-1-thiogalactopyranoside (IPTG; Amresco, Solon, OH, USA); Ni^{2+}-NTA agarose (Qiagen, Valencia, CA, USA); Qubit protein assay kit (Invitrogen, Carlsbad, CA, USA); *pEASY*-E2 vector (TransGen); N,N'-diacetyl chitobiose ($GlcNAc_2$), N,N',N'',N'''-tetraacetyl chitotetraose ($GlcNAc_4$), *p*-nitrophenyl β-N-acetylgalactosaminide (*p*NP GalNAc), and *p*-nitrophenyl β-D-glucopyranoside (*p*NPGlc) (J&K Scientific Ltd., Beijing, China); peptidoglycan from *Bacillus subtilis* (Ekear, Shanghai, China); *p*-nitrophenol (*p*NP), *p*-nitrophenyl-β-D-xylopyranoside (*p*NPXyl), *p*-nitrophenyl α-D-galactopyranoside (*p*NPGal), *p*-nitrophenyl-α-L-arabinofuranoside (*p*NPAra), *p*-nitrophenyl β-N-acety lglucosaminide (*p*NPGlcNAc), mutanolysin from *Streptomy ces globisporus* ATCC 21553, chitin and chitosan (Sigma-Aldrich, St. Louis, MO, USA); and silica gel G plates (Haiyang, Qingdao, China). All other chemicals were of analytical grade.

Gene cloning and sequence analysis

The GlcNAcase-encoding gene, designated *hj5Nag*, was obtained by genome sequencing performed on a Miseq sequencer (Illumina) in our lab. The HJ5 genome sequencing details were described in our previous study [19].

Open reading frames from the draft genome of HJ5 were predicted using the tools described in our previous study [17]. BLASTN and BLASTP (http://www.ncbi.nlm.nih.gov/BLAST/) online tools were used to search homologous sequences and to calculate the identity values. SignalP (http://www.cbs.dtu.dk/services/SignalP/) and InterPro (http://www.ebi.ac.uk/interpro/) online tools were used to predict the signal peptide and protein domain, respectively. Vector NTI 10.3 (InforMax, Gaithersburg, MD, USA) software was used to calculate frequencies of the amino acid residues. Multiple sequences and structures alignments were generated by ESPript [20].

Structure analysis

The homology model of HJ5Nag was predicted using the SwissModel platform (http://swissmodel.expasy.org/). The proportions of secondary structures were analysed by VADAR [21]. The charge distributions on the surfaces of various GlcNAcases were calculated using Discovery Studio 2.5 software (Accelrys, San Diego, CA, USA).

Heterologous expression of HJ5Nag

The coding sequence of HJ5Nag was amplified by PCR using the primer set r*hj5Nag*EF (5'-GGGTGCAGCCC CGCCGC-3') and r*hj5Nag*ER (5'-CTCGGTGGCCCAG TCGATCTCGC-3'). The resulting PCR product was ligated to *pEASY*-E2 vector, which has a single 3′-T overhang at the insertion site according to the manufacturer. *E. coli* BL21 (DE3) competent cells were transformed with the plasmid for recombinant enzyme expression. A positive transformant harbouring the recombinant plasmid was confirmed by DNA sequencing performed by Tsingke (Beijing, China). A seed culture of the positive transformant was grown overnight at 37 °C and then inoculated in 1:100 dilutions into fresh Luria–Bertani medium with the addition of 100 μg mL^{-1} ampicillin. Upon reaching an $OD_{600\ nm}$ of approximately 0.7, IPTG at a final concentration of 0.25 mM was added to the culture to induce enzyme expression at 20 °C for approximately 20 h.

Purification and identification of recombinant GlcNAcase

Cultures containing positive transformant cells were centrifuged and resuspended in ice-cold buffer A containing 20 mM Tris–HCl, 0.5 M NaCl, and 10% (w/v) glycerol (pH 7.2). The cells were disrupted by ultrasonication on ice with 100 short bursts of 4 s each at a power output of 150 W. After removing cell debris by centrifugation, the supernatant was loaded onto Ni^{2+}-NTA agarose gel columns to bond the recombinant enzyme. The target recombinant enzyme was eluted with a linear imidazole gradient of 20–500 mM in buffer A. The protein concentration was determined using a Qubit protein assay kit using a Qubit 2.0 fluorometer (Invitrogen).

Sodium dodecyl sulphate–polyacrylamide gel electrophoresis (SDS–PAGE) was performed to analyse the expression of the recombinant enzyme and the purity of the eluted fractions. The molecular masses of internal peptides from the single band present in the SDS–PAGE gel were analysed via matrix-assisted laser desorption/ionization time-of-flight mass spectrometry (MALDI–TOF MS) and were compared with the molecular masses of the internal peptides from rHJ5Nag.

Enzyme assay and substrate specificity

An enzyme assay of purified rHJ5Nag towards various substrates was performed spectrophotometrically using the *p*NP and 3,5-dinitrosalicylic acid (DNS) methods, as described in our previous study [18]. The *p*NP method was applied to determine the activity of rHJ5Nag towards substrates *p*NPXyl, *p*NPGal, *p*NPGlc, *p*NPAra, *p*NPGlcNAc and *p*NPGalNAc. The DNS method was used for the substrates $GlcNAc_2$, $GlcNAc_4$, colloidal chitin, chitosan, peptidoglycan and muropeptides (degradation products of peptidoglycan by mutanolysin treatment).

Biochemical characterization

The *p*NP method using *p*NPGlcNAc as substrate was applied to determine the biochemical characteristics of purified rHJ5Nag, unless otherwise noted. McIlvaine buffer was used for pH 5.0–8.0, and 0.1 M glycine–NaOH was used for pH 9.0–12.0.

The pH-dependent activity of rHJ5Nag was determined at 37 °C, with pH values ranging from 5.0 to 9.0. The pH-dependent stability of rHJ5Nag was determined by measuring the residual enzyme activity after incubating the enzyme at pH 5.0–12.0 for 1 h.

The temperature-dependent activity of rHJ5Nag was examined at pH 6.0 and temperatures from 0 to 60 °C. The thermostability of rHJ5Nag was determined by measuring the residual activity of the enzyme after preincubation at 30, 45 and 50 °C and pH 6.0 for 1–60 min.

To determine the effects of various metal ions and chemical reagents on purified rHJ5Nag, the enzyme activities were measured after the addition of different metal ions and chemical reagents to the reaction mixture. The metal ions and chemical reagents included 1.0 or 10.0 mM (final concentration) KCl, $AgNO_3$, $CaCl_2$, $CoCl_2$, $HgCl_2$, $Pb(CH_3COO)_2$, $MgSO_4$, $NiSO_4$, $CuSO_4$, $MnSO_4$, $ZnSO_4$, $FeCl_3$, EDTA, β-mercaptoethanol and SDS, and 0.5 or 1.0% (v/v) Tween 80 and Triton X-100. The absorption was measured after removing the precipitate by centrifugation when $AgNO_3$, $HgCl_2$ or SDS was added to the reaction mixture. Furthermore, the enzyme activity in 1.0–30.0% (w/v) NaCl was also examined, and the enzyme stability in NaCl was determined by measuring the residual enzyme activity after preincubating the enzyme in 3.0–30.0% (w/v) NaCl for 1 h.

The K_m, V_{max} and k_{cat} values of purified rHJ5Nag were analysed using 0.1–1.2 mM *p*NPGlcNAc as substrate at 45 °C and pH 6.0. A nonlinear regression Michaelis–Menten fit was applied to analyse the data using GraphPad Prism (GraphPad Software, San Diego, CA, USA). The extent of GlcNAc inhibition on rHJ5Nag activity was determined adding 5.0 and 10.0 mM GlcNAc to the reaction solution. Then, the β-N-acetylglucosaminidase activity was determined according to the aforementioned kinetic assay. The inhibition constant (K_i) was calculated using GraphPad Prism using the nonlinear mixed-model inhibition equation described detailedly in a previous paper [22].

Hydrolytic property of purified rHJ5Nag

The hydrolysis mixture included 160 μL of 0.5% (w/v) $GlcNAc_2$ or $GlcNAc_4$ and approximately 0.1 units of purified rHJ5Nag. After incubating the mixture in McIlvaine buffer (pH 6.0) at 37 °C for 6 h, the hydrolysis products were analysed by thin-layer chromatography (TLC), as previously described [17].

Synergistic action

Commercial chitinase from *Streptomyces griseus*, designated CtnSg, was purchased from YuanYe Bio-Technology (Shanghai, China). The reaction mixture contained colloidal chitin (0.5%, w/v) and either CtnSg (0.01 U mL^{-1} reaction system), rHJ5Nag (0.5 U mL^{-1} reaction system) or both enzymes and was incubated for 120 min. In the time-course study, the reaction mixture containing 0.5% (w/v) colloidal chitin and CtnSg alone was incubated for 30, 60 or 90 min, rHJ5Nag was then added to the reaction and incubated for 90, 60 or 30 min, respectively. According to the manufacturer's instructions for CtnSg and the biochemical characteristics of rHJ5Nag, degradation of colloidal chitin was performed at 25 °C and pH 6.0 (McIlvaine buffer). The DNS method was used to measure the amount of reducing sugars released from the colloidal chitin. A reaction containing heat-inactivated enzyme (incubating in a boiling water bath for 10 min) was used as a control. The degree of synergy was calculated using the following equations:

$$DS = \frac{RS_{rHJ5NagCtnSg}}{RS_{rHJ5Nag} + RS_{CtnSg}}$$

where DS is the degree of synergy, $RS_{\mathrm{rHJ5NagCtnSg}}$ is the amount of reducing sugars released from simultaneous or sequential enzyme reactions, and RS_{rHJ5Nag} and RS_{CtnSg} are the amounts of reducing sugars released from individual enzyme reactions.

Accession numbers

The accession numbers of *hj5Nag* and *Microbacterium* sp. HJ5 16S rDNA in GenBank are KX400857 and KX400858, respectively.

Results

Strain identification

Based on the results of a BLASTN search, the nucleotide identity was 98.8% between the partial 16S rDNA sequence from HJ5 (1375 bp) and the 16S rDNA sequences from *Microbacterium flavescens* (accession no. HQ530520), *Microbacterium takaoensis* (AB201047) and *Microbacterium trichothecenolyticum* (KU179351). Phylogenetic analysis also placed HJ5 in the *Microbacterium* cluster but not in the clusters grouped by the species of other genera of *Microbacteriaceae* (Additional file 1: Figure S1). Therefore, HJ5 belonged to the genus *Microbacterium.*

Gene cloning and sequence analysis

Sequence data of approximately 606.6 Mbp was generated by genome sequencing for HJ5, and these data yielded a draft genomic DNA sequence of approximately 3.6 Mbp after sequence assembly. The analysis of the draft genome predicted an open reading frame designated gene *hj5Nag. hj5Nag* has a length of 1608 bp and encodes the 535-residue GlcNAcase HJ5Nag, with a predicted molecular mass of 55.9 kDa.

The alanine (A) and glycine (G) frequencies of the GlcNAcase were 16.3 and 10.8%, respectively, both of which are the highest values among the GlcNAcases shown in Table 1. The frequency of acidic aspartic acid (D) and glutamic acid (E) of the GlcNAcase was 13.5%, which was the third highest (Table 1). However, frequencies of cysteine (C) and lysine (K) in the GlcNAcase were only 0.2 and 1.1%, respectively, both of which were the second lowest (Table 1). Therefore, the ratio of the total frequency of D, E, A and G to the total frequency of C and K of HJ5Nag (31.2) was the highest among the GlcNAcases shown in Table 1.

The predicted signal peptide of HJ5Nag is cleaved between A25 and A26 to yield a mature polypeptide consisting of the typical N-terminal domain of bacterial GlcNAcases (from V14 to R153; domain signature: IPR015882) and the catalytic domain of GH 20 GlcNAcases (from F154 to V510; IPR015883; Fig. 1, Additional file 1: Figure S2). A BLASTP search for HJ5Nag showed that the enzyme was most similar to multiple putative GH 20 GlcNAcases annotated based on genome analysis. Among these putative GH 20 members, the enzyme from *Microbacterium* sp. Root180 (accession no. WP_056119055) shared the highest identity of 78.9% with HJ5Nag. Based on the alignment of HJ5Nag with other experimentally characterized GlcNAcases (Fig. 1) [23–25], residue D334 probably functioned as the catalytic nucleophile/base, and E335 probably functioned as the catalytic proton donor/acceptor. In addition, residues D162, R165, D194, D195, W365, W430 and W465 might play an

Table 1 Amino acid residue frequencies of the experimentally characterized GH 20 GlcNAcases

GlcNAcase (PDB ID or accession no.)	Frequency (%)						Organism
	D & E	A	G	C	K	DEAG/CK	
ADJ68332	14.3	7.7	6.9	0.0	4.7	6.1	*Vibrio harveyi* [71]
AAC44672	13.7	8.2	5.2	1.6	3.8	5.0	*Vibrio furnissii* [37]
KX400857	13.5	16.3	10.8	0.2	1.1	31.2	*Microbacterium* sp. (This study)
ADJ68333	13.5	7.6	5.6	1.9	4.5	4.2	*V. harveyi* [71]
3RCN (WP_014923268)	13.4	11.2	10.1	0.2	1.7	18.3	*Arthrobacter aurescens* (Unpublished)
3NSM (ABI81756)	13.3	7.5	5.9	2.0	6.1	3.3	*Ostrinia furnacalis* [72]
ABA27426	13.0	4.5	6.5	1.3	5.1	3.8	*Spodoptera frugiperda* [70]
ABA27427	13.0	4.5	6.5	1.3	5.1	3.8	*S. frugiperda* [70]
4PYS	12.9	6.1	7.5	1.6	6.1	3.4	*Bacteroides fragilis* (Unpublished)
BAD00143	12.9	9.5	7.6	0.7	7.1	3.8	*Aeromonas hydrophila* [73]
CAH55822	12.9	9.7	7.6	0.7	7.1	3.9	*Aeromonas caviae* [48]
1QBA (AAB03808)	12.8	9.9	8.4	0.7	6.9	4.1	*Serratia marcescens* [23]
BAC76622	12.8	14.4	8.7	0.5	0.7	29.9	*Streptomyces thermoviolaceus* [42]
AAQ05800	12.7	12.1	8.9	0.0	1.2	28.1	*Cellulomonas fimi* [45]
1 M01 (AAC38798)	12.5	11.9	10.5	0.4	3.6	8.7	*Streptomyces plicatus* [74]
BAC41255	12.5	8.0	5.7	1.2	3.0	6.2	*Aspergillus oryzae* [12]
AKC34129	12.3	9.7	7.1	0.4	7.1	3.9	Uncultured organism [44]
BAA92145	12.2	10.1	9.7	0.7	4.3	6.4	*Aeromonas* sp. [75]
2GJX (AAD13932)	11.9	4.9	6.4	1.5	3.6	4.5	*Homo sapiens* [24]
BAF76001	11.8	9.1	8.3	1.8	2.4	7.0	*A. hydrophila* [76]
CAD10500	11.5	6.9	5.7	1.8	6.2	3.0	*Entamoeba histolytica* [77]
BAM42836	11.2	11.2	6.5	1.1	1.6	10.7	*Lentinula edodes* [78]
AKC34128	10.7	14.4	9.1	0.4	1.6	17.1	Uncultured organism [44]
3GH4 (BAI63641)	10.5	8.0	8.8	0.4	5.1	5.0	*Paenibacillus* sp. [25]
BAA88762	10.4	12.9	10.0	0.4	2.4	11.9	*S. thermoviolaceus* [46]
BAA06136	10.4	10.0	5.3	1.2	6.0	3.6	*Pseudoalteromonas piscicida* (previous *Alteromonas* sp.) [79]
BAB84321	10.4	8.3	5.1	0.4	4.9	4.5	*P. piscicida* (previous *Alteromonas* sp.) [80]
CAE46968	10.3	4.3	5.1	1.4	6.4	2.5	*E. histolytica* [77]
BAA76352	10.1	15.2	5.5	0.0	7.2	4.3	*Lactobacillus casei* [41]
EGR50812	8.9	10.0	6.4	1.0	3.4	5.8	*Trichoderma reesei* [49]

important role in binding the GlcNAcase ligand (Fig. 1) [23–25].

Structure analysis

The model of HJ5Nag showed a GMQE score of 0.72 and a sequence identity of 48.9% when the template was the crystal structure of the GH 20 GlcNAcase from *Streptomyces plicatus* (PDB ID: 1 M01). The homology model of HJ5Nag and the crystal structures of 7 GlcNAcases were used to calculate the proportions of secondary structures and the charge distributions on the surfaces. Among these GlcNAcases, the proportions of amino acids used to build the α-helix and β-sheet structures of HJ5Nag were the second lowest, while the proportion of amino acids used to build random coil structures and the ratio of random coils to α-helices of HJ5Nag were both the highest (Table 2). As shown in Fig. 2, almost the whole surface of HJ5Nag was negatively charged. Compared with other GlcNAcases, HJ5Nag had a much larger negatively charged surface (Fig. 2).

Expression, purification and identification of rHJ5Nag

As shown by an SDS–PAGE gel (Additional file 1: Figure S3), the recombinant GlcNAcase was successfully expressed in *E. coli* BL21 (DE3) and purified by Ni^{2+}-NTA affinity chromatography. The MALDI–TOF MS spectrum of the purified band further confirmed

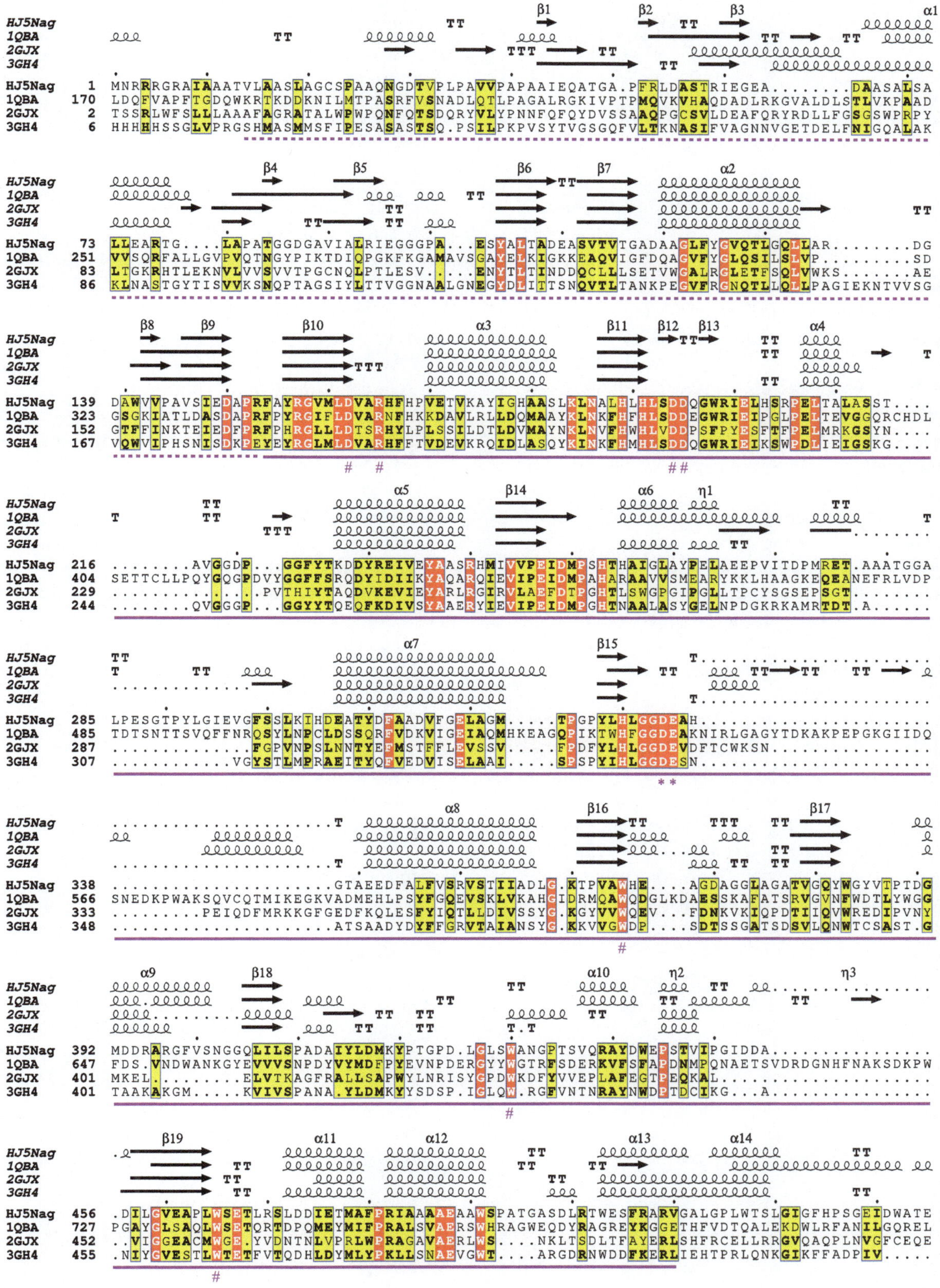

Fig. 1 (See legend on next page.)

(See figure on previous page.)

Fig. 1 Partial amino acid sequences and structures alignments of HJ5Nag with GH 20 GlcNAcases. Sequence names are shown with PDB IDs (except HJ5Nag) as follows: 1QBA (accession no. AAB03808), the GlcNAcase from *S. marcescens* [23]; 2GJX (accession no. AAD13932), the GlcNAcase from *H. sapiens* [24]; and 3GH4 (accession no. BAI63641), the GlcNAcase from *Paenibacillus* sp. [25]. *Asterisks* and *number signs* show the putative catalytic residues and ligand-binding residues, respectively. *Dotted purple line* shows the N-terminal domain of bacterial GlcNAcases (domain signature: IPR015882). *Solid purple line* shows the catalytic domain of GH 20 GlcNAcases (domain signature: IPR015883)

that the purified product was rHJ5Nag (Additional file 1: Figure S4).

Biochemical characterization

At pH 6.0 and 45 °C, purified rHJ5Nag showed specific activities of 1773.1 ± 1.1 (Table 3), 146.0 ± 3.1, 481.4 ± 2.3 and 445.0 ± 1.7 μmol min^{-1} mg^{-1} towards substrates of *p*NPGlcNAc, *p*NPGalNAc, $GlcNAc_2$ and $GlcNAc_4$, respectively. However, no rHJ5Nag activity was detected towards substrates of *p*NPXyl, *p*NPAra, *p*NPGlc, *p*NPGal, peptidoglycan, muropeptides, colloidal chitin or chitosan. Thus, rHJ5Nag is a GlcNAcase that can hydrolyse chitooligosaccharides but cannot participate in cell wall turnover or recycling of bacteria.

Purified rHJ5Nag had activity in the range of pH 5.5–9.0 and 10 – 50 °C, with apparent optima at pH 6.0 and 45 °C (Fig. 3a, c). The enzyme exhibited good stability at 30 °C and maintained more than 55% of its initial activity after incubation in buffers ranging from pH 6.0 to 10.0 for 1 h (Fig. 3b, d).

Purified rHJ5Nag exhibited good salt tolerance. Although the activity of purified rHJ5Nag was completely inhibited by $AgNO_3$, $HgCl_2$ and SDS, the addition of most metal ions and chemical reagents at final concentrations of 1.0 and 10.0 mM or 0.5 and 1.0% (v/v) showed little to no effect (retaining 84.5 – 131.5% activity) on the enzyme activity (Table 4). Furthermore, the enzyme retained more than 80% activity in the presence of 1.0–10.0% (w/v) NaCl and even 53.6% activity in 20.0% (w/v) NaCl (Fig. 3e). The enzyme retained approximately 100% of its initial activity after incubation in 3.0–20.0% (w/v) NaCl for 60 min (Fig. 3f).

Without the addition of GlcNAc, the K_m, V_{max}, k_{cat} and k_{cat}/K_m of purified rHJ5Nag towards *p*NPGlcNAc were 0.52 ± 0.05 mM, 3097 ± 124 μmol min^{-1} mg^{-1}, 2886 ± 116 s^{-1} and 5550 ± 223 mM^{-1} s^{-1}, respectively (Table 3; Fig. 4). When 5.0 and 10.0 mM GlcNAc was added to the reaction solution, the V_{max} of purified rHJ5Nag towards *p*NPGlcNAc decreased to 2732 ± 91 and 2606 ± 87 μmol min^{-1} mg^{-1}, respectively, the k_{cat} decreased to 2546 ± 84 and 2429 ± 81 s^{-1}, respectively, and the k_{cat}/K_m decreased to 4106 ± 135 and 3327 ± 111 mM^{-1} s^{-1}, respectively (Fig. 4). However, the K_m of purified rHJ5Nag towards *p*NPGlcNAc increased to 0.62 ± 0.04 and 0.73 ± 0.05 mM in the presence of 5.0 and 10.0 mM GlcNAc, respectively (Fig. 4). Furthermore, the value of the parameter Alpha in the mixed-model inhibition equation was 3.412 based on the analysis of kinetic data using GraphPad Prism. These results indicate that GlcNAc can inhibit the activity of rHJ5Nag via a mixture of competitive and noncompetitive inhibition. As a result, the enzyme was inhibited by GlcNAc, with a K_i of 14.59 mM.

Hydrolytic properties of purified rHJ5Nag

As shown in Fig. 5, GlcNAc was the end product of $GlcNAc_2$ and $GlcNAc_4$ hydrolysed by purified rHJ5Nag.

Synergistic action

Although purified rHJ5Nag cannot hydrolyse colloidal chitin alone, simultaneous addition of CtnSg and rHJ5Nag improved the degradation of colloidal chitin by 2.02-fold (Table 5). Sequential addition of the two enzymes improved the degradation of colloidal chitin

Table 2 Proportions of secondary structures of GH 20 GlcNAcases

GlcNAcase (PDB ID)	α-Helix	β-Sheet	Coil	Coil/α-Helix
2GJX [24]	176 (35.8%)	133 (27.0%)	183 (37.2%)	1.04
3RCN (Unpublished)	185 (35.2%)	146 (27.8%)	194 (37.0%)	1.05
1 M01 [74]	170 (34.1%)	121 (24.2%)	208 (41.7%)	1.22
3NSM [72]	193 (33.7%)	162 (28.3%)	217 (37.9%)	1.12
4PYS (Unpublished)	165 (33.5%)	134 (27.2%)	193 (39.2%)	1.17
3GH4 [25]	165 (32.5%)	137 (27.0%)	205 (40.4%)	1.24
KX400857 (This study)	152 (30.2%)	126 (25.0%)	226 (44.8%)	1.49
1QBA [23]	232 (27.0%)	282 (32.9%)	344 (40.1%)	1.48

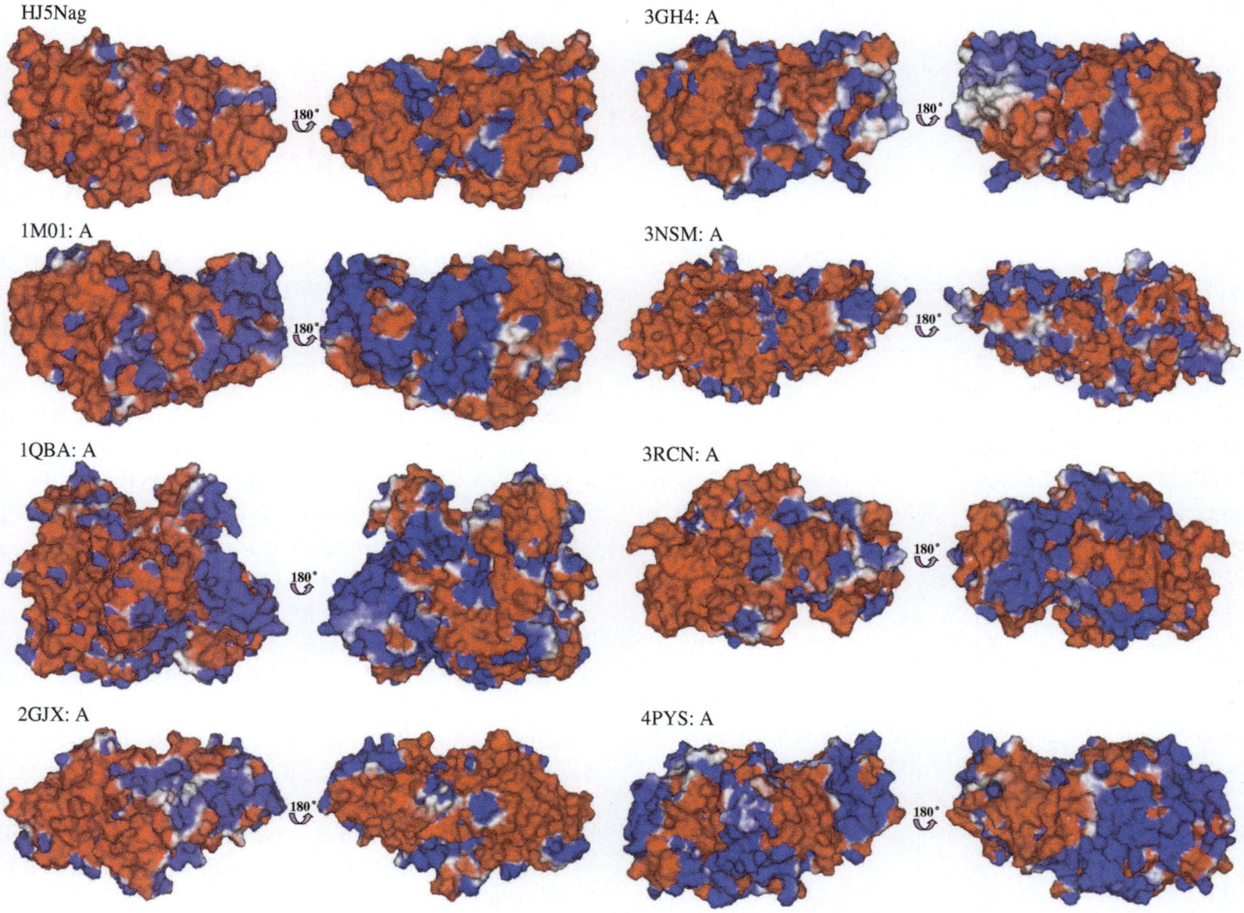

Fig. 2 Charge distributions on the surfaces of the GH 20 GlcNAcases. The A-chain of GlcNAcase is used. PDB IDs are presented as the names of GlcNAcases (except HJ5Nag) as follows: 1 M01, the GlcNAcase from *S. plicatus* [74]; 1QBA, the GlcNAcase from *S. marcescens* [23]; 2GJX, the GlcNAcase from *H. sapiens* [24]; 3GH4, the GlcNAcase from *Paenibacillus* sp. [25]; 3NSM, the GlcNAcase from *O. furnacalis* [72]; 3RCN, the GlcNAcase from *A. aurescens*; 4PYS, the GlcNAcase from *B. fragilis*. The charge distribution on the surface was calculated at pH 7.0. Positive charges are depicted in *blue* and negative charges in *red*

by 1.98-fold, 1.72-fold and 1.29-fold when rHJ5Nag was added and incubated for 90, 60 and 30 min, respectively (Table 5). Furthermore, these results indicate that the synergistic effect benefits from the earlier addition of rHJ5Nag.

Discussion

Microorganisms isolated from extremophilic locations are of great interest and have attracted substantial attention. *Microbacterium* strains have been found in various extremophilic locations, including Antarctic habitats [26], the Atacama Desert [27], sediment of the Indian Ocean [28], uranium-rich soil [29] and salty soil [30]. To be active in these extremophilic locations, enzymes from local *Microbacterium* strains may have to evolve. To the best of our knowledge, although some pioneering studies have revealed the important functions of some enzymes from *Microbacterium* strains in recent years [31–36], few of these enzymes were found to have adapted to extremophilic locations [32]. Furthermore, we found that all the GlcNAcases from *Microbacterium* strains were annotated based on *in silico* analysis without biochemical characterization. Therefore, the study may be the first to report an experimentally characterized GlcNAcase from *Microbacterium*.

Drastic inhibition by Ag^{+} and Hg^{2+} appears to be the common property of GlcNAcases, including HJ5Nag, the GH 20 GlcNAcases from *Paenibacillus* [25] and *V. furnissii* [37], the GH 3 GlcNAcases from *R. miehei* [38] and *Sphingobacterium* sp. [18], the GH 84 GlcNAcase from *C. paraputrificum* [39] and the GlcNAcase from *S. cerradoensis* [40]. Ag^{+} and Hg^{2+} have irreversible effects on sulfhydryl groups, and Hg^{2+} can also oxidize indole rings. Therefore, cysteine or tryptophan residues may be involved in the catalytic activity of GH 20 GlcNAcases. Sequence analysis revealed that W365, W430 and W465

Table 3 Activities of the experimentally characterized GlcNAcases

GlcNAcase	GH Family	Specific activity[a]	V^{a}_{max}	Organism
-	84	1.2	212	*Penicillium chrysogenum* [81]
NagA	20	1926.0	-	*A. oryzae* [12]
HJ5Nag	20	1773.1	3097	*Microbacterium* sp. (This study)
-	20	913.6	-	*P. piscicida* (previous *Alteromonas* sp.) [79]
Nag1	20	319.9	-	*T. reesei* [49]
ExoI	20	200	270	*V. furnissii* [37]
-	20	73.3	-	*L. casei* [41]
Sfhex	20		68.4	*S. frugiperda* [70]
NagC	20	44.7	-	*S. thermoviolaceus* [42]
NAG20A	20	30.0	115	*A. hydrophila* [73]
*Vh*Nag2	20	19.4	-	*V. harveyi* [82]
-	20	3.8	3.8	*E. histolytica* [77]
Hex1	20	-	212	*Paenibacillus* sp. [25]
Hex2	20	-	150	*Paenibacillus* sp. [25]
NagA	3	64.4	24.8	*S. thermoviolaceus* [83]
RmNag	3	21.2	49.3	*Rhizomucor miehei* [38]
Nag3HWLB1	3	19.4	19.9	*Sphingobacterium* sp. [18]
NahA	3	8.8	-	*Symbiobacterium thermophilum* [47]
ExoII	3	0.7	1.1	*V. furnissii* [68]
NagA	3	-	0.3	*Thermotoga maritima* [84]
CbsA	3	-	0.3	*Thermotoga neapolitana* [84]
-	-	521.5	-	*Enterobacter* sp. [85]
-	-	67.2	-	*Vibrio alginolyticus* [65]
-	-	40.5	-	*Talaromyces emersonii* [86]
-	-	27	2.0	*Streptomyces cerradoensis* [40]
NagA	-	5.3	245	*Pseudomonas fluorescens* [3]
	-	0.03	-	*Trichinella spiralis* [64]

[a]These values (μmol min^{-1} mg^{-1}) were determined using *p*NPGlcNAc as substrate

of HJ5Nag might affect ligand binding. As shown in Fig. 1, the conserved W491 may also be essential for the activity of HJ5Nag. SDS is an inhibitor of GlcNAcases, in addition to Ag^{+} and Hg^{2+}. All the GlcNAcases from *Microbacterium* sp. in this study, as well as *R. miehei* [38], *Sphingobacterium* sp. [18] and *L. casei* [41], are strongly inhibited by SDS. Purified rHJ5Nag showed peak activity at pH 6.0 and 45 °C. These properties are the same or similar to those of many GlcNAcases. For example, the GH 20 GlcNAcase from *S. thermoviolaceus* presents maximum activity at pH 6.0 and 50 °C [42], the GH 20 GlcNAcase from *H. latex* presents maximum activity at pH 6.0 and 45–50 °C [43], the GH 3 GlcNAcase from *R. miehei* presents maximum activity at pH 6.5 and 50 °C [38], and the GH 84 GlcNAcase from *Clostridium paraputrificum* presents maximum activity at pH 6.5 and 50 °C [39]. Furthermore, the molecular mass of HJ5Nag (55.9 kDa) is similar to that of some GlcNAcases, such as the GH 20 GlcNAcases from a soil-derived uncultured organism (53.4 kDa) [44], *C. fimi* (54.2 kDa) [45], *Paenibacillus* sp. (57.5 kDa) [25] and *S. thermoviolaceus* (58.3 kDa) [46], and the GH 3 GlcNAcase from *S. thermophilum* (55.9 kDa) [47].

Purified rHJ5Nag exhibited good salt tolerance, as the addition of most salts had little to no effect on the enzyme activity, especially $CuSO_4$ and NaCl. Most GlcNAcases reported to date are strongly inhibited by Cu^{2+}, such as the GH 20 GlcNAcases from *Paenibacillus* [25], *A. caviae* [48] and *L. casei* [41], the GH 3 GlcNAcase from *R. miehei* [38], the GH 84 GlcNAcase from *C. paraputrificum* [39] and the GlcNAcase from *S. cerradoensis* [40]. Furthermore, the GH 20 GlcNAcases from *Paenibacillus* are strongly inhibited by Ni^{2+} [25]; the GH20 GlcNAcase from *A. caviae* is strongly inhibited by Mn^{2+} [48]; the GH20 GlcNAcase from *T. reesei* is inhibited by Fe^{3+} [49]; the GH 84 GlcNAcase from *C. paraputrificum* is inhibited by Ni^{2+}, Co^{2+}, Mn^{2+}, Zn^{2+} and Fe^{3+} [39]; and the GH 3 GlcNAcase from *R. miehei* is inhibited by Ni^{2+}, Co^{2+}, Zn^{2+} and EDTA [38]. Salt tolerance has attracted considerable attention because various salts are present or used in industries [14, 16, 18, 50–60], such as various marine materials and foods in China, with NaCl contents of 3.5–20.0% (w/w) [16, 18]. In addition, material processing and fermentation with a high concentration of NaCl is helpful to reduce the overall cost because sterilization is unnecessary [61].

To reveal the potential catalytic adaptations of rHJ5Nag to high concentrations of salt, sequence and structure comparisons to its homologs were performed. (I). Similar to the results from other salt-tolerant enzymes [50, 51, 53–59], statistical analysis of the amino acid residues showed that HJ5Nag presented higher proportions of acidic D and E along with A and G but lower proportions of C and K than its homologs (Table 1). Carboxyl groups from acidic amino acid residues can bind more water molecules and cations than other groups from other amino acid residues, especially K, and thus enable the protein to form a particle stabilized by cooperative hydrated ion networks [57, 58]. Hydrophobicity must be finely balanced to maintain a protein's flexibility and stability [62]. High NaCl concentrations

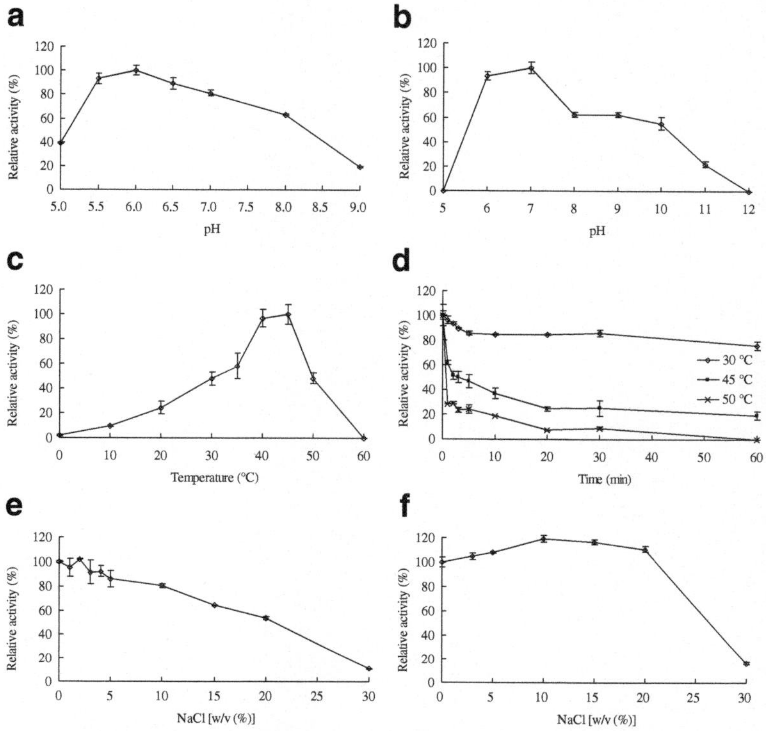

Fig. 3 Enzymatic properties of purified rHJ5Nag. **a** pH-dependent activity profiles. **b** pH-dependent stability profiles. **c** Temperature-dependent activity profiles. **d** Thermostability assay. **e** Effect of NaCl on rHJ5Nag activity. **f** Stability of rHJ5Nag in NaCl. *Error bars* represent the means ± SD ($n = 3$)

strengthen hydrophobic interactions, which can affect protein folding [62]. Salt-tolerant proteins are thought to reduce the size of hydrophobic residues to compensate for the increased hydrophobic effect caused by the high salt concentration [58]. The high proportion of A and G, which are small hydrophobic residues [57], is compatible with that function. In addition, a low abundance of C might give a protein more flexibility in high salt environments due to the decrease in rigid disulphide bridges [56]. (II). At the level of the secondary structure, a higher percentage of random coil structures at the expense of α-helices is observed for salt-tolerant proteins [56, 58], including HJ5Nag (Table 2). Greater protein rigidity is usually accompanied by a higher percentage of α-helix structures, while the opposite is observed for a higher percentage of random coils. (III). At the level of the tertiary structure, an excess of negatively charged surfaces is a typical property of salt-tolerant proteins

Table 4 Effects of metal ions and chemical reagents on the activity of purified rHJ5Nag

Reagent	Relative activity (%)[a]		Reagent	Relative activity (%)[a]	
	1.0 mM	10.0 mM		1.0 mM	10.0 mM
None	100.0 ± 0.4	100.0 ± 2.7	$MnSO_4$	100.9 ± 2.6	99.6 ± 2.4
$FeCl_3$	117.4 ± 1.8	103.1 ± 0.0	$ZnSO_4$	96.8 ± 3.1	84.5 ± 9.9
$MgSO_4$	113.8 ± 5.8	99.3 ± 3.4	$AgNO_3$	0.0	0.0
$CaCl_2$	108.7 ± 5.4	96.6 ± 3.5	$HgCl_2$	0.0	0.0
KCl	105.4 ± 4.8	101.4 ± 1.3	Tween 80	131.5 ± 1.7[b]	108.1 ± 2.1[c]
$CoCl_2$	103.7 ± 3.1	100.4 ± 1.7	EDTA	106.7 ± 4.5	99.5 ± 2.1
$NiSO_4$	101.7 ± 8.7	95.0 ± 0.7	β-Mercaptoethanol	106.1 ± 1.2	100.1 ± 0.6
$CuSO_4$	101.4 ± 4.8	88.6 ± 5.5	Triton X-100	100.5 ± 4.7[b]	93.9 ± 5.2[c]
$Pb(CH_3COO)_2$	101.0 ± 5.8	106.7 ± 3.7	SDS	0.0	0.0

[a]Values represent the means ± SD ($n = 3$) relative to the untreated control sample
[b]Final concentration: 0.5% (w/v)
[c]Final concentration: 1.0% (w/v)

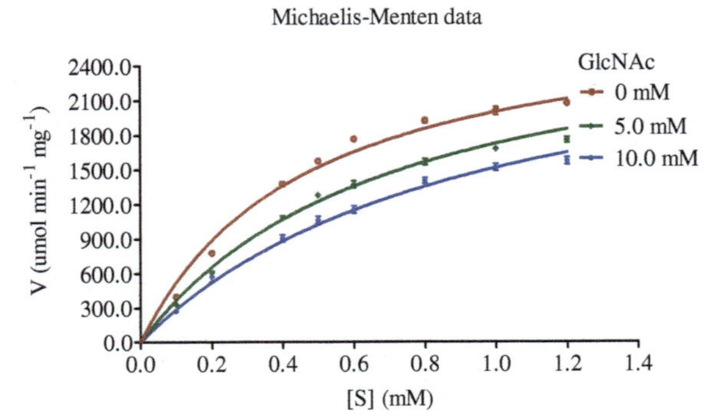

rHJ5Nag	GlcNAc (mM)		
	0	5.0	10.0
K_m (mM)	0.52 ± 0.05	0.62 ± 0.04	0.73 ± 0.05
V_{max} (μmol min^{-1} mg^{-1})	3097 ± 124	2732 ± 91	2606 ± 87
k_{cat} (s^{-1})	2886 ± 116	2546 ± 84	2429 ± 81
k_{cat}/K_m (mM^{-1} s^{-1})	5550 ± 223	4106 ± 135	3327 ± 111
K_i (mM)	14.59		

Fig. 4 Kinetic characterization of purified rHJ5Nag. *Error bars* represent the means ± SD ($n = 3$)

[50, 51, 53, 56–59, 62, 63]. The increase in negative charge on the surface can counteract the lower dielectric constant at high salinity and improve the ability of enzymes to bind water and salt ions, thus improving protein solubility [56–58].

Most characterized GlcNAcases, including GH 3, 20 and 84 members, show specific activities or V_{max} of lower than 1000 μmol min^{-1} mg^{-1} towards *p*NPGlcNAc, as summarized in Table 3. In this study, rHJ5Nag exhibited a specificity activity of 1773.1 ± 1.1 μmol min^{-1} mg^{-1} and a V_{max} of 3097 ± 124 μmol min^{-1} mg^{-1} towards *p*NPGlcNAc. Furthermore, rHJ5Nag showed a specificity activity of 481.4 ± 2.3 μmol min^{-1} mg^{-1} towards $GlcNAc_2$, which is much higher than the values from other GlcNAcases recorded in the literature, such as the values from *Sphingobacterium* sp. (0.3 μmol min^{-1} mg^{-1}) [18], *T. spiralis* (0.79 nmol min^{-1} mg^{-1}) [64], *R. miehei* (1.06 μmol min^{-1} mg^{-1}) [38], *V. alginolyticus* (18.1 μmol min^{-1} mg^{-1}) [65] and *S. thermophilum* (76.8 μmol min^{-1} mg^{-1}) [47]. As discussed above, the structure of HJ5Nag may be more flexible because it has higher proportions of A, G and random coil structures but lower proportions of C and α-helix structures than its homologs. Furthermore, HJ5Nag has a higher proportion of D and E, which are not only negatively charged but also flexible residues. High B-factor values, which are believed to provide high structural flexibility, are usually observed for D and E. The high flexibility is presumed to be a factor in the high activity of the GH 20 GlcNAcase rHJ5Nag.

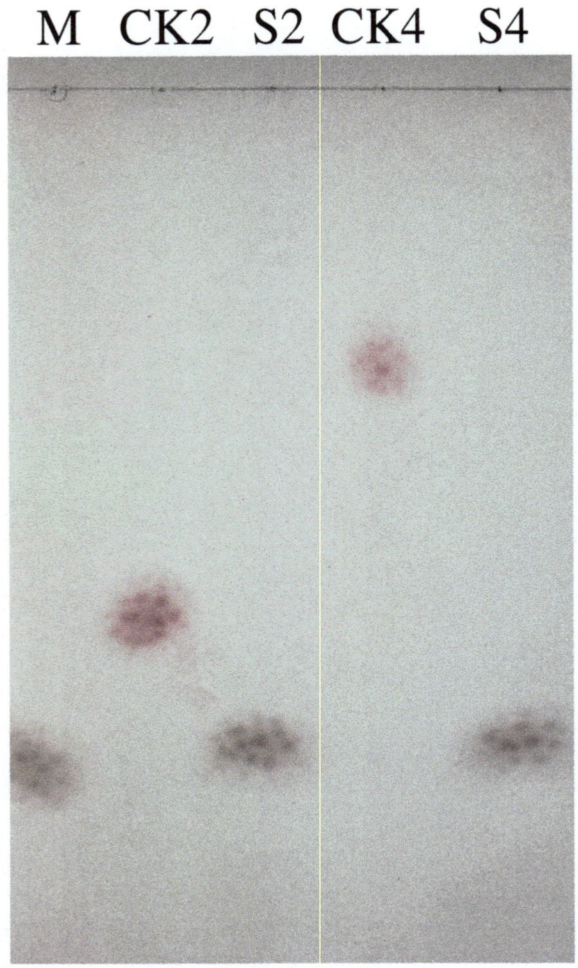

Fig. 5 TLC analysis of the hydrolytic product. Lanes: M, 0.5% (w/v) GlcNAc; CK2 and CK4, inactivated rHJ5Nag (incubating in a boiling water bath for 10 min) with 0.5% (w/v) $GlcNAc_2$ and $GlcNAc_4$, respectively; S2 and S4, hydrolysis of 0.5% (w/v) $GlcNAc_2$ and $GlcNAc_4$ by purified rHJ5Nag, respectively

Although the synergistic action of chitinases and GlcNAcases for complete enzymatic degradation of chitin is attractive, there are few reports [18, 66, 67]. This study may be the first to report a synergistic effect between a chitinase and a GH 20 GlcNAcase. Significant synergy was observed between the commercial CtnSg and rHJ5Nag for chitin degradation, indicating that rHJ5Nag has great potential in the bioconversion of chitin waste. The tolerance of rHJ5Nag to GlcNAc may be one contributor to the significant synergy because GlcNAc can inhibit GlcNAcases and result in the accumulation of chitooligosaccharides, which can inhibit chitinases [38]. The GlcNAcases from *V. furnissii*, *Trichoderma harzianum*, *A. oryzae*, *S. frugiperda*, *R. miehei* and *T. spiralis* were competitively or noncompetitively inhibited by GlcNAc, with K_i values of 0.21 [68],

Table 5 Synergy between the commercial chitinase CtnSg and rHJ5Nag for the degradation of colloidal chitin

Order of enzyme addition and reaction time				Reducing sugars (μmol)	Degree of synergy[a]
First enzyme	Time (min)	Second enzyme	Time (min)		
CtnSg	120	No	0	0.28	1.00
rHJ5Nag	120	No	0	ND[b]	-
CtnSg & rHJ5Nag	120	No	0	0.57	2.02
CtnSg	30	rHJ5Nag	90	0.55	1.98
CtnSg	60	rHJ5Nag	60	0.48	1.72
CtnSg	90	rHJ5Nag	30	0.36	1.29

[a]The degree of synergy was calculated using the following equations:
$DS = \frac{RS_{rHJ5NagCtnSg}}{RS_{rHJ5Nag}+RS_{CtnSg}}$
where DS is the degree of synergy, $RS_{\mathrm{rHJ5NagCtnSg}}$ is the amount of reducing sugars released from simultaneous or sequential enzyme reactions, and RS_{rHJ5Nag} and RS_{CtnSg} are the amounts of reducing sugars released from individual enzyme reactions
[b]Not detected

0.21 [69], 1.6 [13], 2.0 [70], 9.68 [38] and 15.75 mM [64], respectively. In this study, rHJ5Nag showed relatively high GlcNAc tolerance, with a K_i of 14.59 mM.

Conclusions

A GH 20 GlcNAcase was isolated from *Microbacterium* sp. HJ5 harboured in the saline soil of an abandoned salt mine. Biochemical characterization revealed that the GlcNAcase had novel salt–GlcNAc tolerance and high activity. These characteristics suggest that the enzyme has versatile potential in biotechnological applications, such as bioconversion of chitin waste and processing of marine materials and saline foods. Molecular characterization provides an understanding of the molecular–function relationships for the salt tolerance and the high activity of the GH 20 GlcNAcase.

Additional file

Additional file 1: Figure S1. The phylogenetic tree constructed on the basis of the 16S rDNA sequences of various genera of *Microbacteriaceae*. **Figure S2.** Structures of HJ5Nag and the GlcNAcase from *H. sapiens*. **Figure S3.** SDS–PAGE analysis of rHJ5Nag. **Figure S4.** MALDI–TOF MS spectrum of the purified protein. *hj5Nag* (KX400857). HJ5Nag translated from *hj5Nag*. *Microbacterium* sp. HJ5 16S rDNA (KX400858).

Abbreviations

A: Alanine; Buffer A: 20 mM Tris–HCl, 0.5 M NaCl, 10% (w/v) glycerol, pH 7.2; C: Cysteine; D: Aspartic acid; DNS: 3,5-dinitrosalicylic acid; *DS*: Degree of synergy; E: Glutamic acid; G: Glycine; GH: Glycoside hydrolase; GlcNAc: N-acetylglucosamine; $GlcNAc_2$: N,N'-diacetyl chitobiose; $GlcNAc_4$: N,N',N'',N'''-tetraacetyl chitotetraose; GlcNAcases: β-N-acetylglucosaminidases; HJ5Nag: The GlcNAcase from *Microbacterium* sp. HJ5; *hj5Nag*: The GlcNAcase-encoding gene from *Microbacterium* sp. HJ5; IPTG: Isopropyl-β-D-1-thiogalactopyranoside; K: Lysine; MALDI-TOF MS: Matrix-assisted laser desorption/ionization time-of-flight mass spectrometry; *p*NP: *p*-nitrophenol; *p*NPAra: *p*-nitrophenyl-α-L-arabinofuranoside; *p*NPGal: *p*-nitrophenyl α-D-galactopyranoside; *p*NPGalNAc: *p*-nitrophenyl β-N-acetylgalactosaminide; *p*NPGlc: *p*-nitrophenyl β-D-glucopyranoside; *p*NPGlcNAc: *p*-nitrophenyl β-N-acetylglucosaminide; *p*NPXyl: *p*-nitrophenyl-β-D-xylopyranoside; rHJ5Nag: The recombinant HJ5Nag; RS_{CtnSg}: The amount of reducing sugars released by CtnSg from colloidal chitin without the addition of rHJ5Nag; RS_{rHJ5Nag}: The amount of reducing sugars released by rHJ5Nag from colloidal chitin without the addition of CtnSg; $RS_{\mathrm{rHJ5NagCtnSg}}$: The amount of reducing sugars released from simultaneous or sequential enzyme reactions when determining the synergistic action; SDS-PAGE: Sodium dodecyl sulphate-polyacrylamide gel electrophoresis; TLC: Thin-layer chromatography

Acknowledgements

Not applicable.

Funding

This work was supported by the National Natural Science Foundation of China (Grant no. 31260215, 31660445), the Yunling Scholars (Grant no. 2015 56), the Yunling Industry Leading Talents (Grant no. 2014 1782), the Reserve Talents Project for Young and Middle-Aged Academic and Technical Leaders of Yunnan Province (Grant no. 2015HB033) and the Applied and Basic Research Foundation of Yunnan Province (Grant no. 201401PC00224).

Authors' contributions

JZ conceived of the idea, designed the experiments, performed data analysis and wrote the manuscript. ZS and RL performed the experiments. RZ helped perform data analysis and write the manuscript. QW, JL, XT, BX, JD and NH coordinated the study. ZH conceived of the idea, designed the experiments and coordinated the study. All authors read and approved the final manuscript.

Competing interests

The authors declare that they have no competing interests.

Author details

[1]Engineering Research Center of Sustainable Development and Utilization of Biomass Energy, Ministry of Education, Yunnan Normal University, Kunming 650500, People's Republic of China. [2]College of Life Sciences, Yunnan Normal University, No. 768 Juxian Street, Chenggong, Kunming, Yunnan 650500, People's Republic of China. [3]Key Laboratory of Yunnan for Biomass Energy and Biotechnology of Environment, Yunnan, Kunming 650500, People's

Republic of China. [4]Key Laboratory of Enzyme Engineering, Yunnan Normal University, Kunming 650500, People's Republic of China.

References

1. Litzinger S, Duckworth A, Nitzsche K, Risinger C, Wittmann V, Mayer C. Muropeptide rescue in *Bacillus subtilis* involves sequential hydrolysis by β-N-acetylglucosaminidase and N-acetylmuramyl-L-alanine amidase. J Bacteriol. 2010;192(12):3132–43.
2. Patil RS, Ghormade V, Deshpande MV. Chitinolytic enzymes: an exploration. Enzyme Microb Tech. 2000;26(7):473–83.
3. Park JK, Kim WJ, Park YI. Purification and characterization of an exo-type β-N-acetylglucosaminidase from *Pseudomonas fluorescens* JK-0412. J Appl Microbiol. 2011;110(1):277–86.
4. Cody RM, Davis ND, Lin J, Shaw D. Screening microorganisms for chitin hydrolysis and production of ethanol from amino sugars. Biomass. 1990;21(4):285–95.
5. Vyas P, Deshpande M. Enzymatic hydrolysis of chitin by *Myrothecium verrucaria* chitinase complex and its utilization to produce SCP. J Gen Appl Microbiol. 1991;37(3):267–75.
6. Salvatore S, Heuschkel R, Tomlin S, Davies SE, Edwards S, Walker-Smith JA, et al. A pilot study of N-acetyl glucosamine, a nutritional substrate for glycosaminoglycan synthesis, in paediatric chronic inflammatory bowel disease. Aliment Pharm Ther. 2000;14(12):1567–79.
7. Shikhman AR, Amiel D, D'Lima D, Hwang SB, Hu C, Xu A, et al. Chondroprotective activity of N-acetylglucosamine in rabbits with experimental osteoarthritis. Ann Rheum Dis. 2005;64(1):89–94.
8. Suzuki K, Mikami T, Okawa Y, Tokoro A, Suzuki S, Suzuki M. Antitumor effect of hexa-N-acetylchitohexaose and chitohexaose. Carbohyd Res. 1986;151:403–8.
9. Herlihey FA, Moynihan PJ, Clarke AJ. The essential protein for bacterial flagella formation FlgJ functions as a β-N-acetylglucosaminidase. J Biol Chem. 2014;289(45):31029–42.
10. Okada S, Obrien JS. Tay–Sachs disease: generalized absence of a β-D-N-acetylhexosaminidase component. Science. 1969;165(3894):698–700.
11. Li HZ, Morimoto K, Katagiri N, Kimura T, Sakka K, Lun S, et al. A novel β-N-acetylglucosaminidase of *Clostridium paraputrificum* M-21 with high activity on chitobiose. Appl Microbiol Biot. 2002;60(4):420–7.
12. Matsuo I, Kim S, Yamamoto Y, Ajisaka K, Maruyama J, Nakajima H, et al. Cloning and overexpression of β-N-acetylglucosaminidase encoding gene *nagA* from *Aspergillus oryzae* and enzyme-catalyzed synthesis of human milk oligosaccharide. Biosci Biotech Bioch. 2003;67(3):646–50.
13. Rajnochova E, Dvorakova J, Hunkova Z, Kren V. Reverse hydrolysis catalysed by β-N-acetylhexosaminidase from *Aspergillus oryzae*. Biotechnol Lett. 1997;19(9):869–72.
14. Zhou JP, Gao YJ, Dong YY, Tang XH, Li JJ, Xu B, et al. A novel xylanase with tolerance to ethanol, salt, protease, SDS, heat, and alkali from actinomycete *Lechevalieria* sp. HJ3. J Ind Microbiol Biot. 2012;39(7):965–75.
15. Zhou JP, He LM, Gao YJ, Han NY, Zhang R, Wu Q, et al. Characterization of a novel low-temperature-active, alkaline and sucrose-tolerant invertase. Sci Rep. 2016;6:32081.
16. Zhou JP, Liu Y, Lu Q, Zhang R, Wu Q, Li CY, et al. Characterization of a glycoside hydrolase family 27 α-galactosidase from *Pontibacter* reveals its novel salt–protease tolerance and transglycosylation activity. J Agr Food Chem. 2016;64(11):2315–24.
17. Zhou JP, Lu Q, Zhang R, Wang YY, Wu Q, Li JJ, et al. Characterization of two glycoside hydrolase family 36 α-galactosidases: novel transglycosylation activity, lead–zinc tolerance, alkaline and multiple pH optima, and low-temperature activity. Food Chem. 2016;194:156–66.
18. Zhou JP, Song ZF, Zhang R, Ding LM, Wu Q, Li JJ, et al. Characterization of a NaCl-tolerant β-N-acetylglucosaminidase from *Sphingobacterium* sp. HWLB1. Extremophiles. 2016;20(4):547–57.
19. Zhou JP, Shen JD, Zhang R, Tang XH, Li JJ, Xu B, et al. Molecular and biochemical characterization of a novel multidomain xylanase from *Arthrobacter* sp. GN16 isolated from the feces of *Grus nigricollis*. Appl Biochem Biotech. 2015;175(1):573–88.
20. Robert X, Gouet P. Deciphering key features in protein structures with the new ENDscript server. Nucleic Acids Res. 2014;42(W1):W320–W4.
21. Willard L, Ranjan A, Zhang HY, Monzavi H, Boyko RF, Sykes BD, et al. VADAR: a web server for quantitative evaluation of protein structure quality. Nucleic Acids Res. 2003;31(13):3316–9.
22. Meux E, Morel M, Lamant T, Gerardin P, Jacquot JP, Dumarcay S, et al. New substrates and activity of *Phanerochaete chrysosporium* Omega glutathione transferases. Biochimie. 2013;95(2):336–46.
23. Tews I, Perrakis A, Oppenheim A, Dauter Z, Wilson KS, Vorgias CE. Bacterial chitobiase structure provides insight into catalytic mechanism and the basis of Tay–Sachs disease. Nat Struct Mol Biol. 1996;3(7):638–48.
24. Lemieux MJ, Mark BL, Cherney MM, Withers SG, Mahuran DJ, James MNG. Crystallographic structure of human β-hexosaminidase A: interpretation of Tay–Sachs mutations and loss of G(M2) ganglioside hydrolysis. J Mol Biol. 2006;359(4):913–29.
25. Sumida T, Ishii R, Yanagisawa T, Yokoyama S, Ito M. Molecular cloning and crystal structural analysis of a novel β-N-acetylhexosaminidase from *Paenibacillus* sp. TS12 capable of degrading glycosphingolipids. J Mol Biol. 2009;392(1):87–99.
26. Han SR, Kim KH, Ahn DH, Park H, Oh TJ. Complete genome sequence of carotenoid-producing *Microbacterium* sp. strain PAMC28756 isolated from an Antarctic lichen. J Biotechnol. 2016;226:18–9.
27. Mandakovic D, Cabrera P, Pulgar R, Maldonado J, Aravena P, Latorre M, et al. Complete genome sequence of *Microbacterium* sp. CGR1, bacterium tolerant to wide abiotic conditions isolated from the Atacama Desert. J Biotechnol. 2015;216:149–50.
28. Zhang YB, Ren HH, Zhang GY. *Microbacterium hydrothermale* sp. nov., an actinobacterium isolated from hydrothermal sediment. Int J Syst Evol Micr. 2014;64:3508–12.
29. Mondani L, Piette L, Christen R, Bachar D, Berthomieu C, Chapon V. *Microbacterium lemovicicum* sp. nov., a bacterium isolated from a natural uranium-rich soil. Int J Syst Evol Micr. 2013;63:2600–6.
30. Kook M, Son HM, Yi TH. *Microbacterium kyungheense* sp. nov. and *Microbacterium jejuense* sp. nov., isolated from salty soil. Int J Syst Evol Micr. 2014;64:2267–73.
31. Valk V, Eeuwema W, Sarian FD, van der Kaaij RM, Dijkhuizen L. Degradation of granular starch by the bacterium *Microbacterium aurum* strain B8.A involves a modular α-amylase enzyme system with FNIII and CBM25 domains. Appl Environ Microb. 2015;81(19):6610–20.
32. Lu J, Wu XD, Jiang YL, Cai XF, Huang LY, Yang YB, et al. An extremophile *Microbacterium* strain and its protease production under alkaline conditions. J Basic Microbiol. 2014;54(5):378–85.
33. Isotani K, Kurokawa J, Suzuki F, Nomoto S, Negishi T, Matsuda M, et al. Gene cloning and characterization of two NADH-dependent 3-quinuclidinone reductases from *Microbacterium luteolum* JCM 9174. Appl Environ Microb. 2013;79(4):1378–84.
34. Kim DW, Feng JH, Chen HZ, Kweon O, Gao Y, Yu LR, et al. Identification of the enzyme responsible for N-acetylation of norfloxacin by *Microbacterium* sp. strain 4 N2-2. Appl Environ Microb. 2013;79(1):314–21.
35. Ohta Y, Hatada Y, Hidaka Y, Shimane Y, Usui K, Ito T, et al. Enhancing thermostability and the structural characterization of *Microbacterium saccharophilum* K-1 β-fructofuranosidase. Appl Microbiol Biot. 2014;98(15):6667–77.
36. Wang JJ, Zhu YX, Zhao GG, Zhu JG, Wu S. Characterization of a recombinant (+)-γ-lactamase from *Microbacterium hydrocarbonoxydans* which provides evidence that two enantiocomplementary γ-lactamases are in the strain. Appl Microbiol Biot. 2015;99(7):3069–80.
37. Keyhani NO, Roseman S. The chitin catabolic cascade in the marine bacterium *Vibrio furnissii*—molecular cloning, isolation, and characterization of a periplasmic β-N-acetylglucosaminidase. J Biol Chem. 1996;271(52):33425–32.
38. Yang SQ, Song S, Yan QJ, Fu X, Jiang ZQ, Yang XB. Biochemical characterization of the first fungal glycoside hydrolyase family 3 β-N-acetylglucosaminidase from *Rhizomucor miehei*. J Agr Food Chem. 2014;62(22):5181–90.
39. Li HZ, Morimoto K, Kimura T, Sakka K, Ohmiya K. A new type of β-N-acetylglucosaminidase from hydrogen-producing *Clostridium paraputrificum* M-21. J Biosci Bioeng. 2003;96(3):268–74.
40. da Silva Junior Sobrinho I, Bataus LAM, Maitan VR, Ulhoa CJ. Purification and properties of an N-acetylglucosaminidase from *Streptomyces cerradoensis*. Biotechnol Lett. 2005;27(17):1273–6.
41. Senba M, Kashige N, Nakashima K, Miake F, Watanabe K. Cloning of the gene of β-N-acetylglucosaminidase from *Lactobacillus casei* ATCC 27092 and characterization of the enzyme espressed in *Escherichia coli*. Biol Pharm Bull. 2000;23(5):527–31.

42. Kubota T, Miyamoto K, Yasuda M, Inamori Y, Tsujibo H. Molecular characterization of an intracellular β-N-acetylglucosaminidase involved in the chitin degradation system of *Streptomyces thermoviolaceus* OPC-520. Biosci Biotech Bioch. 2004;68(6):1306–14.
43. Sukprasirt P, Wititsuwannakul R. A chitinolytic endochitinase and β-N-acetylglucosaminidase-based system from *Hevea latex* in generating N-acetylglucosamine from chitin. Phytochemistry. 2014;104:5–11.
44. Nyffenegger C, Nordvang RT, Zeuner B, Lezyk M, Difilippo E, Logtenberg MJ, et al. Backbone structures in human milk oligosaccharides: trans-glycosylation by metagenomic β-N-acetylhexosaminidases. Appl Microbiol Biot. 2015;99(19):7997–8009.
45. Mayer C, Vocadlo DJ, Mah M, Rupitz K, Stoll D, Warren RAJ, et al. Characterization of a β-N-acetylhexosaminidase and a β-N-acetylglucosaminidase/β-glucosidase from *Cellulomonas fimi*. FEBS J. 2006;273(13):2929–41.
46. Tsujibo H, Hatano N, Mikami T, Izumizawa Y, Miyamoto K, Inamori Y. Cloning, characterization and expression of β-N-acetylglucosaminidase gene from *Streptomyces thermoviolaceus* OPC-520. BBA-Gen Subjects. 1998; 1425(2):437–40.
47. Ogawa M, Kitagawa M, Tanaka H, Ueda K, Watsuji T, Beppu T, et al. A β-N-acetylhexosaminidase from *Symbiobacterium thermophilum*; gene cloning, overexpression, purification and characterization. Enzyme Microb Tech. 2006;38(3–4):457–64.
48. Lin H, Xiao X, Zeng X, Wang FP. Expression, characterization and mutagenesis of the gene encoding β-N-acetylglucosaminidase from *Aeromonas caviae* CB101. Enzyme Microb Tech. 2006;38(6):765–71.
49. Chen F, Chen XZ, Qin LN, Tao Y, Dong ZY. Characterization and homologous overexpression of an N-acetylglucosaminidase Nag1 from *Trichoderma reesei*. Biochem Bioph Res Co. 2015;459(2):184–8.
50. Liu XS, Huang ZQ, Zhang XN, Shao ZZ, Liu ZD. Cloning, expression and characterization of a novel cold-active and halophilic xylanase from *Zunongwangia profunda*. Extremophiles. 2014;18(2):441–50.
51. Yamaguchi R, Tokunaga H, Ishibashi M, Arakawa T, Tokunaga M. Salt-dependent thermo-reversible α-amylase: cloning and characterization of halophilic α-amylase from moderately halophilic bacterium. *Kocuria varians*. Appl Microbiol Biot. 2011;89(3):673–84.
52. Wang GZ, Wang QH, Lin XJ, Ng TB, Yan RX, Lin J, et al. A novel cold-adapted and highly salt-tolerant esterase from *Alkalibacterium* sp. SL3 from the sediment of a soda lake. Sci Rep. 2016;6:19494.
53. Premkumar L, Greenblatt HM, Bageshwar UK, Savchenko T, Gokhman I, Sussman JL, et al. Three-dimensional structure of a halotolerant algal carbonic anhydrase predicts halotolerance of a mammalian homolog. Proc Natl Acad Sci U S A. 2005;102(21):7493–8.
54. Shen JD, Zhang R, Li JJ, Tang XH, Li RX, Wang M, et al. Characterization of an exo-inulinase from *Arthrobacter*: a novel NaCl-tolerant exo-inulinase with high molecular mass. Bioengineered. 2015;6(2):99–105.
55. Zhou JP, Peng MZ, Zhang R, Li JJ, Tang XH, Xu B, et al. Characterization of *Sphingomonas* sp. JB13 exo-inulinase: a novel detergent-, salt-, and protease-tolerant exo-inulinase. Extremophiles. 2015;19(2):383–93.
56. Paul S, Bag SK, Das S, Harvill ET, Dutta C. Molecular signature of hypersaline adaptation: insights from genome and proteome composition of halophilic prokaryotes. Genome Biol. 2008;9(4):R70.
57. Madern D, Ebel C, Zaccai G. Halophilic adaptation of enzymes. Extremophiles. 2000;4(2):91–8.
58. Warden AC, Williams M, Peat TS, Seabrook SA, Newman J, Dojchinov G, et al. Rational engineering of a mesohalophilic carbonic anhydrase to an extreme halotolerant biocatalyst. Nat Commun. 2015;6:10278.
59. Qin YJ, Huang ZQ, Liu ZD. A novel cold-active and salt-tolerant α-amylase from marine bacterium *Zunongwangia profunda*: molecular cloning, heterologous expression and biochemical characterization. Extremophiles. 2014;18(2):271–81.
60. Shi RR, Li ZM, Ye Q, Xu JH, Liu Y. Heterologous expression and characterization of a novel thermo-halotolerant endoglucanase Cel5H from *Dictyoglomus thermophilum*. Bioresource Technol. 2013;142:338–44.
61. Margesin R, Schinner F. Potential of halotolerant and halophilic microorganisms for biotechnology. Extremophiles. 2001;5(2):73–83.
62. Siglioccolo A, Paiardini A, Piscitelli M, Pascarella S. Structural adaptation of extreme halophilic proteins through decrease of conserved hydrophobic contact surface. BMC Struct Biol. 2011;11:50.
63. Wu GJ, Wu GB, Zhan T, Shao ZZ, Liu ZD. Characterization of a cold-adapted and salt-tolerant esterase from a psychrotrophic bacterium *Psychrobacter pacificensis*. Extremophiles. 2013;17(5):809–19.
64. Bruce AF, Gounaris K. Characterisation of a secreted N-acetyl-β-hexosaminidase from *Trichinella spiralis*. Mol Biochem Parasit. 2006;145(1):84–93.
65. Ohishi K, Murase K, Etoh H. Purification and properties of β-N-acetylglucosaminidase from *Vibrio alginolyticus* H-8. J Biosci Bioeng. 1999;88(1):98–9.
66. Fukamizo T, Kramer KJ. Mechanism of chitin oligosaccharide hydrolysis by the binary enzyme chitinase system in insect moulting fluid. Insect Biochem. 1985;15(1):1–7.
67. Suzuki K, Sugawara N, Suzuki M, Uchiyama T, Katouno F, Nikaidou N, et al. Chitinases A, B, and C1 of *Serratia marcescens* 2170 produced by recombinant *Escherichia coli*: enzymatic properties and synergism on chitin degradation. Biosci Biotech Bioch. 2002;66(5):1075–83.
68. Chitlaru E, Roseman S. Molecular cloning and characterization of a novel β-N-acetyl-D-glucosaminidase from *Vibrio furnissii*. J Biol Chem. 1996;271(52):33433–9.
69. Nieder V, Kutzer M, Kren V, Gallego RG, Kamerling JP, Elling L. Screening and characterization of β-N-acetylhexosaminidases for the synthesis of nucleotide-activated disaccharides. Enzyme Microb Tech. 2004;34(5):407–14.
70. Tomiya N, Narang S, Park J, Abdul-Rahman B, Choi O, Singh S, et al. Purification, characterization, and cloning of a *Spodoptera frugiperda* Sf9 β-N-acetylhexosaminidase that hydrolyzes terminal N-acetylglucosamine on the N-glycan core. J Biol Chem. 2006;281(28):19545–60.
71. Suginta W, Chuenark D, Mizuhara M, Fukamizo T. Novel β-N-acetylglucosaminidases from *Vibrio harveyi* 650: cloning, expression, enzymatic properties, and subsite identification. BMC Biochem. 2010;11:40.
72. Liu TA, Zhang HT, Liu FY, Wu QY, Shen X, Yang Q. Structural determinants of an insect β-N-acetyl-D-hexosaminidase specialized as a chitinolytic enzyme. J Biol Chem. 2011;286(6):4049–58.
73. Lan XQ, Ozawa N, Nishiwaki N, Kodaira R, Okazaki M, Shimosaka M. Purification, cloning, and sequence analysis of β-N-acetylglucosaminidase from the chitinolytic bacterium *Aeromonas hydrophila* strain SUWA-9. Biosci Biotech Bioch. 2004;68(5):1082–90.
74. Mark BL, Wasney GA, Salo TJS, Khan AR, Cao ZM, Robbins PW, et al. Structural and functional characterization of *Streptomyces plicatus* β-N-acetylhexosaminidase by comparative molecular modeling and site-directed mutagenesis. J Biol Chem. 1998;273(31):19618–24.
75. Ueda M, Fujita Y, Kawaguchi T, Arai M. Cloning, nucleotide sequence and expression of the β-N-acetylglucosaminidase gene from *Aeromonas* sp. no. 10S-24. J Biosci Bioeng. 2000;89(2):164–9.
76. Lan XQ, Zhang X, Kodaira R, Zhou Z, Shimosaka M. Gene cloning, expression, and characterization of a second β-N-acetylglucosaminidase from the chitinolytic bacterium *Aeromonas hydrophila* strain SUWA-9. Biosci Biotech Bioch. 2008;72(2):492–8.
77. Riekenberg S, Flockenhaus B, Vahrmann A, Muller MCM, Leippe M, Kiess M, et al. The β-N-acetylhexosaminidase of *Entamoeba histolytica* is composed of two homologous chains and has been localized to cytoplasmic granules. Mol Biochem Parasit. 2004;138(2):217–25.
78. Konno N, Takahashi H, Nakajima M, Takeda T, Sakamoto Y. Characterization of β-N-acetylhexosaminidase (LeHex20A), a member of glycoside hydrolase family 20, from *Lentinula edodes* (shiitake mushroom). AMB Express. 2012;2:29.
79. Tsujibo H, Fujimoto K, Tanno H, Miyamoto K, Kimura Y, Imada C, et al. Molecular cloning of the gene which encodes β-N-acetylglucosaminidase from a marine bacterium, *Alteromonas* sp. strain O-7. Appl Environ Microb. 1995;61(2):804–6.
80. Tsujibo H, Miyamoto K, Yoshimura M, Takata M, Miyamoto J, Inamori Y. Molecular cloning of the gene encoding a novel β-N-acetylhexosaminidase from a marine bacterium, *Alteromonas* sp. strain O-7, and characterization of the cloned enzyme. Biosci Biotech Bioch. 2002;66(2):471–5.
81. Slamova K, Kulik N, Fiala M, Krejzova-Hofmeisterova J, Ettrich R, Kren V. Expression, characterization and homology modeling of a novel eukaryotic GH84 β-N-acetylglucosaminidase from *Penicillium chrysogenum*. Protein Expres Purif. 2014;95:204–10.
82. Meekrathok P, Suginta W. Probing the catalytic mechanism of *Vibrio harveyi* GH20 β-N-acetylglucosaminidase by chemical rescue. PLoS One. 2016;11(2):e0149228.
83. Tsujibo H, Hatano N, Mikami T, Hirasawa A, Miyamoto K, Inamori Y. A novel β-N-acetylglucosaminidase from *Streptomyces thermoviolaceus* OPC-520: gene cloning, expression, and assignment to family 3 of the glycosyl hydrolases. Appl Environ Microb. 1998;64(8):2920–4.

84. Choi KH, Seo JY, Park KM, Park CS, Cha J. Characterization of glycosyl hydrolase family 3 β-N-acetylglucosaminidases from *Thermotoga maritima* and *Thermotoga neapolitana*. J Biosci Bioeng. 2009;108(6):455–9.
85. Matsuo Y, Kurita M, Park JK, Tanaka K, Nakagawa T, Kawamukai M, et al. Purification, characterization and gene analysis of N-acetylglucosaminidase from *Enterobacter* sp. G-1. Biosci Biotech Bioch. 1999;63(7):1261–8.
86. O'Connell E, Murray P, Piggott C, Hennequart F, Tuohy M. Purification and characterization of a N-acetylglucosaminidase produced by *Talaromyces emersonii* during growth on algal fucoidan. J Appl Phycol. 2008;20(5):557–65.

18

Expression and functional analyses of a Kinesin gene *GhKIS13A1* from cotton (*Gossypium hirsutum*) fiber

Yan-Jun Li[1†], Shou-Hong Zhu[1†], Xin-Yu Zhang[1], Yong-Chang Liu[1], Fei Xue[1], Lan-Jie Zhao[2] and Jie Sun[1*]

Abstract

Background: Cotton fiber, a natural fiber widely used in the textile industry, is differentiated from single cell of ovule epidermis. A large number of genes are believed to be involved in fiber formation, but so far only a few fiber genes have been isolated and functionally characterized in this developmental process. The Kinesin13 subfamily was found to play key roles during cell division and cell elongation, and was considered to be involved in the regulation of cotton fiber development.

Results: The full length of coding sequence of *GhKIS13A1* was cloned using cDNA from cotton fiber for functional characterization. Expression pattern analysis showed that *GhKIS13A1* maintained a lower expression level during cotton fiber development. Biochemical assay showed that GhKIS13A1 has microtubule binding activity and basal ATPase activity that can be activated significantly by the presence of microtubules. Overexpression of *GhKIS13A1* in *Arabidopsis* reduced leaf trichomes and the percentage of three-branch trichomes, and increased two-branch and shriveled trichomes compared to wild-type. Additionally, the expression of *GhKIS13A1* in the *Arabidopsis Kinesin-13a-1* mutant rescued the defective trichome branching pattern of the mutant, making its overall trichome branching pattern back to normal.

Conclusions: Our results suggested that *GhKIS13A1* is functionally compatible with *AtKinesin-13A* regarding their role in regulating the number and branching pattern of leaf trichomes. Given the developmental similarities between cotton fibers and *Arabidopsis* trichomes, it is speculated that *GhKIS13A1* may also be involved in the regulation of cotton fiber development.

Keywords: Kinesin13 subfamily, Cotton fiber, Trichome, *GhKIS13A1*

Background

Cotton is one of the most important commercial crops in the world. Cotton fiber, a natural fiber widely used in the textile industry, is differentiated from single cell of ovule epidermis. The development of cotton fiber consists of four overlapping stages: fiber initiation, cell elongation, secondary wall deposition and maturation [1]. A large number of genes are believed to be involved in fiber formation, but so far only a few fiber genes have been isolated and functionally characterized in this developmental process [2–6]. Therefore, further efforts to clone genes involved in fiber development and study their molecular mechanism governing fiber development are needed. The Kinesin family belongs to a class of motor proteins which can move along microtubule filaments by utilizing the energy released from ATP hydrolysis. Kinesins are present in all studied eukaryotic organisms and are implicated in a diverse range of cellular processes including intracellular transport, signal transduction, cell morphogenesis and mitosis [7–11]. Kinesins can be divided into 14 subfamilies from Kinesin-1 to Kinesin-14 based on the motor domain, a highly conserved domain with the ATP-binding and microtubule-binding activities [12]. The motor domain can be located at the C-terminus, N-terminus or in the middle of Kinesins. With the complete sequencing of many plant genomes, much research has been carried

* Correspondence: sunjie@shzu.edu.cn
†Equal contributors
[1]The Key Laboratory of Oasis Eco-agriculture, Agriculture College, Shihezi University, Bei 4 Road, Shihezi 832003, Xinjiang, China
Full list of author information is available at the end of the article

out on the Kinesin family in model plants. Sixty-one and forty-one Kinesin genes have been identified in the *Arabidopsis* and rice genome, respectively [13–15]. To our knowledge, only four Kinesin genes have been reported in cotton (*Gossypium hirsutum*). *GhKCBP*, *GhKCH1* and *GhKCH2* belonging to the Kinesin-14 subfamily were found to decorate microtubules and microfilaments and regulate their alignment and dynamic distribution [16–18]. *GhKinesin-13A* belonging to the Kinesin-13 subfamily was found to be located in the Golgi apparatus of cotton fiber based on immunofluorescent detection [19]. These studies suggested that Kinesin genes may play key roles in cotton fiber growth and developmental processes.

Much more attention was paid to the Kinesin-13 subfamily recently because of its roles in cell division and cell elongation. Mutation of *SRS3* encoding a member of the Kinesin-13 subfamily causes small and round seed phenotype in rice due to a decrease in cell length in the longitudinal direction [20]. T-DNA insertion in the *AtKinesin-13A* gene results in a sharp decrease of the size and number of Golgi vesicles in root cap peripheral cells [21]. Moreover, it was shown that over 70% of leaf trichomes in the *Kinesin-13a-1* mutants bore four branches, while the trichomes of wild-type (Col-0) predominantly bore three branches [19]. Both cotton fibers and *Arabidopsis* trichomes are derived from single cells of epidermis, and likely to be regulated by a similar development mechanism. Therefore, genes that influence the development of *Arabidopsis* trichome may also play key roles in cotton fiber development. In a previous study, we isolated seven putative *G. hirsutum* homologues of *AtKinesin-13A* using primers designed based on the genome sequence of *G. raimondii*, and analyzed the sequence characteristics of these Kinesin genes [22]. Of the seven *Kinesin-13* genes, *GhKIS13A1* was found to be virtually identical to the previously reported *GhKinesin-13A*, a homologue of the *Arabidopsis AtKinesin-13A* [19]. In this study, on the basis of expression pattern and biochemical characterization of *GhKIS13A1*, we showed that *GhKIS13A1* not only influences the number and branching patterns of *Arabidopsis* leaf trichomes but also is able to rescue the mutant phenotype of the *Arabidopsis Kinesin-13a-1* mutant, suggesting that *GhKIS13A1* may play an important role in cotton fiber development and is a potential candidate for improving cotton fiber quality by genetic engineering.

Methods

Materials

Flowers and fibers were collected from the cotton (*Gossypium hirsutum*) variety "Xinluzao 36" plants grown in a normal agronomic field. Flower buds on the anthesis day were tagged and marked as 0 day post anthesis (DPA). Bolls were collected at 3, 6, 9, 12, 15, 18, 21, 24 and 27DPA. Ovules were excised from the bolls and fibers were scraped from the ovules. The seeds of "Xinluzao 36" were sterilizated by 0.1% $HgCl_2$ for 10 min, washed with sterile water 3–4 times and then sown on the Murashige and Skoog (MS) ager medium. The root, hypocotyl and leaf were collected from the tissue-cultured seedlings grown for about 2 weeks. All the collected materials were immediately frozen in liquid nitrogen and stored at -80°C until RNA extraction. The ecotype Columbia (Col-0) of *Arabidopsis thaliana* and the *Kinesin-13a-1* mutant were used in transgenic analysis.

Cloning of *GhKIS13A1*

Total RNAs were extracted from cotton fibers and then reverse transcribed to generate cDNA. Gene-specific primers KIS-1: 5′-ATG GGT GGC CAG ATG CAG CAA AGC-3′ and KIS-2: 5′-ACG AGG AAC TCT TTT CCG ACT C-3′ were designed based on the *GhKIS13A1* sequence (accession number, KP036626) downloaded from the GenBank database. The open reading frame (ORF) of *GhKIS13A1* was amplified by PCR using the gene-specific primers and cotton fibers cDNA as a template. The PCR amplification program was: 94°C for 3 min, 30 PCR cycles (94°C for 1 min, 56°C for 1 min and 72°C for 2.5 min), and 72°C for 10 min. The PCR product was purified and cloned into the pGEM-T easy vector (Promega, USA), and then transformed into *E. coli* DH10B cells for sequencing.

Phylogenetic analysis

Kinesin proteins from *Arabidopsis*, rice, and corn were obtained from the GenBank database (http://www.ncbi.nlm.nih.gov/genbank/). Thirteen cotton Kinesins were obtained from the genome database of *Gossypium hirsutum* (www.cottongen.org/) by using GhKIS13A1 as a probe. MEGA 4.0 was used for the construction of phylogenetic tree, using the neighbor-joining method and the bootstrap test carried out with 1000 replicates.

Expression pattern of *GhKIS13A1* in different cotton organs

Total RNAs from cotton roots, hypocotyls, leaves, flowers, ovules and fibers were extracted according to a previous method [23], and reversely transcribed into cDNAs. Then, the cDNAs were used as templates in Real-time Quantitative PCR (qPCR) reactions with gene-specific primers. The forward primer KIS-5 is 5′-CTG GTC GAA GGG TAG CAG AG-3′, and the reverse primer KIS-6 is 5′-GGC TCG AAG AAC CAC CAT AA-3′. The cotton *UBI* gene was used as an internal reference for normalization of cDNA templates. The forward primer of *UBI* is 5′-CAG ATC TTC GTA AAA CCC T-3′, and the reverse primer is 5′-GAC TCC TTC TGG ATG TTG TA-3′. The qPCR reactions were conducted using a SYBR Green I Master mixture (Roche, Switzerland) on a LightCycler 480IIsystem (Roche, Switzerland) under conditions of an initial denaturation at 95 °C for 2 min followed by 40 cycles of denaturing at 95°C for 15 s, annealing at 55 °C for 20 s, and extending at 72 °C for 15 s. The qPCR reaction was repeated three times. Quantification was performed using the $2^{-\Delta\Delta Ct}$ method.

Expression and purification of the recombinant GhKIS13A1 protein

The *GhKIS13A1* ORF was cloned into plasmid pET28a (+) (with a His tag) using *Eco*R1/*Hin*dIII restriction sites to generate the pET28a-*GhKIS13A1* construct. After sequencing confirmation, the pET28a-*GhKIS13A1* construct was transformed into *E. coli* BL21 (DE3) cells for fusion protein expression. A single colony of *E.coli* BL21 cells harboring the recombinant plasmid pET28a-*GhKIS13A1* was incubated at 37 °C until the OD_{600} reached about 0.6, and protein synthesis was induced by adding 0.8 mM isopropyl thio-β-D-galactoside (IPTG) for up to 6 h at 37°C. Protein purification was performed with His-Binding-Resin following the manufacturer's instructions (Yuekebio, Shanghai, China). The purity of the GhKIS13A1-His fusion protein was loaded on 12% SDS-PAGE gel for western blot experiment with anti-6 × His antibody (Sigma). The protein concentration was determined by the Bradford method with bovine serum albumin (BSA) as the standard.

Assays for Kinesin activity of GhKIS13A1

Microtubule binding assay was performed according to a previous method [24]. To prepare microtubules, the tubulin protein (GenMed Scientifics Inc. USA) was polymerized by incubating with 2 $mmol{\bullet}L^{-1}$ GTP and 30 $\mu mol{\bullet}L^{-1}$ paclitaxel for 30 min at 37 °C. A volume of 50 μl microtubules (2.7 $mg{\bullet}mL^{-1}$) was mixed with 50 μl purified GhKIS13A1 (0.44 $mg{\bullet}mL^{-1}$) in the absence or presence of 10 $mmol{\bullet}L^{-1}$ ATP and 5 $mmol{\bullet}L^{-1}$ 5'-adenylylimidodiphosphate (AMP-PNP; a non-hydrolyzable ATP analogue). After incubation for 15 min at 37 °C, the reactions were centrifuged at 14,000 g for 30 min. The supernatant and pellet were electrophoresed by SDS-PAGE and stained with Coomassie Brilliant Blue R-250.

To analyze the ATPase activity and microtubule-activated ATPase activity, 100 μl purified GhKIS13A1 protein (0.44 $mg{\bullet}mL^{-1}$) was incubated with 100 μl microtubulins (2.7 $mg{\bullet}mL^{-1}$) for 15 min at 37°C. Then 20 μl ATP (20 $mmol{\bullet}L^{-1}$) was added and incubated for another 15 min. The reaction was terminated by adding 20 μl of 100% trichloroacetic acid followed by placing on ice for 10 min. After 10 min of centrifugation at 4°C, the supernatant was collected and used to determine the inorganic phosphate according to the LeBel's method [25].

Vector construction and *Arabidopsis* transformation

The *GhKIS13A1* ORF was amplified using KIS-1 and KIS-2 primers and then constructed into plant expression vector pGWB17 (with a 35S promoter) by the Gateway technology. To ensure the sequence conformity to the entry vector in the right direction in the process of vector construction, the recognition sequence 'CACC' was added to the 5' end of the forward primer KIS-1. The PCR product was integrated into the entry vector pENTR/D-TOPO following the instructions of pENTR™ Directional TOPO® Cloning Kits (Invitrogen, USA). After confirmed by DNA sequencing, the entry vector carrying the *GhKIS13A1* ORF was mixed with expression vector pGWB17 for LR recombination reaction following the instructions of Gateway® LR Clonase™ II Enzyme Mix (Invitrogen, USA). The confirmed plasmid was transformed into *Agrobacterium tumefaciens* GV3101 by the standard transformation method [26]. *Arabidopsis* ecotype Col-0 and the *Kinesin-13a-1* mutant were used in transformation by the floral dip method [27] to generate transgenic lines overexpressing *GhKIS13A1* (*GhKIS13A1*-OX) and complementing the *Kinesin-13a-1* mutation (*GhKIS13A1*-COM), respectively.

PCR and qPCR detection of *Arabidopsis* transgenic plants

Genomic DNA was extracted from *Arabidopsis* leaves and used in PCR analysis for detection of the presence of the inserted *GhKIS13A1* sequence. The gene-specific primers KIS-1 and KIS-2 were used to detect exogenous *GhKIS13A1* transformations. PCR-positive plants harboring the *GhKIS13A1* gene were further analyzed for its expression level by qPCR. Total RNA was extracted from 9-day-old wild-type, transgenic lines and the *Kinesin-13a-1* mutant using TRIZOL Reagent (Tiangen, China), and reversely transcribed into cDNAs. Then, the cDNAs were used as templates in qPCR reactions with gene-specific primers KIS-5 and KIS-6. The *Arabidopsis ACTIN2* gene was used as an internal reference for normalization of cDNA templates. The forward primer of *ACTIN2* is 5′-GGT AAC ATT GTG CTC AGT GGT GG, and the reverse primer is 5′-AAC GAC CTT AAT CTT CAT GCT GC-3′.

Analysis of leaf trichomes of *Arabidopsis* transgenic plants

Six *GhKIS13A1*-OX and *GhKIS13A1*-COM transgenic lines, each with seven randomly selected plants, were used in leaf trichome analysis. The leaf of the same part on each plant was picked for counting the total number of trichomes and calculating the percentages of trichomes with different number of branches.

Results

Analysis of phylogenetic relationships

In the genome database of *Gossypium hirsutum*, thirteen Kinesin13 proteins were identified by using the protein sequence of GhKIS13A1 as a probe. Along with sequences from *Arabidopsis*, rice and corn, a phylogenetic tree was constructed (Fig. 1). These proteins from other plants are members of either the Kinesin-13 subfamily or the other subfamilies, including Kinesin-1, Kinesin-2, Kinesin-7, Kinesin-12 and Kinesin-14 subfamilies. Phylogenetic analysis showed that all cotton Kinesins were clustered to Kinesin-13 subfamily, and divided into Kinesin-13A and Kinesin-13B, which were close to AtKinesin-13A and AtKinesin-13B, respectively (Fig. 1). GhKIS13A1 was close to GhKinesin-13A and GhA09G0604 in Kinesin-13A subclade. BLAST searches revealed that the GhKIS13A1 showed 99.56% identity with previously reported GhKinesin-13A [19], and 99.45% identity with GhA09G0604 identified from cotton database. These suggest that GhKIS13A1 was virtually identical to GhKinesin-13A and GhA09G0604, located on

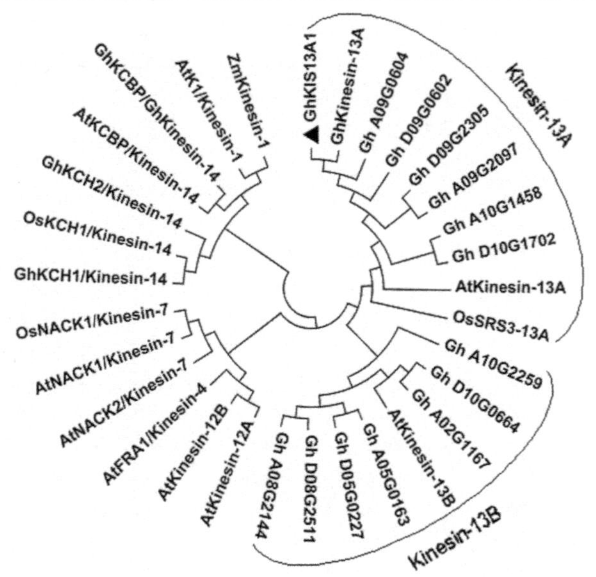

Fig. 1 Phylogenic relationship of GhKIS13A1 and other Kinesin proteins. The phylogenetic tree was constructed by MEGA 4.0 using the neighbor-joining method and the bootstrap test carried out with 1000 bootstrap replicates. The sequence used are GhKIS13A1 (*Gossypium hirsutum*, KP036626), GhKinesin-13A (*Gossypium hirsutum*, AAQ18797), AtKCBP/Kinesin-14 (*Arabidopsis thaliana*, AAC37475), GhKCBP/GhKinesin-14 (*Gossypium hirsutum*, AAP41107), AtKinesin-13A (*Arabidopsis thaliana*, AAL07208), AtKinesin-13B (*Arabidopsis thaliana*, AAK96543), AtKinesin-12A (*Arabidopsis thaliana*, AAF78893), AtKinesin-12B (*Arabidopsis thaliana*, AEE76799), AtK1/Kinesin-1 (*Arabidopsis thaliana*, BAC03248), OsNACK1/Kinesin-7 (*Oryza sativa*, BAB86283), GhKCH1/Kinesin-14 (*Gossypium hirsutum*, AAW03152), GhKCH2/Kinesin-14 (*Gossypium hirsutum*, ABO28522), OsSRS-13A (*Oryza sativa*, EEE62380), ZmKinesin-1 (*Zea mays*, DAA35856), OsKCH1/Kinesin-14 (*Oryza sativa*, NP_001066967), AtFRA1/Kinesin-4 (*Arabidopsis thaliana*, AAN86114), and 13 Kinesins from cotton genome database by using GhKIS13A1 as a probe, including Gh_D09G0602, Gh_A09G0604, Gh_A09G2097, Gh_D09G2305, Gh_D10G1702, Gh_A10G1458, Gh_A10G2259, Gh_D10G0664, Gh_A05G0163, Gh_D05G0227, Gh_D08G2511, Gh_A08G2144, Gh_A02G1167

the A09 chromosome. The sequence differences among GhKIS13A1, GhKinesin-13A and GhA09G0604 may be caused by the difference of cotton varieties and the unavoidable small errors during sequencing or sequence assembly.

Expression analysis of *GhKIS13A1*

The qPCR was performed to investigate the expression pattern of *GhKIS13A1* in different cotton organs. The result showed that *GhKIS13A1* was expressed preferentially in flower and leaf, and at a lower level in hypocotyl, root and fiber. In contrast, *GhKIS13A1* was expressed at the lowest level in 6-15 DPA fiber cells. Although the expression of *GhKIS13A1* was observed at a slightly higher level in 3 DPA ovule and 18-27DPA fiber cells, the level were significantly lower than in the flower and leaf. In addition to 24DPA fiber cell, the expression level of *GhKIS13A1* in fiber cells was lower than in the hypocotyl (Fig. 2).

Biochemical characterization of the motor domain in GhKIS13A1

To carry out the biochemical characterization of GhKIS13A1, the *GhKIS13A1* ORF was amplified and ligated into plasmid pET-28a (+) and expressed in *E.coli* BL21 cells. The GhKIS13A1-His fusion protein prepared using *E.coli* was tested in vitro for its ATPase activity and microtubulin binding activity. SDS-PAGE showed the IPTG induced production of a 110 kDa protein (Fig. 3a), consistent with the predicted molecular mass of GhKIS13A1-His. The recombinant enzyme GhKIS13A1-His was purified by His-Binding-Resin purification column. Western blot experiment showed that GhKIS13A1-His protein could be recognized by his-antibody (Fig. 3b).

To define the ATPase activity and microtubule-activated ATPase activity, the purified GhKIS13A1-His protein was incubated with ATP in the absence or presence of polymerized microtubulins (MTs). As shown

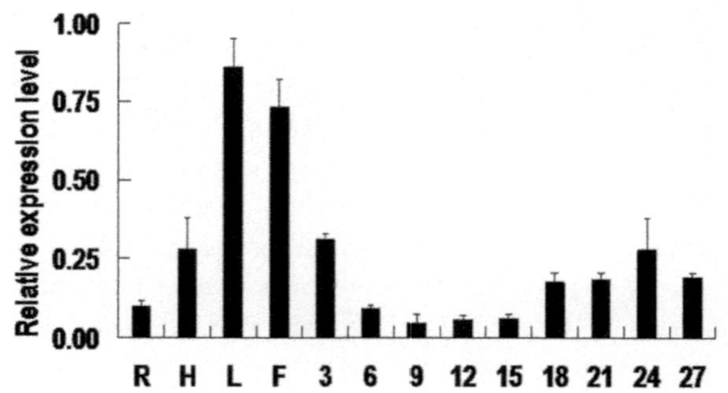

Fig. 2 Expression pattern of *GhKIS13A1* in different cotton organs. All expression levels are relative to the cotton *UBI* gene that was used as an internal control. R: root; H: hypocotyl; L: leaf; F: flower; 3: 3DPA ovule; 6-27: the fiber cells at 6, 9, 12, 15, 18, 21, 24 and 27 DPA, respectively

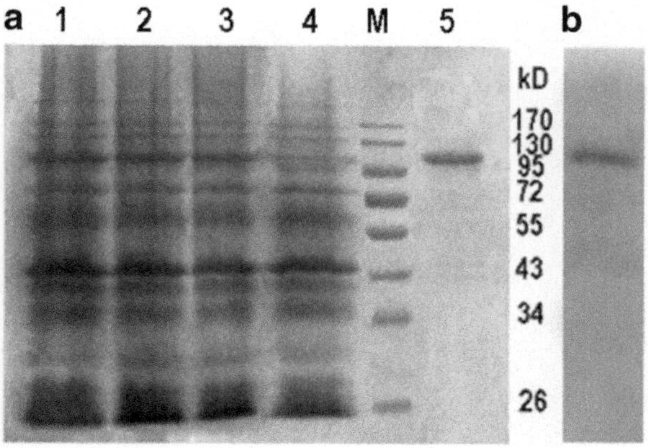

Fig. 3 SDS-PAGE analysis of the induction and purification of GhKIS13A1 expressed in *E.coli* BL21. **a**, SDS-PAGE analysis of induction and purification of GhKIS13A1-His fusion protein. M: protein molecular marker; Lane 1: total proteins of cells after 2 h of IPTG induction; lane 2: total proteins of cells after 4 h of IPTG induction; lane 3: total proteins of cells after 6 h of IPTG induction; lane 4: total proteins of cells without IPTG induction; lane 5: purified GhKIS13A1-His fusion protein. **b**, GhKIS13A1-His fusion protein confirmed by western blot experiment with anti-6 × His antibody

in Fig. 4, the ATPase activity of GhKIS13A1 was 15.42 ± 1.56 nmol•Pi•Min^{-1}•mg^{-1} in the absence of MTs, and increased by approximately 3-fold to 56.48 ± 3.15 nmol•-Pi•Min^{-1}•mg^{-1} in the presence of MTs. No ATPase activity was detected in the reaction system containing MTs and lacking GhKIS13A1-His. The result suggested that GhKIS13A1 has a basal ATPase activity that can be activated significantly by the presence of MTs.

To determine the microtubulin binding activity of GhKIS13A1, the purified GhKIS13A1-His protein was incubated with MTs in the absence or presence of either ATP or AMP-PNP, a non-hydrolyzable ATP analogue. In the absence of MTs, the GhKIS13A1-His protein remained in the supernatant after 30 min of centrifugation (Lanes 1-2 of Fig. 5). In the presence of MTs, the GhKIS13A1-His protein was co-precipitated with MTs only in the presence of AMP-PNP but not in the absence of AMP-PNP (comparing Lanes 5-6 and Lanes 7-8 of Fig. 5). These results suggest that AMP-PNP can promote the binding between GhKIS13A1 and MTs, and that GhKIS13A1 has the nucleotide-dependent microtubulin binding ability.

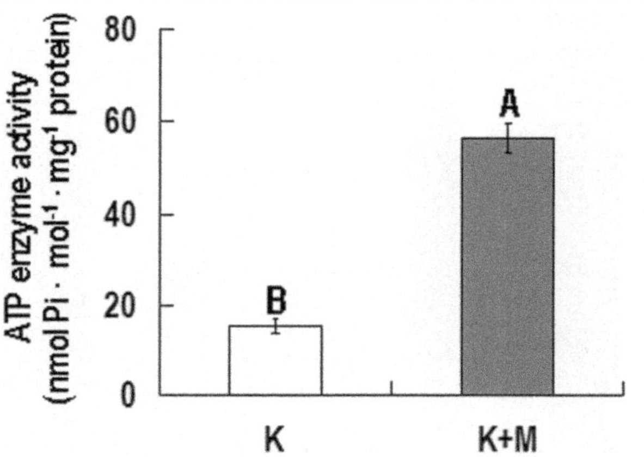

Fig. 4 ATPase activity assays of GhKIS13A1. K: ATPase assays of GhKIS13A1 in the absence of MTs; K + M: ATPase assays of GhKIS13A1 in the presence of MTs. Data were analyzed by Paired-samples *T* test. Different uppercase letters indicate a significant difference between K and K + M at the 1% probability level

Screening for *GhKIS13A1* transgenic *Arabidopsis* plants

The *GhKIS13A1* overexpression vector was constructed and introduced into the *Arabidopsis* wild-type (Col-0) and the *Kinesin-13a-1* mutant plants to generate *GhKIS13A1*-OX and *GhKIS13A1*-COM lines, respectively. Successful transgenic plants were selected using PCR based on the presence of the exogenous *GhKIS13A1* gene. The T2 generation plants of three *GhKIS13A1*-OX lines (L1, L2 and L8) and three *GhKIS13A1*-COM lines (C3, C7 and C10) were selected for leaf trichome investigation. The expression levels of *GhKIS13A1* in these independent transgenic lines were quantified using qPCR. As expected, no expression of *GhKIS13A1* was detected in the wild-type (Col-0) and the *Kinesin-13a-1* mutant, while expression of *GhKIS13A1* was detected in the three *GhKIS13A1*-OX lines (L1, L2 and L3) and three *GhKIS13A1*-COM lines (C3, C7 and C10) (Fig. 6). Among the three *GhKIS13A1*-OX lines, L2 showed the highest transcript level of *GhKIS13A1*, followed by the L1 and L8 lines (Fig. 6b). Among the three *GhKIS13A1*-COM lines, C3 showed the highest transcript level of *GhKIS13A1*, followed by the C7 and C10 lines (Fig. 6c). Exogenous expression of *GhKIS13A1* in *Arabidopsis* had no effect on the overall growth and development of the transgenic plants.

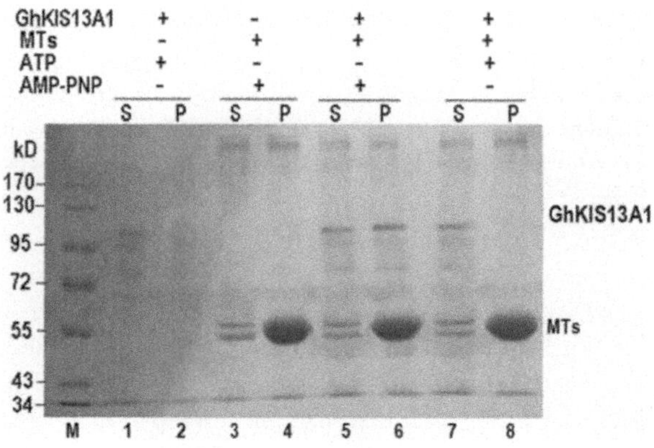

Fig. 5 Microtubule binding assay of GhKIS13A1. M: protein molecular marker; S: supernatant; P: precipitation; lanes 1-2: microtubule binding assays of GhKIS13A1 in the absence of MTs (control test); lanes 3-4: microtubule binding assays of MTs in the absence of GhKIS13A1 (control test); lanes 5-6: microtubule binding assays of GhKIS13A1 in the presence of AMP-PNP; lanes 7-8: microtubule binding assays of GhKIS13A1 in the presence of ATP

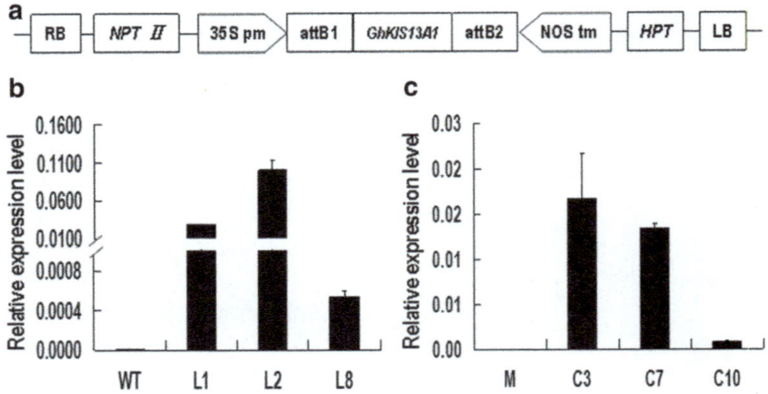

Fig. 6 The PCR and qPCR detection of transgenic plants. **a**, Schematic representation of the expression cassettes in the plant expression vector pGWB17-*GhKIS13A1* used in *Arabidopsis* transformation. 35S pm, CaMV 35S promoter; NOS tm, NOS terminor; *HPT*, hygromycin phosphotralsferase gene; *NPT II*, neomycin phosphotransferase IIgene; attB1 and attB2, recombination site sequences in the Gateway system; *GhKIS13A1*, cDNA sequence of the *GhKIS13A1* gene; RB, right border; LB, left border; (**b**), The expression level of *GhKIS13A1* in wild-type (Col-0) and three *GhKIS13A1*-OX lines (L1, L2 and L8). WT, wild-type; (**c**), The expression level of *GhKIS13A1* in the Kinesin-13a-1 mutant and three *GhKIS13A1*-COM lines (C3, C7 and C10). M, Kinesin-13a-1 mutant. All expression levels are relative to the *Arabidopsis ACTIN2* gene that was used as an internal control

Overexpression of *GhKIS13A1* influences the number and branching of the *Arabidopsis* leaf trichomes

A previous study showed that *AtKinesin-13A* plays a role in trichome morphogenesis, and that its *Kinesin-13a-1* mutant has more trichomes with four branches than wild-type [19]. Therefore, we compared the number and branching pattern of leaf trichomes between the wild-type and the transgenic lines. The eighth rosette leaves taken from seven plants of each transgenic line were used in the comparison.

The number of trichomes of the three *GhKIS13A1*-OX lines were significantly less than that of the wild-type (Fig. 7), and was negatively correlated with the expression level of *GhKIS13A1* in the corresponding *GhKIS13A1*-OX line. Line L2, with the highest *GhKIS13A1* expression level (Fig. 6), produced 52 trichomes per leaf, a 54.9% reduction compared with the wild-type (116 trichomes per leaf); line L8, with the lowest expression level of *GhKIS13A1*, produced 92 trichomes per leaf, which is 20.4% less than that of the wild-type. Regarding the branching pattern, the trichomes of the wild-type predominantly had three branches (94.7%), although trichomes with four branches were occasionally seen (4.2%). In the *GhKIS13A1*-OX lines, although the majority of trichomes also had three branches, their percentages were significantly lower than that of the wild-type (Table 1). Notably, the transgenic lines produced more two-branch trichomes and shriveled-branch trichomes, suggesting that ectopic expression of *GhKIS13A1* in *Arabidopsis* has a negative effect on branching of trichomes with more than two branches. These data collectively indicated that *GhKIS13A1* overexpression can influence the number and branching of the leaf trichomes in *Arabidopsis*.

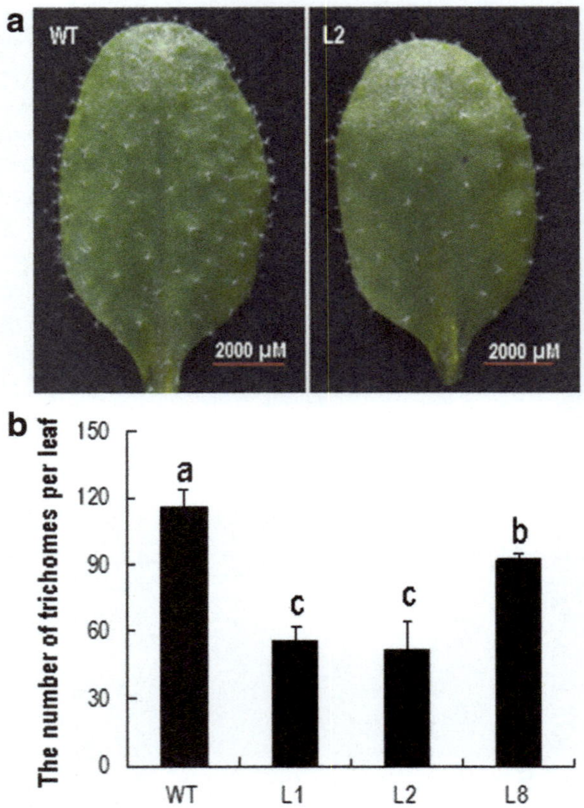

Fig. 7 The morphology and number of leaf trichomes in the wild-type (Col-0) and *GhKIS13A1*-OX lines. **a**, Representative leaf images from the wild-type and the L2 line showing the number and distribution of trichomes. **b**, The number of leaf trichomes in the wild-type and three *GhKIS13A1*-OX lines. Different lower-case letters indicate significant differences among different lines at the 5% probability level (One-way ANOVA, Least Significant Difference test)

Table 1 Percentages of leaf trichomes with different number of branches in the wild-type and the three *GhKIS13A1*-OX lines

	No. of trichome branches			
	5	4	3	2
WT	0	4.2 a	94.7 a	0.8 c
L1	0	2.7 c	85.4 c	11.3 a
L2	0	2.0 c	86.1 c	12.0 a
L8	0	2.7 b	93.3 b	3.4 b

Data represent the mean percentages of trichomes with the indicated number of branches on the eighth rosette leaf from seven plants. Data were analyzed by one-way ANOVA. Different lower-case letters indicate significant differences among different lines at the 5% probability level (One-way ANOVA, Least Significant Difference test)

GhKIS13A1 is able to rescue the defective trichome phenotype of the *Kinesin-13a-1*mutant

To further confirm the function of *GhKIS13A1*, we carried out genetic complementation experiment by transforming the homozygous *Kinesin-13a-1* mutant with *GhKIS13A1* driven by the 35S promoter. In the three *GhKIS13A1*-COM lines (C3, C7 and C10), the number of trichomes was significantly reduced compared with the mutant plants (Fig. 8), and was negatively correlated with their corresponding *GhKIS13A1* expression levels, consistent with the results observed in the *GhKIS13A1*-OX lines. Additionally, unlike the *Kinesin-13a-1* mutant, which had most of its trichomes with four branches, the *GhKIS13A1*-COM lines had majority of their trichomes with three branches, and the percentages of such trichomes (93.0–94.9%) were similar to that of the wild-type (94.7%). As a result, the *GhKIS13A1*-COM lines had an overall branching pattern of trichomes similar to the wild-type (Table 2). These results suggest that *GhKIS13A1* is able to rescue the trichome defects of the *Kinesin-13a-1* mutant. Overall, the data observed in the transgenic plants indicated that *GhKIS13A1* and *AtKinesin-13A* has a similar function in regulating trichome development in *Arabidopsis*, but overexpression of *GhKIS13A1* in *Arabidopsis* has a negative effect on the total number of trichomes.

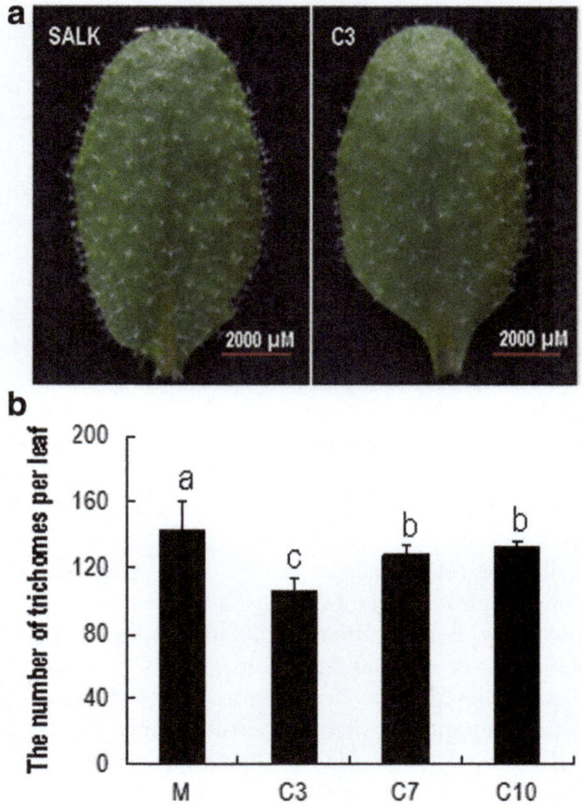

Fig. 8 The morphology and number of leaf trichomes in the *Kinesin-13a-1* mutant and three *GhKIS13A1*-COM lines. **a**, Representative leaf images of the *Kinesin-13a-1* mutant and the C3 line showing the number and distribution of trichomes. **b**, The number of leaf trichomes in the wild-type and three *GhKIS13A1*-COM lines. Different lower-case letters indicate significant differences among different lines at the 5% probability level (One-way ANOVA, Least Significant Difference test)

Discussion

Kinesins are involved in a diverse number of cellular functions. It is expected that Kinesins found in cotton fiber have important cellular roles in cotton fiber development. Four cotton Kinesin genes previously reported to be associated directly or indirectly with cytoskeletons were thought to be related to fiber development [16–19]. We previously cloned 7 cDNAs encoding Kinesins in *G. hirsutum* [22]. Among them, *GhKIS13A1* was chosen for functional analysis. *GhKIS13A1* is a central motor domain Kinesin and its full length cDNA is 2730 bp encoding 909aa. Our results showed that GhKIS13A1 has microtubule-activated ATPase activity and microtubules binding activity similar to conventional Kinesins. GhKIS13A1 exhibited a relatively low basal ATPase activity under the condition of without MTs, and its ATPase activity was enhanced by approximately 3-fold in the

Table 2 Percentages of leaf trichomes with different number of branches in the *Arabidopsis Kinesin-13a-1* mutant and the three *GhKIS13A1*-COM lines

	No. of trichome branches			
	5	4	3	2
Mutant	1.4 a	89.3 a	8.7 b	0.7 a
C3	0 b	3.4 b	94.9 a	1.7 a
C7	0 b	4.0 b	94.4 a	1.7 a
C10	0 b	6.4 b	93.0 a	0.9 a

Data represent the mean percentages of trichomes with the indicated number of branches on the eighth rosette leaf from seven plants. Data were analyzed by one-way ANOVA. Different lower-case letters indicate significant differences among different lines at the 5% probability level (One-way ANOVA, Least Significant Difference test)

presence of MTs. However, the 3-fold was much lower than other prokaryotic-expressed plant Kinesins and many animal Kinesins [18, 28–30]. The ATPase activity of GhKCH2 was found to be enhanced by 10-fold in the presence of MTs [18]. The Kinesins must be firstly attached to microtubules, ATP is then hydrolyzed, phosphate is released, and the Kinesins greatly reduced affinity for microtubules [29]. Therefore, much research found that AMP-PNP, the non-hydrolyzable ATP analogue, could enhance the binding of Kinesins to microtubules [30, 31]. GhKIS13A1 cannot be co-precipitated with microtubules in the presence of ATP, and can be co-precipitated with microtubules in the presence of AMP-PNP. However, some Kinesins can also co-precipitate with microtubules in the presence of ATP or in a nucleotide-free state [18, 32]. To our knowledge, this is the first report describing the biochemical characterization of a cotton Kinesin.

A number of genes with a role in the regulation of trichome number and branches have been isolated in *Arabidopsis*. Loss-of-function mutants of some of these genes showed glabrous leaves or a decease number of leaf trichomes [33–35], but for some genes, their loss-of-function mutants showed an increased clustering of trichomes [36, 37]. In addition, disruption of some genes caused trichomes with abnormal number of branches [38–40]. In this study, we showed that *Arabidopsis* plants overexpressing *GhKIS13A1* exhibited less number of trichomes than the wild-type. We also showed that the *Kinesin-13a-1* mutant harboring *GhKIS13A1* also had less number of trichomes but had the majority of its trichomes with a restored three-branch morphology. These data suggest that the expression of *GhKIS13A1* can inhibit the tichome growth of *Arabidopsis*, and can be a potential candidate for synchronously regulating the number and branching pattern in plants.

Cotton fibers are single-celled, non-glandular hairs of epidermal. Fiber initiation and fiber elongation are fairly synchronous on each ovule and progresses for 20-30 days [1]. Expression pattern analysis showed that *GhKIS13A1* was expressed at a lower level in cotton fiber (containing only trichomes), and maintained a relatively lower level during fiber initiation and elongation stages (Fig. 2). The result suggested that the lower expression of *GhKIS13A1* may be beneficial to the fiber initiation and elongation. The hypothesis is in line with the results that the lower expression level of *GhKIS13A1* produced more trichomes per leaf in six transgenic lines of *Arabidopsis*. It is speculated that *GhKIS13A1* may negatively regulate the cotton fiber development.

Both *Arabidopsis* trichomes and cotton fibers are single-celled, non-glandular hairs of epidermal. A number of studies have suggested that there are developmental similarities between cotton fibers and *Arabidopsis* trichomes [41]. *Arabidopsis* has been served as a useful model plant for dissecting the mechanisms controlling cotton fiber development. Many *Arabidopsis* homologous genes have been isolated from cotton, and some of them have been shown to play a similar role in trichome formation in *Arabidopsis* [42–44]. In this study, the expression pattern analysis and the investigation of the trichome phenotype of *Arabidopsis* harboring *GhKIS13A1* may shed insight into the possible roles of *GhKIS13A1* in cotton fiber development. The *GhKinesin13A1,* whose sequence is identical to the *GhKIS13A1,* is localized to Golgi stacks, which are frequently associated with microtubules and actin microfilaments [19]. *GhKIS13A1* may thus play a similar regulatory role in trichome morphogenesis through its effects on the organization of Golgi stacks. This paper firstly functionally characterized a Kinesin gene from cotton that may be pivotal in fiber development.

Conclusions

In this study, biochemical characterization showed that GhKIS13A1 has microtubule binding activity, basal ATPase activity and microtubule-activated ATPase activity. *Arabidopsis* overexpressing *GhKIS13A1* exhibited less trichomes than the wild-type, and the *Kinesin-13a-1* mutant harboring *GhKIS13A1* showed less trichomes and had majority of trichomes with restored three-branch morphology, and the percentage of such trichome was matched with that in wild-type. These data suggest that *GhKIS13A1* is involved in the trichome development by regulating the total trichome number and branching. It is proposed that *GhKIS13A1* may also involve in the regulation of cotton fiber development.

Abbreviations

AMP-PNP: 5′-adenylylimidodiphosphate; ATP: Adenosine triphosphate; BSA: Bovine serum albumin; DPA: Day post anthesis; GTP: Guanosine triphosphate; IPTG: Isopropyl thio-β-D-galactoside; MS: Murashige and Skoog; MTs: Microtubules; ORF: Open reading frame; Pi: Inorganic phosphate; qPCR: Real-time quantative PCR; SDS-PAGE: Sodium dodecyl sulfate-polyacrylamide gel electrophoresis

Acknowledgments

The seeds of the *Kinesin-13a-1* mutant of *A. thaliana* were kindly provided by Dr Bo Liu from University of California at Davis. We thank Prof. Wang YZ and Zhu QH for insightful comments on the manuscript.

Funding

This research was supported by the National Natural Science Foundation of China (Grant No: 31301394) and the National Key Research and Development Program (Grant No: 2016YFD0101900).

Authors' contributions

YJL acquired the funding, performed partial experiments and wrote the manuscript. SHZ performed partial experiments and analyzed data. JS supervised the research group. XYZ and LJZ contributed to preparation for the experiment. YCL and FX revised and enhanced the paper. All authors reviewed and approved the final manuscript.

Competing interests
The authors declare that they have no competing interests.

Author details
[1]The Key Laboratory of Oasis Eco-agriculture, Agriculture College, Shihezi University, Bei 4 Road, Shihezi 832003, Xinjiang, China. [2]College of Life Sciences, Shihezi University, Bei 4 Road, Shihezi 832003, Xinjiang, China.

References

1. Kim HJ, Triplett BA. Cotton fiber growth in plant and in vitro: models for plant cell elongation and cell wall biogenesis. Plant Physiol. 2001;127(4):1361–6.
2. Ruan YL, Llewellyn DJ, Furbank RT. Suppression of sucrose synthase gene expression represses cotton fiber cell initiation, elongation, and seed development. Plant Cell. 2003;15(4):952–64.
3. Li XB, Fan XP, Wang XL, Cai L, Yang WC. The cotton ACTIN1 gene is functionally expressed in fibers and participates in fiber elongation. Plant Cell. 2005;17(3):859–75.
4. Zhang M, Zheng XL, Song SQ, Zeng QW, Hou L, Li DM, Zhao J, Wei Y, Li XB, Luo M, Xiao YH, Luo XY, Zhang JF, Xiang CB, Pei Y. Spatiotemporal manipulation of auxin biosynthesis in cotton ovule epidermal cells enhances fiber yield and quality. Nat Biotechnol. 2011;29(5):453–8.
5. Xu B, Gou JY, Li FG, Shangguan XX, Zhao B, Yang CQ, Wang LJ, Yuan S, Liu CJ, Chen LY. A cotton BURP domain protein interacts with α-expansin and their co-expression promotes plant growth and fruit production. Mol Plant. 2013;6(3):945–58.
6. Han LB, Li YB, Wang HY, Wu XM, Li CL, Luo M, Wu SJ, Kong ZS, Pei Y, Jiao GL, Xia GX. The dual functions of WLIM1a in cell elongation and secondary wall formation in developing cotton fibers. Plant Cell. 2013;25(11):4421–38.
7. Schliwa M, Woehlke G. Molecular motors. Nature. 2003;422(6933):759–65.
8. Hirokawa N, Noda Y, Tanaka Y, Niwa S. Kinesin superfamily motor proteins and intracellular transport. Nat Rev Mol Cell Biol. 2009;10(10):682–96.
9. Verhey KJ, Meyer D, Deehan R, Blenis J, Schnapp BJ, Rapoport TA, Margolis B. Cargo of Kinesin identified as JIP scaffolding proteins and associated signaling molecules. J Cell Biol. 2001;152(5):959–70.
10. Sakai T, Honing HV, Nishioka M, Uehara Y, Takahashi M, Fujisawa N, Saiji K, Seki M, Shinozaki K, Jones MA, Smirnoff N, Okada K, Wasteneys GO. Armadillo repeat-containing Kinesins and a NIMA-related kinase are required for epidermal-cell morphogenesis in Arabidopsis. Plant J. 2008;53(1):151–71.
11. Vanstraelen M, Inze D, Geelen D. Mitosis-specific Kinesins in Arabidopsis. Trends Plant Sci. 2006;11(4):167–75.
12. Miki H, Okada Y, Hirokawa N. Analysis of the Kinesin superfamily: insights into structure and function. Trends Cell Biol. 2005;15(9):467–76.
13. Reddy AS, Day IS. Kinesins in the Arabidopis genome: a comparative analysis among eukaryotes. BMC Genomics. 2001;2(1):2.
14. Vale RD. The molecular motor toolbox for intracellular transport. Cell. 2003; 112(4):467–80.
15. Richardson DN, Simmons MP, Reddy AS. Comprehensive comparative analysis of Kinesins in photosynthetic eukaryotes. BMC Genomics. 2006; 7(1):18.
16. Preuss ML, Delmer DP, Liu B. The cotton Kinesin-like calmodulin-binding protein associates with cortical microtubules in cotton fibers. Plant Physiol. 2003;132(1):154–60.
17. Preuss ML, Kovar DR, Lee YR, Staiger CJ, Delmer DP, Liu B. A plant-specific Kinesin binds to actin microfilaments and interacts with cortial microtubules in cotton fibers. Plant Physiol. 2004;136(4):3945–55.
18. Xu T, Qu Z, Yang X, Qin X, Xiong J, Wang Y, Ren D, Liu G. A cotton Kinesin GhKCH2 interacts with both microtubules and microfilaments. Biochem J. 2009;421(2):171–80.
19. Lu L, Lee YR, Pan R, Maloof JN, Liu B. An internal motor Kinesin is associated with the Golgi apparatus and plays a role in trichome morphogenesis in Arabidopsis. Mol Biol Cell. 2005;16(2):811–23.
20. Kitagawa K, Kurinami S, Oki K, Abe Y, Ando T, Kono I, Yano M, Kitano H, Iwasaki Y. A novel Kinesin 13 protein regulating rice seed length. Plant Cell Physiol. 2010;51(8):1315–29.
21. Wei L, Zhang W, Liu Z, Li Y. AtKinesin-13A is located on Golgi-associated vesicle and involved in vesicle formation/budding in Arabidopsis root-cap peripheral cells. BMC Plant Biol. 2009;9(1):138.
22. Zhao LJ, Xue F, Zhu SH, Li YJ, Liu YC, Sun J. Cloning and expression analysis of GhKinesin13 subfamily genes in Gossypium hirsutum. Acta Agron Sin. 2015;41(4):539–47 (in Chinese with English abstract).
23. Jiang JX, Zhang TZ. Extraction of total RNA in cotton tissues with CTAB-acidic phenolic method. Cotton Sci. 2003;15(3):166–7 (in Chinese with English abstract).
24. Li XY, Wang HQ, Xu T, Cao QH, Ren DT, Liu GQ. Molecular cloning, expression and biochemical property analysis of AtPK1, a Kinesin gene from Arabidopsis thaliana. Chinese Sci Bull. 2007;52(10):1338–46.
25. LeBel D, Poirier GG, Beaudoin AR. A conveninent method for the ATPase Assay. Anal Biochem. 1978;85(1):86–9.
26. Wise AA, Liu Z, Binns AN. Three Methods for the Introduction of Foreign DNA into Agrobacterium. Methods Mol Biol. 2006;343:43–53.
27. Clough SJ, Bent AF. Floral dip: a simplified method for Agrobacterium-mediated transformation of *Arabidopsis thaliana*. Plant J. 1998;16(6):735–43.
28. Umeki N, Mitsui T, Umezu N, Kondo K, Maruta S. Preparation and characterization of a novel rice plant-specific Kinesin. J Biochem. 2006;139(4):645–54.
29. Matthies HJG, Baskin RJ, Hawley RS. Orphan Kinesin NOD lacks motile properties but does possess a microtubule-stimulated ATPase activity. Mol Biol Cell. 2001;12(12):4000–12.
30. Kuznetsov SA, Gelfand VI. Bovine brain Kinesin is a microtubule-activivated ATPase. Proc Natl Acad Sci U S A. 1986;83(22):8530–4.
31. Vale RD, Reese TS, Sheetz MP. Identification of a novel force-generating protein, Kinesin, involved in microtule-based motility. Cell. 1985;42(1):39–50.
32. Alonso MC, Drummond DR, Kain S, Hoeng J, Amos L, Cross RA. An ATP gate controls tubulin binding by the tethered head of Kinesin-1. Science. 2007; 316(5821):120–3.
33. Zimmermann IM, Heim MA, Weisshaar B, Uhrig JF. Comprehensive identification of Arabidopsis thaliana MYB transcription factors interacting with R/B-like BHLH proteins. Plant J. 2004;40(1):22–34.
34. Payne CT, Zhang F, Lloyd AM. GL3 encodes a bHLH protein that regulates trichome development in arabidopsis through interaction with GL1 and TTG1. Genetics. 2000;156(3):1349–62.
35. Zhang F, Gonzalez A, Zhao M, Payne CT, Lloyd A. A network of redundant bHLH proteins functions in all TTG1-dependent pathways of Arabidopsis. Development. 2003;130(20):4859–69.
36. Wada T, Tachibana T, Shimura Y, Okada K. Epidemal cell differentiation in Arabidopsis determined by a Myb homolog, CPC. Science. 1997;277(5329): 1113–6.
37. Hulskamp M, Misra S, Jurgens G. Genetic dissection of trichome cell development in Arabidopsis. Cell. 1994;76(3):555–66.
38. Buschmann H, Hauptmann M, Niessing D, Lloyd CW, Schaffner AR. Helical growth of the Arabidopsis mutant tortifolia2 does not depend on cell division pattens but involves handed twisting of isolated cells. Plant Cell. 2009;21(7):2090–106.
39. Nakamura M, Hashimoto TA. Mutation in the Arabidopsis gamma-tubulin-containing complex causes helical growth and abnormal microtubule branching. J Cell Sci. 2009;122(Pt 13):2208–17.
40. Kong Z, Hotta T, Lee YR, Horio T, Liu B. The {gamma}-tubulin complex protein GCP4 is required for organizing functional microtubule arrays in Arabidopsis thaliana. Plant Cell. 2010;22(1):191–204.
41. Guan XY, Yu N, Shangguan XX, Wang S, Lu S, Wang LJ, Chen XY. Arabidopsis trichome research sheds light on cotton fiber development mechanisms. Chinese Sci Bull. 2007;52(13):1734–41.
42. Shangguan XX, Yang CQ, Zhang XF, Wang LJ. Functional characterization of a basic Helix-Loop-Helix (bHLH) transcription factor GhDEL65 from cotton (Gossypium hirsutum). Physiol Plant. 2016;158(2):200–12.
43. Guan X, Pang M, Nah G, Shi X, Ye W. miR828 and miR858 regulate homoeologous MYB2 gene functions in Arabidopsis trichome and cotton fibre development. Nut Commun. 2014;5(1):3050.
44. Wang S, Wang JW, Yu N, Li CH, Luo B, Gou JY, Wang LJ, Chen XY. Control of plant trichome development by a cotton fiber MYB gene. Plant Cell. 2004;16(9):2323–34.

19

Enhancement of thermoalkaliphilic xylanase production by *Pichia pastoris* through novel fed-batch strategy in high cell-density fermentation

Tingting Shang[1†], Dayong Si[1†], Dongyan Zhang[1], Xuhui Liu[1], Longmei Zhao[1,2], Cong Hu[1], Yu Fu[1] and Rijun Zhang[1*]

Abstract

Background: Xylanase degrades xylan into monomers of various sizes by catalyzing the endohydrolysis of the 1,4-β-D-xylosidic linkage randomly, possessing potential in wide industrial applications. Most of xylanases are susceptible to be inactive when suffering high temperature and high alkaline process. Therefore, it is necessary to develop a high amount of effective thermoalkaliphilic xylanases. This study aims to enhance thermoalkaliphilic xylanase production in *Pichia pastoris* through fermentation parameters optimization and novel efficient fed-batch strategy in high cell-density fermentation.

Results: Recombinant xylanase activity increased 12.2%, 7.4%, 12.0% and 9.9% by supplementing the *Pichia pastoris* culture with 20 g/L wheat bran, 5 mg/L L-histidine, 10 mg/L L-tryptophan and 10 mg/L L-methionine in shake flasks, respectively. Investigation of nutritional fermentation parameters, non-nutritional fermentation parameters and feeding strategies in 1 L bioreactor and 1 L shake flask revealed that glycerol and methanol feeding strategies were the critical factors for high cell density and xylanase activity. In 50 L bioreactor, a novel glycerol feeding strategy and a four-stage methanol feeding strategy with a stepwise increase in feeding rate were developed to enhance recombinant xylanase production. In the initial 72 h of methanol induction, the linear dependence of xylanase activity on methanol intake was observed ($R^2 = 0.9726$). The maximum xylanase activity was predicted to be 591.2 U/mL, while the actual maximum xylanase activity was 560.7 U/mL, which was 7.05 times of that in shake flask. Recombinant xylanase retained 82.5% of its initial activity after pre-incubation at 80 °C for 50 min (pH 8.0), and it exhibited excellent stability in the broad temperature (60–80 °C) and pH (pH 8.0–11.0) ranges.

Conclusions: Efficient glycerol and methanol fed-batch strategies resulting in desired cell density and xylanase activity should be applied in other *P. pastoris* fermentation for other recombinant proteins production. Recombinant xylanases with high pH- and thermal-stability showed potential in various industrial applications.

Keywords: Thermoalkaliphilic xylanase, *Pichia pastoris*, Fermentation parameter optimization, Bioprocess, Fed-batch strategy, High cell-density fermentation

* Correspondence: feedbiotech@yahoo.com
†Equal contributors
[1]Laboratory of Feed Biotechnology, State Key Laboratory of Animal Nutrition, College of Animal Science & Technology, China Agricultural University, No.2 Yuanmingyuan West RoadHaidian District, Beijing 100193, China
Full list of author information is available at the end of the article

Background

Xylan is the major component of hemicellulose, and it is the second most abundant renewable resource [1]. Xylanase (β-1,4-Endoxylanases, EC 3.2.1.8) degrades xylan into monomers by catalyzing the endohydrolysis of the 1,4-β-D-xylosidic linkage between molecules of xylose in the main chain randomly [2]. Lignocellulosic matter consists of ca. 40% cellulose, 33% hemicellulose, and 23% lignin by dry weight [3]. Endoxylanases play an important role in bioconversion of lignocellulose in feed, food, biofuel, pulp and paper industry and are widely used as raw materials in lots of industrial processes [2, 4].

Thermoalkaliphilic xylanases are required in detergent, textile, paper and pulp industries. With them, enzymes hydrolysis will be performed at elevated temperatures without cooling substrate down, thus process time will be shortened, energy will be saved and saccharification yield can be improved, leading to a more economical and feasible process [5]. Up to now, a large number of xylanases have been produced and investigated, however, most of xylanases are susceptible to denaturation when exposed to the high temperature and high alkaline process [6]. Therefore, the development of a high amount of effective thermoalkaliphilic xylanases is highly desirable.

With the development of recombinant protein engineering, genetic engineering strains are employed to produce recombinant xylanase effectively. The methylotrophic yeast *Pichia pastoris* (*P. pastoris*), one of the most extensively used expression systems for heterologous proteins production, has many advantages: (1) generally strong and tightly regulated promoters, (2) available molecular manipulation tools, (3) high cell-density fermentation on inexpensive substrate, (4) protein production in secretory fashion [7, 8]. Unlike extremozyme production by extremophile under unusual condition, the bioprocess of extremozyme production by mesophilic host organisms such as *P. pastoris* can be optimized for high biomass and enzyme activity. Therefore, expression of thermoalkaliphilic xylanases in *P. pastoris* is attractive.

Different micro-organisms have different physiological phenotype and biosynthetic capacity, thus fermentation parameters including growth medium composition and cultivation conditions (pH, temperature, dissolved oxygen (DO), etc.) affect producing cells growth and recombinant xylanase production [9]. The levels of recombinant protein production in *P. pastoris* in bioreactor cultures are typically much higher than those in shake-flask cultures [10], since these factors including pH, temperature, oxygen transfer and substrate addition can be monitored and controlled simultaneously. In bioreactor cultures, *P. pastoris* can grow to high cell density (exceeding 100 g/L dry cell weight (DCW)) [11, 12], and a large amount of recombinant proteins production is based on high concentration of producing cells in the culture [13]. For *P. pastoris* expression system, alcohol oxidase 1 (AOX1) promoter has been the most widely reported and utilized, which can be regulated by methanol for recombinant proteins production [14]. Therefore, cultivation of *P. pastoris* is normally separated into three phases: batch phase, where cells grow until initial carbon source has been exhausted; fed-batch phase, where the same carbon source is fed for high cell density; induction phase, recombinant protein is induced to produce by feeding inducer after initial carbon source exhaustion [15–17]. Nutrient feeding strategy is critical for recombinant proteins production by *P. pastoris* in high cell-density fermentation. Glycerol is a common initial carbon source, while it may repress the AOX1 promoter and lower recombinant proteins production [18, 19]. The inducer methanol excessing a certain level is cytotoxic and inhibits cells growth [20]. To prevent overfeeding of nutrients, feedback-controlled feeing systems are developed, in which the nutrient is added according to a change in DO, pH, substrate concentration or by-product (often has negative effects on targeted recombinant protein production) concentration [21–24].

In previous study, the *xynA* xylanase gene from *Thermobiafida fusca* YX was cloned and expressed in *P. pastoris* X-33. *Thermobifida fusca* is a thermophilic actinomycete and a major degrader of plant cell walls in heated organic materials. This recombinant xylanase was optimally active at 80 °C and pH 8.0 [25]. In the present study, we optimized fermentation parameters and developed novel glycerol and methanol fed-batch strategies to enhance recombinant thermoalkaliphilic xylanase production in *P. pastoris* in high cell-density fermentation in 50 L bioreactor.

Methods

Strains and reagents

To obtain the secreted thermoalkaliphilic xylanase, a *P. pastoris* strain X-33 transformed with pPICZα-A vector bearing the construct *xynA* of *Thermobifida fusca* YX with a C-terminal histidine tag locates at the downstream of AOX1 promoter was used, and the xylanase was optimally active at 80 °C and pH 8.0 [25]. Yeast extract and peptone were obtained from OXOID (Basingstoke, Hampshire, UK), yeast nitrogen base (YNB) without amino acids was purchased from BD (Sparks, MD, USA), biotin and phenylmethylsulfonyl fluoride (PMSF) were obtained from Amresco (Solon, OH, USA). Xylan from beechwood was purchased from Sigma Chemical Company (St. Louis, MO, USA). Xylose was from Wako (Osaka, Japan). Corn bran, corncob, cottonseed hull and wheat bran were purchased from Xiangchi (Shandong, China), Chunjing (Guangdong, China), Hengfeng Mining (Hebei, China) and Zhongnongzhenya (Beijing, China), respectively. All of the lignocellulosic materials passed through 0.2-mm screen sieve. Other chemicals were obtained from Sinopharm Chemical Reagent Co., Ltd (Shanghai, China).

Cultivation of recombinant *P. pastoris* in shake flask

Yeast extract peptone dextrose (YPD) plates for seed cultivation contained 10 g/L yeast extract, 20 g/L peptone, 20 g/L dextrose and 20 g/L agar. A single colony was inoculated in 50 mL buffered glycerol-complex medium (BMGY, 10 g/L yeast extract, 20 g/L tryptone, 13.4 g/L YNB, 4×10^{-4} g/L biotin, 100 mM potassium phosphate (pH 6.0), 1% (v/v) glycerol), and incubated at 30 °C, 220 rpm for 24 h in a 500 mL baffled shake flask, then 5 mL of the culture was transferred into 100 mL BMGY medium in 1 L baffled shake flask and cultured at the same condition. To select the optimal carbon source, xylanase activity and DCW were determined with 1% (w/v) glucose or 1% (w/v) maltose as carbon source to replace 1% (v/v) glycerol in the BMGY medium. After incubation for 60 h, 0.5% (v/v) methanol was added every 24 h. The culture was sampled at 0 h and 24 h of fermentation, and then every 12 h. We also investigated the xylanase activity and DCW with 1% (v/v) glycerol or 1% (w/v) glucose as the initial carbon source, and 0.5% (v/v) methanol was added every 24 h after inoculation. To obtain highly active xylanase, 2% (w/v) tryptone in the BMGY medium was replaced by various nitrogen sources (2% $(NH_4)_2SO_4$, 2% $(NH_2)_2CO$, 2% soya peptone, 2% (v/v) ammonia solution (pH 6.0)). To study the effect of amino acid, protease inhibitor and lignocellulosic material on xylanase production, recombinant *P. pastoris* was cultured in BMGY medium supplemented with 5 mg/L L-histidine, 10 mg/L L-tryptophan, 10 mg/L L-methionine, 1 mM PMSF, 1 mM ethylenediaminetetraacetic acid (EDTA), 20 g/L corn bran, 20 g/L corncob, 20 g/L cottonseed hull, 20 g/L wheat bran, 10 g/L corn bran + 10 g/L wheat bran, 10 g/L corncob + 10 g/L wheat bran, and 10 g/L cottonseed hull + 10 g/L wheat bran for 96 h, respectively. Methanol (0.5% (v/v)) was added every 24 h to maintain induction. Samples were collected at 96 h for xylanase activity analysis.

Cultivation of recombinant *P. pastoris* in 1 L bioreactor

Seed culture was prepared in BMGY medium at 30 °C, 220 rpm in baffled shake flask for 24 h. Seed culture (30 mL) was inoculated to 600 mL BMGY medium supplemented with 3.0 mg L-histidine, 6.0 mg L-tryptophan, 6.0 mg L-methionine and 12.0 g wheat bran in 1 L bioreactor. Cultivation was conducted at 30 °C, 800 rpm. The pH was maintained at 6.0 with 28% (w/v) NH_4OH and DO was kept above 20%. Different treatments were performed: I) for 1 L bioreactors, the glycerol growth phase was maintained for 24 h by feeding 1% (v/v) glycerol at 9 h and 17 h, the methanol induction phase was maintained for following 120 h by feeding 0.5% (v/v) methanol at inoculated time of 24 h, 32 h, 48 h, then every 8 h to the end; II) for the other 1 L bioreactors, 0.5% (v/v) methanol was added every 24 h after inoculation; III) *P. pastoris* expressed xylanase in 100 mL BMGY medium without supplementation of amino acids and wheat bran in 1 L shake flasks at 30 °C, 220 rpm, but having the same glycerol and methanol feeding strategies with treatment II.

Cultivation of recombinant *P. pastoris* in 50 L bioreactor

Seed culture (1.25 L) was inoculated to 25 L basal salt medium as Li et al. described, and $(NH_4)_2SO_4$ was replaced by $(NH_2)_2CO$ in a 50 L bioreactor (Baoxing Co., Shanghai, China) [26]. In addition, we added 0.125 g L-histidine, 0.25 g L-tryptophan, 0.25 g L-methionine and 500 g wheat bran into the 50 L bioreactor. Temperature was maintained at 30 °C by water cooling, pH was automatically adjusted to 6.0 with 28% (w/v) NH_4OH, the agitation rate ranged from 200 to 350 rpm, and the tank pressure was 0.7 bar. A total cultivation time of 120 h included glycerol growth stage (0–24 h) and methanol induction stage (25–120 h). Glycerol cultivation was divided into three phases: batch phase (0–19 h): cells grew until the initial carbon source glycerol was depleted, which was indicated by a sudden increase in DO; fed-batch phase (20–22 h): 50% (v/v) glycerol containing 1.2% (v/v) PTM1 trace salt solution was fed at the rate of 10.4 mL h^{-1} L^{-1}; starvation phase (23–24 h): glycerol feeding was stopped, and depletion of glycerol was indicated by DO rose to 100%. For methanol induction, feeding manner of 100% (v/v) methanol containing 1.2% (v/v) PTM1 trace salt solution was shown in the Table 1. Samples were collected every 12 h.

Determination of DCW

Five milliliters of culture broth was placed in pre-weighed tube and centrifuged at 12,000 rpm for 5 min. The supernatant was removed and stored at −20 °C for xylanase activity analysis. The tube was dried at 100 °C to a constant weight, and then DCW was calculated and reported as g/L (g DCW per L of culture).

Determination of xylanase activity

The xylanase activity was determined by measuring the reducing sugar released from beechwood xylan with dinitrosalicylic acid (DNS) method [27, 28]. The reaction

Table 1 Methanol feeding rate in methanol induction phase

	Methanol feeding rate (mL h^{-1} L^{-1})			
	0.8	1.2	1.6	1.8
Induction time (h)	0–12	13–24	25–48	49–96

mixture contained 20 μL of appropriately diluted crude enzyme solution and 180 μL of 1% (w/v) beechwood xylan in 50 mM phosphate buffer (pH 8.0). After being incubated at 80 °C for 20 min, the reaction was terminated by chilling the mixture on ice and adding 200 μL of DNS solution. After boiling sample for 10 min, the absorbance of 200 μL sample at 540 nm was measured by iMark Microplate Reader (Bio-Rad, Hercules, CA, USA). One unit of xylanase activity was defined as the amount of xylanase releasing 1 μmole of reducing sugar per minute under the assay condition (with xylose as a standard).

Characterization of the recombinant xylanase

To determine the effect of pH on xylanase stability, xylanase was pre-incubated at 50 °C in different pH buffers (pH 8.0 using sodium phosphate buffer, pH 9.0–11.0 using glycine-NaOH buffer) for 0, 1, 2, 3, 5, 7, 9, 11 h. Samples were taken at different time intervals and xylanase residual activity was measured according to the above method after cooling.

To evaluate thermal stability, xylanase was pre-incubated at 60 °C, 70 °C, 80 °C and 90 °C for 0, 5, 10, 15, 20, 25, 30, 40, 50 and 60 min. Samples were taken at different time intervals. After cooling, residual activity of treated xylanase was assayed according to the above method.

Statistical analysis

The SPSS 17.0 statistical software package and software Microsoft Excel 2010 were used to analyze and plot the experimental data. The data were expressed as the means ± standard deviations. The differences among the groups were analyzed by one-way analysis variance (ANOVA) followed by Duncan's method. A *p*-value less than 0.05 was considered statistically significant.

Results and discussion

Effect of carbon source on DCW and xylanase activity in shake flask

Carbon source provides energy and carbon skeletons for *P. pastoris* cells growth and recombinant protein production. The xylanase activity was almost zero before induction with glycerol or glucose as initial carbon source (Fig. 1a). It meant that *P. pastoris* cells growth could be supported by glycerol or glucose, however, xylanase gene expression driven by AOX1 promoter can't be induced in glycerol- or glucose-containing medium, which was in agreement with Hellwig's report [29]. When 0.5% (v/v) methanol was added at 60 h, the xylanase activity increased in the medium whose initial carbon source was glucose or glycerol, and it was always higher than that in the medium with maltose as initial carbon source. There was almost no change in DCW after fermentation for 24 h in glycerol- or glucose-grown *P. pastoris* cultivation, while the DCW was lower at the beginning of fermentation and increased with the time in maltose-grown *P. pastoris* cultivation (Fig. 1a). We also investigated the DCW and xylanase activity in glycerol- and glucose-grown *P. pastoris* cultivation when 0.5% (v/v) methanol was added every 24 h after inoculation (Fig. 1b), and the maximum xylanase activity in both glycerol-grown and glucose-grown *P. pastoris* cultivation detected at 84 h was about 75.0 U/mL and kept more or less the same until 108 h. After fermentation for 24 h, the DCW ranged from 13.53 g/L to 14.43 g/L with glycerol as initial carbon source, and it ranged from 10.75 g/L to 13.23 g/L with glucose as initial carbon source (Fig. 1b). Glucose, one of fermentative carbon sources, tends to be avoid in favor of the non-fermentative carbon source – glycerol, because the by-product ethanol may repress the AOX1 promoter for recombinant protein production [10, 30]. In this study, glycerol was treated as the optimal initial carbon source.

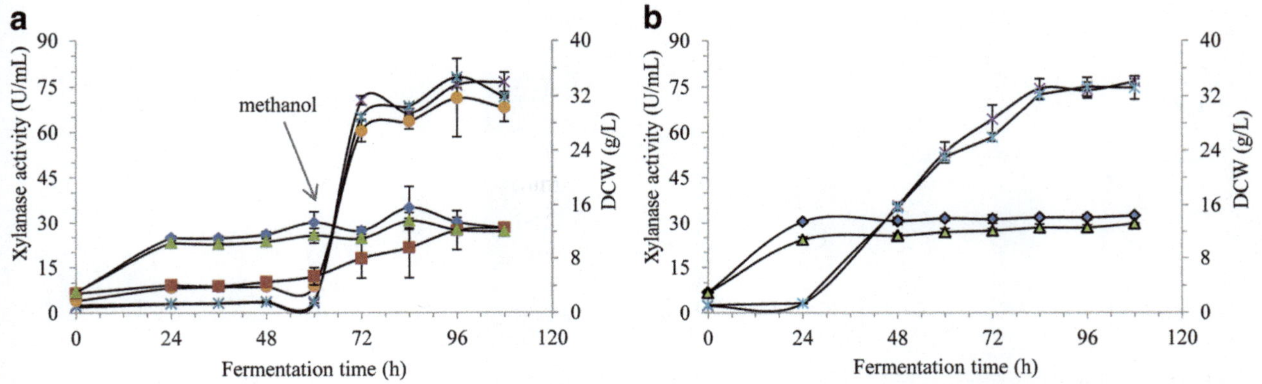

Fig. 1 DCW and xylanase activity profiles with xylanase expression at different carbon sources in shake flask. **a** methanol was added after fermentation for 60 h, **b** methanol was added every 24 h. Circles, xylanase activity (maltose); asterisks, xylanase activity (glucose); crosses, xylanase activity (glycerol); triangles, DCW (glucose); squares, DCW (maltose); diamonds, DCW (glycerol)

Supplementation of wheat bran improved xylanase activity in shake flask

Lignocellulosic material contains a plenty of xylan and other nutrients. It may serve as substrate for cells growth or inducer for hemicellulose enzyme production, [31]. BMGY medium supplemented with lignocellulosic materials (corn bran, wheat bran, cottonseed hull and corncob) was used for xylanase production in *P. pastoris*. The xylanase activity was significantly improved by supplementing 2% (w/v) wheat bran (Fig. 2a). Wheat bran can provide high quality proteins, minerals, and phenolic and bioacitve carbohydrate after enzymatical hydrolysis [32–34], which may be benefit for *P. pastoris* cells growth and recombinant xylanase production. In previous studies, corncob could be used by filamentous fungi (*Rhizopus stolonifer* JS-1008, *Streptomyces sp.* CS802, *Aspergillus niger* CECT 2700, etc.) for xylanase production [35–37]. However, the xylanase activity was decreased by 70.31% with adding 2% (w/v) corncob to *P. pastoris* culture in this study (Fig. 2a).

Effect of nitrogen source on xylanase activity in shake flask

Nitrogen source plays an important role in the growth of micro-organisms and metabolite production. To study the effect of nitrogen source on xylanase activity, tryptone in BMGY medium was replaced by one organic nitrogen source – soya peptone and three inorganic nitrogen sources – $(NH_2)_2CO$, $(NH_4)_2SO_4$ and NH_4OH (pH 6.0), respectively. As shown in Fig. 2b, when tryptone was used as nitrogen source, the highest xylanase activity reached 71.3 U/mL at 96 h, and the second highest xylanase activity was 61.3 U/mL with $(NH_2)_2CO$ as nitrogen source. No obvious difference in xylanase activity was observed when NH_4OH (pH 6.0), $(NH_4)_2SO_4$ or soya peptone was used. For economical and commercial purpose, $(NH_2)_2CO$ has potential application in large-scale fermentation.

Supplementation of amino acid and protein inhibitor improved xylanase activity in shake flask

When adding PMSF to the medium, xylanase activity was increased from 102.3 U/mL to 106.7 U/mL ($P < 0.05$), it can be inferred that extracellular proteases were produced during xylanase production in *P. pastoris*. Some researchers indicated that supplementation of amino acids improved recombinant protein production [38], and addition of casamino acids could protect protein expressed by *P. pastoris* from proteolytic degradation [39]. The xylanase activity was

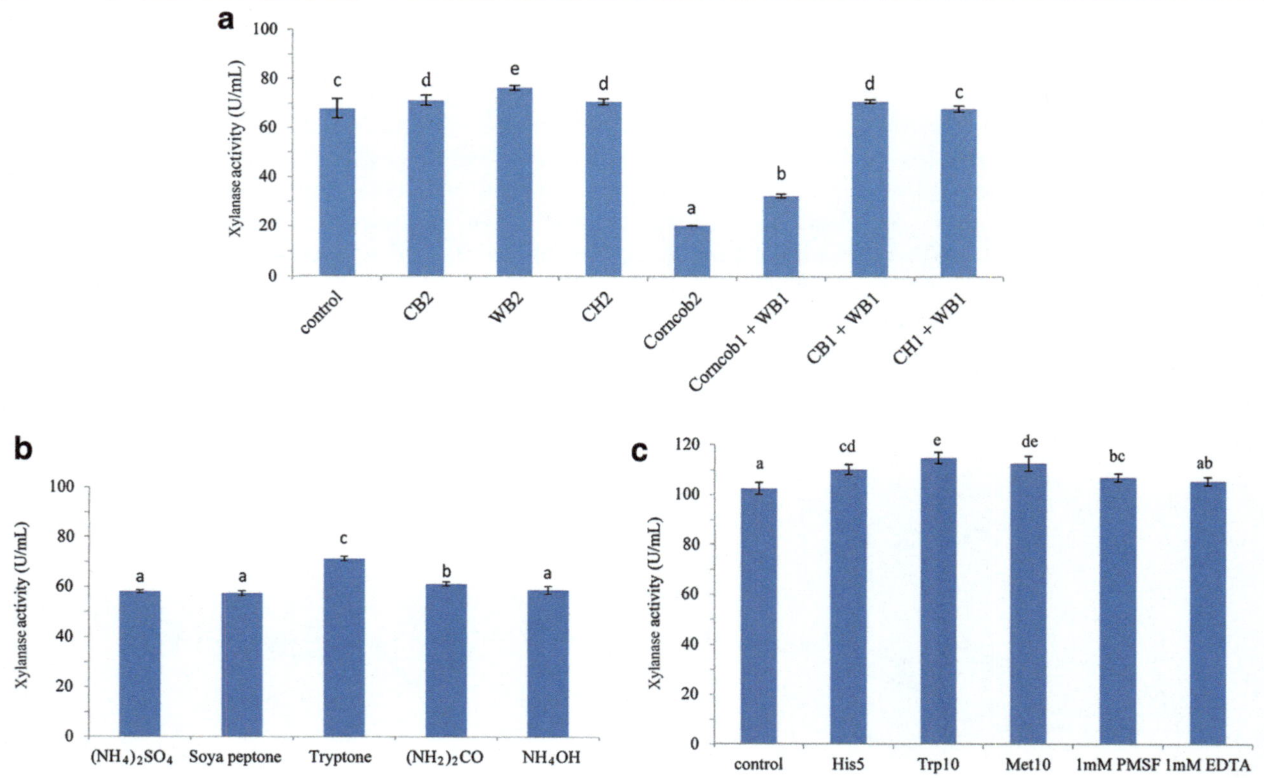

Fig. 2 Effect of fermentation parameters on xylanase activity in shake flask. **a** lignocellulosic materials, **b** nitrogen sources, **c** amino acids and protease inhibitors. CB2, corn bran (20 g/L); WB2, wheat bran (20 g/L); CH2, cottonseed hull (20 g/L); Corncob2, corncob (20 g/L); Corncob1 + WB1, corncob (10 g/L) and wheat bran (10 g/L); CB1 + WB1, corn bran (10 g/L) and wheat bran (10 g/L); CH1 + WB1, cottonseed hull (10 g/L) and wheat bran (10 g/L); His5, L-histidine (5 mg/L); Trp10, L-tryptophan (10 mg/L); Met10, L-methionine (10 mg/L)

increased by 7.4%, 12.0% and 9.9% when supplemented BMGY medium with 5 mg/L L-histidine, 10 mg/L L-tryptophan and 10 mg/L L-methionine, respectively (Fig. 2c). Besides acting as alternative and competing substrates to decrease hydrolysis of target protein, sufficiently available exogenous amino acids in nitrogen-metabolism can be incorporared directly into biomass [40].

Xylanase production in *P. pastoris* in 1 L bioreactor and 1 L shake flask

Singh et al. reported that the recombinant protein produced in bioreactor was significantly higher than that in shake flask [41]. As shown in Fig. 3, DCW in all of the treatments was stable after 48 h, however, DCW in treatment I was 45.05 g/L at 96 h, significantly higher than those in treatment II (26.73 g/L) and treatment III (14.48 g/L). It can be inferred that DCW was highly correlated to glycerol content. The maximum xylanase activity reached 114.1 U/mL in treatment II and 108.8 U/mL in treatment III at 72 h, however, the maximum xylanase activity was 224.4 U/mL at 96 h in treatment I. These results obviously suggested that glycerol and methanol feeding strategies were the critical factors in high DCW and xylanase activity. In treatment I, glycerol had already been depleted at 24 h of fermentation, which was indicated by a spike in DO, and then pulsed addition of 6 ml of methanol (50% (v/v)) in the 600 mL BMGY medium was performed at 24 h, 32 h, 48 h, and then every 8 h to the end of fermentation in 1 L bioreactors. When *P. pastoris* cells were growing with sufficient carbon source, DO was strongly demanded for carbon source assimilation, as a result the DO tension fell and became very low; while the DO tension increased sharply once carbon source was depleted. According to these phenomena, the methanol depletion time from the time point pulsed addition of methanol to the time point when DO rose to 50% again was recorded and shown in Fig. 4. In general, methanol consumption rate increased in the initial 64 h and then showed a tendency of decrease, and the result also suggested methanol was insufficient in *P. pastoris* cultivation (Fig. 4).

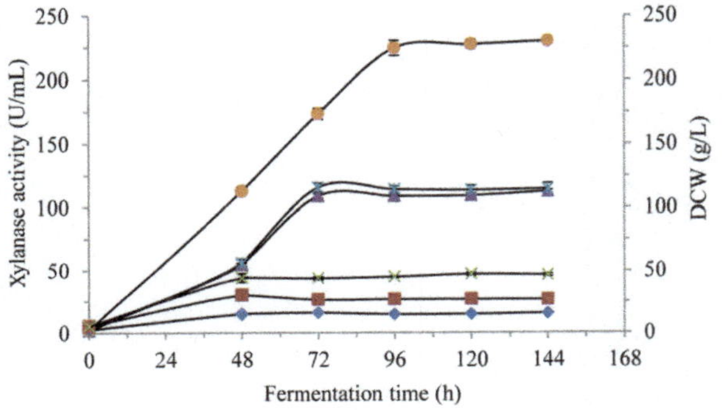

Fig. 3 Recombinant xylanase production in *P.pastoris* in 1 L bioreactor and 1 L shake flask. Circles, xylanase activity in treatment I; asterisks, xylanase activity in treatment II; triangles, xylanase activity in treatment III; crosses, DCW in treatment I; squares, DCW in treatment II; diamonds, DCW in treatment III

Development of novel glycerol and methanol fed-batch strategies in 50 L bioreactor

Based on the above findings, novel glycerol and methanol fed-batch strategies were adopted to improve DCW and xylanase activity in 50 L bioreactor, where glycerol and methanol feeding rates, DO, pH and temperature could be easily controlled and monitored. Glycerol enters glycolysis and requires respiration to oxidize NADH to serve as energy source [18]. A lower initial glycerol concentration and fed-batch operation led to a higher cell growth rate and desired biomass [23], and the concentration of the producing cells in the medium affects the volumetric productivity of recombinant enzymes [13]. In this study, glycerol cultivation (total 24 h) consisted of batch phase (0–19 h), fed-batch phase (20–22 h) and starvation phase (23–24 h). After culturing *P. pastoris* in glycerol-containing medium for 24 h, depletion of glycerol was indicated by DO rose to 100%, and the DCW reached 41.8 g/L, but almost no xylanase activity was detected (Fig. 5), which indicated AOX1 promoter could be tightly regulated by methanol, and glycerol catabolic process was relatively independent of the methanol metabolism pathways [42]. Weinhandl et al. also reported glycerol supported *P. pastoris* cells growth without inducing the AOX promoter [43], however, Bara et al. observed leaky expression from AOX promoter, and functional recombinant protein was detected in bioreactor prior to methanol induction in glycerol-grown *P. pastoris* cultivation [20]. In any case, pre-induction phase cultivation should be optimized to maximize recombinant protein production.

The methanol not only induces the AOX1 promoter in *P. pastoris* for recombinant protein production, but it is also the carbon/energy source. The methanol concentration seriously affects cell uptake rate and specific production rate. Notably, excess of methanol led to accumulation of formaldehyde to toxic level and reduced biomass yield [44]. Methanol is assimilated vigorously with DO consumption. The oxygen demand not only associates with the cellular electron transport, but alcohol oxygenase also requires molecular oxygen as a substrate [45]. In this study, methanol feeding rate was adjusted to response to online measured fermentation parameter – DO, which was kept fluctuating around fixed value. Methanol intake (feeding) rate gradually increased as described in Materials and Methods. And as shown in Fig. 6a,

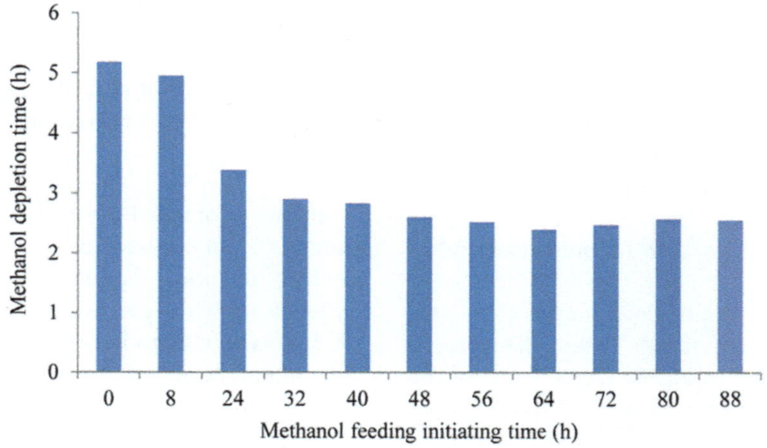

Fig. 4 Methanol depletion time with different methanol feeding initiating time in 1 L bioreactor

in the initial 72 h of methanol induction, methanol intake was adapted to the equation:

$$MI = 0.1984 m_{time}^{2} + 23 m_{time} - 30.286 \left(R^2 = 0.9989\right) \quad (1)$$

where MI was methanol intake, m_{time} was methanol induction time. The xylanase activity increased with m_{time} and finally reached a plateau (Fig. 5). It was fitted for equation:

$$Xa = -0.0015\, m_{time}^{3} + 0.2184\, m_{time}^{2} - 1.285\, m_{time} + 28.034 \left(R^2 = 0.9842\right) \quad (2)$$

where Xa was xylanase activity. The predicted maximum xylanase activity was 591.2 U/mL when methanol was fed for 94.0 h according to Eq. 2 and extreme value theorem, and the actual maximum xylanase activity was 560.7 U/mL, which was 7.05 times of that in shake flask. The linear dependence of xylanase activity on methanol intake in initial 72 h of methanol induction was observed (Fig. 6b), and the resulting equation:

$$Xa = 0.1906\, MI + 1.5033 \left(R^2 = 0.9726\right) \quad (3)$$

At the end of total cultivation, the DCW reached 53.48 g/L in 50 L bioreactor, far more than that in shake flask (Fig. 5). Both xylanase activity and DCW were significantly improved through novel glycerol and methanol fed-batch strategies in high cell-density fermentation.

Characterization of xylanase activity under high pH and temperature

The optimal temperature of xylanase has been proved to be 80 °C at the optimal pH 8.0 [25]. We evaluated the residual xylanase activity at 50 °C with different incubation time in a range of pH 8.0–11.0. It showed

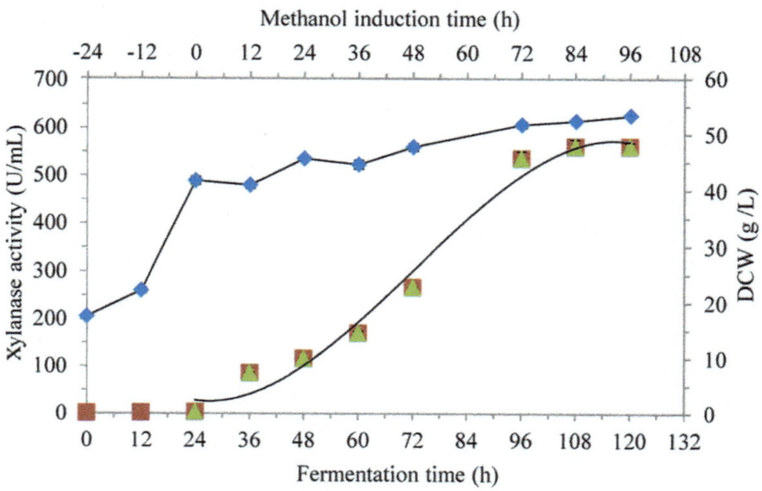

Fig. 5 Recombinant xylanase production in *P. pastoris* in 50 L bioreactor

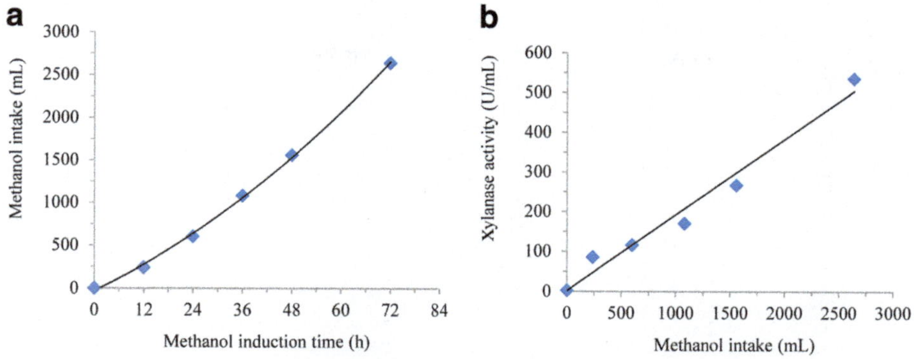

Fig. 6 Time-course of methanol intake and dependence of xylanase activity on methanol intake in 50-L bioreactor. **a** time-course of the methanol intake, **b** linear dependence of xylanase activity on methanol intake

excellent pH stability. After pre-incubating for 11 h, residual xylanase activity was over 95.0% at pH 8.0, and more than 85.0% of xylanase activity was retained even at pH 11.0 (Fig. 7a). According to Ergun and Calik, the most promising recombinant thermoalkaliphilic xylanase of *B. halodurans* TSEV1 expressed in *P. pastoris* retained 50.0% of its initial activity at 80 °C for 45 min (pH 9.0), whose xylanase activity was 502.0 U/mL [5, 46]. In this study, residual xylanase activity was still over 80.0% after pre-incubation at 80 °C for 50 min (pH 8.0); about 90.0% of its initial activity was retained at 70 °C for 60 min (pH 8.0) (Fig. 7b). This xylanase with high pH- and temperature tolerance showed potential in industry, especially in the pulp and paper industry [6]. The bioprocess will be more economical and feasible when the labor intensive pH and temperature readjustment step is trimmed down [47].

Long-term storage stability

Enzyme stability is a critical factor for industrial application [48]. The fermentation broth was stored at −20 °C for 1 year to analyze xylanase activity. The residual xylanase activity was 99.6% after 1 year, providing important support for xylanase to apply widely.

Conclusions

Recombinant thermoalkaliphilic xylanase production in *P. pastoris* was enhanced by fermentation parameters optimization. Glycerol and methanol feeding strategies were the critical factors leading to the difference of DCW and xylanase activity between in bioreactor and in shake flask. Novel efficient glycerol and methanol fed-batch strategies were developed to improve xylanase activity to be 560.7 U/mL in high cell-density fermentation in 50 L bioreactor, and should be applied in other *P. pastoris* fermentation for recombinant proteins production. Recombinant xylanase with high stability in the broad temperature (60–80 °C) and pH (pH 8.0–11.0) ranges was suitable for various industrial applications such as detergent, textile, and paper and pulp industry.

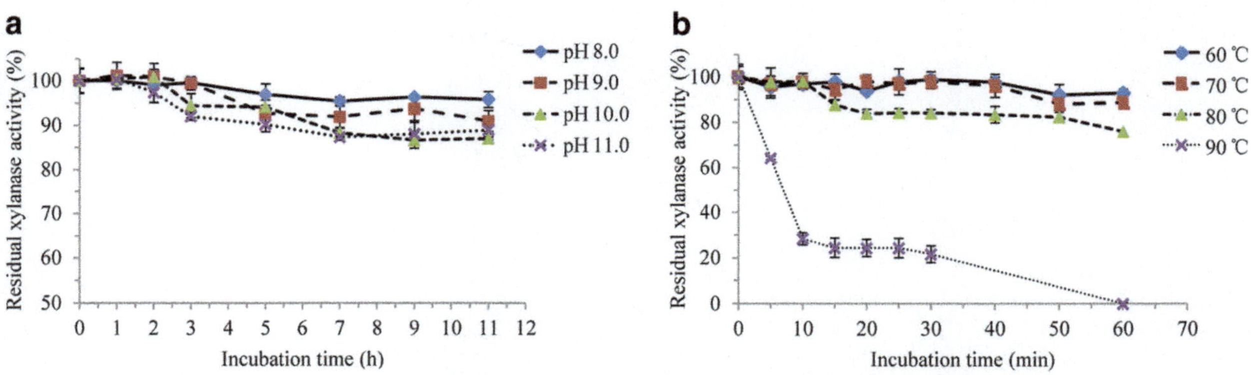

Fig. 7 The properties of recombinant xylanase. **a** effect of pH on the stability of recombinant xylanase, **b** effect of temperature on the stability of recombinant xylanase

Abbreviations
AOX1: Alcohol oxidase 1; BMGY: Buffered glycerol-complex medium; DCW: Dry cell weight; DNS: Dinitrosalicylic acid; DO: Dissolved oxygen; *E. coli*: *Escherichia coli*; EDTA: Ethylenediaminetetraacetic acid; NSPs: Nonstarch polycaccharides; *P. pastoris*: *Pichia pastoris*; PMSF: Phenylmethylsulfonyl fluoride; YNB: Yeast nitrogen base; YPD: Yeast extract peptone dextrose medium

Acknowledgements
Not applicable.

Funding
This research was supported by the 'Five-twelfth' National Science and Technology Support Program of China (2011BAD26B0403).

Authors' contributions
TS and DS designed and conducted the experiment, analyzed the data and drafted this manuscript. DZ and XL supported data analysis and read the proof manuscript. LZ participated in shake flask fermentation. CH and YF participated in 1 L bioreactor fermentation. RZ is the leader of the group and led the work. All authors read and approved the final manuscript.

Competing interests
The authors declare that they have no competing interests.

Author details
[1]Laboratory of Feed Biotechnology, State Key Laboratory of Animal Nutrition, College of Animal Science & Technology, China Agricultural University, No.2 Yuanmingyuan West RoadHaidian District, Beijing 100193, China. [2]College of Animal Science and Technology, Henan University of Science and Technology, Luoyang 471003, China.

References

1. Cayetano-Cruz M, de los Santos AIP, Garcia-Huante Y, Santiago-Hernandez A, Pavon-Orozco P, Lopez VELY, Hidalgo-Lara ME. High level expression of a recombinant xylanase by *Pichia pastoris* cultured in a bioreactor with methanol as the sole carbon source: Purification and biochemical characterization of the enzyme. Biochem Eng J. 2016;112:161–9.
2. Van Dyk JS, Pletschke BI. A review of lignocellulose bioconversion using enzymatic hydrolysis and synergistic cooperation between enzymes-Factors affecting enzymes, conversion and synergy. Biotechnol Adv. 2012;30:1458–80.
3. Sa-Pereira P, Paveia H, Costa-Ferreira M, Aires-Barros MR. A new look at xylanases - An overview of purification strategies. Mol Biotechnol. 2003;24: 257–81.
4. Martin-Sampedro R, Rodriguez A, Ferrer A, Garcia-Fuentevilla LL, Eugenio ME. Biobleaching of pulp from oil palm empty fruit bunches with laccase and xylanase. Bioresour Technol. 2012;110:371–8.
5. Ergun BG, Calik P. Lignocellulose degrading extremozymes produced by *Pichia pastoris*: current status and future prospects. Bioprocess Biosyst Eng. 2016;39:1–36.
6. Lin XQ, Han SY, Zhang N, Hu H, Zheng SP, Ye YR, Lin Y. Bleach boosting effect of xylanase A from *Bacillus halodurans* C-125 in ECF bleaching of wheat straw pulp. Enzyme Microb Tech. 2013;52:91–8.
7. Ahmad M, Hirz M, Pichler H, Schwab H. Protein expression in *Pichia pastoris*: recent achievements and perspectives for heterologous protein production. Appl Microbiol Biotechnol. 2014;98:5301–17.
8. Gmeiner C, Saadati A, Maresch D, Krasteva S, Frank M, Altmann F, Herwig C, Spadiut O. Development of a fed-batch process for a recombinant *Pichia pastoris* Delta och1 strain expressing a plant peroxidase. Microbial Cell Factories. 2015;14:1.
9. Hahn-Hagerdal B, Karhumaa K, Larsson CU, Gorwa-Grauslund M, Gorgens J, van Zyl WH. Role of cultivation media in the development of yeast strains for large scale industrial use. Microbial cell factories. 2005;4:31.
10. Macauley-Patrick S, Fazenda ML, McNeil B, Harvey LM. Heterologous protein production using the *Pichia pastoris* expression system. Yeast. 2005;22:249–70.
11. Jahic M, Veide A, Charoenrat T, Teeri T, Enfors SO. Process technology for production and recovery of heterologous proteins with *Pichia pastoris*. Biotechnol Prog. 2006;22:1465–73.
12. Jahic M, Rotticci-Mulder JC, Martinelle M, Hult K, Enfors SO. Modeling of growth and energy metabolism of *Pichia pastoris* producing a fusion protein. Bioprocess Biosyst Eng. 2002;24:385–93.
13. Salehmin MNI, Annuar MSM, Chisti Y. High cell density fed-batch fermentations for lipase production: feeding strategies and oxygen transfer. Bioprocess Biosyst Eng. 2013;36:1527–43.
14. Wang JR, Li YY, Liu DN. Improved Production of *Aspergillus usamii* endo-beta-1,4-xylanase in *Pichia pastoris* via Combined Strategies. Biomed Res Int. 2016;2016:3265895.
15. Cereghino GPL, Cereghino JL, Ilgen C, Cregg JM. Production of recombinant proteins in fermenter cultures of the yeast *Pichia pastoris*. Curr Opin Biotech. 2002;13:329–32.
16. Cos O, Ramon R, Montesinos JL, Valero F. Operational strategies, monitoring and control of heterologous protein production in the methylotrophic yeast *Pichia pastoris* under different promoters: A review. Microbial cell factories. 2006;5:17.
17. Zhou XY, Yu Y, Tao JJ, Yu L. Production of LYZL6, a novel human c-type lysozyme, in recombinant *Pichia pastoris* employing high cell density fed-batch fermentation. J Biosci Bioeng. 2014;118:420–5.
18. Zhang WH, Potter KJH, Plantz BA, Schlegel VL, Smith LA, Meagher MM. *Pichia pastoris* fermentation with mixed-feeds of glycerol and methanol: growth kinetics and production improvement. J Ind Microbiol Biot. 2003;30:210–5.
19. Fang ZG, Xu L, Pan DJ, Jiao LC, Liu ZM, Yan YJ. Enhanced production of *Thermomyces lanuginosus* lipase in *Pichia pastoris* via genetic and fermentation strategies. J Ind Microbiol Biot. 2014;41:1541–51.
20. Bawa Z, Routledge SJ, Jamshad M, Clare M, Sarkar D, Dickerson I, Ganzlin M, Poyner DR, Bill RM. Functional recombinant protein is present in the pre-induction phases of *Pichia pastoris* cultures when grown in bioreactors, but not shake-flasks. Microbial cell factories. 2014;13:127.
21. Zheng J, Zhao W, Guo N, Lin FL, Tian J, Wu LS, Zhou HB. Development of an industrial medium and a novel fed-batch strategy for high-level expression of recombinant beta-mananase by *Pichia pastoris*. Bioresource Technol. 2012;118:257–64.
22. Zhao W, Wang JW, Deng RQ, Wang XZ. Scale-up fermentation of recombinant *Candida rugosa* lipase expressed in *Pichia pastoris* using the GAP promoter. J Ind Microbiol Biot. 2008;35:189–95.
23. Markošová K, Weignerová L, Rosenberg M, Křen V, Rebroš M. Upscale of recombinant alpha-L-rhamnosidase production by *Pichia pastoris* Mut(S) strain. Front Microbiol. 2015;6:1140.
24. Ding J, Gao MJ, Hou GL, Liang KX, Yu RS, Li Z, Shi ZP. Stabilizing porcine interferon-alpha production by *Pichia pastoris* with an ethanol on-line measurement based DO-Stat glycerol feeding strategy. J Chem Technol Biot. 2014;89:1948–53.
25. Zhao LM, Geng J, Guo YQ, Liao XD, Liu XH, Wu RJ, Zheng ZJ, Zhang RJ. Expression of the *Thermobifida fusca* xylanase Xyn11A in *Pichia pastoris* and its characterization. Bmc Biotechnol. 2015;15:18.
26. Li YY, Zhong KX, Hu AH, Liu DN, Chen LZ, Xu SD. High-level expression and characterization of a thermostable xylanase mutant from *Trichoderma reesei* in *Pichia pastoris*. Protein Expres Purif. 2015;108:90–6.
27. Bailey MJ, Biely P, Poutanen K. Interlaboratory testing of methods for assay of xylanase activity. J Biotechnol. 1992;23:257–70.
28. Miller GL. Use of dinitrosalicylic acid reagent for determination of reducing sugar. Anal Chem. 1959;31:426–8.
29. Hellwig S, Emde F, Raven NPG, Henke M, van der Logt P, Fischer R. Analysis of single-chain antibody production in *Pichia pastoris* using on-line methanol control in fed-batch and mixed-feed fermentations. Biotechnol Bioeng. 2001;74:344–52.
30. Inan M, Meagher MM. The effect of ethanol and acetate on protein expression in *Pichia pastoris*. J Biosci Bioeng. 2001;92:337–41.
31. Delabona PD, Farinas CS, Lima DJD, Pradella JGD. Experimental mixture design as a tool to enhance glycosyl hydrolases production by a new *Trichoderma harzianum* P49P11 strain cultivated under controlled bioreactor submerged fermentation. Bioresource Technol. 2013;132:401–5.
32. Maes C, Delcour JA. Structural characterisation of water-extractable and water-unextractable arabinoxylans in wheat bran. J Cereal Sci. 2002;35:315–26.
33. Majzoobi M, Pashangeh S, Farahnaky A, Eskandari MH, Jamalian J. Effect of particle size reduction, hydrothermal and fermentation treatments on phytic acid content and some physicochemical properties of wheat bran. J Food Sci Tech Mys. 2014;51:2755–61.

34. Wood IP, Cook NM, Wilson DR, Ryden P, Robertson JA, Waldron KW. Ethanol from a biorefinery waste stream: Saccharification of amylase, protease and xylanase treated wheat bran. Food Chem. 2016;198:125–31.
35. Perez-Rodriguez N, Oliveira F, Perez-Bibbins B, Belo I, Agrasar AT, Dominguez JM. Optimization of xylanase production by filamentous fungi in solid-state fermentation and scale-up to horizontal tube bioreactor. Appl Biochem Biotechnol. 2014;173:803–25.
36. Simkhada JR, Yoo HY, Choi YH, Kim SW, Yoo JC. An extremely alkaline novel xylanase from a newly isolated *streptomyces* strain cultivated in corncob medium. Appl Biochem Biotechnol. 2012;168:2017–27.
37. Zhang ZC, Li JS, Feng F, Liu D, Pang QX, Li M, Chen KP. Optimization of nutrition constituents for xylanase activity by *Rhizopus stolonifer* under solid-state fermentation on corncob. Bioresources. 2013;8:2018–32.
38. Gorgens JF, van Zyl WH, Knoetze JH, Hahn-Hagerdal B. Amino acid supplementation improves heterologous protein production by *Saccharomyces cerevisiae* in defined medium. Appl Microbiol Biotechnol. 2005;67:684–91.
39. Werten MWT, Van den Bosch TJ, Wind RD, Mooibroek H, De Wolf FA. High-yield secretion of recombinant gelatins by *Pichia pastoris*. Yeast. 1999;15: 1087–96.
40. Albers E, Larsson C, Liden G, Niklasson C, Gustafsson L. Influence of the nitrogen source on *Saccharomyces cerevisiae* anaerobic growth and product formation. Appl Environ Microb. 1996;62:3187–95.
41. Singh S, Gras A, Fiez-Vandal C, Ruprecht J, Rana R, Martinez M, Strange PG, Wagner R, Byrne B. Large-scale functional expression of WT and truncated human adenosine A(2A) receptor in *Pichia pastoris* bioreactor cultures. Microbial cell factories. 2008;7:28.
42. Sibirny AA, Titorenko VI, Gonchar MV, Ubiyvovk VM, Ksheminskaya GP, Vitvitskaya OP. Genetic-control of methanol utilization in yeasts. J Basic Microb. 1988;28:293–319.
43. Weinhandl K, Winkler M, Glieder A, Camattari A. Carbon source dependent promoters in yeasts. Microbial cell factories. 2014;13:5.
44. Khatri NK, Hoffmann F. Impact of methanol concentration on secreted protein production in oxygen-limited cultures of recombinant *Pichia pastoris*. Biotechnol Bioeng. 2006;93:871–9.
45. Lee CY, Lee SJ, Jung KH, Katoh S, Lee EK. High dissolved oxygen tension enhances heterologous protein expression by recombinant *Pichia pastoris*. Process Biochem. 2003;38:1147–54.
46. Kumar V, Satyanarayana T. Generation of xylooligosaccharides from microwave irradiated agroresidues using recombinant thermo-alkali-stable endoxylanase of the polyextremophilic bacterium *Bacillus halodurans* expressed in *Pichia pastoris*. Bioresource Technol. 2015;179:382–9.
47. Kumar V, Satyanarayana T. Biochemical and thermodynamic characteristics of thermo-alkali-stable xylanase from a novel polyextremophilic *Bacillus halodurans* TSEV1. Extremophiles. 2013;17:797–808.
48. Soozanipour A, Taheri-Kafrani A, Isfahani AL. Covalent attachment of xylanase on functionalized magnetic nanoparticles and determination of its activity and stability. Chem Eng J. 2015;270:235–43.

20

Pigment fingerprint profile during extractive fermentation with *Monascus anka* GIM 3.592

Kan Shi, Rui Tang, Tao Huang, Lu Wang and Zhenqiang Wu*

Abstract

Background: Traditional submerged fermentation mainly accumulates intracellular orange pigments with absorption maxima at 470 nm, whereas extractive fermentation of *Monascus spp.* with Triton X-100 can promote the export of intracellular pigments to extracellular broth, mainly obtaining extracellular yellow pigments with absorption maxima at approximately 410 nm. In this study, a strain of *Monascus* (*M. anka* GIM 3.592) that produces high yields of pigments was employed to investigate the differences in pigment fingerprint profiles between submerged and extractive fermentations.

Results: Using extractive fermentation with this high-yield strain, the extracellular pigments exhibited an absorption maximum at 430 nm, not 410 nm, as previously observed. By comparing the pigment fingerprint profiles between submerged and extractive fermentations, extractive fermentation was found to not only export intracellular pigments to the extracellular broth, but also to form four other yellow pigments (**Y1-Y4**) that accounted for a large proportion of the extracellular pigments and that were not produced in submerged fermentation. The yields of **Y1-Y4** were closely related to the concentration and feeding time point of Triton X-100. **Y1-Y4** presented identical UV-Vis spectra with absorption maxima at 430 nm and fluorescence spectra with absorption maxima (emission) at 565 nm. HPLC-MS and the spectral analysis showed that the four pigments (**Y1-Y4**) had not been previously reported. The results indicated that these pigments may rely on the bioconversion of orange pigments (rubropunctatin and monascorubrin).

Conclusions: Using extractive fermentation with *M. anka* led to a high yield of extracellular yellow pigments ($AU_{410\ nm} = 114$), and the pigment fingerprint profile significantly differed compared to the results of traditional submerged fermentation. These results provide information and a detailed view of the composition and variation of pigments in extractive fermentation and could also contribute to characterizing pigment metabolism during extractive fermentation.

Keywords: *Monascus anka*, Extractive fermentation, Pigment profile, New yellow pigments, Bioconversion

Background

Monascus pigments are fungal azaphilone metabolites that consist of three groups of pigment components (i.e., yellow, orange and red pigments) that have been used as food colourants in Eastern and Southeastern Asian countries for a long time. The *Monascus* pigment composition mainly consists of two yellow pigments (monascin and ankaflavin), two orange pigments (rubropunctatin and monascorubrin) and two red pigments (rubropunctamine and monascorubramine), all of which are intracellular pigments. By 2012, more than 50 pigments produced by *Monascus. spp* were found and identified [1]. Some of these pigments possess specific bioactivities, such as anti-microbial [2–4], anti-tumour [5–8], anti-oxidative stress [9], anti-obesity [10, 11], anti-inflammation [6, 12, 13] and anti-diabetes [9, 14] properties.

Submerged fermentation (SF) at a low pH can rapidly accumulate a large amount of intracellular orange pigments (rubropunctatin and monascorubrin), independent of the nitrogen sources that are employed [15, 16].

* Correspondence: btzhqwu@scut.edu.cn
School of Bioscience and Bioengineering, South China University of Technology, Guangzhou 510006, China

Additionally, SF can avoid generating the mycotoxin, citrinin [17, 18]. In contrast to SF at low pH, extractive fermentation (EF) with Triton X-100 (TX-100) facilitates export of intracellular pigments directly to the fermentation broth, preventing intracellular product inhibition and further degradation of the pigments to improve their yield [19, 20]. In SF, the intracellular pigments mainly consist of orange pigments (rubropunctatin and monascorubrin) with absorption maxima at 470 nm. However, yellow pigments (monascin and ankaflavin) were found to dominate both the intracellular and extracellular pigments, which had the same absorption maxima at 410 nm throughout the EF. This phenomenon was due to the export of intracellular yellow pigments and the subsequent blockage of yellow pigment conversion to orange pigments [21].

In this study, a strain of *Monascus* with a high pigment yield was selected and used for pigment production via EF. The differences in pigment fingerprint profiles between SF and EF were investigated in detail.

Methods

Microorganisms and culture

Monascus anka (GIM 3.592, Guangdong Culture Collection Centre of Microbiology (GDMCC/GIMCC), China), which was maintained on potato dextrose agar plates and preserved at 4 °C, was used for pigment production. A subculture was carried out at 30 °C for 5 days per month.

The inoculum medium consisted of 20 g of glucose, 10 g of peptone, 3 g of yeast extract, 0.5 g of KCl, 4 g of KH_2PO_4, 0.01 g of $ZnSO_4 \cdot 7H_2O$, and 0.01 g of $FeSO_4 \cdot 7H_2O$ in 1 L of double-distilled water. For inoculum cultivation, a suspension of spores (approximately 3×10^6 spores) was inoculated in 50 mL of the inoculum medium in a 250-mL Erlenmeyer flask. The inoculum was cultivated at 30 °C and 180 rpm for 24 h.

The SF medium consisted of 50 g of glucose, 5 g of $(NH_4)_2SO_4$, 0.5 g of KCl, 4 g of KH_2PO_4, 0.01 g of $ZnSO_4 \cdot 7H_2O$, 0.5 g of $MgSO_4 \cdot 7H_2O$, and 0.01 g of $FeSO_4 \cdot 7H_2O$ per litre of double-distilled water. For EF, a certain amount of TX-100 was feed to the fermentation medium at a specific time as indicated. Before inoculation, the pH of all the fermentation media was adjusted to 4.0 with 10% (V/V) hydrochloric acid. For fermentation cultures, an aliquot of the inoculum culture (5 mL) was inoculated into 50 mL of fermentation medium. The fermentation was carried out at 30 °C and 180 rpm for 6.5 days.

For direct extraction of the intracellular pigments using 50 mL of 50 g/L TX-100 aqueous solution (pH was adjusted to 2.0 with 10% (V/V) hydrochloric acid) was made in a 250-mL Erlenmeyer flask, and it was added to 3.0 g aliquots of wet mycelia after 6 days of SF, which were collected under a vacuum through a 0.8 mm mixed cellulose esters membrane and then washed thoroughly with double-distilled water. Meanwhile, as a control, the fermentation supernatant from 6.5 days of SF without mycelia was also collected, and TX-100 was added to a final concentration of 50 g/L (the pH of this solution was approximately 2.0). Then, 50 mL of this solution was added to 3.0 g aliquots of wet mycelia after 6 days of SF in a 250-mL Erlenmeyer flask. The Erlenmeyer flasks were incubated at 30 °C and 180 rpm.

Analytical methods

Pigment, biomass and residual glucose analyses

The fermentation broth in the Erlenmeyer flasks was filtered under vacuum through a 0.8 mm mixed cellulose ester membrane. The collected filtrate (extracellular broth) was normalized to a volume of 50 mL for analysis of extracellular pigments. The mycelia remaining on the membrane were thoroughly washed with double distilled water until the runoff was almost colourless or foamless, and they were then resuspended in 50 mL of acidic aqueous ethanol (70%, v/v, in aqueous hydrochloric acid, pH 2.0) [15, 22] and incubated for 1 h at 30 °C with intermittent shaking. The acidic aqueous ethanol was used to prevent pigment shifting and amination reactions of orange pigments with amino compounds. The suspension was then filtered with pre-weighed filter paper. The filtrate (intracellular extract) was subjected to analysis of the intracellular pigments, and the mycelia remaining on the filter were dried to a constant weight at 60 °C to determine their biomass (dry cell weight, DCW). The comprehensive spectra of the extracellular pigments (i.e., extracellular broth) and intracellular pigments (i.e., intracellular extract) were recorded using a UV/VIS spectrophotometer (Unico, Dayton, OH, USA) from 350 nm to 550 nm at 1 nm intervals. The absorbance units (AU) at specific wavelengths (410, 470 and 510 nm) were used as an index of the extracellular pigment concentration. The residual glucose was quantified using a standard 3,5-dinitrosalicylic acid (DNS) method.

Pigment analysis using HPLC-PDAD, HPLC-FLD and TLC

Analytical HPLC was performed with an Alliance e2695 system (Waters, Milford, CT, USA) using a 2998 photodiode array detector (PDAD) (Waters, Milford, CT, USA) and a 2475 multi-wavelength fluorescence detector (FLD) (Waters, Milford, CT, USA). A Sunfire C18 column (250 × 4.6 mm, 5 μm, Waters, Milford, CT, USA) was used and the column temperature was 30 °C. Mobile phase A (aqueous H_3PO_4, pH 2.5) and mobile phase B (acetonitrile) were used according to the following gradients at a 1 mL/min flow rate: 0 min, 20% B; 25 min, 80% B; 35 min, 80% B; 36 min, 20% B; and 41 min, 20% B. The PADD was set to 200–600 nm for recording the UV-Vis spectra of the pigments. When recording the pigment profiles and concentrations, the detection wavelength was set to 310 nm. With the FLD, the pigment excitation spectra were recorded from 300 nm to 500 nm at 10 nm intervals using a 565 nm emission wavelength and emission spectra of pigments were recorded from 400 nm to 700 nm at 15 nm intervals using a 331 nm excitation wavelength. Thin-layer chromatography (TLC) analysis was conducted using a Silica gel 60 F_{254} TLC plate (Merck, Germany) with n-hexane/ethyl acetate/petroleum ether (30:17:5) as the developing solvent.

Determination of the pigment composition using LC-MS

Liquid chromatography-mass spectrometry (LC-MS) was performed using an HP1100 HPLC system (Agilent, Palo Alto, CA, USA) and a microTOF-QII mass spectrometer (Bruker, Rheinstetten, Germany). The C18 column and chromatographic conditions were the same as the conditions mentioned above, except for mobile phase A (0.1% formic acid in water). The extracellular broth containing high concentrations of TX-100 was processed in advance to remove as much TX-100 as possible before LC-MS analysis. During processing, 6.5-day EF extracellular broth was mixed with isometric diethyl ether to form a Winsor I microemulsion at 4 °C to remove TX-100 [23]. The organic solvent phase (diethyl ether) was collected and diethyl ether was removed by rotary evaporation at 36 °C. The residue was redissolved in an aqueous solution (the pH was adjusted to 2.0 with 10% (V/V) hydrochloric acid), and then the aqueous solution with residue was subjected to microemulsion extraction again. After 3 successive microemulsion extractions, the residue was redissolved in chloroform and a type of post-alkali treated macroporous resin was added to the chloroform. Then, the pigments that were absorbed in the resin were eluted with methanol (0.1% (w/v) hydrochloric acid). The pigments collected after elution were redissolved in chloroform and then submitted to macroporous resin absorption again. In total, the macroporous resin absorption and elution of pigments was successively conducted 4 times to further remove TX-100. The residual pigments were then redissolved in acidic aqueous ethanol (70%, v/v, in aqueous hydrochloric acid, pH 2.0) for LC-MS analysis.

Results

Characteristics of pigment metabolism in EF

As shown in Fig. 1a and b, in traditional submerged fermentation without Triton X-100 addition (SF), the pigments mainly accumulated intracellularly, and the AU_{470} can reach 135 after 6.5 days of fermentation. Compared to the production of intracellular pigments, the production of extracellular pigments were extremely low, and the AU_{470} was approximately only 1. On day 6.5, the intracellular AU_{470}/extracellular AU_{470} ratio was 138.32, and the intracellular AU_{410}/extracellular AU_{410} was 12.98. Both the intracellular and extracellular pigment metabolisms were coupled to cell growth. The logarithmic phase of cell growth began on the 1st day and lasted until nearly the 5th day. The highest biomass (DCW) was 8.9 g/L on the 5th day. Cell growth and pigment production during EF (50 g/L Triton X-100) showed notable differences compared to SF in the beginning (Fig. 1c and d). The highest production of intracellular pigments were obtained on the 4th day when the AU at 470 nm reached 123. However, they decreased sharply after this day. The production of extracellular pigments kept increasing during the entire fermentation, and the final AU at 410 nm reached 114. The extracellular accumulation of pigments was significant after 6.5 days of fermentation. The intracellular AU_{470}/extracellular AU_{470} ratio was 0.58 and the intracellular AU_{410}/extracellular AU_{410}/ extracellular AU_{410} ratio was 0.4. The metabolism of the intracellular pigments was coupled to cell growth, but the metabolism of extracellular pigments was not. It was clear that intracellular pigment production lagged for one day. The logarithmic phase of cell growth was shorter and began on the 1st day and stopped on the 3rd day. Compared to SF, EF had much faster glucose consumption. However, cell growth was actually inhibited to some extent, and the highest biomass was only 6.5 g/L.

The comprehensive spectra of the pigments produced during SF and EF showed significant differences (Fig. 1e and f). The extracellular pigment spectra from SF had absorption maxima at approximately 350 nm. However, the pigments from EF showed absorption maxima at 430 nm. The intracellular pigment spectra from SF presented absorption maxima at 470 nm, which indicated that orange pigments were the dominant intracellular pigments. However, in EF, in addition to absorption maxima at 470 nm, the intracellular pigment spectra also had a higher shoulder at 405 nm, which indicated yellow pigments (monascin and ankaflavin) also accounted for a large proportion.

Pigment fingerprint profiles and the production of four new yellow pigments during EF

Because the yield of extracellular water-soluble pigments was extremely low, the pigment analysis was focused on low-polarity pigments. As shown in Fig. 2a, the HPLC pigment fingerprint profile indicated that the intracellular pigments from SF and EF, and the extracellular pigments in SF were mainly low-polarity compounds. It was obvious that compounds **1**, **2**, **3** and **4** were found in the intracellular pigments in SF (Fig. 2A-c) and EF (Fig. 2A-b) and the extracellular pigments in EF (Fig. 2A-a). However, interestingly, besides the four original intracellular pigments (**1–4**) produced by *Monascus* strains, EF also produced a large amount of the

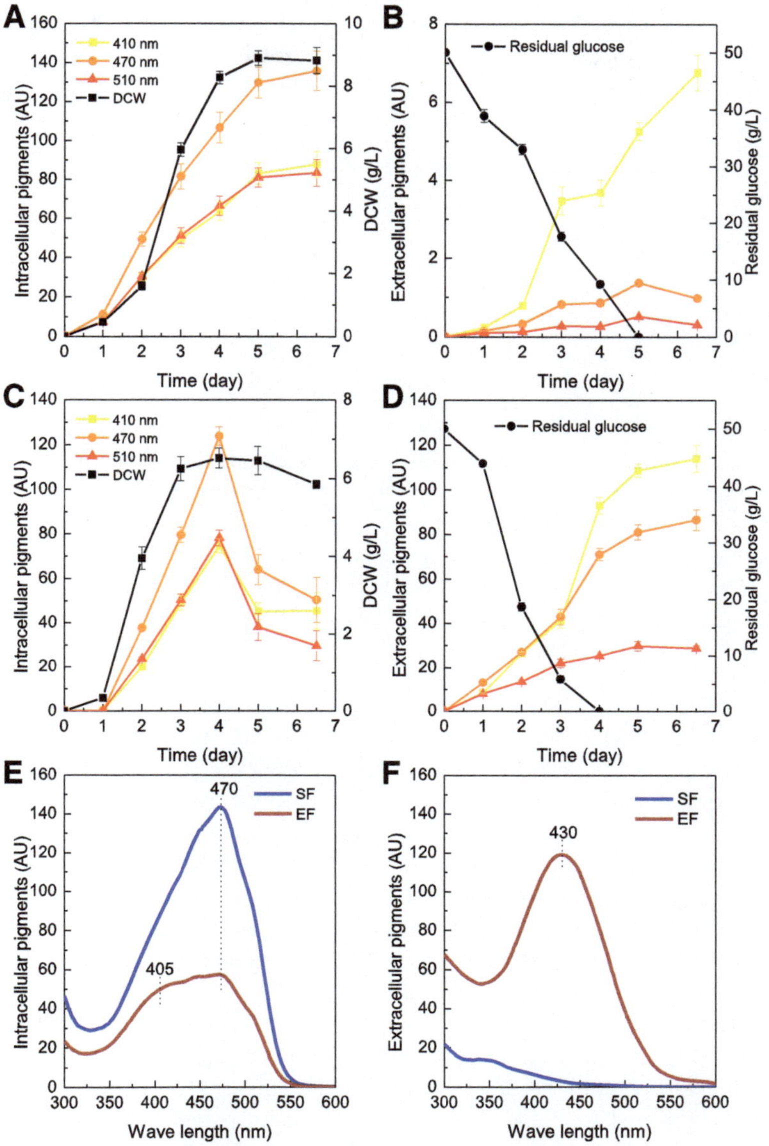

Fig. 1 Cell growth, glucose consumption, and intracellular and extracellular pigment production during SF (**a**, **b**) or EF (**c**, **d**). Comprehensive spectra of intracellular (**e**) and extracellular (**f**) pigments after 6.5 days of SF or EF

other four compounds **Y1**, **Y2**, **Y3** and **Y4** (Fig. 2A-a). These four compounds were not produced as extracellular or intracellular pigments in SF.

Compounds **1** and **3** were identified as monascin and ankaflavin (the two major yellow pigments accumulated intracellularly in most *Monascus* strains) by comparison of their UV-Vis spectra and molecular weights with data available in the literature (Table 1) [22, 24, 25]. Compounds **2** and **4** were identified as rubropunctatin and monascorubrin (the two major orange pigments accumulated intracellularly in most *Monascus* strains), respectively, by comparison with data available in the literature (Table 1) [22, 24, 25]. **Y1**-**Y4** had nearly identical UV-Vis spectra with three absorption maxima at approximately 219, 310 and 430 nm (Fig. 3), indicating that they were pigments with a yellow colour, as shown through TLC (Fig. 2b). In addition, as measured by FLD, the results also showed the same fluorescence excitation (Ex) spectra with two absorption maxima at approximately 310 and 420 nm as well as the same fluorescence emission (Em) spectra with absorption maxima at approximately 565 nm (Fig. 3). The fluorescence spectra absorption maxima at 565 nm indicated that the four pigments had orange fluorescence under specific excitation wavelengths; these results were also confirmed by the presence of orange fluorescent bands on the TLC plate under ultraviolet light (365 nm) (Fig. 2b). The molecular weights of **Y1**, **Y2**, **Y3** and **Y4** were 481, 509,

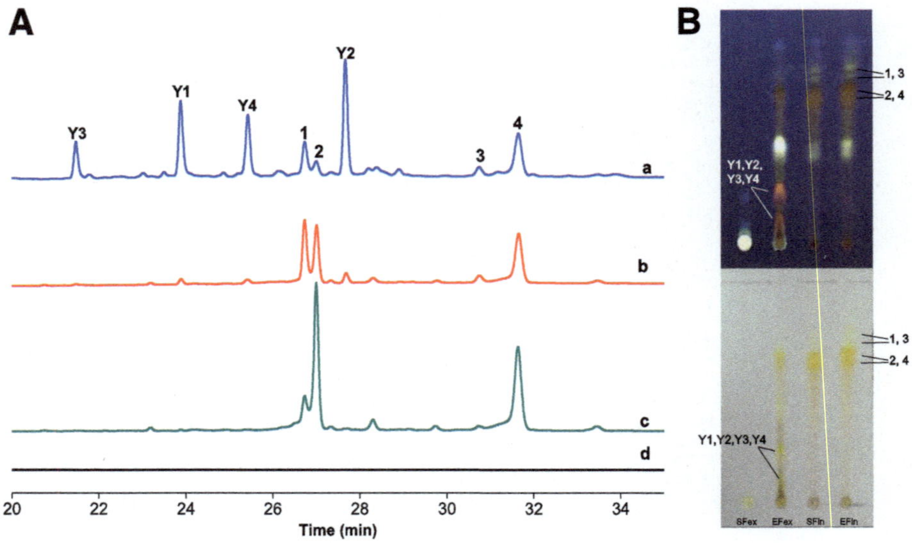

Fig. 2 HPLC profiles (**A**) of extracellular pigments of EF (**a**) or SF (**d**), and intracellular pigments of EF (**b**) or SF (**c**) on the 6.5th day. The Y axis for chromatogram c was adjusted appropriately to make the peaks were shown completely. TLC analysis (**B**) of extracellular pigments of SF (SFex) or EF (EFex) and intracellular pigments of SF (SFin) or EF (EFin) on the 6.5th day. The upper image of TLC was taken under ultraviolet light (365 nm). The lower image of TLC was taken under natural light

468 and 496, respectively. Thus, based on their UV-Vis spectra, fluorescence spectra and molecular weights, the four compounds have not been described or reported before.

The time course for pigment production during EF was recorded based on the pigment peak area variations (Fig. 4). These peak area time courses were mainly used to describe the relative variation of the pigment amount. As shown in Fig. 4a and b, extracellular **Y1**, **Y2**, **Y3** and **Y4**, as well as extracellular monascin and ankaflavin, increased rapidly at the early and middle stage of fermentation. **Y2** and **Y4** kept increasing until the 5th day, **Y1** and **Y3** increased rapidly in the first four days and then increased slightly in the next 2.5 days. At the later stage of fermentation the concentration of intracellular monascin and rubropunctatin was almost 1.4 and 3.6 times higher than the extracellular concentration of monascin and rubropunctatin. During the last 2.5 days, intracellular rubropunctatin and monascorubrin sharply decreased, and their amount on day 6.5 was less than 1/4 of the amount on the 4th day (Fig. 4c). Extracellular monascorubrin increased until the 5th day, while extracellular rubropunctatin increased during the first 3 days and decreased in the following 3.5 days. Both the intracellular and extracellular amounts of monascorubrin were maintained at higher levels than rubropunctatin.

Effect of the concentration and feeding time point of TX-100 on pigment fingerprint profiles

Different TX-100 concentrations impact the content of extracellular and intracellular pigments (Fig. 5). The content of extracellular pigments increased significantly with increases of the TX-100 concentration (Fig. 5b, Additional file 1: Figure S1-A and B), and when the TX-100 concentration was 100 g/L, the AU_{410} reached as high as 170. However, **Y1**, **Y2**, **Y3** and **Y4** were detected and accounted

Table 1 Chemical characteristics of extracellular *Monascus* pigments during EF

No.	Pigments	Peaks position of spectra (nm)[a]	HRESIMS m/z $[M+H]^{+b}$	Molecular formula[b]
1	Monascin	230, 291, 388	359.1835	$C_{21}H_{26}O_5$
2	Rubropunctatin	214, 247, 286, 471	355.1554	$C_{21}H_{22}O_5$
3	Ankaflavin	230, 293, 388	387.2179	$C_{23}H_{30}O_5$
4	Monascorubrin	214, 274, 285, 471	383.1865	$C_{23}H_{26}O_5$
Y1	Not reported	219, 310, 430	481.1878	–
Y2	Not reported	219, 310, 430	509.2193	–
Y3	Not reported	218, 310, 430	468.2041	–
Y4	Not reported	218, 310, 430	496.2111	–

[a]detected by HPLC-PDAD; [b]detected by LC-MS

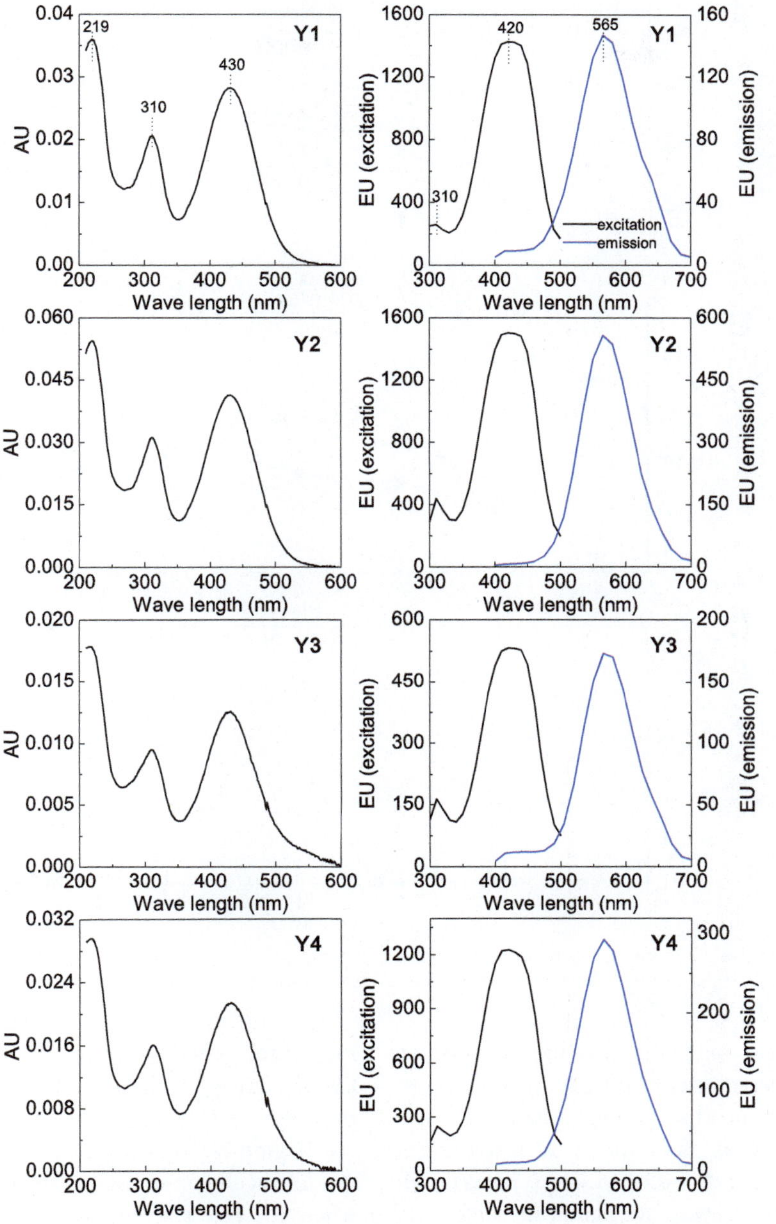

Fig. 3 UV-Vis spectra of **Y1**, **Y2**, **Y3** and **Y4** measured by HPLC-PDAD (*left*). Fluorescence excitation and emission spectra of **Y1**, **Y2**, **Y3** and **Y4** measured by HPLC-FLD (*right*)

for a large proportion of the extracellular pigment fraction despite the different TX-100 concentrations (Fig. 5a), which resulted in the same absorption maxima of the extracellular pigments at approximately 430 nm (Fig. 5b). In contrast to the extracellular pigment content, the intracellular pigment content, especially for rubropunctatin and monascorubrin, decreased significantly with increases of the TX-100 concentration (Fig. 5d and Additional file 1: Figure S1-C). When TX-100 concentration increase from 10 g/L to 100 g/L, the content of intracellular rubropunctatin and monascorubrin (peak area) decreased from 8.02×10^6 and 7.15×10^6, respectively, to 1.61×10^5 and 3.98×10^5, respectively. The intracellular pigments mainly consisted of the two major orange pigments (**2** and **4**) and the two major yellow pigments (**1** and **3**) (Fig. 5c). However, the proportion of each pigment was very different. The intracellular pigments were dominated by the orange pigments (**2** and **4**) when the TX-100 concentration was 10 g/L. When the TX-100 concentration was 50 g/L, the proportion of yellow pigments (**1** and **3**) in the intracellular pigment fraction increased substantially, and the intracellular pigment spectra had absorption maxima at 470 nm with a higher shoulder at approximately 410 nm (Fig. 5d). When the concentration was 100 g/L,

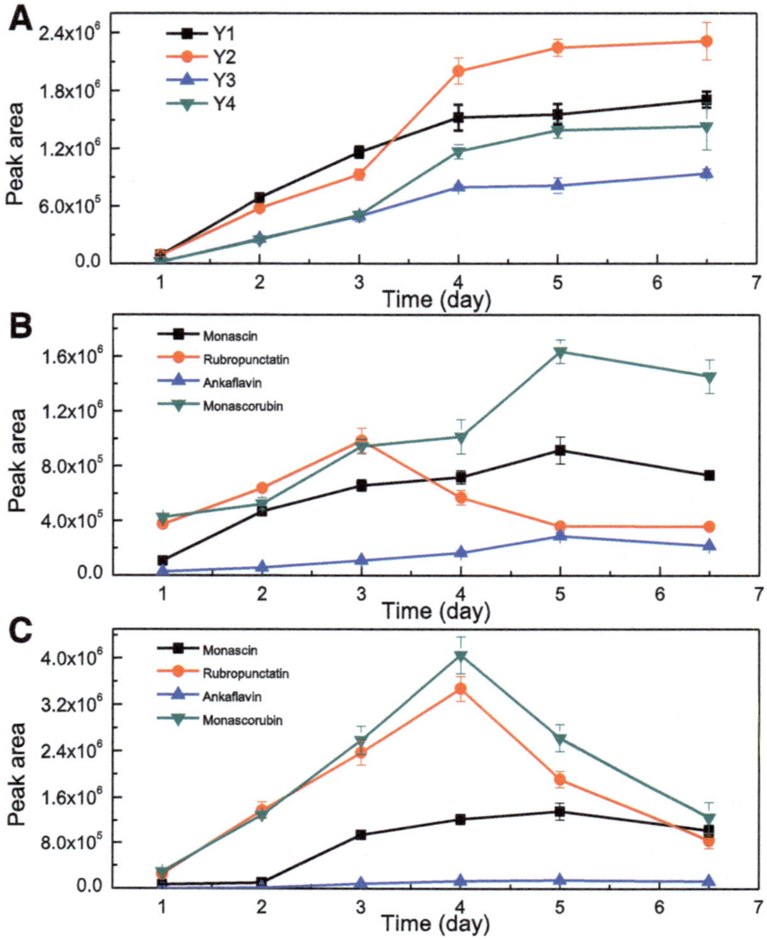

Fig. 4 Variation of the peak area of extracellular (**a**, **b**) and intracellular (**c**) pigments during EF measured by HPLC-PDA

the proportion of yellow pigments was higher, and the intracellular pigment spectra had absorption maxima at approximately 420 nm with no absorption maxima at 470 nm (Fig. 5d). The final DCW for EF with 10, 50 and 100 g/L TX-100 was 6.32 ± 0.27, 5.86 ± 0.10, 6.40 ± 0.27 g/L, respectively. Therefore, there was no significant difference among DCW under different concentration of TX-100.

For testing the effect of TX-100 feeding time point on the pigment profiles, TX-100 was feed at different fermentation times at a concentration of 50 g/L. The different TX-100 feeding time points had an impact on both the extracellular and intracellular pigment content (Fig. 6 and Additional file 2: Figure S2). Feeding TX-100 on the 6th day gave the highest intracellular pigment content and the lowest extracellular pigment content (Fig. 6b and d). The highest extracellular pigment content was obtained when TX-100 was feed on the 4th day. Feeding TX-100 at an early fermentation stage was shown to give lower amounts of intracellular pigments. The earlier feeding time points would lead to the more thorough exportation of intracellular pigments, especially for the intracellular orange pigments. As shown in Additional file 2: Figure S2, the content of two intracellular orange pigments were around 1×10^6 when TX-100 was feed at the beginning and the 2nd day, whereas the content of two intracellular orange pigments were around 5.5×10^6 when TX-100 was feed at the 4th and 6th day. Whenever TX-100 was feed, the extracellular pigments, which had spectra with the same absorption maxima at 430 nm, were also dominated by **Y1**, **Y2**, **Y3** and **Y4** (Fig. 6a and b). The intracellular pigments were dominated by the orange pigments when TX-100 was feed at a late fermentation stage (4th and 6th day). However, at an early fermentation stage (the beginning and 2nd day), the yellow pigments were a larger proportion of the intracellular pigments (Fig. 6c and d).

Pigment fingerprint profiles in a TX-100 aqueous solution extraction of mycelia from submerged fermentation

Y1, **Y2**, **Y3** and **Y4** were found when TX-100 was feed on the 6th day. Similar to the organic solvent, the TX-100 aqueous solution could also directly extract intracellular

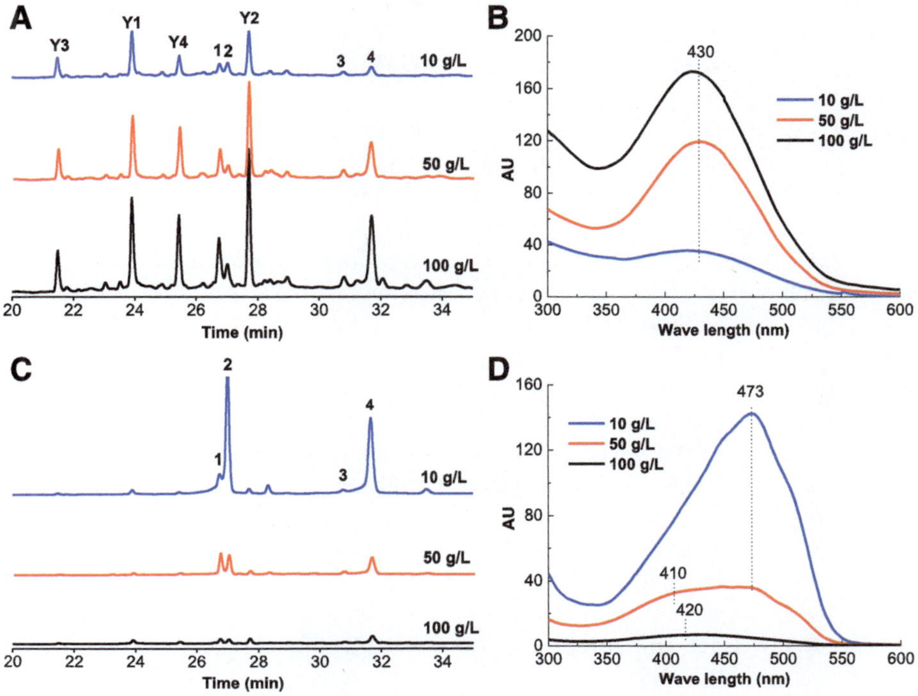

Fig. 5 HPLC profiles of extracellular (**a**) and intracellular (**c**) pigments, and comprehensive spectra of extracellular (**b**) and intracellular (**d**) pigments in EF with different TX-100 concentrations (10 g/L, 50 g/L, 100 g/L). The total fermentation time was 6.5 days

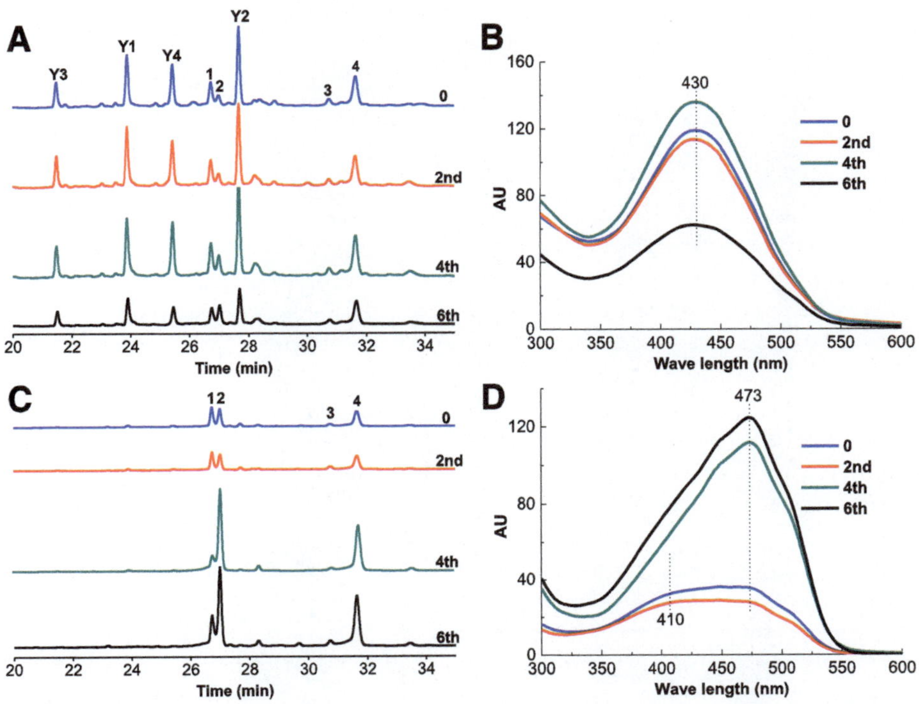

Fig. 6 HPLC profiles of extracellular (**a**) and intracellular (**c**) pigments, and comprehensive spectra of extracellular (**b**) and intracellular (**d**) pigments in EF with different TX-100 (50 g/L) feeding time point (the beginning, 2rd day, 4th day or 6th day). The total fermentation time was 6.5 days

pigments from the mycelia. It was mentioned before that SF can only produce the two yellow (**1** and **3**) and two orange pigments (**2** and **4**) (Fig. 2A-c). To determine if direct extraction with TX-100 aqueous solution could yield **Y1**, **Y2**, **Y3** and **Y4**, wet mycelia from 6 days of SF were placed into 50 g/L TX-100 aqueous solution (pH 2.0) and incubated at 30 °C and 180 rpm. Then, the extracellular extract profiles were analysed using HPLC-PDAD and UV-Vis spectroscopy (Fig. 7a and b). Meanwhile, a control group was examined by placing wet mycelia from 6 days of SF into the fermentation supernatant from 6 days of SF without mycelia (adding TX-100 to 50 g/L), and the extracellular extract profiles of this group were shown in Fig. 7c. It was very obvious that a direct extraction with a TX-100 aqueous solution could not produce **Y1**, **Y2**, **Y3** and **Y4** and could only extract the intracellular pigments (**1**, **2**, **3** and **4**) into the extracellular broth. The spectra of the extracellular extract presented an absorption maximum at approximately 447 nm and a high shoulder at approximately 474 nm (Fig. 7b). Additionally, when this direct extracellular extract was mixed with isometric broth from 6 days of SF (without mycelia), the profile of the mixture showed no differences between different incubation times. However, large amounts of **Y1**, **Y2**, **Y3** and **Y4** were produced in the control group, and the amount increased substantially over time (Fig. 7c). These results indicated that the production of **Y1**, **Y2**, **Y3** and **Y4** was closely related to the cellular metabolism of *Monascus*.

Discussion

In our previous study, EF with TX-100 successfully exported intracellular pigments to the extracellular broth, leading to the accumulation of high amounts of extracellular and intracellular pigments with the same absorption maxima at 410 nm that mainly consisted of yellow pigments (monascin (**1**) and ankaflavin (**3**)) [21]. However, under the same fermentation conditions, SF mainly produces intracellular orange pigments (rubropunctatin (**2**) and monascorubrin (**4**)) with absorption maxima at 470 nm. This interesting phenomenon may be related to the export of intracellular yellow pigments to the extracellular broth by TX-100 and the subsequent prevention of the conversion of yellow pigments into orange pigments. However, in this study, a strain with a high pigment yield (*Monascus anka* GIM 3.592) showed substantial differences with EF. During SF, this strain still mainly accumulated intracellular orange pigments (**2** and **4**) with absorption maxima at 470 nm (Figs. 1e and 2A-c). However, during EF, the intracellular pigments were dominated by neither the yellow pigments nor the orange pigments. Both the yellow and orange pigments were produced at high levels in the intracellular pigment

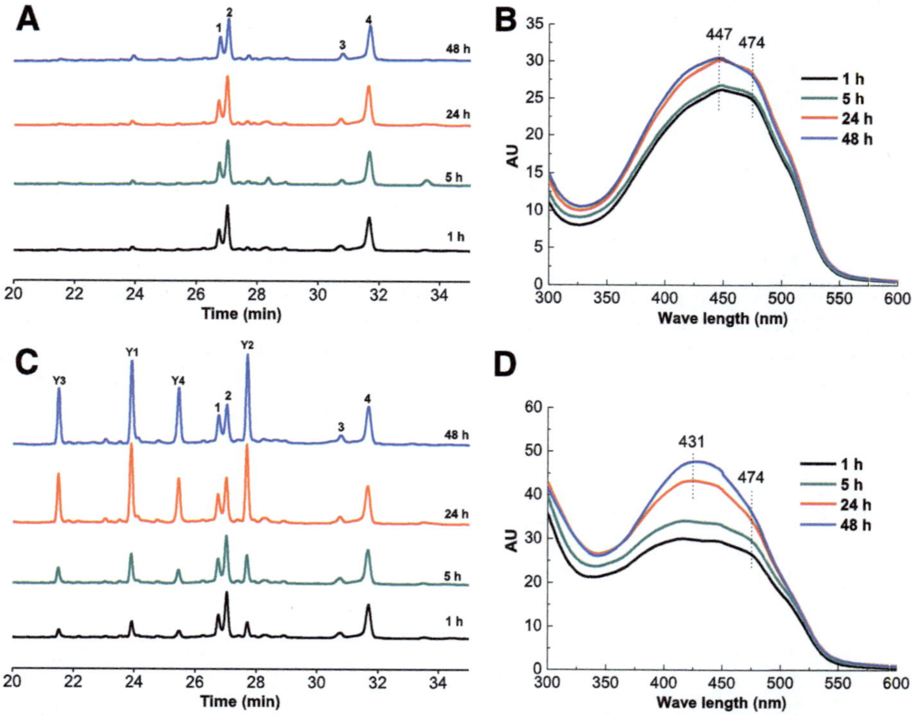

Fig. 7 HPLC profiles (**a**) and comprehensive spectra (**b**) of extracellular extract from 6-day SF wet mycelia with a 50 g/L TX-100 aqueous solution (pH 2.0); HPLC profiles (**c**) and comprehensive spectra (**d**) of extracellular extract from 6-day SF wet mycelia with the fermentation supernatant from 6 days of SF without mycelia (adding TX-100 to 50 g/L)

fraction (Fig. 2A-b), resulting in intracellular pigment spectra having an absorption maximum at 470 nm and a high shoulder at 405 nm (Fig. 1e). In addition to the yellow and orange pigments (**1**, **2**, **3** and **4**), EF produced four new yellow pigments with strong orange fluorescence (**Y1**, **Y2**, **Y3** and **Y4**), which all had the same absorption maximum at approximately 430 nm (Fig. 2A-a, B and 3). The four new pigments were a large proportion of the extracellular pigments (Fig. 2A-a), resulting in the comprehensive spectrum of the extracellular pigments having an absorption maximum at approximately 430 nm (Fig. 1f).

The TX-100 aqueous solution can directly extract the intracellular pigments (Fig. 7a). From the 1st to 4th day of EF, the biosynthesis rate of intracellular pigments was higher than the extraction rate of intracellular pigments with TX-100. Therefore, both the extracellular and intracellular pigments accumulated rapidly (Fig. 1c and d). Approaching the end of EF, the biosynthesis rate slowed down or even ceased, but the extraction rate with TX-100 remained high. Therefore, the intracellular pigments decreased sharply in the last 2.5 days. However, as shown in Fig. 7a and c, the accumulation of extracellular pigments during EF cannot be attributed to only the direct extraction of intracellular pigments with TX-100 because direct extraction with TX-100 aqueous solution can only export the intracellular pigments (**1**, **2**, **3** and **4**) but cannot give rise to the four new yellow pigments (**Y1**, **Y2**, **Y3** and **Y4**). Bioconversion among the pigments must exist, resulting in the formation of the four new yellow pigments.

An interesting phenomenon during the last 2.5 days of EF was that the reduction of intracellular rubropunctatin and monascorubrin was not equivalent to the incremental increase of extracellular rubropunctatin and monascorubrin (Fig. 4). The reduction of intracellular monascorubrin was more than 7 times greater than the extracellular monascorubrin increase. Furthermore, both intracellular and extracellular rubropunctatin decreased. Therefore, this remarkable phenomena indicated that the significant missing rubropunctatin and monascorubrin might be the sources of **Y1**, **Y2**, **Y3** and **Y4**. Meanwhile, the accumulation of **Y1**, **Y2**, **Y3** and **Y4** were closely related to the level of the exportation of intracellular orange pigments. The experiments investigating the effect of TX-100 concentration also support this deduction (Fig. 5 and Additional file 1: Figure S1). The different TX-100 concentration had no effect on the final DCW. However, with increases of the TX-100 concentration, intracellular rubropunctatin and monascorubrin decreased significantly (Additional file 1: Figure S1-C), and extracellular **Y1**, **Y2**, **Y3** and **Y4** increased significantly (Additional file 1: Figure S1-A). These results are consistent with a previous study, which found that the exportation of intracellular pigments intensified as TX-100 concentrations increased [26]. Higher TX-100 concentrations helped to export more intracellular pigments and thus left a low level of intracellular pigments. Especially, when TX-100 concentration increase from 10 g/L to 100 g/L, the reduction of intracellular rubropunctatin and monascorubrin (peak area) was approximately 7.85×10^6 and 6.77×10^6, respectively (Additional file 1: Figure S1-C), however, the increase of extracellular rubropunctatin and monascorubrin (peak area) was approximately 4.15×10^5 and 2.66×10^5, respectively (Additional file 1: Figure S1-B). The reduction of intracellular rubropunctatin and monascorubrin was almost than 19 and 26 times greater than the extracellular rubropunctatin and monascorubrin increase when TX-100 concentration increase from 10 g/L to 100 g/L. What is more, when TX-100 concentration increase from 10 g/L to 100 g/L, the reduction of intracellular monascin (1.80×10^6) and ankaflavin (1.96×10^5) was almost equivalent to the incremental increase of extracellular monascin (1.18×10^6) and ankaflavin (2.52×10^5). This results also strongly supported the aforementioned deduction that the missing rubropunctatin and monascorubrin might be the sources of **Y1**, **Y2**, **Y3** and **Y4**.

Y1, **Y2**, **Y3** and **Y4** had the same UV-Vis and fluorescence spectra (Fig. 4), indicating that they share a unique chromophore. The difference in the molecular weight between these pigments also indicated that rubropunctatin and monascorubrin might be the sources of **Y1**, **Y2**, **Y3** and **Y4**. The difference in molecular weight between **Y1** and **Y2** and between **Y3** and **Y4** was 28.0315 and 28.0070, respectively (Table 1). The molecular weight difference between rubropunctatin and monascorubrin is 28.0311, which indicated that **Y1** and **Y2** might be formed form the conversion of rubropunctatin and monascorubrin, respectively. The difference of molecular weight between **Y1** and rubropunctatin (126.0324) was almost the same as that between **Y2** and monascorubrin (126.0328). This indicated that **Y1** and **Y2**might be formed by the same modification of rubropunctatin and monascorubrin, respectively. However, the difference of molecular weight between **Y3** and rubropunctatin (113.0487) and between **Y4** and monascorubrin (113.0246). Therefore, **Y3** and **Y4** might be due to the different modification of rubropunctatin and monascorubrin.

Conclusions

In summary, EF with *Monascus anka* GIM 3.592 not only caused the export of intracellular pigments to the extracellular broth, which led to a high yield of extracellular yellow pigments ($AU_{410} = 114$), but also altered the composition of pigments. The pigment fingerprint profile from EF showed significant differences from that of traditional SF. In addition to the four original intracellular pigments (**1-4**), four new yellow pigments (**Y1-Y4**) were detected, and these pigments dominated the extracellular pigment fraction. Furthermore, the results indicated that **Y1-Y4** may be formed by bioconversion of orange pigments (rubropunctatin and monascorubrin). These results provide information and a detailed view of the composition and variation of pigments from EF and could also contribute to the elucidation of pigment metabolism during EF.

Additional files

Additional file 1: Figure S1. Peak area of each pigment in extracellular (A and B) and intracellular (C) pigment fractions in EF with different TX-100 concentrations (10 g/L, 50 g/L, 100 g/L). The total fermentation time was 6.5 days.

Additional file 2: Figure S2. Peak area of each pigment in extracellular (A and B) and intracellular (C) pigment fractions in EF with different Triton X-100 (50 g/L) feeding time point (the beginning, 2rd day, 4th day or 6th day). The total fermentation time was 6.5 days.

Abbreviations

SF: Submerged fermentation; EF: Extractive fermentation; TX-100: Triton X-100; DCW: Dry cell weight; AU: Absorbance units; PDAD: Photodiode array detector; FLD: Fluorescence detector; TLC: Thin-layer chromatography; LC-MS: Liquid chromatography-mass spectrometry

Acknowledgements

The authors gratefully acknowledge the financial support of the funding.

Funding

This work was supported by the National Natural Science Foundation of China [grant number 31271925], the Special Project on the Integration of Industry, Education and Research of Guangdong Province, China [grant number 2013B090600015], and the Science and Technology Program of Guangzhou, China [grant number 2014 J4100192].

Authors' contributions

ZW conceived and oversaw the study and revised the manuscript. KS designed and carried out the experiments and wrote the manuscript. RT and TH assisted to carry out the experiments and prepare figures. LW assisted the data analysis. All authors read and approved the final manuscript.

Competing interests

The authors declare that they have no competing interests.

References

1. Feng Y, Shao Y, Chen F. *Monascus* pigments. Appl Microbiol Biot. 2012;96:1421–40.
2. Kim C, Jung H, Kim YO, Shin CS. Antimicrobial activities of amino acid derivatives of *Monascus* pigments. FEMS Microbiol Lett. 2006;264:117–24.
3. Martínková L, J Zlová P, Veselý D. Biological activity of polyketide pigments produced by the fungus *Monascus*. J Appl Bacteriol. 1995;79:609–16.
4. Vendruscolo F, Tosin I, Giachini AJ, Schmidell W, Ninow JL. Antimicrobial activity of *Monascus* pigments produced in submerged fermentation. J Food Process Pres. 2014;38:1860–5.
5. Akihisa T, Tokuda H, Ukiya M, Kiyota A, Yasukawa K, Sakamoto N, Kimura Y, Suzuki T, Takayasu J, Nishino H. Anti-tumor-initiating effects of monascin, an azaphilonoid pigment from the extract of *Monascus pilosus* fermented rice (red-mold rice). Chem Biodivers. 2005;2:1305–9.
6. Hsu LC, Hsu YW, Liang YH, Kuo YH, Pan TM. Anti-tumor and anti-inflammatory properties of ankaflavin and monaphilone A from *Monascus purpureus* NTU 568. J Agric Food Chem. 2011;59:1124–30.
7. Su N, Lin Y, Lee M, Ho C. Ankaflavin from *Monascus*-fermented red rice exhibits selective cytotoxic effect and induces cell death on Hep G2 cells. J Agr Food Chem. 2005;53:1949–54.
8. Zheng Y, Xin Y, Shi X, Guo Y. Anti-cancer effect of rubropunctatin against human gastric carcinoma cells BGC-823. Appl Microbiol Biotechnol. 2010;88:1169–77.
9. Shi Y, Liao VH, Pan T. Monascin from red mold dioscorea as a novel antidiabetic and antioxidative stress agent in rats and Caenorhabditis elegans. Free Radical Bio Med. 2012;52:109–17.
10. Choe D, Lee J, Woo S, Shin CS. Evaluation of the amine derivatives of *Monascus* pigment with anti-obesity activities. Food Chem. 2012;134:315–23.
11. Lee C, Wen J, Hsu Y, Pan T. *Monascus*-fermented yellow pigments monascin and ankaflavin showed antiobesity effect via the suppression of differentiation and lipogenesis in obese rats fed a high-fat diet. J Agr Food Chem. 2013;61:1493–500.
12. Hsu W, Lee B, Liao T, Hsu Y, Pan T. *Monascus*-fermented metabolite monascin suppresses inflammation via PPAR-γ regulation and JNK inactivation in THP-1 monocytes. Food Chem Toxicol. 2012;50:1178–86.
13. Lin C, Lin Y, Huang P, Tsai H, Chen Y. Inhibition of endothelial adhesion molecule expression by *Monascus purpureus*-fermented rice metabolites, monacolin K, ankaflavin, and monascin. J Sci Food Agr. 2011;91:1751–8.
14. Lee B, Hsu W, Liao T, Pan T. The *Monascus* metabolite monascin against TNF-α-induced insulin resistance via suppressing PPAR-γ phosphorylation in C2C12 myotubes. Food Chem Toxicol. 2011;49:2609–17.
15. Kang B, Zhang X, Wu Z, Qi H, Wang Z. Effect of pH and nonionic surfactant on profile of intracellular and extracellular *Monascus* pigments. Process Biochem. 2013;48:759–67.
16. Shi K, Song D, Chen G, Pistolozzi M, Wu Z, Quan L. Controlling composition and color characteristics of *Monascus* pigments by pH and nitrogen sources in submerged fermentation. J Biosci Bioeng. 2015;120:145–54.
17. Chen G, Shi K, Song D, Quan L, Wu Z. The pigment characteristics and productivity shifting in high cell density culture of *Monascus anka* mycelia. BMC Biotechnol. 2015;15:72.
18. Kang B, Zhang X, Wu Z, Wang Z, Park S. Production of citrinin-free *Monascus* pigments by submerged culture at low pH. Enzyme Microb Technol. 2014;55:50–7.
19. Hu Z, Zhang X, Wu Z, Qi H, Wang Z. Export of intracellular *Monascus* pigments by two-stage microbial fermentation in nonionic surfactant micelle aqueous solution. J Biotechnol. 2012;162:202–9.
20. Hu Z, Zhang X, Wu Z, Qi H, Wang Z. Perstraction of intracellular pigments by submerged cultivation of *Monascus* in nonionic surfactant micelle aqueous solution. Appl Microbiol Biotechnol. 2012;94:81–9.
21. Xiong X, Zhang X, Wu Z, Wang Z. Accumulation of yellow *Monascus* pigments by extractive fermentation in nonionic surfactant micelle aqueous solution. Appl Microbiol Biotechnol. 2015;99:1173–80.
22. Shi K, Chen G, Pistolozzi M, Xia F, Wu Z. Improved analysis of *Monascus* pigments based on their pH-sensitive Uv-Vis absorption and reactivity properties. Food Addit Contam A. 2016;33:1396–401.
23. Shen L, Zhang X, Liu M, Wang Z. Microemulsion extraction of *Monascus* pigments from nonionic surfactant using high polarity of diethyl ether as excess oil phase. Sep Sci Technol. 2014;49:279–300.
24. Teng SS, Feldheim W. Analysis of anka pigments by liquid chromatography with diode array detection and tandem mass spectrometry. Chromatographia. 1998;47:529–36.
25. Zheng Y, Xin Y, Guo Y. Study on the fingerprint profile of *Monascus* products with HPLC–FD, PAD and MS. Food Chem. 2009;113:705–11.
26. Kang B, Zhang X, Wu Z, Qi H, Wang Z. Solubilization capacity of nonionic surfactant micelles exhibiting strong influence on export of intracellular pigments in *Monascus* fermentation. Microb Biotechnol. 2013;6:540–50.

Permissions

All chapters in this book were first published in BIOTECHNOLOGY, by BioMed Central; hereby published with permission under the Creative Commons Attribution License or equivalent. Every chapter published in this book has been scrutinized by our experts. Their significance has been extensively debated. The topics covered herein carry significant findings which will fuel the growth of the discipline. They may even be implemented as practical applications or may be referred to as a beginning point for another development.

The contributors of this book come from diverse backgrounds, making this book a truly international effort. This book will bring forth new frontiers with its revolutionizing research information and detailed analysis of the nascent developments around the world.

We would like to thank all the contributing authors for lending their expertise to make the book truly unique. They have played a crucial role in the development of this book. Without their invaluable contributions this book wouldn't have been possible. They have made vital efforts to compile up to date information on the varied aspects of this subject to make this book a valuable addition to the collection of many professionals and students.

This book was conceptualized with the vision of imparting up-to-date information and advanced data in this field. To ensure the same, a matchless editorial board was set up. Every individual on the board went through rigorous rounds of assessment to prove their worth. After which they invested a large part of their time researching and compiling the most relevant data for our readers.

The editorial board has been involved in producing this book since its inception. They have spent rigorous hours researching and exploring the diverse topics which have resulted in the successful publishing of this book. They have passed on their knowledge of decades through this book. To expedite this challenging task, the publisher supported the team at every step. A small team of assistant editors was also appointed to further simplify the editing procedure and attain best results for the readers.

Apart from the editorial board, the designing team has also invested a significant amount of their time in understanding the subject and creating the most relevant covers. They scrutinized every image to scout for the most suitable representation of the subject and create an appropriate cover for the book.

The publishing team has been an ardent support to the editorial, designing and production team. Their endless efforts to recruit the best for this project, has resulted in the accomplishment of this book. They are a veteran in the field of academics and their pool of knowledge is as vast as their experience in printing. Their expertise and guidance has proved useful at every step. Their uncompromising quality standards have made this book an exceptional effort. Their encouragement from time to time has been an inspiration for everyone.

The publisher and the editorial board hope that this book will prove to be a valuable piece of knowledge for researchers, students, practitioners and scholars across the globe.

List of Contributors

Jianfeng Yuan, Mianbin Wu, Jianping Lin and Lirong Yang
Key Laboratory of Biomass Chemical Engineering of the Ministry of Education,College of Chemical and Biological Engineering, Zhejiang University, Hangzhou 310027, China

Felix Bracharz, Veronika Redai, Kathrin Bach, Farah Qoura and Thomas Brück
Industrial Biocatalysis Group, Technische Universität München, Lichtenbergstraße 4, 85748 Garching, Germany

Bin Tang, Ping Wei, Lina Zhao, Zuokun Shi, Qida Shen, Mengmeng Yang, Guoqiang Xie and Shigui Wang
Hangzhou Key Laboratory of Animal Adaptation and Evolution, College of Life and Environmental Sciences, Hangzhou Normal University, Hangzhou, Zhejiang, China

Qinhong Wang
Department of Strategic and Integrative Research, Tianjin Institute of Industrial Biotechnology, Chinese Academy of Sciences, 300308 Tianjin, China

Wenqin Bai, Yufan Cao and Jun Liu
Department of Strategic and Integrative Research, Tianjin Institute of Industrial Biotechnology, Chinese Academy of Sciences, 300308 Tianjin, China
College of Life Science, Shanxi Normal University, Linfen 041004, China

Zhenhu Jia
College of Life Science, Shanxi Normal University, Linfen 041004, China

Nan Wang, Kai Ren, Rong Jia, Wenting Chen and Ruirui Sun
School of Life Science, Anhui University, 111 Jiulong Road, Economic and Technology Development Zone, Hefei, Anhui 230601, People's Republic of China

Ying Zhao, Fei-Fei Li, Zhen-Ning Liu and Guang-Rong Zhao
Department of Pharmaceutical Engineering, School of Chemical Engineering and Technology, Tianjin University, Tianjin 300072, China
Key Laboratory of Systems Bioengineering, Ministry of Education, Tianjin 300072, China
Syn Bio Research Platform, Collaborative Innovation Center of Chemical Science and Engineering, Tianjin 300072, China

Chang-Song Wang
Department of Pharmaceutical Engineering, School of Chemical Engineering and Technology, Tianjin University, Tianjin 300072, China
Key Laboratory of Systems Bioengineering, Ministry of Education, Tianjin 300072, China
Syn Bio Research Platform, Collaborative Innovation Center of Chemical Science and Engineering, Tianjin 300072, China
PPG Coating (Tianjin) Co., Ltd. Tianjin Economic Technological Development Area (TEDA), 192 Huanghai Road, Tianjin 300457, China

Jun-Zheng Wu, Qin Liu, Xiao-Shan Geng, Li-Juan Luo and Jin-Ping Liu
Hainan Key Laboratory for Sustainable Utilization of Tropical Bioresources, College of Agriculture, Hainan University, Haikou, Hainan Province 570228, China

Kai-Mian Li
The Institute of Tropical Bioscience and Biotechnology, Chinese Academy of Tropical Agricultural Sciences, Haikou, Hainan Province 571101, China

Xiuling Shang, Shuwen Liu, Yong Liang and Yun Zhang
CAS Key Laboratory of Pathogenic Microbiology and Immunology, Institute of Microbiology, Chinese Academy of Sciences, Beijing 100101, People's Republic of China

Xin Chai and Yu Zhang
CAS Key Laboratory of Pathogenic Microbiology and Immunology, Institute of Microbiology, Chinese Academy of Sciences, Beijing 100101, People's Republic of China
University of Chinese Academy of Sciences, Beijing 100049, People's Republic of China

Tingyi Wen
CAS Key Laboratory of Pathogenic Microbiology and Immunology, Institute of Microbiology, Chinese Academy of Sciences, Beijing 100101, People's Republic of China
Savaid Medical School, University of Chinese Academy of Sciences, Beijing 100049, People's Republic of China

Ting Zhou and Dion Lepp
Guelph Food Research Centre, Agriculture and Agri-Food Canada, 93 Stone Road West, Guelph, Ontario N1G 5C9, Canada

Rafiq Ahad
Guelph Food Research Centre, Agriculture and Agri-Food Canada, 93 Stone Road West, Guelph, Ontario N1G 5C9, Canada
Department of Plant Agriculture, University of Guelph, Guelph, Ontario N1G 2W1, Canada

K. Peter Pauls
Department of Plant Agriculture, University of Guelph, Guelph, Ontario N1G 2W1, Canada

Hongwei Gao, Min Sun and Xizhi Xiao
Shandong Entry-Exit Inspection and Quarantine Bureau of People's Republic of China, Qingdao, China

Xiaofan Yu and Ronggui Li
Qingdao University Medical College, Qingdao, China

Tingting Deng, Xin Huang and Ying Chen
Chinese Academy of Inspection and Quarantine Institute, Beijing, China

Thomas S. Lane, Caroline S. Rempe and C. Neal Stewart Jr.
The Graduate School of Genome Science and Technology, University of Tennessee, Knoxville, TN Department of Plant Sciences, University of Tennessee, Knoxville, TN 37996, USA

Jack Davitt and Margaret E. Staton
Department of Entomology and Plant Pathology, University of Tennessee, Knoxville, TN 37996, USA

Yanhui Peng
Department of Plant Sciences, University of Tennessee, Knoxville, TN 37996, USA

Douglas Edward Soltis
Department of Biology, University of Florida, Gainesville, FL 32611, USA
Florida Museum of Natural History, Gainesville, FL 32611, USA
Genetics Institute, University of Florida, Gainesville, FL 32611, USA

Michael Melkonian
Department of Biology, Cologne Biocenter, University of Cologne, Cologne 50674, Germany

Michael Deyholos
Department of Biology, University of British Columbia, Kelowna, BC V1V 1V7, Canada

James H. Leebens-Mack
Department of Biological Sciences, University of Georgia, Athens, GA 30602, USA

Mark Chase
Jodrell Laboratory, Royal Botanic Gardens Kew, Richmond, Surrey TW9 3AE, UK
Department of Biology, Duke University, Durham, NC 27708, USA

Carl J. Rothfels
Department of Integrative Biology, University of California, University Herbarium, Berkeley, CA 94720-2465, USA

Dennis Stevenson
New York Botanical Garden, Bronx, New York, NY 10458, USA

Sean W. Graham
Department of Botany, University of British Columbia, Vancouver, BC, Canada

Jun Yu
CAS Key Laboratory of Genome Sciences and Information, Beijing Key Laboratory of Genome and Precision Medicine Technologies, Beijing Institute of Genomics, Chinese Academy of Sciences, Beijing 100101, People's Republic of China

Tao Liu
College of Marine Life Sciences, Ocean University of China, Qingdao 266003, People's Republic of China

J. Chris Pires
Bond Life Sciences Center, Division of Biological Sciences, University of Missouri, Columbia, MO, USA

Patrick P. Edger and Ying Zhu
BGI-Shenzhen, Beishan Industrial Zone, Yantian District, Shenzhen 518083, China

Yinlong Xie
BGI-Shenzhen, Beishan Industrial Zone, Yantian District, Shenzhen 518083, China
Department of Computer Science, HKU-BGI Bioinformatics Algorithms and Core Technology Research Laboratory, University of Hong Kong, Pokfulam, Hong Kong
School of Bioscience and Bioengineering, South China University of Technology, Guangzhou, China

Gane Ka-Shu Wong
BGI-Shenzhen, Beishan Industrial Zone, Yantian District, Shenzhen 518083, China
Department of Biological Sciences, University of Alberta, Edmonton, AB T6G 2E9, Canada
Department of Medicine, University of Alberta, Edmonton, AB T6G 2E1, Canada

Yong Zhang
Department of Computer Science, HKU-BGI Bioinformatics Algorithms and Core Technology Research Laboratory, University of Hong Kong, Pokfulam, Hong Kong

Eric Carpenter
Department of Biological Sciences, University of Alberta, Edmonton, AB T6G 2E9, Canada

Martinus J.A. Daas, Antonius H.P. van de Weijer, John van der Oost and Willem M. de Vos
Laboratory of Microbiology, Wageningen University, Stippeneng 4, Wageningen 6708 WE, The Netherlands

Richard van Kranenburg
Laboratory of Microbiology, Wageningen University, Stippeneng 4, Wageningen 6708 WE, The Netherlands
Corbion, Arkelsedijk 46, Gorinchem 4206 AC, The Netherlands

Patricia Murciano Martínez and Mirjam A. Kabel
Laboratory of Food Chemistry, Wageningen University, Bornse Weilanden 9, Wageningen 6708 WG, The Netherlands

Hanna Amelina, Vera Moiseeva, Laura Catharine Collopy, Siân Rosanna Pearson, Christine Anne Armstrong and Kazunori Tomita
Chromosome Maintenance Group, UCL Cancer Institute, University College London, Paul O'Gorman Building, Huntley Street, London WC1E 6DD, UK

Wending Mei, Lu Wang and Zhaojuan Zheng
College of Chemical Engineering, Nanjing Forestry University, Nanjing 210037, People's Republic of China

Jia Ouyang
College of Chemical Engineering, Nanjing Forestry University, Nanjing 210037, People's Republic of China
Key Laboratory of Forest Genetics & Biotechnology of the Ministry of Education, Nanjing, People's Republic of China

Ying Zang
College of Forestry, Nanjing Forestry University, Nanjing 210037, People's Republic of China

Sandra Ziemons, Katerina Koutsantas, Kordula Becker, Tim Dahlmann and Ulrich Kück
Lehrstuhl für Allgemeine und Molekulare Botanik, Ruhr-Universität Bochum, ND7/131, Universitätsstraße 150, 44780 Bochum, Germany

Changfu Li and Yansheng Zhang
CAS Key Laboratory of Plant Germplasm Enhancement and Specialty Agriculture, Wuhan Botanical Garden, Chinese Academy of Sciences, Wuhan 430074, China

Chen Zhou and Jing Li
CAS Key Laboratory of Plant Germplasm Enhancement and Specialty Agriculture, Wuhan Botanical Garden, Chinese Academy of Sciences, Wuhan 430074, China
University of Chinese Academy of Sciences, Beijing 100049, China

Junpei Zhou, Rui Zhang, Qian Wu, Junjun Li, Xianghua Tang, Bo Xu, Junmei Ding, Nanyu Han and Zunxi Huang
Engineering Research Center of Sustainable Development and Utilization of Biomass Energy, Ministry of Education, Yunnan Normal University, Kunming 650500, People's Republic of China

College of Life Sciences, Yunnan Normal University, No. 768 Juxian Street, Chenggong, Kunming, Yunnan 650500, People's Republic of China
Key Laboratory of Yunnan for Biomass Energy and Biotechnology of Environment, Yunnan, Kunming 650500, People's Republic of China
Key Laboratory of Enzyme Engineering, Yunnan Normal University, Kunming 650500, People's Republic of China

Zhifeng Song and Rui Liu
College of Life Sciences, Yunnan Normal University, No. 768 Juxian Street, Chenggong, Kunming, Yunnan 650500, People's Republic of China

Yan-Jun Li, Shou-Hong Zhu, Xin-Yu Zhang, Yong-Chang Liu, Fei Xue and Jie Sun
The Key Laboratory of Oasis Eco-agriculture, Agriculture College, Shihezi University, Bei 4 Road, Shihezi 832003, Xinjiang, China

Lan-Jie Zhao
College of Life Sciences, Shihezi University, Bei 4 Road, Shihezi 832003, Xinjiang, China

Tingting Shang, Dayong Si, Dongyan Zhang, Xuhui Liu, Cong Hu, Yu Fu and Rijun Zhang
Laboratory of Feed Biotechnology, State Key Laboratory of Animal Nutrition, College of Animal Science & Technology, China Agricultural University, No.2 Yuanmingyuan West RoadHaidian District, Beijing 100193, China

Longmei Zhao
Laboratory of Feed Biotechnology, State Key Laboratory of Animal Nutrition, College of Animal Science & Technology, China Agricultural University, No.2 Yuanmingyuan West RoadHaidian District, Beijing 100193, China
College of Animal Science and Technology, Henan University of Science and Technology, Luoyang 471003, China

Kan Shi, Rui Tang, Tao Huang, Lu Wang and Zhenqiang Wu
School of Bioscience and Bioengineering, South China University of Technology, Guangzhou 510006, China

Index